how to know the
insects

D. Crouch

The **Pictured Key Nature Series** has been published since 1944 by the Wm. C. Brown Company. The series was initiated in 1937 by the late Dr. H. E. Jaques, Professor Emeritus of Biology at Iowa Wesleyan University. Dr. Jaques' dedication to the interest of nature lovers in every walk of life has resulted in the prominent place this series fills for all who wonder **"How to Know."**

John F. Bamrick and Edward T. Cawley
Consulting Editors

The Pictured Key Nature Series

How to Know the
  AQUATIC INSECTS, Lehmkuhl
  AQUATIC PLANTS, Prescott
  BEETLES, Arnett-Downie-Jaques, Second Edition
  BUTTERFLIES, Ehrlich
  ECONOMIC PLANTS, Jaques, Second Edition
  FALL FLOWERS, Cuthbert
  FERNS AND FERN ALLIES, Mickel
  FRESHWATER ALGAE, Prescott, Third Edition
  FRESHWATER FISHES, Eddy-Underhill, Third Edition
  GILLED MUSHROOMS, Smith-Smith-Weber
  GRASSES, Pohl, Third Edition
  IMMATURE INSECTS, Chu
  INSECTS, Bland-Jaques, Third Edition
  LAND BIRDS, Jaques
  LICHENS, Hale, Second Edition
  LIVING THINGS, Jaques, Second Edition
  MAMMALS, Booth, Third Edition
  MARINE ISOPOD CRUSTACEANS, Schultz
  MITES AND TICKS, McDaniel
  MOSSES AND LIVERWORTS, Conard-Redfearn, Third Edition
  NON-GILLED FLESHY FUNGI, Smith-Smith
  PLANT FAMILIES, Jaques
  POLLEN AND SPORES, Kapp
  PROTOZOA, Jahn, Bovee, Jahn, Third Edition
  SEAWEEDS, Abbott-Dawson, Second Edition
  SEED PLANTS, Cronquist
  SPIDERS, Kaston, Third Edition
  SPRING FLOWERS, Cuthbert, Second Edition
  TREMATODES, Schell
  TREES, Miller-Jaques, Third Edition
  TRUE BUGS, Slater-Baranowski
  WATER BIRDS, Jaques-Ollivier
  WEEDS, Wilkinson-Jaques, Third Edition
  WESTERN TREES, Baerg, Second Edition

how to know the insects

Third Edition

Roger G. Bland
Central Michigan University

H.E. Jaques
formerly Iowa Wesleyan College

The Pictured Key Nature Series
Wm. C. Brown Company Publishers
Dubuque, Iowa

Copyright © 1947 by H.E. Jaques

Copyright © 1978 by Wm. C. Brown Company Publishers

Library of Congress Catalog Card Number: 77-88344

ISBN 0—697—04752—0 (Paper)
ISBN 0—697—04753—9 (Cloth)

All rights reserved. No part of this publication may be reproduced, stored in a retrieval system, or transmitted in any form or by any means, electronic, mechanical, photocopying, recording, or otherwise, without the prior written permission of the publisher.

Printed in the United States of America
20 19 18 17 16 15 14 13

Contents

Preface vii
Acknowledgments ix
Introduction xi

How to Use This Book 1
How to Collect Insects 3
How to Preserve and Mount Insects 13
How to Organize and Maintain an Insect Collection 22
How to Observe and Rear Insects 25
Where to Look for Insects 27
Insect Structure and Development 30
General References on Insect Biology 36
Classification and Nomenclature 38
Key to the Orders of Adult Insects 40
Subclass Apterygota 57
    Order Protura—Proturans 57
    Order Diplura—Diplurans 58
    Order Collembola—Springtails 59
    Order Thysanura—Bristletails 62
Subclass Pterygota 64
    Order Ephemeroptera—Mayflies 64
    Order Odonata—Dragonflies and Damselflies 69
    Order Plecoptera—Stoneflies 82
    Order Orthoptera—Grasshoppers, Katydids, and Crickets 89
    Order Phasmida—Walkingsticks 101
    Order Dictyoptera—Cockroaches and Mantids 102
    Order Grylloblattodea—Rock Crawlers or Icebugs 107
    Order Dermaptera—Earwigs 108
    Order Embioptera—Webspinners 111

Order Isoptera—Termites 112
Order Zoraptera—Zorapterans 117
Order Psocoptera—Psocids 118
Order Mallophaga—Chewing Lice 122
Order Anoplura—Sucking Lice 126
Order Thysanoptera—Thrips 129
Order Hemiptera—True Bugs, Cicadas, Leafhoppers, Planthoppers, Aphids, Scale Insects, and Others 133
Order Coleoptera—Beetles 175
Order Strepsiptera—Twistedwinged Parasites 237
Order Neuroptera—Dobsonflies, Alderflies, Snakeflies, Lacewings, Antlions, Mantispids, and Others 239
Order Mecoptera—Scorpionflies 247
Order Trichoptera—Caddisflies 249
Order Lepidoptera—Butterflies and Moths 257
Order Diptera—Flies 315
Order Siphonaptera—Fleas 355
Order Hymenoptera—Ants, Bees, Sawflies, Wasps, and Allies 361

Index and Glossary 393

Preface

"How to Know the Insects" has occupied a unique niche as an early entomological handbook and its continuing use affirms the talent of Professor Jaques to create a book of great utility. Many changes have occurred in the classification of insects since the last revision in 1947 and the time arrived to revise the text and incorporate the changes.

This new edition is completely revised and provides a more complete account of insect natural history to accompany the keys and examples of species. The book may be used at three levels. The most elementary level includes the key to orders and the general information on the 29 insect orders. At the second level, collectors who wish to classify insects into families and learn more about them will find that the keys to common families and the descriptive biology of the families will serve these purposes. Other readers may have specimens they would like to identify to species, and the descriptions of selected common species will be of assistance at this third level.

I have added many features useful in the identification and natural history portions of a beginning or advanced class in college entomology. Teachers of high school biology classes and the layman interested in the world of insects will also continue to find the book a valuable source of reference in the classroom, field, or personal library. New and revised features include the following:

1. Revised keys to orders and common families. Keys to families are more accurate and usable and include additional common families.

2. Hundreds of new illustrations. All drawings are larger and improved in clarity. Many show the variance in insect form within families and others illustrate wing venation and various body features needed to use the keys.

3. A description of the basic biology of each order and of most families. Characteristics used to identify immatures are included for many orders and certain families.

4. A description of selected common species or genera in most families.

5. General literature references for future reading at the end of each chapter on the insect orders. References are also given for general textbooks in entomology and for books on rearing insects, and names and addresses are listed for biological supply companies.

6. Directions for finding, collecting, and preserving insects are rewritten and expanded. Chapters on insect classification, structure and development, and rearing, are new or revised. The glossary includes many additional terms used in the text.

Roger G. Bland

Acknowledgments

The chapters on insect orders were reviewed by specialists at the Smithsonian Institution, the U.S. Department of Agriculture Systematic Entomology Laboratory, universities, and other institutions. I am very grateful for their corrections and additions, and for updating the nomenclature as needed. The reviewers are as follows: *Anoplura:* Ke C. Kim; *Coleoptera:* Donald M. Anderson, Terry L. Erwin, Robert D. Gordon, John M. Kingsolver, Paul J. Spangler, Theodore J. Spilman, Donald R. Whitehead; *Collembola:* Kenneth Christiansen, Richard J. Snider; *Dermaptera, Dictyoptera, Grylloblattodea, Phasmida,* and *Zoraptera:* Ashley B. Gurney; *Diptera;* Richard H. Foote, Raymond J. Gagne, Robert H. King, Lloyd V. Knutson, Wayne N. Mathis, Curtis W. Sabrosky, George C. Steyskal, F. Christian Thompson, Willis W. Wirth (particular thanks go to Dr. Thompson for his many suggestions); *Embioptera:* Edward S. Ross (who also supplied a key and other data); *Ephemeroptera:* George F. Edmunds, Jr.; *Hemiptera (Heteroptera):* Richard C. Froeschner, Carl W. Schaefer, James A. Slater (Drs. Froeschner's and Slater's assistance is especially appreciated); *Hemiptera (Homoptera):* Richard C. Froeschner, James P. Kramer, Douglass R. Miller, Manya B. Stoetzel (Dr. Miller's assistance is especially appreciated); *Hymenoptera:* Suzanne Batra, Robert W. Carlson, Gordon Gordh, Paul M. Marsh, Arnold S. Menke, David R. Smith (Drs. Carlson and Smith were particularly helpful with their corrections); *Isoptera:* William L. Nutting (who made extensive suggestions and supplied necessary data); *Lepidoptera:* Don R. Davis, W. Donald Duckworth; *Mallophaga:* K.C. Emerson (who was especially helpful in providing information); *Mecoptera:* George W. Byers (who supplied improved identification data); *Neuroptera* and *Trichoptera:* Oliver S. Flint, Jr.; *Odonata:* Donald J. Borror, Harold B. White (both made extensive improvements and supplied useful data); *Orthoptera:* Irving J. Cantrall, Ashley B. Gurney (both were especially helpful in suggesting and supplying references and other data); *Plecoptera:* Richard W. Baumann; *Psocoptera:* Edward L. Mockford (who provided additional useful data); *Siphonaptera:* Robert Traub; *Strepsiptera:* Richard M. Bohart; *Thysanoptera:* Lewis J. Stannard. Informative data also were received from Pedro Wygodzinsky *(Diplura),* David C. Rentz *(Orthoptera),* Alexander B. Klots *(Lepidoptera),* Robert E. Lewis *(Siphonaptera),* and Glenn B. Wiggins *(Trichoptera).*

The choice of insects to illustrate and the quality of illustrations is my responsibility and does not reflect the judgment of the reviewers.

Many people generously contributed their time and talent to produce the new illustrations of insect species. Without this assistance the book would have been far less useful. The illustrators are as follows: John R. Cassani, Stephanie Chemis, Penny I. Foldenauer, Thomas W. Ford, David A. Ginnebaugh, Dennis J. Goebel, Barbara Saddler Henry, Mary K. Hunt, John Kelly, Donald B. Lewis, Susan Marquand, James S. McEwan, Kim I. McGuire, Theresa M. Nelson, Priscilla Olson, David Pearson, Brian Perry, Janet A. Rice, Frank R. Rogala, Bradley V. Sias, Michael A. Simmons (frontispiece), Michele A. Szok, Peggy Whiting, and Mildred G. Wujek. Illustrations of Trichoptera were modified from Ross and the Illinois Natural History Survey.

The index was compiled expertly by my wife, Kathryn. She also edited the text to improve its clarity and syntax, typed the index and many of the chapters, and typed many letters to correspondents. Louise K. Barrett, of Wm. C. Brown Company Publishers, provided the competent editorial supervision.

Roger G. Bland
Mt. Pleasant, Michigan

Introduction

An insect is a small invertebrate, or animal without a backbone, with the following external, adult characteristics: (1) hardened external skeleton; (2) three distinct (usually) body regions (head, thorax, and abdomen); (3) one pair of segmented antennae; (4) one pair of compound eyes in most cases; (5) three pairs of segmented legs with one pair on each of the three thoracic segments; and (6) usually one or two pairs of wings although some adults are wingless. Immature insects may appear like miniature adults without fully developed wings but most do not resemble the adult and may be wormlike or assume other shapes.

The first insects evolved approximately 340 million years ago and today insects form about 5/6 of the known animal life. Approximately 1 million insect species are catalogued in the world and many more continue to be discovered. North American species (north of Mexico) number over 98,000. Insects range in size from parasitic wasps about 0.2 mm long to tropical species such as moths (300 mm wingspread), walkingsticks (330 mm long), and goliath beetles (75 mm wide). The colors vary from nearly transparent individuals to luminous or iridescent species. Upon close inspection their body segments, legs, eyes, antennae, wings, etc., are endlessly modified and often bizarre. The following chapters examine the collecting and identification of common North American insects.

A Mud Dauber, *Sceliphron caementarium* (Sphecidae).

How to Use This Book

KEYS

Dichotomous keys are used to identify orders and common families of insects in North America, north of Mexico. This type of key gives the user two choices in each couplet followed by a number which directs the person to the next part of the key. By careful and accurate interpretation of the insect characters and their associated illustrations in the key or text, the user proceeds through the key and narrows down the alternatives until the unknown specimen matches a description of an order or family. The number in parentheses indicates the part of the previous couplet that directed the user to the current number. The numbers in parentheses are especially useful in backtracking at any point or taking a specimen in a suspected order or family and working it backwards through the key to confirm the identification.

ILLUSTRATIONS

Illustrations of whole insects represent selected common species and in some cases show the diversity of shapes occurring in the orders and families. The reader should not assume that a specimen identified to a family is the same species as a similar insect shown in an illustration unless the specimen in hand matches the written description of the illustrated species.

DESCRIPTION OF ORDERS AND FAMILIES

The orders in this book are in general phylogenetic (evolutionary) arrangement, based on structural and developmental characteristics, from the most primitive to the most advanced. The families within each order are also listed phylogenetically in the same manner except for the Lepidoptera which are listed in reverse order (most to least advanced). General information is provided for quick identification and a basic understanding of the natural history of insects in the orders and families. Occasionally a commonly used alternate name or an older name still used in textbooks will be placed in parentheses next to the family heading.

DESCRIPTIONS OF GENERA AND SPECIES

In some families one to several genera are described rather than species because the families contain too many similar species or the species are identifiable only with complex keys.

Species are best identified through the use of keys. However, some common species can be identified accurately without a key if various features are described. Once the specimen has been identified to family, one may check the species' descriptions to see if a match occurs. All the characteristics given for a species must match (color is sometimes an exception) because in some cases a single feature may eliminate an incorrect but very similar species. For most families only a few of the common species are provided and if the collector's specimens are quite common but do not match any description very well, then it is best to assume that the specimens are not described in the book. What is common and abundant in one part of a county, state, or region may be rare or absent in another area.

Dimensions given in millimeters represent the total body length from the front of the head to the tip of the abdomen unless otherwise specified. Wingspread is the span of the front wings from tip to tip; wing length is the length of a single front wing. A millimeter rule is reproduced at the front and back of the book for measuring insects.

The ranges given for species cover general regions and a certain amount of overlapping occurs. The U.S. regions are as follows:

1. Eastern U.S.—Wisconsin south along the Mississippi River to approximately northern Mississippi; east to South Carolina and north to Maine.
2. Southern U.S.—Eastern Texas to Florida; north to southern Missouri and southern Virginia.
3. Great Plains—Iowa west to the Rocky Mts.; southwest to northern Texas.
4. Midwest—Similar to Great Plains but shifted slightly eastward to include Missouri, Minnesota, Illinois, and Indiana.
5. Southwestern U.S.—Texas north to Kansas; west to southern Utah, southern Nevada, and Arizona.
6. Northwestern U.S.—Washington, Oregon, and northern California east to Idaho and western Montana.
7. Far West—Washington to California, western Idaho and Nevada.
8. Western U.S.—west of a line from approximately western Texas north to Montana.

The numbers of North American species given in this book represent those found north of Mexico. Data for the large orders are approximate values as no precise information on the numbers of species is available for North America or the world.

How to Collect Insects

EQUIPMENT AND SUPPLIES

Minimum Needed
1. Killing jar if insects are not studied alive.
2. Toxic chemical for the killing jar.
3. Insect net for flying or hopping insects.

Supplementary Materials
1. One or two small, flat, tissue-lined containers to store dead insects periodically removed from the killing jar.
2. Extra killing jar.
3. Vials filled with alcohol or other killing solution; soft-bodied insects (e.g., larvae, aphids, termites, etc.) are placed in these vials.
4. Forceps and syringe to handle small insects on land and in water, respectively.
5. Envelopes to store butterflies, moths, dragonflies and damselflies.
6. Knife to pry bark from trees or dig into rotting wood.
7. Collecting bag for the above materials.

Specialized Equipment
1. Beating tray or umbrella for insects that fall to the ground from trees or shrubs.
2. Aspirator to pick up minute insects.
3. Sifters and funnels to separate insects from debris.
4. Nets and screens to collect aquatic insects.
5. Lights to attract insects at night.

DESCRIPTION OF EQUIPMENT AND SUPPLIES

Killing Jars
Killing jars or bottles may be made easily or otherwise purchased from a biological supply company. Wide-mouth jars (6-16 oz.) with air-tight lids are commonly used. A large test tube or slender bottle with a cork or rubber stopper is a handy, supplemental container that fits easily into a pocket for carrying.

Ethyl Acetate Jar (Fig. 1)—Make a thick mixture of plaster of paris and water. Pour the mixture into a clean jar to a height of 20-30 mm and allow to air dry at room temperature (requires several days) or under low heat (light bulb or lowest oven temperature). When completely dry add enough ethyl acetate to saturate the plaster of paris, pour back any excess, and replace the lid or stopper. There should not be any standing liquid. Ethyl acetate acts as a fumigant and is obtained primarily from biological or chemical supply companies (see the list of companies at the end of this section). It is safer to use than cyanide or carbon tetrachloride. The bottom and lower sides of the jar can be taped to reduce potential breakage. A label with "poison" written on it can be adhered to the jar. Cotton or pieces of rubber tubing covered by a tight layer of cardboard may be substituted for plaster of paris although they are less suitable.

chemical supply company and occasionally an insecticide dealer.

Figure 2 Cyanide killing jar. a, hard plaster of paris; b, dry plaster of paris or sawdust; c, cyanide crystals; d, tape.

The killing jar is made by placing a layer about 5-10 mm thick of the finely granular or powdered cyanide compound on the bottom of a clean and dry jar. The cyanide is covered with a slightly thicker layer of dry plaster of paris (or fine sawdust) followed by a wet plaster of paris layer of similar thickness. After the plaster has set and dried (a day or two outdoors or in a garage), the jar is capped or corked and ready to use in about two more days.

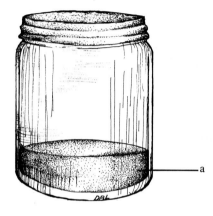

Figure 1 Ethyl acetate killing jar. a, hard plaster of paris.

The killing jar must be recharged with ethyl acetate after several hours of collecting depending on how often and how long the lid was removed. The lid should always be left on unless the collector is actually placing an insect in or removing one from the jar.

Cyanide Jar (Fig. 2)—Cyanide is much more hazardous than ethyl acetate but if handled with care it is reasonably safe to use. Cyanide is preferred by many collectors because it lasts for months to a year or more and kills quickly. The jars must be taped to reduce the chance of breakage and marked with a "poison" label. Potassium and sodium cyanide are the least dangerous to work with and last the longest. They may be obtained from a biological or

Other Toxic Chemicals—Substitutes for ethyl acetate include carbon tetrachloride, chloroform, and cigarette lighter fluid. All are more toxic to humans than ethyl acetate. Alcohol does not work well in a killing jar and should be avoided.

Insect Nets

Various types of insect nets may be homemade or purchased from a biological supply company.

Aerial Net (Fig. 3)—This type of net has an open mesh bag with a reinforced rim of muslin. The net is swung easily, causes the least damage to the insect, and enables the collector to view the specimen through the bag. The netting portion of the bag may be made of nylon or nylon marquisette, scrim, organdy, or fine-mesh bolting cloth. A pattern is shown in Figure 3A. The net's rim is generally 0.3 m (12 inches) in diameter and made of number six or eight wire. The wooden or aluminum handle is about 1 m long with two grooves and holes cut to receive the ends of the wire hoop (Fig. 3B). A metal collar, hose clamps, or a wrapping of wire, heavy cord, or friction tape will hold the rim to the handle.

Collapsible nets can be devised by using hinges that divide the handle and rim into two folding parts. The frame of a collapsible, fish-landing net also serves this purpose.

Combination Aerial-Sweeping Net—The net bag consists of muslin except for the bottom third which is netting (Fig. 3C). The bag can be used for sweeping with minimal damage from tearing and the insects at the bottom can be seen. This type of net is good for all-purpose collecting.

Sweeping Nets—The net bag is made of strong muslin or fine-mesh bolting cloth to withstand damage when sweeping vegetation. The rim is reinforced with canvas.

Aquatic Nets and Related Equipment—Aerial and combination aerial-sweeping nets work satisfactorily when collecting aquatic insects in water unless continuous collecting is planned. In this case more specialized nets are available.

Standard and Heavy-duty Nets—Standard aquatic nets have a metal rod for a rim, the netting is very strong, and heavy muslin or canvas reinforces the rim to help withstand the weight of debris, mud, and water that may be brought to shore. The net bag is usually much shallower than terrestrial bags. Heavy-duty nets have very stout handles and heavier bags and rims. Aquatic nets also are made with a triangular, square, or "D"-shaped rim to help scrape the bottom of ponds and streams.

Dip Nets—The shallow, heavy-duty bag of these nets is usually no deeper than the diameter of the rim. The strong handle may be 1.7-2 m long to aid in reaching out into the water.

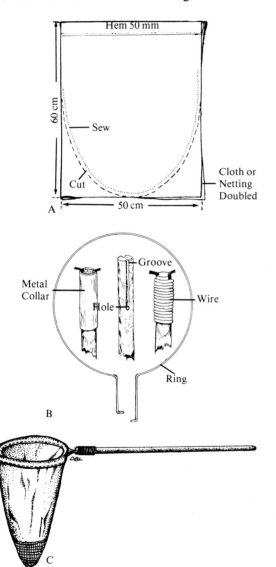

Figure 3 Insect net. A, pattern for making a 0.3 m (12 inch)-diameter aerial or sweeping net; B, ring, grooved handle, and two types of collars; C, combination aerial-sweeping net bag.

Kitchen strainers 100-150 mm in diameter will work as dip nets. A long-handled, white enamel dipper is a standard device used to sample mosquito larvae.

Apron Nets—These are triangular, scoop-shaped nets with a long handle (Fig. 4). The bottom is made of fine screening and the top is covered with coarse mesh screening which keeps heavy vegetation from entering. A trap door opens at the back.

Screens and Drift Nets—A piece of 20 mesh wire or plastic screening about 45 cm long and 30 cm high can be attached at each end to a wooden handle (Fig. 5). The screen is held vertically in a stream as another person disturbs the streambed while walking about 2 m upstream, or the collector may walk backward upstream while holding the screen. Drift nets are usually made of netting rather than screening. They can be used as a large handscreen with a person holding each end, or smaller ones with square rims can be tied to a tree or rock and faced upstream (Fig. 6). Drifting insects are caught in the net bag.

Figure 6   Drift net for stream insects.

Dredges and Miscellaneous Samplers—Ekman and Peterson dredges are used to sample a specific area of a lake bottom. The traplike jaws are opened, the dredge lowered on a rope to the bottom, and a metal messenger-weight dropped along the rope to close the jaws on the top layer of mud. These devices and others which sample specific areas of streambeds or lake beds are described in references for insect collecting on pages 24 (Peterson 1964) and 36 (Merritt and Cummins 1978; Usinger 1956).

Beating Trays and Umbrellas

Many insects fall to the ground when vegetation is disturbed rather than fly away. To collect these specimens a slightly concave frame can be made in the fashion of a kite frame with one of the two crossmembers left longer to serve as a handle (Fig. 7). The four ends of the crossmembers insert into pockets sewed in each corner of a square piece of sheeting or canvas. An old umbrella will serve the same purpose although the handle should be hinged to lay horizontally. The tray or umbrella is carefully placed under the bush or tree and the plant shaken or beaten. Best results are often ob-

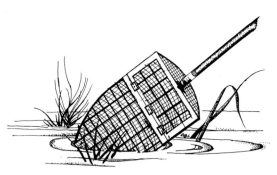

Figure 4   Apron net for specialized aquatic collecting.

Figure 5   Handscreen for stream insects.

tained at night or while the air is still cool in the morning and the insects are inactive.

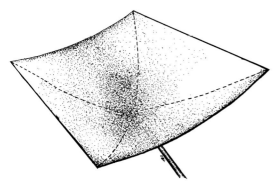

Figure 7 Beating tray to catch insects dislodged from trees and shrubs.

Aspirators

Small insects may be captured alive by sucking them into a glass or plastic vial or tube (Fig. 8). The collector sucks on a mouthpiece inserted into one end of a rubber tube. The other end surrounds a glass or metal tube which inserts through a cork or stopper into the vial. The end of the tube in the vial is covered with a piece of fine netting or screening to prevent the inhalation of insects. A bent glass or metal tube protrudes from the vial and is placed against the desired insect. A direct sucking-type aspirator should not be used to remove insects from bird or mammal nests; aspirators where the collector blows to obtain a sucking action are available for this purpose.

Sifters and Funnels

Sifters—Small insects in trash and other debris may be captured by sifting the material through a box with a screen bottom, or a series of screens of increasing mesh, and onto a white sheet or shallow enamel pan. Bottom samples from streams, ponds, and lakes are flushed through a series of sieves of increasing mesh that are placed on top of each other. The contents of each sieve are sorted under a little water in a shallow enamel pan.

Funnels—The Berlese funnel is a standard device used to extract small insects from soil, plant debris, and moss (Fig. 9). The material is

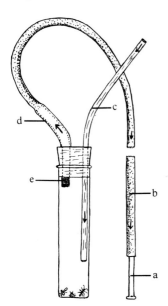

Figure 8 Aspirator for very small insects. a, glass or plastic mouthpiece; b, rubber tubing; c and d, metal, plastic, or glass tubes; e, wire or cloth screening.

Figure 9 Berlese funnel for minute insects in debris. a, light bulb; b, ventilated cover (optional); c, screen; d, large funnel; e, container of alcohol.

placed on a coarse screen platform which has been inserted into a large funnel. The tip of the funnel extends into a jar of 70% alcohol. Heat from a light bulb suspended above the funnel slowly dries the debris after several hours to a day and drives the insects downward into the alcohol.

Lights

Many kinds of insects are attracted to lights and can be captured by hand or netted. Incandescent lights and camping lanterns work satisfactorily although fluorescent lights and especially blacklights are usually superior. Blacklight lamps are easily obtainable; a 15 watt lamp with the correct fixture is a good size and not too expensive. A DC-AC converter may be purchased so the AC unit can be operated by using a car battery (the converter can be plugged into the cigarette lighter). Very low-wattage blacklights also can be purchased to replace the fluorescent lamps of portable, battery-operated camping lanterns.

A white sheet suspended vertically behind the blacklight acts as a light reflector and as a resting site for insects. The sheet can be oriented to face into woods, a lake, etc.

Traps

Many types of traps are described in Peterson (1964) which is referenced on page 24.

Light Traps—The simplest trap consists of an incandescent or blacklight bulb suspended over a funnel (Fig. 10). The funnel sits on the rim of a large can that contains an open killing jar. The space between the rim and funnel should be made nearly airtight by using weatherstripping, etc. Another design uses a vertical blacklight with three or four baffles placed parallel to the lamp (Fig. 11). Specimens captured in light traps are not always in good condition and often are covered with scales from trapped moths. Fewer insects are collected on moonlit nights.

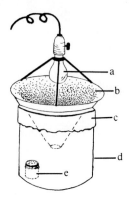

Figure 10   Light trap for nocturnal insects. a, light bulb (regular or blacklight); b, funnel; c, rubberstrip or weatherstripping to hold fumigant inside; d, can or other container; e, open killing jar.

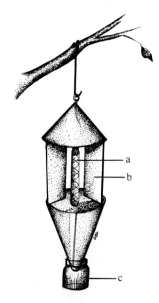

Figure 11   Light trap with baffles. a, fluorescent blacklight; b, baffle; c, killing jar.

Pitfall Traps—Ground-inhabiting insects are captured by various types of sunken traps. A tin or plastic container with holes in the bottom for water drainage is sunk in the ground until the rim is at ground level (Fig. 12A). Bait such as molasses, candy, fruit, meat, fish, manure, fermented solids or liquids, 3% formalin, etc., is placed in the bottom and the attracted insects

8   How to Collect Insects

will fall in and generally be unable to leave. To reduce damage from rain and large animals, a roof can be made from a flat rock or board that rests on supports 2-3 cm high placed around the opening. Traps located in fields, woods, sandy trails, shorelines, etc., all yield different types of insects.

Colored Trays and Panels—Shallow, colored trays filled with water and a little detergent will attract aphids and other insects. Bright yellow trays are especially attractive. Colored panels smeared with a very thin layer of sticky adhesive or petroleum jelly also are used.

Fly Traps—Flies are attracted to bait and then crawl or fly upward into a holding chamber when trying to leave (Figs. 12 B, C).

Malaise Traps—These tentlike, square or rectangular traps consist of panels or baffles of fine netting supported vertically by one or more stakes and guy ropes (Fig. 13). The trap may be 2 m or more in height and several meters long depending on the modification. The trap takes advantage of the tendency of many flying insects to crawl or fly upward when hitting an obstacle. By doing so the insects become trapped at the highest point inside the netting and move into a killing jar or jar of alcohol. Fast-flying

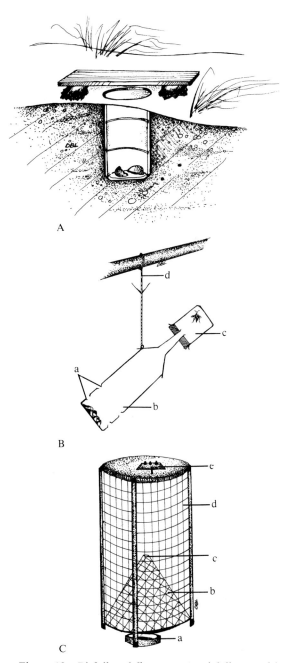

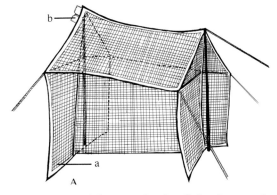

Figure 12 Pitfall and fly traps. A, pitfall trap with bait in container. B, suspended fly trap. a, openings; b, bait chamber; c, capture chamber (without poison); d, greased suspension wire. C, self-supporting fly trap; a, bait; b, inverted funnel; c, opening; d, screen cage; e, entry door.

Figure 13 Malaise trap for fast-flying insects. A, netting portion; a, baffle, b, killing jar.

How to Collect Insects   9

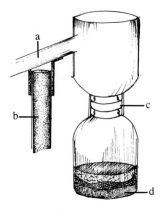

Figure 13 (cont.) Malaise trap for fast-flying insects. B, killing jar; a, entryway at highest point of netting; b, supporting pole; c, connecting collar; d, poison.

wasps, flies, and other insects not often collected are trapped.

COLLECTING TECHNIQUES

Terrestrial Insects

Adults—When using a net, the collector may look for particular insects and swing at them or walk rapidly through a field sweeping the net sideways through vegetation in order to obtain a larger number and variety of specimens. Once insects are in the net, their escape should be prevented by twisting the net sideways so the bag folds over the rim (Fig. 14). The net then should be placed on the ground. If the insects

Figure 14 Twisting the net handle to retain captured specimens.

are to be killed, a killing jar freshly charged with the poison should be used. Most adult insects, with the exception of butterflies and moths, can also be killed in alcohol or other solutions and no great color loss or other damage occurs; however, killing jars are usually preferred.

There are various ways to remove the insects from a net.

A. Work the uncapped killing jar or other container into the bag and intercept the insects on the walls. This method is often used to collect butterflies and moths since the scales (powdery material) are easily rubbed off if the insect is handled. Many collectors of butterflies and moths will grasp the specimen through the net (never by the wings) and gently pinch the thorax to stun the insect before placing it in the killing jar. This method reduces wing damage and scale loss from fluttering.

B. Swing the net rapidly a few times to force the insects to the end of the bag and insert the end into the killing jar. Cap the jar and after the insects are stunned they may be removed from the net and replaced in the jar. This method works well with stinging insects such as bumble bees, honey bees, and wasps.

C. Invert the whole net into a large killing jar or cage.

D. Open the net bag and select the desired insects by picking them up with the fingers or with forceps, scraping them into a killing jar, or using an aspirator.

Most insects will die within 10 minutes in a newly charged killing jar. Dead insects should be removed within an hour or two or discoloration may occur (especially in cyanide jars). A hot killing jar also causes discoloration. In addition, moisture may form on the inside walls of hot jars and cause damage to many insects (especially butterflies). The walls can be wiped periodically or some tissue paper added while collecting to absorb the moisture and also cushion the insects. Large living specimens like grasshoppers and beetles should not be placed in the jar with smaller insects (dead or alive) or damage to the latter may result. Delicate butterflies and moths should be killed in a separate, dry jar.

Killing jars should never be allowed to fill up with dead insects. Many collectors use small boxes or paper or celluloid envelopes to transfer dead specimens while collecting. The insects are placed between layers of tissue paper in the boxes. Envelopes work well for butterflies, moths, dragonflies, and other large-winged specimens. Celluloid stamp and coin envelopes may be used. Paper envelopes of various sizes are made easily by cutting rectangular strips of smooth, white paper and folding them in triangles (Fig. 15). Write the collecting data on the outside of the envelope and place the insect inside with the wings together over the body.

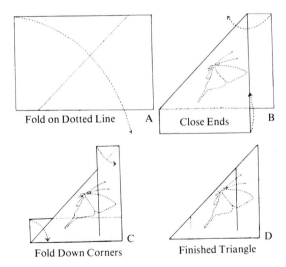

Figure 15   Procedure for making paper triangles, A-D.

Minute insects are collected using an aspirator or forceps, or by dipping a small watercolor brush in alcohol and dabbing it on the insect. A jackknife or sheath knife is very useful in cutting open fruit, galls and stems, peeling bark, and cutting into rotting wood.

Soft-bodied adults and minute adults that will be mounted on slides or preserved in alcohol may be placed directly into alcohol or another killing solution. These insects include aphids, bristletails, caddisflies, fleas, some katydids and crickets, lice, mayflies, springtails, termites, and thrips.

Immatures—Most immature insects are soft-bodied and thus killed in liquids to prevent shrinkage. The few that are hard-bodied may be killed in a killing jar if desired. Liquid killing agents and methods include the following:

A. Drop immatures directly into vials of 70-90% ethyl or isopropyl (rubbing) alcohol carried in the field. Although this method is the easiest, large larvae (e.g., moth caterpillars, beetle grubs) may turn brown or black due to bacterial decay from poor alcohol penetration. Methods B-D will usually prevent discoloration of large larvae.

B. Place living immatures in near-boiling water for up to five minutes and transfer them to 70-90% ethyl or isopropyl alcohol. Small insects are left in the water only a minute or so.

C. Drop immatures directly into vials of a 95% alcohol-xylene mixture (1:1). Transfer the specimens to 70-90% alcohol within a few hours.

D. Drop immatures directly into KAAD (7-10 parts 95% ethyl alcohol, 2 parts acetic acid, 1 part kerosene, and 1 part dioxane). Transfer the specimens to 70-90% alcohol within a few hours.

Aquatic Insects

Adults and Immatures—Once the specimens and associated mud and debris are in a net or other collecting device, the contents are poured into a white enamel pan, sieve or, if necessary, on the bare ground. Select and pick up the desired specimens with the fingers, forceps, or a syringe. The hard-bodied adults are placed in a killing jar or in liquid and the soft-bodied adults, minute adults, and immatures are dropped into alcohol or another liquid killing solution for immatures as described above.

How to Preserve and Mount Insects

Adult insects that are not placed temporarily or permanently in alcohol are mounted on insect pins or, if too small to pin, mounted on points or minuten pins. Immature insects are preserved in alcohol or alcoholic solutions. Adults should be mounted preferably the same day they are collected because once dry they are brittle and easily broken. Refer to the chapter on dragonflies and damselflies for an alternate method to pinning these insects. Insects mounted at a later date are stored dry in boxes or envelopes and require relaxing (softening) before mounting. Specimens may be kept in a pliable condition by refrigerating them in a plastic container and adding chlorocresol. Even after they are relaxed insects are subject to more damage than if they had been pinned the same day as collected.

RELAXING METHODS

A relaxing chamber can be made from a plastic freezer tray with a snap-on lid, a wide-mouthed jar or can, or any broad, airtight container. The bottom of the container is covered with wet cotton, sand, or cloth. The specimens are put in an open, shallow box which is placed in the container. Alternatively, a platform can be used to support the specimens over standing water in the bottom of a tall container. With any method used, a tablespoon of moth flakes, ethyl acetate, or a minute amount of carbolic acid (phenol) should be added in a separate container to prevent molds from growing on the insects. The chamber is closed tightly for one to several days until the insects are pliable. If the insects are left inside too long or get too damp, they will discolor and disintegrate.

Barber's fluid will also relax specimens when they are dipped into it for a few minutes. The fluid is made from 95% alcohol and water (25 parts each), ethyl acetate (10 parts), and benzene (3.5 parts).

PINNING SMALL TO LARGE INSECTS

Adults that are hard-bodied and sufficiently large are generally pinned. Most retain their colors and shrivel little, if at all.

Insect Pins

Only steel insect pins should be used; sewing pins and other short types are undesirable. Insect pins are sold in sizes from 00 (extremely slender) to 8 (large diameter). Sizes 1 to 3 are the best for general use. Size 0 may be used to pin very slender insects but the pin tends to bend easily.

Pinning the Insect

Most insects are pinned vertically through the body so that when the pin is vertical the insect is horizontal and not tilted (Fig. 16). The insect may be held in the air or on a flat surface (cardboard, styrofoam, old table) with the thumb

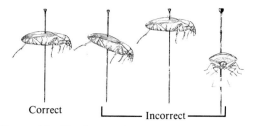

Figure 16 Pinning an insect correctly and incorrectly.

and forefinger or with forceps while pinning. All insects and labels should be pinned at a uniform height above the pin point—about 25 mm for insects. A pinning block may be made or purchased to assist in uniformity (Fig. 17). To use the block, pin the insect and insert the point into the deepest (25 mm) hole. Gently push the insect up to the correct height. The middle hole is used to position the label bearing the collecting data and the bottom hole is used for an additional label if needed.

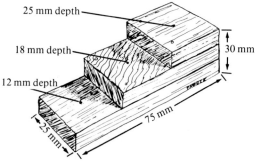

Figure 17 Pinning block to position insects and labels on a pin.

The legs and antennae do not need to be extended into a lifelike position although they should project somewhat to be easily visible. If spreading is desired or if the abdomen sags, the pinned specimens can be pushed into a soft surface such as styrofoam, cork, balsa wood, etc., and the appendages and abdomen arranged. A sagging abdomen can be supported with a piece of cardboard pinned beneath the abdomen or the pinned insect inserted into a vertical surface with the abdomen hanging downward until dry.

Insects are pinned in the following body locations:

1. **Thorax between the bases of the front wings** (Fig. 18A, B)—Butterflies, moths, mantids, cockroaches, flies, wasps, bees, and other insects not included below. Mayflies are sometimes pinned but identical specimens should also be preserved in alcohol. Insert the pin slightly to the right of the midline on flies, wasps, and bees if they are large enough.

2. **Posterior part of the pronotum** (Fig. 18C)—Grasshoppers, katydids, and crickets; treehoppers, leafhoppers, planthoppers, and spittlebugs if they are large enough.

3. **Scutellum** (Fig. 18D)—Stink bugs and other large true bugs (Heteroptera). Pin slightly to the right of the midline.

4. **Right front wing (elytron) toward the base** (Fig. 18E, G)—Beetles. Many collectors pin true bugs (Heteroptera), leafhoppers, planthoppers, and spittlebugs in this position.

5. **Left side of the thorax below the wing base** (Fig. 18F)—Dragonflies and damselflies. The left side faces up and the wings are folded over the back before pinning. The specimens may need to be placed in an envelope for a day to hold the wings in place before pinning (see the chapter on Odonata). These insects can also be pinned as in part 1.

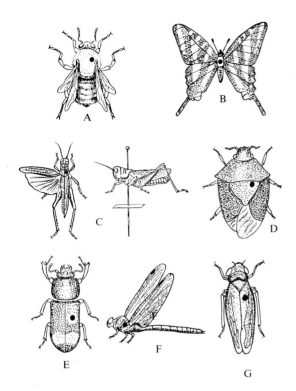

moved up to the same height as for a pinned insect. Insects are generally glued with their right side against the tip of the point which requires that the tip be bent downward with forceps (Fig. 19A). The tip is touched to a minute amount of a fast-drying glue (e.g., household cement, clear nail polish, white glue) and pressed against the insect. Wasps and flies are often mounted on unbent points with the left side of the insect up and the legs or abdomen pointing toward the pin (Fig. 19B). Very flat insects (e.g., Heteroptera) are glued ventrally to the tip of the point (Fig. 19C). A label is added and turned parallel to the point (Fig. 19D) so both extend to the left.

Figure 18 Pinning areas on insects. A, bees and flies; B, butterflies and moths; C, grasshoppers and crickets; D, stink bugs; E, beetles; F, dragonflies and damselflies; G, leafhoppers. See the section on pinning insects for locations on other insects.

MOUNTING VERY SMALL INSECTS

Small or fragile insects likely to be broken or disfigured by pinning are usually mounted on a card point, minuten pin, or microscope slide. Small insects can also be preserved in alcohol or other liquid mixtures. Refer to the sections on Microscope Slide Preparation and Preserving in Fluids for further information on these methods.

Points are elongated triangles cut with a point punch or scissors from light, white cardboard such as file cards. The triangles are 3-4 mm wide at the base and 8-10 mm long. A pin is inserted through the base and the triangle

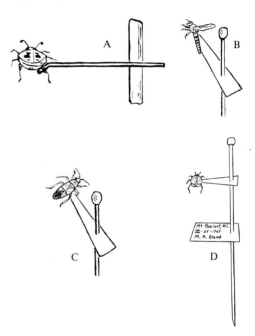

Figure 19 Mounting small insects on points. A, common method with tip of point bent; B, wasps and flies with left side up; C, bugs and other flat insects on unbent tip; D, point and label placement on a pin.

Minuten pins are very fine steel pins. One end is pushed through the specimen and the other end inserted into a piece of cork or other spongy material. A regular insect pin is inserted through the cork (Fig. 20).

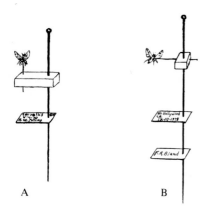

Figure 20 Mounting small insects on minuten pins. A, one label; B, two labels.

PRESERVING IN LIQUIDS

Most immature insects and soft-bodied adults are usually preserved in alcohol to prevent shrinkage. Hard-bodied adult insects may also be preserved in fluid although the wings of butterflies and moths are damaged. Generally 70% ethyl alcohol is used although some collectors prefer higher concentrations for certain insects. Isopropyl (rubbing) alcohol (70%) is also satisfactory. The alcohol should be changed once several days after the immatures are first immersed. Alcohol combined with other chemicals kills immatures better than alcohol alone (see the previous chapter on collecting immatures) but these solutions must be replaced by alcohol within a few hours. Colors like green and yellow will always fade when larvae are preserved in liquids.

Storage containers such as vials or bottles need rubber or other synthetic stoppers if available. Cork stoppers tend to shrink and crack, resulting in fluid evaporation. Screw-cap vials work well if the cap fits tightly or has a tapered insert that is self-sealing. Stoppered, corked, or capped vials can also be placed in a large jar filled with alcohol and tightly capped to prevent fluid evaporation. Some collectors add a little glycerin to the vials to help keep specimens moist in case the alcohol completely evaporates accidentally.

MICROSCOPE SLIDE PREPARATION

Very small insects or insect parts including wings are often viewed more conveniently and more accurately when mounted on a glass slide (Fig. 21) and magnified with a microscope or projected on a screen. Temporary or permanent mounts may be made. Glass microscope slides, cover glasses (coverslips), and a mounting medium are needed. Biological stains are sometimes used for transparent specimens.

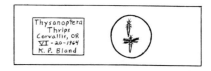

Figure 21 Microscope slide and label.

Temporary Mounts

Water, alcohol, or glycerin serve as a medium for temporary mounts. Specimens may be placed directly into these media from the alcohol preservative. Water and alcohol may evaporate within minutes but glycerin remains for months. A ring of nail polish or other "ringing" cement can be applied to the edges of the cover glass to make a semipermanent slide of glycerin.

Permanent Mounts

Hoyer's Solution—This is a common, water-base medium. It bleaches the specimens slightly. Living insects or dead specimens in alcohol are first dipped into water for about 30 seconds and then placed directly into the drop of Hoyer's solution on the slide. The specimen is oriented with forceps and a cover glass is placed on its edge and gently lowered over the medium. The medium should flow to the edges with light pressure applied to the cover glass as it pushes out most of the air bubbles. A ringing solution may be used to seal the mount so the medium will not dry out.

Resin-based Mounting Media—The most common medium used is Canada balsam although other satisfactory materials are available. The following procedures are used:

A. Dehydrate the specimen in freshly prepared 70% ethyl alcohol for 30 minutes to several hours, depending on the insect's size.
B. Transfer with forceps to pure 95% ethyl alcohol for 15-30 minutes. Highest quality slides are made by transferring specimens to 100% ethyl alcohol but this addition is not needed except for critically important specimens.
C. Transfer to xylene (xylol) for 10 minutes. If the xylene turns cloudy (water is still present) repeat step B.
D. Transfer to several drops of the mounting medium that have been placed on a clean slide. Leave room on the slide for a label.
E. Orient the specimen with forceps under a dissecting microscope if needed and add a cover glass. Attach a label.
F. Keep the slide flat and allow to dry. Drying requires several weeks at room temperature but a slide warmer or a very low-temperature oven will reduce the drying time to several days. A light bulb may be used for heat but the light should not directly strike the specimen. Do not stack slides on each other.

If the specimen is too thick and the cover glass rests on it instead of the medium, the edges of the medium can be built up in layers by allowing them to thicken. A microscope slide with a central concavity also can be used.

SPREADING THE WINGS

The wings of butterflies and moths are commonly spread on a pinning board or similar flat surface to aid in identification and to exhibit their full color. If the colored wings or wing venation of other insects are useful in identification, or if the wings are large and do not fold in a neat manner, spread the wings of at least a few specimens (e.g., grasshoppers with colored hind wings, dragonflies, and damselflies).

Pinning Board

Insect wings are spread on a pinning board unless the specimen is to be displayed under glass, in which case an ungrooved flat surface is often used (see the next section). The board may be purchased or constructed (Fig. 22). Styrofoam, cork, balsa wood, or other sheets of material may be substituted for a board since it is not critical to have sloping sides. A groove should be cut for the body and legs of the pinned insect.

Spreading Procedure

1. Pin the insect through the dorsal side and adjust to the correct height. The specimen must have been recently killed or previously relaxed. Insert the pin near one end of the board's groove until the base of the wings are even with the top of the board. If the wings are folded together they may be opened by squeezing the sides of the thorax with forceps just below the wing bases. A pin on each side of the abdomen can be in-

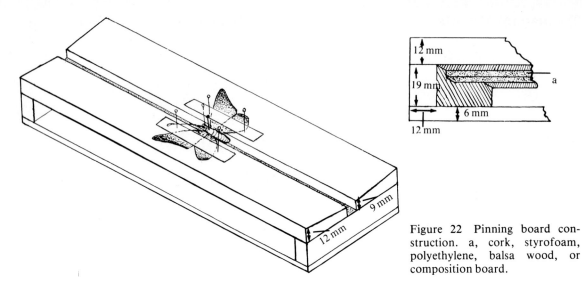

Figure 22 Pinning board construction. a, cork, styrofoam, polyethylene, balsa wood, or composition board.

serted into the groove to prevent the abdomen from twisting. One strip of paper is placed parallel to each side of the groove and pinned at one end.

2. Using a pin, gently push against the base of a large, front wing vein to force the wing forward. Hold the wing in place by gently applying pressure from the strip of paper and insert a pin close to the anterior edge of the front wing (Fig. 23A). *The posterior margin of butterfly and moth front wings should be at right angles to the body.* Front wings of other insects do not need to be moved quite so far forward but should not cover the hind wings. Repeat the procedure for the other front wing.
3. Move the hind wing forward in the same manner and place a pin through the paper strip just behind the wing (Fig. 23B). The anterior edge of the hind wing of butterflies and moths goes under the posterior edge of the front wing. Repeat the procedure for the other hind wing. Some collectors cover the front and hind wings with microscope slides to prevent wing curling and others use broad paper strips for the same result.
4. Orient the antennae and hold them in position with pins or underneath the strips of paper (Fig. 22). If the abdomen sags it will need to be supported with pins. Pin the completed collecting label on the board next to the insect.

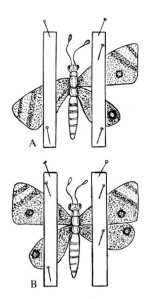

Figure 23 Spreading butterfly and moth wings. A, front wing on one side moved forward until posterior margin is at a right angle to the body; B, hind wing moved forward until anterior margin is under the front wing.

5. Drying time varies with the size of the insect (larger moths may take a week) and the prevailing humidity and temperature. Placing the specimen in an oven at the lowest temperature or under a light (preferably not direct light) will speed drying.

Ungrooved Pinning Board

If insects are to be mounted against a surface and under glass (e.g., Riker mount), a pin is inserted from below and the wings are spread with the insect upside down. The procedures are identical to those used on a pinning board. Once spread, forceps are used to hold the specimen down while the pin is removed.

PREPARING LABELS

All specimens must be labeled. The locality and date of collection are especially important as a future reference for collecting and as a source of basic scientific data on the species. Additional data such as the habitat and name of the collector add to the value of the specimens. Small and neatly written labels are also important in preparing an attractive collection.

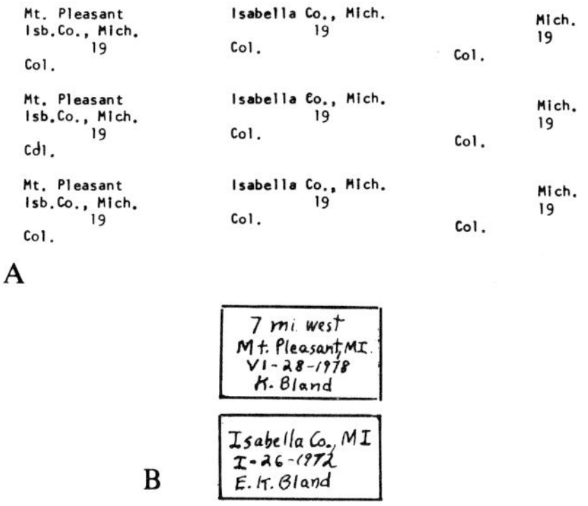

Labels are made easily from white file cards or stiff white paper and should be cut no larger than 8 × 15 mm. Labels printed with a designated locality and collector's name may be ordered from a biological supply company or cut from a photographic reduction of a typed page (Fig. 24A). The locality (town, county, or range/section, and the state), date (VI-28-78 or 28-VI-78 for June 28, 1978), and collector's name are written on a single label in black India ink with a quill pen or other fine-pointed pen (Fig. 24B). A single label is placed about 15 mm above the pin point (Fig. 18C). If the insect is on a point or minuten pin the label is oriented parallel to them (Figs. 19D; 20A).

A second label, if used, could contain the name of the collector (if it did not appear on the first label) and the habitat, or the name of the species. If the label bears the collector's name and/or the habitat, it goes below and parallel to the first label (Fig. 20B). If the label is used to name the species, it would be large (25 × 25 mm) and rest on the pinning surface. This label contains the scientific name of the species, the author, the name of the person who identified the specimen ("det." is an abbreviation for determination, that is, who identified the specimen), and the date of the determination (Fig. 25).

Figure 24 Insect labels. A, pre-printed labels; B, hand-printed labels.

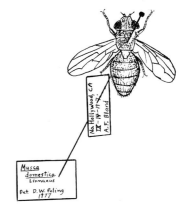

Figure 25 Species label showing the scientific name, author, and person who identified the specimen.

How to Preserve and Mount Insects

Labels for insects preserved in liquids are inserted into the container with the specimen. Microscope slide labels should contain all the data described above and often the name of the family and order are also included. A label may be used on each side of the cover glass if needed.

MISCELLANEOUS PREPARATION AND PRESERVATION TECHNIQUES

Clearing Specimens

Dark specimens often need to be made more transparent before mounting on microscope slides. Fleas and dark-colored lice and thrips are often cleared to varying degrees of transparency. The most readily available solution is 10-15% potassium hydroxide (KOH). The specimen is soaked in warm or cold KOH, the latter temperature requiring several hours to days to clear the insect. Gently warming the specimen in KOH may clear it in a few minutes but damage may occur quickly if the process is not carefully monitored. The insect is then placed in distilled water (sometimes a small amount of acetic acid is also added) and a standard mounting procedure followed for microscope slides.

Butterfly and Moth Wing Mounting

The venation of lepidopteran wings often must be visible for identification to family. If scraping scales from small areas or adding a few drops of xylene to the wing does not work satisfactorily, bleaching and mounting may be necessary. The procedure is as follows:

A. Using forceps to grasp the wing bases, carefully remove the front and hind wings together from one side.
B. Immerse in a shallow container of 95% alcohol for several seconds.
C. Immerse in 10% hydrochloric acid for several seconds.
D. Immerse in a 1:1 solution of table salt (sodium chloride) and chlorine bleach (sodium hypochlorite) for a few minutes or until all color is gone.
E. Dip in water to remove excess solution.
F. Place the wings on a slide by picking them up with forceps or floating them on water and bringing the slide up from underneath. Orient the wet wings with the tips to the right. Allow to air dry.
G. Place a large cover glass (50 × 50 mm) over the wings and wrap it with tape. A paper border is sometimes placed around thick wings to give a better seating for the cover glass before the latter is added. Label the mask or adhere a label to the slide.

Inflating Larvae

Artificially inflated larvae may be pinned in a collection along with the pinned adults of the same species or family. If inflation is carefully done, the caterpillars present a lifelike appearance, retaining most of their spines, markings, and colors. Various types of apparatus may be purchased from supply houses or the collector can assemble the needed materials (Fig. 26A). The inflating procedure is as follows:

1. Kill the larva in hot water and place it on a piece of blotting paper. Make an incision around the anus and press out the body contents using a pencil or other similar object. Starting from the thorax and continuing to the tip of the abdomen, roll the pencil with just enough pressure to force out most of the body contents but not enough to rupture the skin or break the hairs.
2. Insert a piece of glass tubing, which has been drawn to a long point, into the anal opening. Fasten the end of the caterpillar

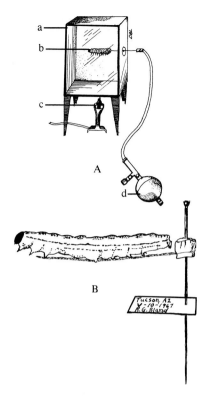

Figure 26 A, Inflation apparatus for larva; B, mounted larva. a, oven; b, inflated larva; c, heat source; d, constant-pressure inflating bulb.

to the tube using a hooked wire clip on the tube. Some collectors attach the skin by heating the point of the glass tube before inserting it and the wet skin adheres to the tube.

3. Inflate the caterpillar by forcing air steadily through the tube, using the mouth, a constant-pressure rubber bulb, or some other small air pump. While inflation is maintained, the insect is dried quickly in an oven hot enough to bake thoroughly without scorching. A rectangular can on an inverted electric iron or supported over a light bulb or an alcohol lamp makes a good oven.
4. When dry the larva is removed carefully from the glass rod and mounted with a drop of glue on some wire, a pinned straw, etc. (Fig. 26B).

Larvae also can be injected (after the contents are removed) with melted beeswax or a beeswax-paraffin wax combination using a syringe, and baked in an oven. The specimens are not as fragile as inflated larvae and the wax can be dyed to add any lost color.

How to Organize and Maintain an Insect Collection

INSECT BOXES AND DISPLAY MOUNTS

Pinned insects should be kept in boxes that are as airtight as possible. Tight-fitting lids exclude dust and insects that feed on dry specimens, and they hold a fumigant longer. Cigar boxes or cardboard boxes lined with sheet cork, styrofoam, soft composition board, polyethylene, etc., serve as inexpensive but not especially airtight containers. Black cardboard boxes with white-lined interiors and inexpensive redwood storage boxes are attractive and commonly available from biological supply companies (Fig. 27A). They are moderately airtight and about 23 × 30 × 6 cm in size. More expensive wooden insect boxes with hinged, tight-fitting lids are available although the best (Schmitt boxes) are quite expensive.

Large collections are generally housed in glass-topped museum drawers that fit into cabinets holding 12 to 48 drawers (Fig. 27B). The drawers usually contain cardboard trays which hold specimens of a single species. Drawers and trays are readily available from supply companies but are expensive. Various sizes and designs of glass-lined drawers for storage or display can be made at home with the proper woodworking equipment. The glass or plexiglass can be part of the lid like the commercial drawers or the glass can slide in grooves along the top edge of the drawer. The drawer bottom is lined with a pinning surface unless separate boxes are used to fill the drawer. In all cases, a tight-fitting unit is necessary to exclude pests and prevent excessive leakage of a fumigant.

Riker mounts are also used to display insects (Fig. 27C). These square or rectangular mounts are shallow cardboard boxes of various sizes (up to about 0.4 m long) that are lined with cotton. The center of the lid is cut out and replaced with glass leaving only the lid edges to fit over the box. The box and lid are covered with black binding tape to strengthen them. Shallow depressions are made in the cotton layer for the insects and labels. Insects are not pinned if they are to go into a Riker mount.

All-glass or plastic mounts are sometimes used to display butterflies and moths. The wings are placed on glued layers of glass or clear plastic supports. Top and bottom layers are cemented to the supports and the "sandwich" bound on the sides with tape. Insects can also be imbedded in bioplastic available from biological supply companies or hobby shops.

Arranging the Insects

Arrange the specimens by order and family (Fig. 27D). A large label indicating the Latin and common names of an order can be pinned

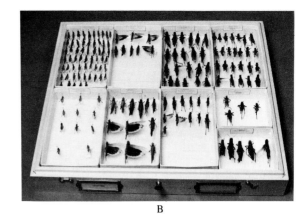

Figure 27 Insect containers. A, inexpensive cardboard or wooden box; B, glass-topped museum drawer with small unit trays; C, glass-topped cardboard Riker mount with a cotton-filled interior; D, labels for families and orders.

into the bottom lining of the box. Smaller labels, each with the Latin and common names of a family in the order, are arranged and pinned below the order label. Specimens belonging to each family are pinned beneath the appropriate label. Because only one label is made for a family, each insect does not need a family label. Labels can be color coordinated for display purposes (e.g., order labels are one color and family labels another).

Insects in vials or other containers may be included with the pinned insects by placing the containers in a horizontal position. They must be securely fastened to the pinning bottom and have leak-proof lids. If the box is sufficiently deep, the vials can be arranged vertically along the sides. Otherwise insects in liquids are stored vertically in separate containers (see also section on Preserving in Liquids).

CARE OF THE COLLECTION

Dried insects are fed upon by dermestid beetles and other museum pests. Sawdustlike material below the insect usually indicates a dermestid larva is feeding inside. Naphthalene and paradichlorobenzene (PDB) are sold as moth flakes or moth balls and are commonly used to repel pests. PDB is more toxic to pests but vaporizes quicker than naphthalene. A combination of both works well. The repellent is placed in a small box located in the corner of the insect box or under the cotton of a Riker mount.

Once an infestation is present, concentrated treatment with PDB or ethyl acetate or heating the insect box to 150°F (66°C) or higher for several hours is necessary. The collection should be checked several times a year and the repellent replaced as needed because a collection can be ruined in a matter of months by a heavy pest infestation.

REFERENCES ON COLLECTING AND PRESERVING INSECTS

Beirne, B.P. 1955. Collecting, Preparing, and Preserving Insects. Canada Dept. Agric. Ent. Div. Pub. 932. 133 pp.

Borror, D.J.; DeLong, D.M.; and Triplehorn, C.A. 1976. *An Introduction to the Study of Insects.* 4th ed. Holt, Rinehart and Winston, N.Y. 852 pp.

British Museum (Natural History). 1974. *Insects. Instructions for Collectors No. 4a.* 5th ed. British Museum (Natural History), London. 169 pp.

Nicholls, C.F. 1970. Some Entomological Equipment. Research Institute Canada Dept. of Agric., Belleville, Ontario. 118 pp.

Oldroyd, H. 1958. *Collecting, Preserving, and Studying Insects.* Macmillan, N.Y. 327 pp.

Oman, P.W., and Cushman, A.D. 1948. Collection and Preservation of Insects. USDA Misc. Pub. 601. 42 pp.

Peterson, A. 1964. *A Manual of Entomological Techniques.* 10th ed. Edwards Bros., Ann Arbor, Mich. 435 pp.

BIOLOGICAL SUPPLY COMPANIES

This is a partial list of companies that sell entomological collecting supplies, equipment, and living or preserved insects. An asterisk denotes a company that sells only living insects and not supplies or equipment.

American Biological Supply Co., 1330 Dillon Heights Ave., P.O. Box 3149, Baltimore, Md. 21228

Bio Quip Products, P.O. Box 61, Santa Monica, Calif. 90406

Bio-Serv, Inc., P.O. Box 100-B, Frenchtown, N.J. 08825 (artificial diets)

Carolina Biological Supply Co., Burlington, N.C. 27215

Clair Armin, 191 W. Palm Ave., Reedley, Calif. 93654 (insect boxes and pins only)

Connecticut Valley Biological Supply Co., Valley Rd., Southampton, Mass. 01073

Dahl Co., P.O. Box 566, Berkeley, Calif. 94710

Entomology Research Institute, Lake City, Minn. 55041

Entomological Supplies, Inc., 5655 Oregon Ave., Baltimore, Md. 21227

Frey Scientific Co., 465 So. Diamond St., Mansfield, Ohio 44903

*Insect Control and Research, Inc., 1330 Dillon Heights Ave., Baltimore, Md. 21228

LaPine Scientific Co., 920 Barker St., Berkeley, Calif. 94710

*Nasco-Steinhilber, Fort Atkinson, Wis. 53538

*Rincon Vitova Insectaries, Inc., P.O. Box 95, Oak View, Calif. 93022

Science Kit Inc., 777 East Park Dr., Tonawanda, N.Y. 14150

Turtox/Cambosco Products, General Biological Supply House, Inc., 8200 So. Hoyne Ave., Chicago, Ill. 60620

Ward's Natural Science Establishment, Inc., P.O. Box 1712, Rochester, N.Y. 14603

Ward's of California, P.O. Box 1749, Monterey, Calif. 93940

How to Observe and Rear Insects

The behavior of insects has interested humans for centuries. The seemingly complex yet stereotyped activity of so many of these tiny, primitive animals has occupied the attention of professional biologists and amateur naturalists alike. Careful observation and interpretation of an insect's behavior throughout its life cycle and of the role of insect castes in colonies form a major part of a meaningful study of insects.

When searching for insects (see the section on Where to Look for Insects) it is often worthwhile to watch them in their natural habitats as they feed and build nests. When collected alive insects can be brought indoors and placed in a habitat created to resemble their natural environment. Pond insects are easily maintained temporarily in aquariums and provide examples of diverse swimming methods, air bubble breathing, and predation. Stream insects require an aerator and a cold water temperature. Ants and bees provide well-known examples of group behavior through the use of ant "farms" and display hives.

Immature and adult insects may be reared by first determining and collecting the fresh food that they require. Moderate (not high) humidity and room temperature (or slightly higher as provided by the heat from light bulbs) are usually satisfactory for rearing. Water, available from a wet cotton wick or sponge inserted in a small vial of water, is needed by insects that eat dry food (exceptions are flour beetles and mealworms which require no direct water). A screened cage or jar with ample ventilation will suffice as a container and materials are added to simulate natural conditions. A screened cylindrical cage can be placed over individual potted plants indoors or small plants outdoors. Insect larvae may need soil to dig into or twigs to climb in order to pupate.

Larvae or pupae (including pupae inside silken cocoons) that hibernate during the winter but are brought indoors in the fall to await their emergence as adults often need to be subjected first to cold temperatures. Place them in a loosely sealed container of clean, moist sand in a refrigerator (not in the freezer) for about six weeks and then return them to a warm room. The insects can also be placed outdoors in a container for the same length of time. This treatment is often necessary for mantid egg cases in the northern states.

Many insects are reared on substitute foods or artificial diets. The following are examples of foods used for some commonly reared insects:

1. **Crickets**—Dry powdered milk mixed with crushed, dry dog food; bits of apple, banana, or lettuce suffice for very young crickets.

2. **Grasshoppers**—Lettuce; powdered dry milk (2 parts), dry alfalfa meal (2 parts), and dry brewer's yeast (1 part).

3. **Cockroaches**—Crushed, dry dog or rat food mixed with dry powdered milk.

4. **Mantids**—Fruit flies and other flies, aphids (leave them on the plant), crickets, and grasshoppers. Mealworms and uncooked hamburger or other meat will be eaten if held with forceps and touched to the mantid's mouth.

5. **Large Milkweed Bug**—Milkweed seeds (collect in the fall and store seeds for later use). A sunflower-feeding strain is also available from supply companies.

6. **Mealworm (beetle)**—Bran with an occasional apple core added.

7. **Confused Flour Beetle**—White flour with whole-wheat flour and cornmeal added.

8. **Butterflies and Moths**—Sugar water.

9. **Greater Wax Moth Larvae**—Mixture of 7 parts granular dog meal, 2 parts honey, and 1 part water.

10. **Mosquito Larvae**—Finely crushed, dry dog food sprinkled very lightly on the water which contains the larvae.

Numerous insect larvae are now reared in laboratories on various gellike artificial diets and some are available commercially for rearing moth and beetle larvae. One well known diet for certain moth larvae is made of casein, wheat germ, sugar, vitamins, minerals, agar, water, and a few other additives.

REFERENCES

Borden, J.H., and Herrin, B.D. 1972. *Insects in the Classroom.* British Columbia Teacher's Federation, Vancouver, B.C. 147 pp.

Cummins, K.W.; Miller, L.D.; Smith, N.A.; and Fox, R.M. 1965. *Experimental Entomology.* Reinhold, N.Y. 176 pp.

Ford, R.L.E. 1973. *Studying Insects.* Frederick Warne, N.Y. 150 pp.

Kalmus, H. 1960. *101 Simple Experiments with Insects.* Doubleday, Garden City, N.Y. 194 pp.

Needham, J.G. (chm.); Galtsoff, P.S.; Lutz, F.E.; and Welch, P.S. 1959. *Culture Methods for Invertebrate Animals.* Dover, N.Y. 590 pp. (1937 reprint).

Peterson, A. 1964. *A Manual of Entomological Techniques.* 10th ed. Edwards Bros., Ann Arbor, Mich. 435 pp.

Sadlers, D. 1971. *Studying Insects.* McGraw-Hill, N.Y. 128 pp.

Singh, P. 1977. *Artificial Diets for Insects, Mites, and Spiders.* Plenum, N.Y. 606 pp.

Siverly, R.E. 1962. *Rearing Insects in Schools.* Wm. C. Brown Company Publishers, Dubuque, Iowa. 113 pp.

Smith, C.N. (ed.). 1966. *Insect Colonization and Mass Production.* Academic Press, N.Y. 618 pp.

Tipton, V.J. 1973. *Toward More Effective Teaching. Entomology.* Brigham Young Univ. Press, Provo, Utah. 101 pp.

———. 1976. *Entomology. Catalog of Instructional Materials.* Brigham Young Univ. Press, Provo, Utah. 440 pp.

Villiard, P. 1969. *Moths and How to Rear Them.* Funk and Wagnalls, N.Y. 242 pp.

———. 1973. *Insects as Pets.* Doubleday, N.Y. 143 pp.

Washington State University Cooperative Extension Service. 1970. A Guide for 4-H Entomology. Washington State Univ., Pullman. 59 pp.

Whitten, R.H. 1973. *Use, Care and Culture of Invertebrates in the Classroom.* Carolina Biological Supply Co., Burlington, N.C. 23 pp.

Wilcox, J.A. 1972. Entomology Projects for Elementary and Secondary Schools. Bull. 422 New York State Museum and Science Service. State Univ. of New York, Albany. 44 pp.

Where to Look for Insects

1. **Plants**
   Inspect and collect from flower and vegetable gardens, grasses in lawns and fields, weeds, bushes, aquatic plants, fruit and shade trees, trees in woodlands and forests, etc.
   A. Flowers—Look on the flower as well as between the petals and in the throat. Thump the flower on a piece of white paper to dislodge thrips. The many kinds of flowers attract many different insects. Willow catkins in spring are very attractive to insects as are blossoms of many trees.
   B. Leaves—Look on both sides as well as inside (leaf miners). The underside of leaves is often a preferred resting site. Leaf damage such as chewed areas may indicate that the insects are still present or may return at night to feed.
   C. Stems—Split stems for larvae where damage occurs (holes, swellings, dead portions).
   D. Roots—Root-feeding soil insects are often near, attached to, or inside roots. Cabbages and radishes are examples of vegetables that commonly have root maggots.
   E. Fruits, nuts, and vegetables—Cut inside these items for larvae if damage is apparent (holes, discoloration). Decaying fruits and vegetables on or in the ground yield many insects.
   F. Mosses and lichens—Carefully inspect these for tiny insects.
   G. Galls—Split open or place in closed containers until the insects emerge as adults.
   H. Shelf (bracket) fungi and mushrooms—Inspect the outside and inside for insects.
   I. Fresh sap on trees attracts insects.
   J. Tree holes filled with water hold mosquito larvae and other insects.
   K. Beneath loose bark of dead or dying trees and stumps.
   L. Inside rotting logs and stumps and on wood piles. The odors emanating from freshly cut wood are especially attractive.

2. **Air**
   A. Sunny, calm days are usually best to find and capture flying insects with a net.
   B. Certain flies and winged ants and termites swarm in the spring and fall (especially if it is warm and sunny after a light rain).

C. Check the radiator or front grill of a car for usable specimens.
D. Insect nets attached to a car traveling about 25-30 mph (40-48 km/hr) may capture specimens if insects are abundant.

3. **Water**
   A. Collect on top of and beneath the water's surface.
   B. Scrape the mud along shorelines (down to 4 cm deep) and for some distance into the center of the body of water.
   C. Inspect rocks (especially the underside) above and beneath the water.
   D. Inspect floating and rooted aquatic plants. Work a net between them and scrape them to collect clinging insects. Some are found inside stems.
   E. Inspect debris that accumulates on the bottom of streams and ponds, on rocks, bushes, overhanging branches, and shorelines. Debris piled up on shores after floods or storms is a good source of insects.
   F. The edges of mud puddles on hot, sunny days attract thirsty insects.

4. **Soil**
   A. Rich humus-containing garden topsoil, mulches, compost piles, and forest litter—Slowly sift through the soil and plant material or heat in a Berlese funnel (see Collecting Methods section).
   B. Digging down about 15 cm under grasses and in garden soils may yield June beetle grubs and adults, click beetle larvae, cutworms, and other insects.

5. Look under stones, logs, boards, and cardboard boxes. Turn them back to the original position when through to reestablish the habitat.

6. Lights at night near wooded areas, in fields, shorelines, other habitats.
   A. Incandescent and fluorescent lights along the sides of buildings.
   B. Street and parking lot lights.
   C. Blacklight in front of a suspended sheet.
   D. Camping lights.

7. Roadside ditches filled with weeds and plant debris frequently yield great numbers and types of insects.

8. Walk through a yard or field at night with a flashlight, camping lantern, or head lamp to locate singing insects.

9. Sweep fields at night to obtain many resting insects not present during the day.

10. Sandy, wet shorelines of ponds and lakes. Water poured along the shoreline may bring out certain insects from burrows.

11. Sandy lake and ocean beaches and sand dunes with clusters of flowering shrubs and beach grasses.

12. Inside and beneath animal manure in pastures (both dry and semi-dry manure).

13. Inside and beneath dead animals (wild birds, fowl, fish, and small mammals hit by cars). A recent road-kill can be staked to the ground or covered with a board and checked daily for insects.

14. Inspect skin and feathers of birds and fur of mammals (cows, dogs, cats, rodents, etc.) for lice and fleas. Caution: fleas on rats and certain ground squirrels in the western states may carry diseases transferable to humans. During the winter lice congregate behind the ears and other protected body areas of cattle and horses. Parasitic bot fly larvae occur in skin lumps along the back of cattle (especially calves). Freshly killed game birds should be placed in a

sealed plastic or paper bag to hold the parasites which will soon drop off the birds.

15. Flies are especially attracted to cows, horses, etc., and the associated barns.

16. Sunny windows in buildings: garages, homes (especially in spring and fall), stables, poultry houses.

17. Basements near furnaces and where boxes and papers are stored may contain silverfish and firebrats.

18. Flour bins, granaries, and stored cereals and flour in houses (especially if the contents are opened and spilled).

19. Picnics or any activity outdoors where food is exposed.

20. Sunlit forest openings. Stand on the shaded edge to wait for insects to fly in.

21. Hilltops attract certain butterflies.

22. Surfaces and edges of melting snow patches in winter may have tiny, jumping springtails and scorpionflies.

Insect Structure and Development

STRUCTURE

Head

The head is the hardened (sclerotized) anterior region of the body that bears the eyes, antennae, and mouthparts (Fig. 28). The large eyes occurring on most adults are called compound eyes because they consist of a few to several thousand individual eye units. Single eyes, or ocelli (usually three), are located on top of the head of adults. Immature insects without compound eyes usually have a cluster of ocelli on each side of the head.

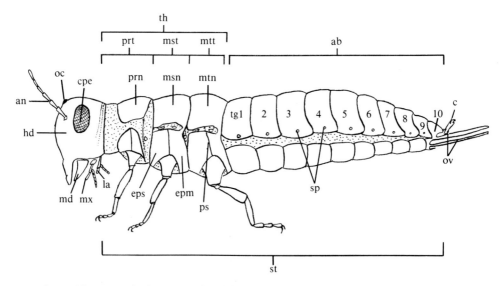

Figure 28 Insect body parts. ab, abdomen; an, antenna; cpe, compound eye; cr, cercus; epm, epimeron; eps, episternum; hd, head; la, labium; md, mandible; msn, mesonotum; mst, mesothorax; mtn, metanotum; mtt, metathorax; mx, maxilla; oc, ocellus; ov, ovipositor; prn, pronotum; prt, prothorax; ps, pleural suture; sp, spiracles; st, sternum; tg 1-10, terga; th, thorax.

Insects have one pair of segmented, flexible antennae that assume a variety of shapes (Fig. 29). Antennae have a sensory function and, depending on the insects, are used to detect odors and sounds and to touch and taste objects.

The mouthparts consist of an upper lip or labrum, a pair of mandibles, a pair of maxillae, sometimes a very short tonguelike hypopharynx, and a lower lip or labium. The mandibles may be stout, curved, and toothed for chewing (Fig. 30A) (includes cutting, crushing, or grind-

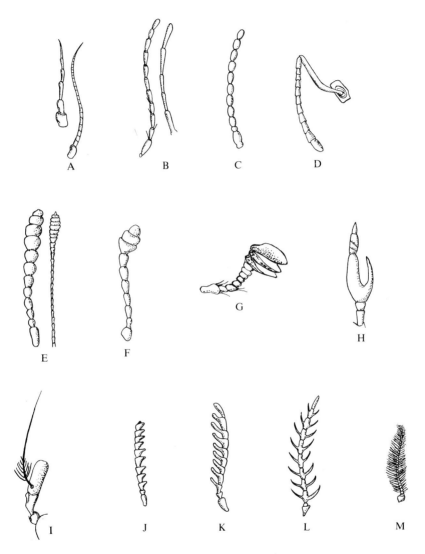

Figure 29 Types of antennae. A, setaceous; B, filiform; C, moniliform; D, elbowed; E, clavate; F, capitate; G, lamellate; H, ringed; I, aristate; J, serrate; K, pectinate, L, bipectinate; M, plumose.

Insect Structure and Development

ing); long and needlelike for piercing and sucking (Fig. 30B); or otherwise modified. The maxillae also may be modified to pierce and suck in combination with the mandibles (e.g., mosquitoes, stink bugs, aphids) or they may be the only sucking organ (e.g., coiled proboscis of butterflies and moths). If the mandibles are used for chewing, the maxillae usually have short antennalike appendages (maxillary palpi) for touching and tasting materials. The labium may be somewhat flattened, bear short sensory palpi, and be used to guide food into the mouth opening. The labium also can be elongated and thickened to surround and hold the piercing mandibles and maxillae. The elongated mouthparts together are called a proboscis or beak.

Thorax

The thorax is divided into three segments: the prothorax, mesothorax, and metathorax (Fig. 28). Each segment has four hardened areas or sclerites. The upper area is the notum, the lower area is the sternum, and the region on each side is the pleuron. Thus the dorsal (upper) part of the prothorax is commonly called the pronotum, the ventral (lower) area of the mesothorax is the mesosternum, etc. The mesonotum has a triangular area, the scutellum, which is a conspicuous feature on true bugs (Heteroptera) (Fig. 124) and, to a lesser degree, beetles. Grooves and ridges often are present on the thorax.

Legs—Each thoracic segment bears a pair of legs on its lower edges. The legs are modified into many forms depending on their function (running, jumping, grasping, swimming, etc.) but are typically divided into six main segments (Fig. 31). The first segment is the short coxa which inserts into the body. The small trochanter (it may be divided into two parts) follows the coxa. Third is the femur which is usually long and sometimes thickened and spined. The tibia is fourth and generally long and slender. Fifth is the tarsus consisting of one to five segments and last is a pair of claws, the pretarsus. Some larval insects (e.g., caterpillars) bear various numbers of fleshy prolegs on the abdomen that have a different segmentation than the thoracic legs (Fig. 234).

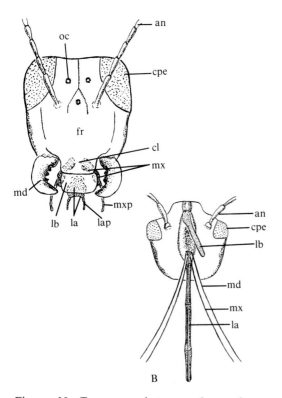

Figure 30 Two general types of mouthparts. A, chewing; B, piercing-sucking (opened). an, antenna; cl, clypeus; cpe, compound eye; fr, frons; la, labium; lap, labial palpus; lb, labrum; md, mandible; mx, maxilla; mxp, maxillary palpus; oc, ocellus.

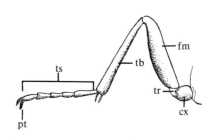

Figure 31 General insect leg. cx, coxa; fm, femur; pt, pretarsus (claws); tb, tibia; tr, trochanter; ts, tarsus.

Wings—Adult insects may bear a pair of wings on both the meso- and metathorax (Fig. 28), on the mesothorax only (flies [Diptera], male scales), or the wings may be absent. The wings vary greatly in size, shape, color, thickness, and venation. The vein pattern is used widely in identification. The names and general position of the major veins are shown in Figure 32 along with the terms used in this book to designate the wing edges. Closed cells are regions cut off from the wing margin by crossveins; open cells extend to the wing margins.

Abdomen

The abdomen is the softer, more flexible, posterior region of an insect (Fig. 28). It may be visible or hidden under the wings. The abdomen usually consists of eleven segments although those most posterior are small or modified and not easily visible. The dorsal (upper) region is the tergum and the ventral (lower) region is the sternum. The slightly hardened plate (sclerite) of each segment is called a tergite if dorsal or sternite if ventral.

Most insects lack appendages on the abdomen except near the tip where cerci may appear as stubs, narrow plates, or filaments (Figs. 28; 65A; 106). Other terminal appendages are associated with or form the external genitalia. Genitalia of adult males may be external or withdrawn into the body. Adult females of some groups bear an egg-laying device, the ovipositor (Fig. 28), that may be stout for digging, bladelike or sawlike for cutting (Fig. 95D), or needlelike for piercing and/or stinging (Fig. 356).

DEVELOPMENT

Eggs

Insect eggs vary greatly in shape and color (Fig. 33). Only a few may be produced or masses totaling a thousand or more eggs are deposited by some insects. The embryo usually develops from a fertilized egg but in some cases (especially wasps and bees) an unfertilized egg produces a male.

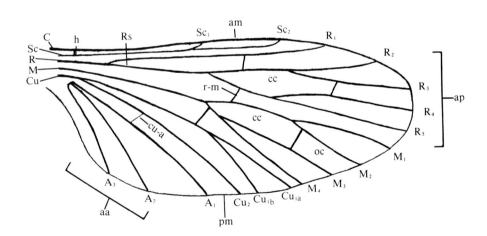

Figure 32 General wing. A, anal vein; aa, anal angle; am, anterior or costal margin; ap, apex; C, costa; cc, closed cell; Cu, cubitus; cu-a, cubito-anal crossvein; h, humeral crossvein; M, media; oc, open cell; pm, posterior margin; R, radius; r-m, radio-medial crossvein; Rs, radial sector; Sc, subcosta.

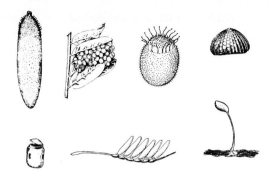

Figure 33   Examples of eggs.

Immatures

Immature insects grow in size by shedding periodically their rigid, outer skin layer (the cuticle), and expanding their newly produced tissues before the new cuticle hardens. Molting is the process involved in loosening the old cuticle and producing the new, larger replacement. The shedding of the old cuticle usually takes a few hours. The number of molts varies from about four to eight but can be much more for some insects. Adults do not molt with the exceptions of mayflies and bristletails. Between molts the insect is called an instar.

The change in form during development is termed metamorphosis. There are various classifications of metamorphosis and this book uses two common divisions: hemimetabolous (incomplete metamorphosis) and holometabolous (complete metamorphosis). Members of the subclass Apterygota (bristletails, springtails, etc.) undergo such a minimal change in appearance that they are often described as having ametabolous development, a type of incomplete metamorphosis.

Hemimetabolous insects are characterized by immatures that resemble the adults and change principally by a size increase and the development of wings and genitalia (Fig. 34). The immatures, usually called nymphs, have compound eyes and mouthparts like adults and the wings develop externally as pads. Immatures generally feed on the same type of food as adults; exceptions are the aquatic immatures of mayflies, dragonflies, damselflies, and stoneflies. These insects do not resemble the adults and live in a different habitat (Figs. 68; 73C; 83B).

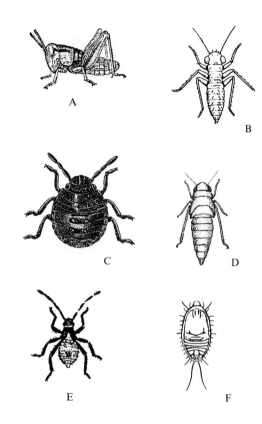

Figure 34   Examples of nymphs. A, grasshopper; B, thrips; C, stink bug; D, leafhopper; E, aphid; F, mealybug.

Immatures of holometabolous insects do not resemble the adults and have one additional developmental stage, the pupa. The instars are called larvae, a term which refers to a worm-like form although many are not this shape (Fig. 35). A series of simple eyes (ocelli) occur on the head of most larvae, along with chewing or chewing-sucking mouthparts and very short antennae. The wings develop beneath the larval skin and are not visible until the pupal stage.

34   Insect Structure and Development

pupal stage. These cells help form the wings, legs, antennae, and mouthparts. The preceding processes temporarily halt if the pupa is dormant during the winter or at other times of the year.

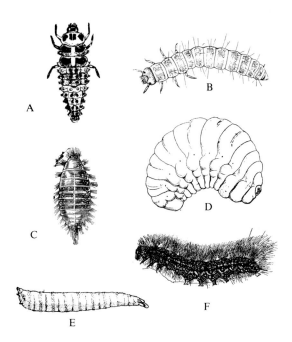

Figure 35 Examples of larvae. A, lady beetle; B, ground beetle; C, dermestid beetle; D, weevil; E, fly; F, moth.

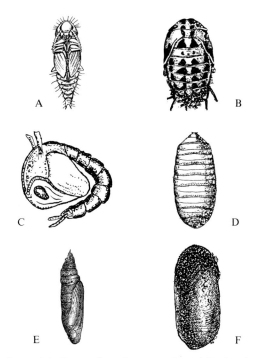

Figure 36 Examples of pupae. A and B, beetles; C, fly (mosquito); D, fly; E, moth; F, wasp.

The mature larva molts into a stage called the pupa (Fig. 36). The pupa does not feed and usually is unable to move (the active mosquito pupae are notable exceptions). Pupae of many moths, some wasps, and various other insects are covered by a silken cocoon. The cocoon is spun by the mature larva just before it pupates (molts into the pupal stage) and may be composed partly of body hairs and debris. Many fly pupae are encased in a puparium which consists of the hardened cuticle of the next-to-last larval instar. Most of the structural changes necessary for the transition from a larva to an adult occur during the pupal stage. Tissue breakdown, reorganization, growth, and differentiation occur to varying degrees to form the adult's digestive system, reproductive organs, and other structures. Many adult tissues are formed from small groups of cells that are present during the larval stages but are limited in growth until the

Adults

After a period of internal activity ranging from several days to weeks, the adult emerges by pushing and crawling out of the pupal skin. The wings are crumpled and the insect body is soft. Within minutes to hours depending on the species, the adult's body dries, hardens, and becomes more pigmented. The wings expand from air and liquid pressure and form a rigid framework. The adult is now ready to assume normal activity and will live from as little as 1 1/2 hours (some mayflies) to perhaps twenty years (tropical termite queens). Generally adults live only a few weeks.

General References on Insect Biology

Annual Review of Entomology. 1956 (vol. 1) and continuing. Annual Reviews, Palo Alto, Calif.

Atkins, M.D. 1978. *Insects in Perspective.* MacMillan, N.Y. 513 pp. Daly, H.V.; Doyen, J.T.; and Ehrlich, P.R. 1978. *An Introduction to Insect Biology and Diversity.* McGraw-Hill, N.Y. 512 pp.

Baker, W.L. 1972. Eastern Forest Insects. USDA Misc. Pub. No. 1175. U.S. Govt. Printing Office, Washington, D.C. 642 pp.

Belkin, J.N. 1972. *Fundamentals of Entomology. A Manual for Introductory Courses* (in two parts). Bio-Rand Foundation, Baltimore, MD. 220 pp.

Borror, D.J.; DeLong, D.M.; and Triplehorn, C.A. 1976. *An Introduction to the Study of Insects.* 4th ed. Holt, Rinehart and Winston. 852 pp.

Borror, D.J., and White, R.E. 1970. *A Field Guide to the Insects of America North of Mexico.* Houghton Mifflin, Boston. 404 pp.

Chapman, R.F. 1971. *The Insects. Structure and Function.* 2nd ed. American Elsevier, N.Y. 819 pp.

Chu, H.F. 1949. *How to Know the Immature Insects.* Wm. C. Brown Company Publishers, Dubuque, Iowa. 234 pp.

Davidson, R.H., and Peairs, L.M. 1966. *Insect Pests of Farm, Garden and Orchard.* 6th ed. Wiley, N.Y. 675 pp.

Dennis, C.J. 1974. *Laboratory Manual for Introductory Entomology.* 3rd ed. Wm. C. Brown Company Publishers, Dubuque, Iowa. 112 pp.

Ebeling, W. 1959. *Subtropical Fruit Pests.* Division of Agric. Sci., Univ. Calif., Richmond, Calif. 436 pp.

———. 1975. *Urban Entomology.* Division of Agric. Sci., Univ. Calif. Richmond, Calif. 695 pp.

Elzinga, R.J. 1978. *Fundamentals of Entomology.* Prentice-Hall, Englewood Cliffs, N.J. 325 pp.

Essig, E.O. 1958. *Insects and Mites of Western North America.* Macmillan, N.Y. 1050 pp.

Fox, R.M., and Fox, J.W. 1964. *Introduction to Comparative Entomology.* Reinhold, N.Y. 450 pp.

Frost, S.W. 1959. *Insect Life and Insect Natural History.* Dover, N.Y. 526 pp.

Graham, S.A., and Knight, F.B. 1965. *Principles of Forest Entomology.* 4th ed. McGraw-Hill, N.Y. 417 pp.

Horn, D.J. 1976. *Biology of Insects.* Saunders, Philadelphia. 439 pp.

Hutchins, R.E. 1966. *Insects.* Prentice-Hall, Englewood Cliffs, N.Y. 324 pp.

Imms, A.D.; Richards, O.W.; Davies, R.G. 1977. *Imms' General Textbook of Entomology.* 10th ed. Vol. 1, 430 pp. Vol. 2, 910 pp. Halsted Press, N.Y.

James, M.T., and Harwood, R.F. 1969. *Herms's Medical Entomology.* 6th ed. Macmillan, N.Y. 484 pp.

Johnson, W.T., and Lyon, H.H. 1976. *Insects That Feed on Trees and Shrubs. An Illustrated Practical Guide.* Comstock Publ. Associates of Cornell Univ. Press, Ithaca, N.Y. 464 pp.

Lanham, U. 1964. *The Insects.* Columbia Univ. Press, N.Y. 292 pp.

Leftwich, A.W. 1976. *A Dictionary of Entomology.* Crane Russak, N.Y. 360 pp.

Linsenmaier, W. 1972. *Insects of the World.* McGraw-Hill, N.Y. 392 pp.

Little, V.A. 1972. *General and Applied Entomology.* 3rd ed. Harper and Row, N.Y. 527 pp.

Merritt, R.W., and Cummins, K.W. (eds.). 1978. *An Introduction to the Aquatic Insects of North America.* Kendall/Hunt, Dubuque, Iowa. 512 pp.

Metcalf, C.L.; Flint, W.P.; and Metcalf, R.L. 1962. *Destructive and Useful Insects.* 4th ed. McGraw-Hill, N.Y. 1087 pp.

Oldroyd, H. 1968. *Elements of Entomology.* Weidenfeld and Nicolson, London. 312 pp.

Peterson, A. 1948. *Larvae of Insects. Part I. Lepidoptera and Plant Infesting Hymenoptera.* Edwards Bros., Ann Arbor, Mich. 315 pp.

———. 1951. *Larvae of Insects. Part II. Coleoptera, Diptera, Neuroptera, Siphonaptera, Mecoptera, Trichoptera.* Edwards Bros., Ann Arbor, Mich. 416 pp.

Pfadt, R.E. (ed.). 1971. *Fundamentals of Applied Entomology.* 2nd ed. Macmillan, N.Y. 693 pp.

Price, P.W. 1975. *Insect Ecology.* Wiley, N.Y. 514 pp.

Rockstein, M. (ed.). 1973-1974. *The Physiology of Insecta.* 2nd ed. Vols. 1-6. Academic Press, N.Y. 3,103 pp.

Romoser, W.S. 1973. *The Science of Entomology.* Macmillan, N.Y. 449 pp.

Ross, H.H. 1965. *A Textbook of Entomology.* 3rd ed. Wiley, N.Y. 539 pp.

Smith, R.F.; Mittler, T.E.; and Smith, C.N. (eds.). 1973. *History of Entomology.* Annual Reviews Inc., Palo Alto, Calif. 517 pp.

Snodgrass, R.E. 1935. *Principles of Insect Morphology.* McGraw-Hill, N.Y. 667 pp.

Swan, L.A., and Papp, C.S. 1972. *The Common Insects of North America.* Harper and Row, N.Y. 750 pp.

Tipton, V.J. 1976. *Entomology. Catalog of Instructional Materials.* Brigham Young Univ. Press, Provo, Utah. 440 pp.

Torre-Bueno, J.R. de la. 1937. *A Glossary of Entomology.* Brooklyn Entomological Society, Brooklyn, N.Y. 336 pp. 1962 printing includes Supplement A.

Tweedie, M. 1974. *Atlas of Insects.* John Day, N.Y. 128 pp.

Usinger, R.L. (ed.). 1956. *Aquatic Insects of California, with Keys to North American Genera and California Species.* Univ. Calif. Press, Berkeley. 508 pp.

Wigglesworth, V.B. 1964. *The Life of Insects.* Weidenfeld and Nicolson, London. 359 pp. (Paperback: World Publishing, N.Y. 383 pp.)

———. 1965. *The Principles of Insect Physiology.* Dutton, N.Y. 741 pp.

Wilson, E.O. 1971. *The Insect Societies.* Harvard Univ. Press, Cambridge, Mass. 548 pp.

Wilson, M.C.; Broersma, D.B.; and Provonsha, A.V. 1977. *Fundamentals of Applied Entomology.* Waveland Press, Prospect Heights, Ill. 166 pp.

Zim, H.S., and Cottam, C. 1956. *Insects: A Guide to Familiar American Insects.* Western, Racine, Wis. 160 pp.

Classification and Nomenclature

The animal kingdom is divided into numerous major groups called phyla and each phylum is separated further into classes. Insects are in the class Insecta in the phylum Arthropoda. Arthropods are invertebrates (animals without a backbone) with jointed appendages. The Latin word Insecta is translated as: *in,* in or into; *sect,* cut. This refers to the segmented or "cut" body. Other classes of arthropods are Crustacea (crabs, shrimp, crayfish, sowbugs, etc.), Arachnida (spiders, mites, ticks, scorpions, etc.), Chilopoda (centipedes), Diplopoda (millipedes), and several smaller classes.

The categories commonly used for insects are listed below in decreasing rank:

Phylum—Arthropoda
  Subphylum—Mandibulata
    Class—Insecta
      Subclass—Apterygota and Pterygota
        Order—Protura to Hymenoptera
            (29 orders)
          Suborder
            Superfamily
              Family
                Subfamily
                  Tribe
                    Genus
                      Subgenus
                        Species
                          Subspecies

Nomenclature refers to the scientific names given to organisms. These names are generally derived from Latin or latinized Greek words although words (including names of people and places) of any language can be used if latinized. Common names are given to certain common species and higher categories of insects but the names will vary from language to language and place to place.

The scientific name used to classify an insect above the species level is a uninomial (one word) name such as Hymenoptera (Order), Apidae (Family), and *Apis* (Genus). Standard word endings are used for certain categories. Order names of winged insects usually end in -ptera (e.g., Diptera), superfamily names in -oidea (e.g., Chalcidoidea), family names in -idae (e.g., Acrididae), subfamily names in -inae (e.g., Troginae), and tribe names end in -ini (e.g., Bembicini).

The scientific name of a species is a binomial (two word) name and consists of the name of the genus *(Apis)* and the specific name *(mellifera).* The species then is *Apis mellifera,* the honey bee. The generic name and those of higher categories begin with a capital letter but the specific name does not. The name of the species is underlined or written in italics. The abbreviation "sp." refers to a single, unnamed species (e.g., *Apis* sp.). More than one species is abbreviated "spp." (e.g., *Apis* spp.).

In some cases a trinomial (three word) name is used to designate a subspecies (e.g., *Papilio polyxenes asterius,* a swallowtail butterfly). Subspecific names are applied to populations of a species that vary in appearance or other features and the populations are usually separated geographically from each other.

The author is the person who first named and published a description of the species and the name of this person follows the species name. For example, *Apis mellifera* Linnaeus indicates that the famous biologist, Linnaeus, was responsible for naming the honey bee. An author's name is often in parentheses which signifies that the author originally placed the species in another genus. For example, *Melanoplus bivittatus* (Say) was described by Say as being in a genus other than *Melanoplus.* This species of grasshopper subsequently was reclassified as *Melanoplus* and the specific name retained.

Common names of orders, families, species, and subspecies used in this book are taken from the "Common Names of Insects" (1975 Revision) published by the Entomological Society of America (ESA). Except for butterflies and a few miscellaneous species, insects not listed in the ESA publication were not given a common name although often one or more could be found in various books. If a family has no common name the Latin name is sometimes used as an adjective (e.g., Otitidae are called Otitid Flies). The name "fly" is written as a separate word (e.g., horse fly) if the fly belongs to the order of flies, Diptera. The name of a fly which is not a two-winged fly (Diptera) is written as one word (e.g., sawfly, dragonfly, and butterfly). The name "bug" is written separately if the insect belongs to the suborder of true bugs, Heteroptera (e.g., stink bug), but otherwise the word is not separated (e.g., ladybug).

Key to the Orders of Adult Insects

1a	Wings present and well developed although sometimes short. 2	4a(3b)	Pronotum very long and extends over or beyond abdomen (easily mistaken for front wings) (Fig. 43A); miniature-sized grasshopper . . . . (p. 89) *Orthoptera* (in part)
1b	Wings absent, or very small with only inconspicuous remnants present . . 26		
2a(1a)	One pair of wings present (Figs. 37A; 38; 40E; 43A; 46A). 3	4b	Pronotum not greatly elongated or extending over abdomen; does not resemble grasshopper 5
2b	Two pairs of wings present 7	5a(4b)	One to 3 slender tails on tip of abdomen (Figs. 40E ♂ 46A); mouthparts poorly developed and inconspicuous. 6
3a(2a)	Hind wing broad and fanlike; minute, slender, often clubbed appendages replace front wings (Fig. 37A); 1 or more antennal segments with a long side branch; less than 4 mm long; uncommon (male) . . . . . . (p. 237) *Strepsiptera* (in part)	5b	No tails on tip of abdomen; mouthparts usually elongated into a proboscis (Fig. 38) . . . . (p. 315) *Diptera* (in part)
3b	Not with above combination of characteristics 4		

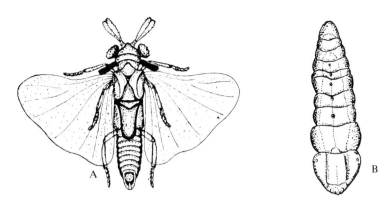

Figure 37 Strepsiptera (Twistedwinged Parasites). A, male; B, female.

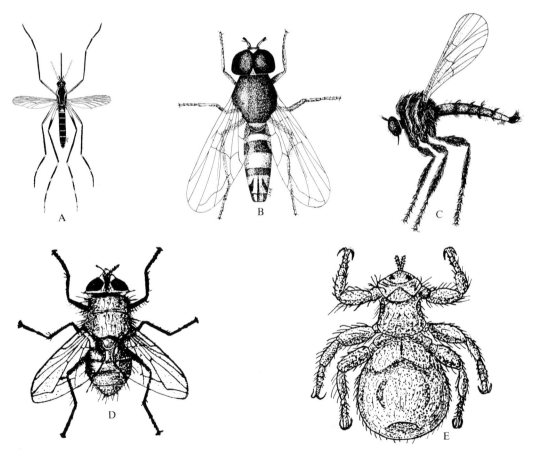

Figure 38 Diptera (Flies). A, mosquito; B, flower fly; C, robber fly; D, blow fly; E, louse fly (a wingless species).

Key to Orders of Adult Insects   41

6a(5a) Antennae long and easily visible (Fig. 40E ♂); wings with only 1 vein (usually forked); minute, slender appendages (halteres) in place of hind wings; 1 spinelike appendage sometimes on tip of abdomen; uncommon (suborder Homoptera)..........
...... (p. 133) *Hemiptera* (in part)

6b Antennae very short and bristlelike (Fig. 46A); wings with many veins; halteres absent; 2-3 long, slender tails on tip of abdomen ....
.... (p. 64) *Ephemeroptera* (in part)

7a(2b) Front and hind wings dissimilar in structure; front wings leathery (thickened and toughened) at their base (Fig. 39) or for their full length (Figs. 40C, D; 41; 43; 44), or distinctly hard (Fig. 42); hind wings membranous if present.......... 8

7b Front and hind wings similar in structure, all being membranous (Figs. 40A, B) or covered with hairs or colored powdery scales (Figs. 45A, B; 53), but not leathery or hard... 13

8a(7a) Mouthparts almost always elongated into a slender beak (Figs. 49A, B).. 9

8b Mouthparts not elongated into a beak, but are a chewing type with mandibles and palpi (Fig. 49C)... 10

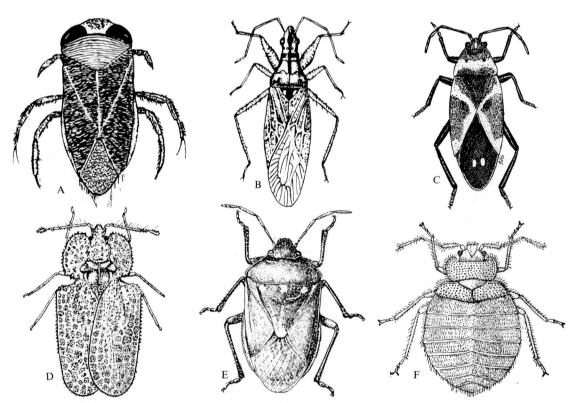

Figure 39 Hemiptera (Suborder Heteroptera—True Bugs). A, water boatman; B, damsel bug; C, seed bug; D, lace bug; E, stink bug; F, bed bug.

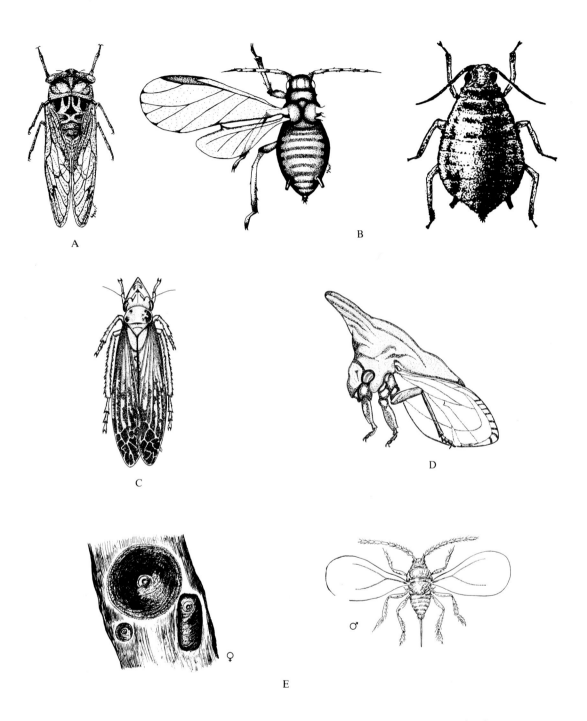

Figure 40 Hemiptera (Suborder Homoptera—Cicadas, Leafhoppers, Planthoppers, Aphids, Scale Insects, others). A, cicada; B, aphids; C, leafhopper; D, treehopper; E, scale insects.

9a(8a) Beak arises below and toward the front of head (Fig. 49A); basal portion of front wings hardened and outer portion membranous (Fig. 39), or if outer portion not membranous then entire wing is sculptured and lacelike (Fig. 39D); wings at rest held flat over body (suborder Heteroptera) . . . (p. 133) *Hemiptera* (in part)

9b Beak arises below and toward the rear of head (Fig. 49B) or appears to arise between front coxae; front wings uniformly and slightly hardened (Figs. 40C, D); wings at rest are held rooflike or vertically over body (suborder Homoptera) . . . . (p. 133) *Hemiptera* (in part)

10a(8b) Pincerlike appendages at tip of abdomen (Fig. 41); front wings short . . . . . . (p. 108) *Dermaptera* (in part)

10b No pincerlike appendages at tip of abdomen; front wings variable in length 11

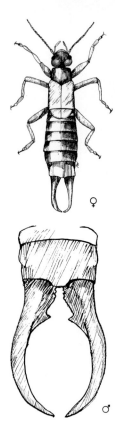

Figure 41 Dermaptera (Earwigs).

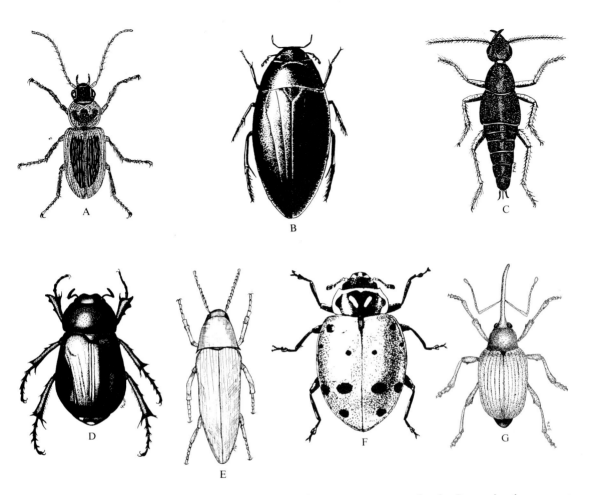

Figure 42 Coleoptera (Beetles). A, ground beetle; B, water scavenger beetle; C, rove beetle; D, scarab beetle; E, click beetle; F, lady beetle; G, weevil.

11a(10b) Front wings leathery or hard, without veins, and meet in a straight line down middle of back (Fig. 42) .. .. (p. 175) *Coleoptera*

11b Front wings leathery, veined, and either overlap or are held rooflike over body (Figs. 43; 44) 12

12a(11b) Tarsi 3- or 4-segmented; hind legs enlarged and thickened for jumping (Fig. 43) .... (p. 89) *Orthoptera* (in part)

12b Tarsi 5-segmented; hind legs long and slender for running (Fig. 44); front legs may be thickened, spined, and used for grasping (Fig. 44B).... ...... (p. 102) *Dictyoptera* (in part)

Key to Orders of Adult Insects   45

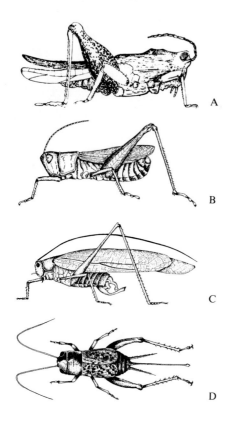

Figure 43 Orthoptera (Grasshoppers, Katydids, Crickets). A, pygmy grasshopper or grouse locust; B, grasshopper; C, katydid; D, crickets.

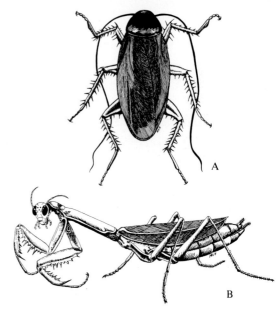

Figure 44 Dictyoptera (Cockroaches, Mantids). A, cockroach; B, mantid.

13a(7b)   Wings completely or partly covered with scales (often powdery) (Fig. 45); mouthparts usually appear as a coiled tube (proboscis, Fig. 49D) . . . . (p. 257) *Lepidoptera* (in part)

13b   Wings not covered with scales; mouthparts not a coiled tube 14

14a(13b)   Mouthparts elongated into a slender beak attached underneath and to the rear of the head (Fig. 49B), sometimes appearing to arise between the front legs; wings held rooflike over body (Figs. 40A, B) (suborder Homoptera) . . . . (p. 133) *Hemiptera* (in part)

14b   Not as above. 15

15a(14b)   Front wings triangular and much larger than the small hind wings (Fig. 46B); wings at rest held vertically above body; 2 or 3 very long, threadlike tails at tip of abdomen . . . . (p. 64) *Ephemeroptera* (in part)

46   Key to Orders of Adult Insects

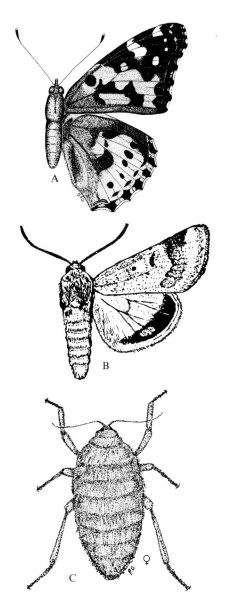

Figure 45 Lepidoptera (Butterflies, Moths). A, brushfooted butterfly; B, noctuid or owlet moth; C, geometrid moth (a wingless species).

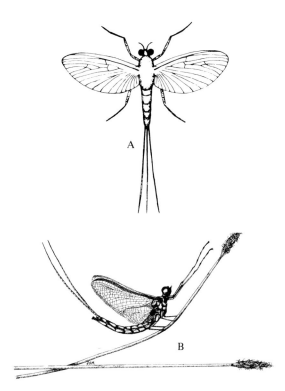

Figure 46 Ephemeroptera (Mayflies). A, two-winged species; B, four-winged species.

Figure 47 Thysanoptera (Thrips).

Key to Orders of Adult Insects 47

Figure 48 Odonata (Dragonflies, Damselflies).

15b Not with above combination of characteristics. 16

16a(15b) Wings very narrow and margined with a fringe of long hairs (Fig. 47); usually less than 5 mm long . . . . (p. 129) *Thysanoptera* (in part)

16b Wings not unusually narrow and fringe of hairs absent 17

17a(16b) Tarsi with 4 or fewer segments (Fig. 31) 18

17b Tarsi with 5 segments 23

18a(17a) Front and hind wings long and similar in length and shape (Fig. 48); wings at rest held out to side or above body; abdomen very long and slender. . . . . (p. 69) *Odonata*

18b Not with above combination of characteristics. 19

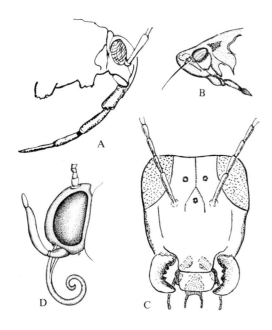

Figure 49 Mouthparts. A, beak toward front of head (Hemiptera, Suborder Heteroptera); B, beak toward rear of head (Hemiptera, Suborder Homoptera); C, chewing; D, siphoning proboscis (Lepidoptera).

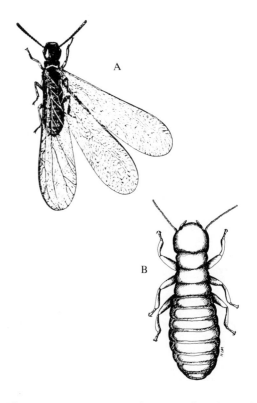

Figure 50 Isoptera (Termites). A, winged; B, wingless.

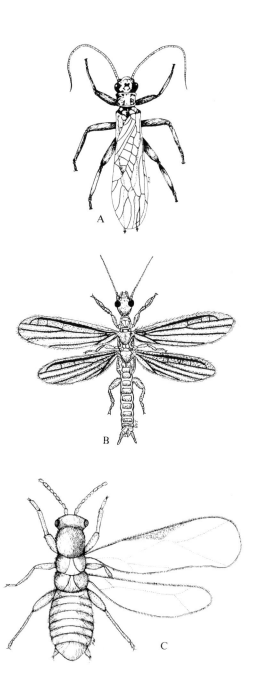

Figure 51 A, Plecoptera (Stoneflies); B, Embioptera (Webspinners); C, Zoraptera (Zorapterans).

19a(18b) Wings nearly equal in size, shape, and venation (Fig. 50A); tarsi 4-segmented . . . (p. 112) *Isoptera* (in part)

19b Front wings longer than hind wings; tarsi with 3 or fewer segments 20

20a(19b) Hind wings very broad, anal area greatly expanded and folded lengthwise at rest (Fig. 51A); body generally flattened; tip of abdomen usually with 2 short tails. . . . .
. . . . (p. 82) *Plecoptera*

20b Hind wings not unusually broad, anal area not expanded or folded at rest; body not flattened; short abdominal tails present or absent . . . 21

21a(20b) Base of front tarsi very stout (Fig. 51B); tarsi 3 segmented; uncommon. . . (p. 111) *Embioptera* (in part)

21b Base of front tarsi not enlarged; tarsi 2- or 3-segmented 22

Key to Orders of Adult Insects   49

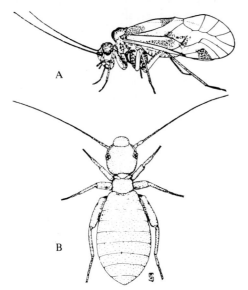

Figure 52 Psocoptera (Psocids). A, winged; B, wingless.

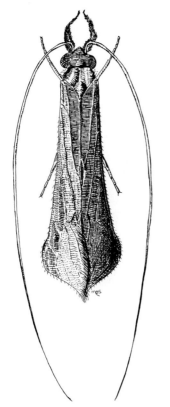

Figure 53 Trichoptera (Caddisflies).

22a(21b) Tip of abdomen with 2 very short tails (Fig. 51C); tarsi 2-segmented; antennae 9-segmented and beadlike; uncommon. . . . . (p. 117) *Zoraptera* (in part)

22b Tip of abdomen without short tails; tarsi 2- or 3-segmented; antennae not beadlike, but long and thin (Fig. 52A) . . . . (p. 118) *Psocoptera* (in part)

23a(17b) Front wings hairy (Fig. 53), and with few crossveins along anterior margin; threadlike antennae nearly as long as, to much longer than body; mouthparts indistinct except for palpi; mothlike . . . . . . (p. 249) *Trichoptera*

23b Front wings not hairy, sometimes many crossveins along anterior margin (Fig. 56); antennae distinctly shorter than body; mandibles well developed; generally not mothlike 24

Figure 54 Hymenoptera (Ants, Bees, Sawflies, Wasps, others). A, sawfly; B, ichneumon; C, ant; D, wasp; E, yellowjacket; F, wasp; G, bumble bee.

Key to Orders of Adult Insects

24a(23b) Front wings longer and with more veins than hind wings; body stout (Fig. 54A) or constricted juncture between thorax and abdomen (Figs. 54B-G); needlelike (sting) or sawlike appendage on tip of female's abdomen ....
..... (p. 361) *Hymenoptera* (in part)

24b Front and hind wings very similar in size and number of veins; body not stout, no constriction between thorax and abdomen; needlelike or sawlike appendage on tip of abdomen absent although tip of abdomen may be greatly enlarged and curved upward...................... 25

25a(24b) Head extended downward to form a stout beak (Fig. 55) ....
.... (p. 247) *Mecoptera* (in part)

25b Head does not form a stout beak (Fig. 56)....... (p. 239) *Neuroptera*

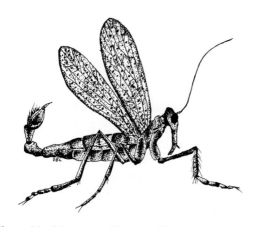

Figure 55 Mecoptera (Scorpionflies).

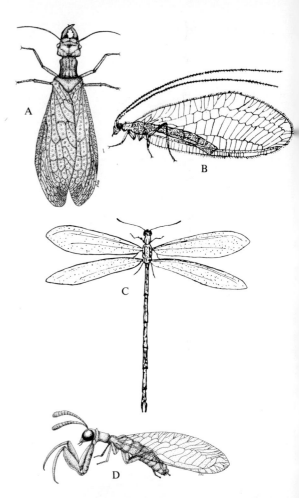

Figure 56 Neuroptera (Dobsonflies, Alderflies, Snakeflies, Lacewings, Antlions, Mantispids, others). A, dobsonfly; B, lacewing; C, antlion; D, mantispid.

26a(1b) Head and legs indistinct (Figs. 37B; 40E ♀); covered with a soft or hard, waxy coating and attached to plants or, if no covering, it occurs as an internal parasite of other insects ... 27

26b Head and legs distinct; only a few with threadlike or powdery, waxy coverings 28

27a(26a)	Round to elongated (Fig. 40E ♀); soft or hard, waxy coating; usually attached in clusters to plants (suborder Homoptera) . . . . (p. 133) *Hemiptera* (in part)	31b	Head equal in width to, or wider than, prothorax (Fig. 57B); 1 or 2 small tarsal claws; tibia without a thumblike process . . . . (p. 122) *Mallophaga*
27b	Elongated (Fig. 37B); no coating on body; internal parasite of bees, wasps, leafhoppers, others . . . . (p. 237) *Strepsiptera* (in part)	32a(29b)	Body flattened from top to bottom (Fig. 38E); does not jump; primarily on birds and sheep, rarely on bats and honey bees. . . . . (p. 315) *Diptera* (in part)
28a(26b)	Body very flattened either from the side (Fig. 57C) or from top to bottom (Figs. 38E; 39F; 57A, B); external parasites of birds, mammals, or honey bees 29	32b	Body flattened from the side (Fig. 57C); jumping insects with long legs . . . . (p. 355) *Siphonaptera*
28b	Body usually not flattened, but if so then 2 or 3 short to long tails are present on the tip of the abdomen (Figs. 58A, B); not parasitic 33		
29a(28a)	Tarsi with 3 or fewer segments . . . 30		
29b	Tarsi with 5 segments 32		
30a(29a)	Antennae longer than head; tarsi 3-segmented (Fig. 39F) (suborder Heteroptera) . . . . (p. 133) *Hemiptera* (in part)		
30b	Antennae equal to or shorter than head; tarsi 1- or 2-segmented 31		Figure 57 A, Anoplura (Sucking Lice); B, Mallophaga (Chewing Lice); C, Siphonaptera (Fleas).
31a(30b)	Head generally narrower than prothorax and often relatively small (Fig. 57A); 1 large tarsal claw; thumblike process on tibia (Fig. 57A) . . . . (p. 126) *Anoplura*		

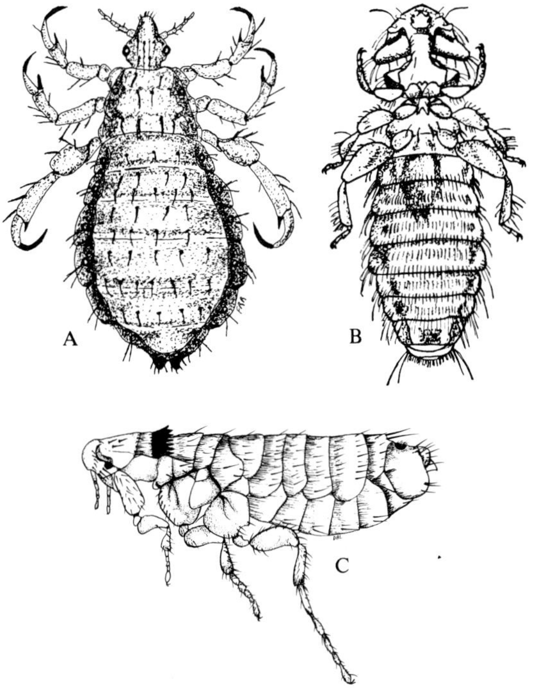

Key to Orders of Adult Insects

33a(28b) Tip of abdomen with 2 or 3 short to long, slender tails (Figs. 58A, B) .. 34

33b Tip of abdomen without slender tails 35

34a(33a) Tip of abdomen with 2 tails; body not scaly (Fig. 58A) ....
.... (p. 58) *Diplura* (in part)

34b Tip of abdomen with 3 tails; body often scaly (Fig. 58B) ....
.... (p. 62) *Thysanura*

35a(33b) Tip of abdomen with 2 pincerlike appendages (Fig. 41)........... 36

35b Tip of abdomen without 2 pincerlike appendages.................. 37

36a(35a) Tarsi 3-segmented; eyes present and distinct (Fig. 41)................
...... (p. 108) *Dermaptera* (in part)

36b Tarsi 1-segmented; eyes absent.....
.... (p. 58) *Diplura* (in part)

37a(35b) Tarsi 1- to 3-segmented......... 38

37b Tarsi 4- or 5-segmented......... 46

38a(37a) Antennae absent (Fig. 58C); less than 2 mm in length; uncommon ....
.... (p. 57) *Protura*

38b Antennae present; size variable... 39

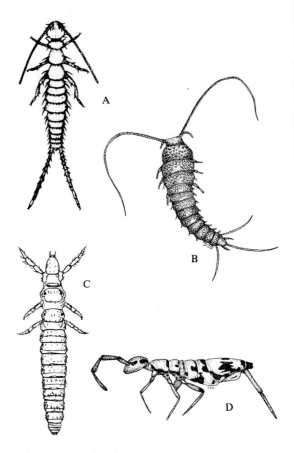

Figure 58 Suborder Apterygota. A, Diplura (Diplurans); B, Thysanura (Bristletails); C, Protura (Proturans); D, Collembola (Springtails).

39a(38b) Body very slender (Fig. 47); mouthparts cone-shaped; tarsi 1- or 2-segmented and often with a pad instead of claws at tip; 5 mm or less in length......................
..... (p. 129) *Thysanoptera* (in part)

54   Key to Orders of Adult Insects

39b	Body oval or elongate-oval, if very slender then body is more than 5 mm long; mouthparts not cone-shaped although they may be long and slender; tarsi 2- or 3-segmented or appear to be absent, claws instead of pad present at tip.................... 40		43b	Base of front tarsi not stout; tarsi 2- or 3-segmented or may appear to be absent..................... 44
40a(39b)	Mouthparts form a slender beak (Figs. 49A, B)............... 41		44a(43b)	Antennae threadlike and with 13 or more segments (Fig. 52B)......... .... (p. 118) *Psocoptera* (in part)
40b	Mouthparts do not form a beak .. 42		44b	Antennae beadlike or relatively stout with up to 9 segments 45
41a(40a)	Beak arises underneath toward the rear of head (Fig. 49B); a pair of tubes may be present near the end of the abdomen (Fig. 40B); usually more than 5 antennal segments (suborder Homoptera).............. .... (p. 133) *Hemiptera* (in part)		45a(44b)	Antennae with 9 segments (Fig. 51C); abdomen with 2 short tails; uncommon .... (p. 117) *Zoraptera* (in part)
			45b	Antennae with 6 or fewer segments; abdomen without tails but a forked, ventral appendage often present (Fig. 58D) ...... (p. 59) *Collembola*
41b	Beak arises underneath toward the front of head (Fig. 49A); abdominal tubes absent; usually 4 or 5 antennal segments; may be antlike (suborder Heteroptera)................... .... (p. 133) *Hemiptera* (in part)		46a(37b)	Distinct constriction or joint between thorax and abdomen (Fig. 54C); body may be very fuzzy and antlike ..... (p. 361) *Hymenoptera* (in part)
			46b	No distinct constriction between thorax and abdomen; if body is fuzzy it is not antlike 47
42a(40b)	Hind legs thickened and lengthened for jumping (Fig. 43) .... (p. 89) *Orthoptera* (in part)		47a(46b)	Head prolonged into a stout ventral beak (Fig. 55).................. .... (p. 247) *Mecoptera* (in part)
42b	Hind legs not thickened and lengthened..................... 43		47b	Head not prolonged into a stout ventral beak 48
43a(42b)	Base of front tarsi very stout (Fig. 51B); tarsi 3-segmented; uncommon...................... ...... (p. 111) *Embioptera* (in part)		48a(47b)	Body densely covered with hairs or scales; coiled proboscis usually present (Fig. 45C)................. ...... (p. 257) *Lepidoptera* (in part)

Key to Orders of Adult Insects

48b	Body never hairy or scaly; proboscis absent and replaced by chewing mouthparts. 49	50a(49a)	Usually very elongate, slender, and sticklike body (Fig. 59). . . . . (p. 101) *Phasmida*
		50b	Not as above. 51
49a(48b)	Pronotum usually smaller than mesonotum (Figs. 50B; 59; 60) . . . 50	51a(50b)	Tarsi 4-segmented; prothorax narrower than head; tip of abdomen without long tails although short stubs may occur (Fig. 50B). . . . . (p. 112) *Isoptera* (in part)
49b	Pronotum larger than mesonotum, may be very broad and partly or completely covering head, curved and sometimes pointed, or very long and slender (Figs. 43; 44) 52	51b	Tarsi 5-segmented; prothorax as wide or wider than head; tip of abdomen with 2 long tails (Fig. 60); uncommon . . . (p. 107) *Grylloblattodea*

Figure 59 Phasmida (Walkingsticks).

		52a(49b)	Femora of hind legs thickened for jumping (Fig. 43). . . . . (p. 89) *Orthoptera* (in part)
		52b	Femora of hind legs not thickened 53
		53a(52b)	Body flattened (Fig. 44A) or front legs thickened, spined, and used for grasping (Fig. 44B). . . . . (p. 102) *Dictyoptera* (in part)
		53b	Body cylindrical and not flattened; front legs not thickened or spiny. . . . . (p. 89) *Orthoptera* (in part)

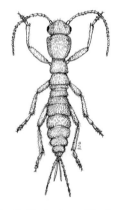

Figure 60 Grylloblattodea (Rock Crawlers or Icebugs).

56 Key to Orders of Adult Insects

Subclass Apterygota[1]

The Apterygota differ from Pterygota by lacking wings and undergoing a very simple metamorphosis (ametabolous). Some adult pterygotes lack wings, but this condition is thought to have evolved from their winged ancestors whereas the apterygotes have never been winged. The thorax of apterygotes is less developed and the abdomen usually bears style-like appendages not found in adult Pterygota.

The Protura, Collembola, and Diplura differ from the Thysanura and Pterygota by having mouthparts that are withdrawn inside the head capsule (entognathous) rather than protruding (ectognathous). The antennae, eyes, segmentation, and specialization of the body and legs of these three orders also differ from the Thysanura. These differences have led many taxonomists to classify proturans, collembolans, and diplurans as offshoots from the class Insecta, placing each of them in a separate class and leaving Thysanura as the only order in the insect subclass Apterygota.

ORDER PROTURA[2]
Proturans

Proturans are minute (0.5-2.0 mm), whitish insects that lack wings, compound eyes, and antennae (Fig. 61). The cone-shaped head contains piercing-sucking mouthparts and the front legs are held up, simulating the appearance of antennae. Styli occur on the undersides of the first three abdominal segments. The immatures have an ametabolous type of development and are unique in adding one abdominal segment during each of three molts (anamorphosis). Some authorities separate proturans from insects and elevate them to the Class Protura.

Proturans are moderately rare. They occur in moist soil and humus, along the edge of woods, under bark, and in rotting logs. These insects are collected by sifting debris or using a Berlese funnel, and are preserved in 80% alcohol or mounted on slides.

Species: North America, 20; world, 152. Families: North America, 3.

---

1. Apterygota: *a*, without; *pterygota*, wings.
2. Protura: *prot*, first; *ura*, tail (refers to the pointed, terminal segment of the abdomen).

KEY TO COMMON FAMILIES OF PROTURA

1a   Tracheae present; 1 pair of spiracles on both the meso- and metathorax (Fig. 61); styli on first 3 abdominal segments have terminal vesicles .... *Eosentomidae*

1b   Tracheae and spiracles absent; only styli on the first abdominal segment have terminal vesicles............ .... *Acerentomidae*

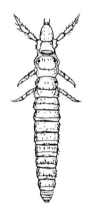

Figure 61   Proturan (Eosentomidae).

GENERAL REFERENCES

Ewing, H.E. 1940. The Protura of North America. Ann. Ent. Soc. Amer. 33:495-551.

Imms, A.D.; Richards, O.W.; Davies, R.G. 1977. *Imms' General Textbook of Entomology.* 10th ed. Vol. 2. Halsted Press, N.Y. 910 pp.

Tuxen, S.L. 1964. *The Protura. A Revision of the Species of the World with Keys for Determination.* Hermann, Paris. 360 pp.

ORDER DIPLURA[3]
Diplurans

Diplurans are small (usually less than 7 mm), pale insects that possess two caudal filaments (Fig. 62) and 1-segmented tarsi. These insects have no wings, compound eyes, or scales, and the mouthparts are a chewing type. The name Entotrophi is used for this order by some authorities and others elevate the group to the Class Diplura. Diplurans live in damp, concealed areas under stones and bark, in soil, and in rotting wood and other debris. Specimens are captured by sifting through debris or using a Berlese funnel, and are preserved in 80% alcohol or mounted on slides.

Species: North America, 75; world, 500. Families: North America, 3.

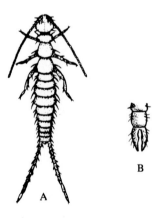

Figure 62   Diplurans. A, Campodeidae; B, forceps-like cerci of Japygidae.

3. Diplura: *dipl*, two; *ura*, tail.

KEY TO COMMON FAMILIES OF DIPLURA

1a   Cerci forcepslike (Fig. 62B), 1-segmented, and shorter than antennae .... (p. 59) *Japygidae*

1b   Cerci not forcepslike; cerci many-segmented and as long as antennae (Fig. 62A) .... (p. 59) *Campodeidae*

Family Campodeidae

This is the most commonly encountered family. *Campodea* (Fig. 62A) is the common genus in North America.

Family Japygidae

Although widely distributed, this family is most frequently collected in the southern U.S. *Japyx* is the common genus in North America.

GENERAL REFERENCES

Condé, B. 1956. Materiaux pour une monographie des Diploures Campodéidés. Mem. Mus. nat. Hist. Nat. (Paris) A, Zool. 12:1-202.

Imms, A.D.; Richards, O.W.; Davies, R.G. 1977. *Imms' General Textbook of Entomology.* 10th ed. Vol. 2. Halsted Press, N.Y. 910 pp.

Smith, L.M. Series of Publications on Diplura Including: (1959) Ann. Ent. Soc. Amer. 52:363-368; (1960) ibid. 53:137-143, 575-583.

ORDER COLLEMBOLA[4]
Springtails

Collembola are small (less than 6 mm), wingless insects that have a unique ventral tube (collophore) on the first abdominal segment and often a forked, springing organ (furcula) on the fourth abdominal segment (Fig. 63A). The collophore acts as an adhesive organ, helps absorb water, and is probably involved in respiration and secretion. The furcula is folded forward under the abdomen when at rest and held in place by a structure (tenaculum) on the third abdominal segment. When the furcula is released, it is forced downward by the release of cuticular tension and the contraction of its muscles. The furcula strikes the ground and enables the insect to spring as high as 100 mm into the air. Collembola have chewing mouthparts and an ametabolous type of development. Some authorities consider the springtails to be separate from insects and have elevated them to the Class Collembola.

Springtails prefer the cool, humid, concealed microhabitats of soil, leafmold, fungi, moss, decaying logs, and ant and termite nests. However some are found on vegetation or even trees. Others occur on the surface of pools, snow fields, and in the intertidal areas of the seashore. These insects range throughout the world from the Arctic to the Antarctic regions. The food of springtails generally consists of decaying materials, algae, lichens, pollen, and fungal spores. Collembolans often mass together in enormous numbers and some species cause damage by feeding on germinating seeds and seedling plants in greenhouses and gardens.

Collembola are collected by sifting or floating debris from the habitats described above, and by using a Berlese funnel. A small brush dipped in alcohol or an aspirator may be used to pick up specimens from masses of springtails. They are best preserved in isopropyl alcohol (although 80% ethyl alcohol is satisfactory) or mounted on slides.

Species: North America, 650; world, 3,600.
Families: North America, 7.

[4]. Collembola: *coll,* glue; *embola,* wedge or peg (refers to the collophore).

KEY TO FAMILIES OF COLLEMBOLA

1a Body spherical or broadly oval (Fig. 63A); first 4 abdominal segments distinctly separated. 2

1b Body elongated; abdomen with 6 distinct segments 3

2a(1a) Antennae much shorter than head; eyes absent. . . . . (p. 61) *Neelidae*

2b Antennae equal or longer than head (Fig. 63A); eyes usually present . . . . (p. 61) *Sminthuridae*

3a(1b) First thoracic segment similar in appearance to other thoracic segments although slightly smaller (Figs. 63B, E), and with at least a few setae; body without markings. 4

3b First thoracic segment different in appearance from other thoracic segments, greatly reduced in size (Figs. 63F, G), and without setae; body often with pigmented markings (Fig. 63F) 6

4a(3a) Furcula and eyes absent (Fig. 63B); skin pores (pseudocelli) on head and body (Fig. 63B); body usually white . . . . (p. 61) *Onychiuridae*

4b Furcula and eyes present or absent; no skin pores; body usually pigmented. 5

5a(4b) Furcula extends beyond end of abdomen (Fig. 63D) . . . . (p. 61) *Poduridae*

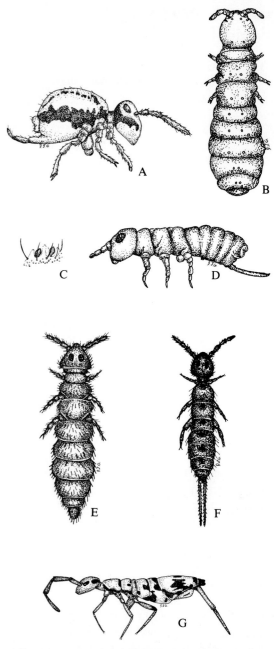

Figure 63 Springtails. A, Sminthuridae; B, Onychiuridae; C, third antennal segment of *Onychiurus folsomi*; D, *Podura aquatica* (Poduridae); E, *Hypogastrura nivicola* (Hypogastruridae); F, *Entomobrya multifasciata* (Entomobryidae); G, *Isotoma viridis* (Isotomidae).

5b   Furcula, when present, never extends beyond end of abdomen.......... .... (p. 62) *Hypogastruridae*

6a(3b)   Body covered with scales or dense setae giving a furry appearance; third and fourth abdominal segments often much longer than others, or fourth at least twice as long as third along middorsal line (Fig. 63F)..... .... (p. 62) *Entomobryidae*

6b   Body not covered with scales or dense furry setae; third and fourth abdominal segments not unusually long and about equal in length along middorsal line (Fig. 63G) .... (p. 62) *Isotomidae*

Family Neelidae

Members of this family are similar to those of Sminthuridae but the antennae are shorter than the head. *Neelus minimus* Willem is a white, eyeless, globular species about 0.5 mm long and common in soil.

Family Sminthuridae

These are globular or oval springtails which often have black eyes.

Common Species

Bourletiella hortensis (Fitch)—1.5 mm; dark purple to blackish; body often dull green dorsally; many small white spots; 8 eyes on each side of head are surrounded by black patches ringed with white; dense setae on body are short and curved; throughout most of North America (in cultivated land; spring and early summer).

Sminthurinus elegans (Fitch) (Fig. 63A)—1 mm; yellow to yellow-green with blackish markings; 2 irregular longitudinal black stripes on each side of body; most of U.S., southern Canada (soil of lawns and pastures).

Family Onychiuridae

Members of this family have skin pores on the head and body and the body is usually white.

Common Species

Onychiurus folsomi (Schäffer) (Fig. 63B)—1.8-2.1 mm; white; body stout; no anal spines; 4 papillae and 2 ovate, erect clubs on 3rd antennal segment (Fig. 63C); throughout U.S., southern Canada (in organic soils, mushroom farms).

Tullbergia granulata Mills—1 mm; white; very elongated body; no eyes; antennae shorter than head; body more heavily granulated on dorsal than ventral side; irregular rows of coarse granules across dorsum of 6th abdominal segment; 2 anal horns; throughout North America (in soil).

Family Poduridae

The only species in this family is the often abundant *Podura aquatica* Linnaeus (Fig. 63D) which occurs throughout most of North America. It is 1.5 mm long and is blue, black, or reddish brown; the furcula extends beyond the abdomen and the head and body lack skin pores. This species is found on the water's surface along the edges of streams and ponds. Some authorities classify the Hypogastruridae as part of Poduridae.

Family Hypogastruridae

The furcula does not extend beyond the tip of the abdomen in this family. *Anurida maritima* (Guérin), a blue, black, or dark-gray species with no furcula, occurs along the seashore, on the surface of small tidal pools, on sand and seaweed, and under rocks between high and low tides.

Common Species

Hypogastrura nivicola (Fitch) (Fig. 63E). Snow flea—2.0-2.7 mm; blue to bluish black; 4 tubercles posterior to antennae; 4th antennal segment with blunt setae; tarsus with 1 large claw curved slightly inward and 1 smaller claw 1/2 its length; 2 short anal spines; northern 1/2 U.S. southeast to NC, southern Canada (on snow surface and pools in forests after a thaw).

Family Entomobryidae

The fourth abdominal segment is at least twice the length of the third in this family. Scales or clubbed setae occur on the body.

Common Species

Entomobrya multifasciata (Tullberg) (Fig. 63F)—2.5 mm; yellow with purplish broken bands on each body segment; eyes on dark patches; body with scales and hairs; throughout North America.

Tomocerus flavescens (Tullberg)—2 mm; silvery gray or cream color; 6 ocelli; 3rd antennal segment almost body length and often coiled; 4th antennal segment shorter than 3rd and sometimes missing; scales on body; throughout North America (under bark, boards, logs).

Family Isotomidae

In this family the third and fourth abdominal segments are about equal in length along the middorsal line and the body lacks scales and clubbed setae.

Common Species

Folsomia elongata (MacGillivray)—1.6 mm; grayish or greenish gray to white; white beneath; anterior portions of body segments pale; abdominal segments 4-6 fused and capsulelike making segmentation indistinct; Rocky Mts. eastward (in soil litter, under stones and bark).

Isotoma viridis (Bourlet) (Fig. 63G)—4 mm; color variable but not black or blue-black; 4th antennal segment without blunt setae; North America, primarily eastern 2/3 U.S.

> GENERAL REFERENCES
> Christiansen, K. 1964. Bionomics of Collembola. Ann. Rev. Ent. 9:147-178.
> Christiansen, K., and Bellinger, P. *The Collembola of North America North of the Rio Grande.* Entomological Reprint Specialists, Los Angeles. Publ. expected 1978.
> Maynard, E.A. 1951. *A Monograph of the Collembola or Springtail Insects of New York State.* Comstock Publ., Ithaca, N.Y. 339 pp.
> Mills, H.B. 1934. *A Monograph of the Collembola of Iowa.* Iowa St. Coll. Press, Monog. 3. Ames, Iowa. 143 pp.
> Salmon, J.T. 1964-1965. An Index to the Collembola. Bull. Roy. Soc. New Zealand, no. 7. vol. 1-3. 651 pp.

ORDER THYSANURA[5]
Bristletails

Bristletails are small- to moderate-sized wingless insects with three slender, taillike appendages and long, slim antennae (Figs. 64A, B). The body is gray, brown, or white and usually covered with scales. These insects are able to jump or run rapidly. Thysanurans have chewing mouthparts and the nymphs undergo an ametabolous type of development.

Eggs are laid singly in cracks and other secluded places. Nymphs mature slowly (up to two years) and they may molt dozens of times.

---

5. Thysanura: *thysan,* bristle or fringe; *ura,* tail.

Most species occur outdoors and feed on plant materials; the house-inhabiting forms eat cereals, glue, starch, and paper.

Outdoor species may be collected under the bark of rotten logs, leaf debris, and stones. Indoor species occur in damp or dry, warm areas such as bathrooms and basements. A box baited with crushed crackers and made accessible by a ramp will trap individuals if the inside walls of the box are coated with petroleum jelly. Specimens are preserved in 80% alcohol.

Species: North America, 50; world, 700. Families: North America, 4.

KEY TO COMMON FAMILIES OF THYSANURA

1a Small compound eyes are widely separated; no styli on middle and hind coxae; abdominal styli on segments 7-9; running insects .... (p. 63) *Lepismatidae*

1b Large compound eyes usually touching each other; styli on middle and hind coxae and abdominal segments 2-9 (Fig. 64B); jumping insects .... (p. 63) *Machilidae*

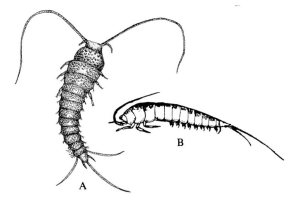

Figure 64 Bristletails. A, silverfish, *Lepisma saccharina* (Lepismatidae); B, *Machilis variabilis* (Machilidae).

Lepismatidae—Firebrats and Silverfish

The two best known members (see Common Species) occur in buildings and may become pests. They are nocturnal and feed on paper, bookbindings, paste, cereals and flour, and starched clothing and curtains. Other species live under debris, in caves, and in ant nests.

Common Species

Lepisma saccharina Linnaeus (Fig. 64A). Silverfish—8-13 mm; silvery gray with mottled dark markings above; throughout North America (in moderately cool, damp areas of buildings).

Thermobia domestica (Packard). Firebrat—8-13 mm; tan or yellowish brown; throughout North America (near warm areas, e.g., furnaces, steampipes).

Machilidae—Machilids or Jumping Bristletails

These brownish insects occur under leaves, bark, and rocks in woods and grassy areas. They are not as flattened as Lepismatidae and jump when disturbed. They are sometimes classified as a separate order, Microcoryphia.

Common Species

Machilis variabilis Say (Fig. 64B)—9-11 mm; grayish brown, antennae extend to tip of longest abdominal filament; North America (under stones and rotten logs).

GENERAL REFERENCES

Imms, A.D.; Richards, O.W.; Davies, R.G. 1977. *Imms' General Textbook of Entomology.* 10th ed. Vol. 2. Halsted Press, N.Y. 910 pp.

Sweetman, H.L. 1938. Physical Ecology of the Firebrat, *Thermobia domestica* (Packard). Ecol. Monogr. 8:285-311.

Wygodzinsky, P. 1972. A Revision of the Silverfish (Lepismatidae, Thysanura) of the United States and the Caribbean Area. Amer. Mus. Novit., No. 2481. 26 pp.

Subclass Pterygota[1]

The vast majority of insects are in the subclass Pterygota and most of the adults are winged. Those without wings probably evolved from winged ancestors that developed a specific mode of life (e.g., parasitic, living underground or in caves or microhabitats, etc.) and eventually lost the need for wings. The thorax of pterygotes is enlarged and strengthened for wing support. Immatures develop via a hemimetabolous (incomplete metamorphosis) or holometabolous (complete metamorphosis) route.

ORDER EPHEMEROPTERA[2]
Mayflies

Adult mayflies are slender, winged insects with two or three long, threadlike filaments projecting from the tip of the abdomen (Fig. 65A). These insects range from about 2 to 32 mm in length. Most species have four membranous wings and the hind wings are much smaller than the front wings or are absent in some species. The wings are held together upright over the body and cannot be folded flat (Fig. 65A).

Adults live from as little as 1 1/2 hours to a week or two although most live only two to three days. They are unable to feed with their nonfunctional mouthparts. Mating flights of males occur over or near water usually near dusk and are characterized by upward and forward movements followed by downward drifting. These swarms may contain tremendous numbers of individuals and attract much attention. Females enter the swarm, are grasped by males with their unusually long front legs, and mating ensues. Females typically land on the water's surface after mating to release eggs and die while still in the water. Some species drop eggs from above, dip the abdomen into the water, or crawl underwater.

Nymphs usually have three (sometimes only two) filaments or tails projecting from the tip of the abdomen. Breathing is by means of gills located on the sides of the abdomen and, in some cases, gill filaments located beneath the head or thorax. Most species develop in fresh water although a few occur in brackish water of estuaries. Nymphs feed primarily on diatoms,

---

1. Pterygota: wings (singular, *pterygo*)
2. Ephemeroptera: *ephemero,* short-lived, for a day; *ptera,* wings.

algae, other aquatic plants, and organic debris; a few species are predaceous. Some species burrow in mud but others are more active. A hemimetabolous type of development occurs and generally there are one or two generations per year. Developmental time is highly dependent on water temperature and some species take two or more years to develop and may molt 50 times. After the last molt a winged form, the subimago, emerges at the water surface and flies to a nearby resting site. The subimago has duller wings and shorter tails than the adult (imago) and it molts once (usually by the following day) to the adult stage. Mayflies are the only insects known to molt after functional wings have developed. Both nymphs and adults are important members of food chains.

Adults are collected with a net or picked off of surfaces by hand. Lights will attract some species. Swarms of males occur along lake and stream margins as well as over vegetation away from water. A long-handled net is needed for high-flying swarms. Adults may be pinned or preserved in 70-80% alcohol. Nymphs are collected in small or large streams and lakes using the standard aquatic collecting methods. Hand screens and drift nets work well in streams and apron and dip nets are effective in lakes. Nymphs are preserved in 70-80% alcohol.

Species: North America, 622; world, 2,200.
Families: North America, 18.

KEY TO COMMON FAMILIES OF EPHEMEROPTERA

A more detailed key that includes most or all of the families will be found in Borror *et al.* (1976) and Edmunds *et al.* (1976).

1a		Veins M and Cu_1 in front wing greatly divergent at base, and base of M_2 greatly bent toward Cu_1 (Fig. 65B) . . . . . (p. 68) *Ephemeridae*
1b		Veins M and Cu_1 weakly divergent at base, and base of M_2 weakly bent toward Cu_1 (Figs. 65C, D) 2
2a(1b)		Hind tarsi 5-segmented. . . . . (p. 67) *Heptageniidae*
2b		Hind tarsi 3- or 4-segmented. 3
3a(2b)		Front wing with 1 or 2 long intercalary veins between veins M_2 and Cu_1 (Fig. 65C) . . . . (p. 68) *Ephemerellidae*
3b		Front wing without long intercalary veins between veins M_2 and Cu_1 . . . 4
4a(3b)		Vein Cu_2 in front wing bent downward toward hind margin (Fig. 65D); 3 tails on tip of abdomen . . . . (p. 67) *Leptophlebiidae*
4b		Vein Cu_2 straight or slightly and smoothly curved; 2 (rarely 3) tails on abdomen 5
5a(4b)		Hind tarsi 3-segmented; hind wings very small or absent; vein M_2 in front wing not attached at its base to M . . . . . . (p. 67) *Baetidae*
5b		Hind tarsi 4-segmented; hind wings normal in size; vein M_2 attached at base to vein M . . . . (p. 66) *Siphlonuridae*

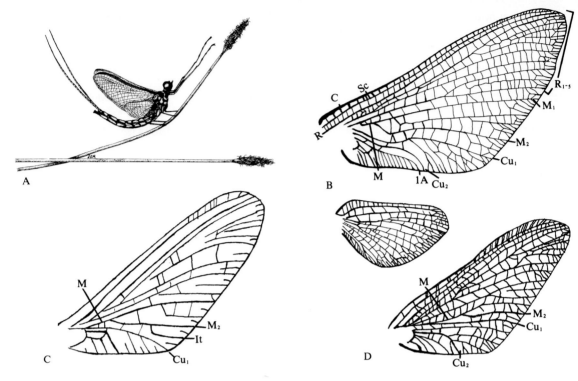

Figure 65 Mayflies. A, lateral view; B, wings of Ephemeridae; C, front wing of Ephemerellidae; D, front wing of Leptophlebiidae. It, intercalary vein.

Family Siphlonuridae

Adults in this family have relatively large hind wings and the upper part of the compound eye has larger facets than the lower part. Nymphs are streamlined and minnowlike in shape (Fig. 66A). The antennae are shorter than the head width and each abdominal segment usually has a sharp dorsolateral projection that points posteriorly. Nymphs are found primarily in rapidly flowing streams and rivers although some occur in swamps and forest pools.

Common Species

Siphlonurus alternatus (Say) (Fig. 66B)—12 mm; head whitish; thorax yellowish brown with wide reddish brown middorsal stripe; abdomen pinkish white with reddish brown oval and triangular markings and a band on posterior margins of tergites, ventral surface marked with distinct bands and 2 black dots near center of each sternite; eastern 1/2 North America.

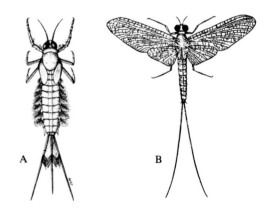

Figure 66 Siphlonuridae. A, nymph; B, *Siphlonurus alternatus*.

Siphlonurus occidentalis Eaton—11-13 mm; yellowish brown and reddish brown; white bands across ventral side of thorax; abdomen with darker dorsal markings, ventral side with dark "U"-shaped markings on each segment; western 1/2 U.S., southwest Canada. *S. spectabilis* Traver, 14 mm, speckled wings, hind wing orange-tinted (♂) CA to NM.

Family Baetidae

The hind wings of these small mayflies are very reduced or absent. The eyes of males are divided into a large upper and smaller lower layer giving a turbanlike effect (Fig. 67A). The family is the largest in the order with about 140 species in North America. Nymphs are streamlined in form and may resemble the Siphlonuridae, but the antennae are usually at least twice the head width and the abdomen lacks posterolateral projections (Fig. 67C). Immatures occur in a variety of aquatic habitats. The largest genus, *Baetis* (Fig. 67B), contains about 62 species that are often difficult to identify. Adults are characterized by the tiny, narrow hind wings with one to three longitudinal veins. Abdominal segments two to six in males are clear, white, or light brown, and the terminal segments are dark.

Family Heptageniidae

These medium-sized mayflies have 5-segmented hind tarsi. The dark-colored nymphs are distinctly flattened and have two or three filaments on the tip of the abdomen (Fig. 68A). Nymphs frequent the undersides of rocks in fast-moving streams but some also occur in ponds and sandy bottoms of rivers.

Common Species

Heptagenia flavescens (Walsh)—9-13 mm; yellowish; top of thorax reddish brown (♂); broad reddish brown middorsal stripe on abdomen; southeastern and central U.S., southcentral Canada.

Stenacron interpunctatum (Say)—8-10 mm; head yellow or white with green tinge; pronotum white with oblique black streak on each side, rest of thorax brown with yellow scutellum; legs yellow with dark bands; anterior margin of front wing light brown, black dash in upper center of wing between 2nd and 3rd longitudinal veins; east of Rocky Mts.

Stenonema tripunctatum (Banks)—9-11 mm; light brown to reddish brown; front wing clear except for shaded area along outer 1/3 of anterior margin; hind wing may have dark apical margin; each abdominal segment with 3 transverse streaks near posterior margin; east of Rocky Mts.

Family Leptophlebiidae

Adults in this family are usually recognized by the strongly curved Cu_2 veins in the front wing (Fig. 65D). The compound eyes of males have large facets in the upper portion and smaller facets below. There are three filaments on the tip of the abdomen. Nymphs have abdominal gills that are narrow and deeply forked, broad with long fringes on the margins, or broad with one or more narrow points on the tips

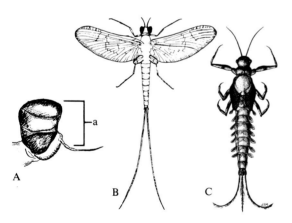

Figure 67 Baetidae. A, head of adult; B, *Baetis* sp; C, nymph. a, turbanlike compound eye.

(Fig. 68B). The habitats of nymphs range from ponds, lakes, and quiet areas of streams and rivers to shallow, rapid streams. Nymphs often prefer areas where silt and debris accumulate and may cling to the underside of rocks.

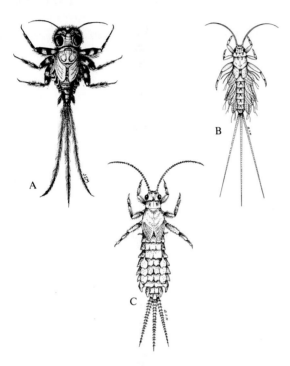

Figure 68 Nymphs. A, Heptageniidae; B, Leptophlebiidae; C, Ephemerellidae (*Ephemerella* sp.).

Common Species

Leptophlebia cupida (Say)—9-11 mm; blackish; wings brown along outer anterior margin; tip of abdomen yellowish, underside of abdomen tan; eastern 1/2 U.S., southern Canada (quiet areas of streams, isolated pools). *L. nebulosa* (Walker), similar, apical 1/3 front wings brownish, abdominal tails ringed, U.S. except Southeast, southeastern and southwestern Canada.

Paraleptophlebia debilis (Walker)—8 mm; deep brown with abdomen white on segments 2-7; tip of wing with small spots; most of North America (adults occur March to November depending on locality).

Family Ephemerellidae

Members of this common family have one or two long, intercalary veins in the front wing between veins M_2 and Cu_1 (Fig. 65C). The compound eyes of males meet dorsally for most of their length. Nymphs may be slender or flattened, some have dorsal tubercles on the body, and most have gills on abdominal segment three (gills are small or absent on segments one and two) (Fig. 68C). Habitats include turbulent shorelines of lakes, slow-moving or stagnant streams, and streams with rapids. All North American species (approximately 76) are in the genus *Ephemerella*.

Family Ephemeridae

Adults, ranging up to 32 mm in length, are among the largest mayflies. Some species emerge in great numbers during May and June. Nymphs have long, upcurved mandibles and the abdominal gills are distinctly fringed (Fig. 69B). The immatures burrow in sand and silt bottoms in quiet areas of streams of all sizes as well as along the shorelines of lakes.

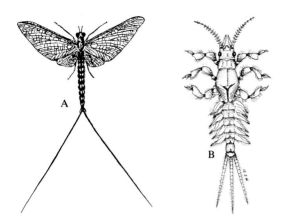

Figure 69 Ephemeridae. A, *Hexagenia limbata;* B, *H. limbata* nymph.

Common Species

Ephemera simulans Walker—10-14 mm; blackish brown or dark reddish brown; wing with many dark blotches and dark crossveins; abdomen yellowish to light brown, tails dark yellow or reddish brown; most of North America.

Hexagenia bilineata (Say)—18-21 mm; reddish brown; outer margin of front wing with dark border, anterior margin faintly darkened, occasional spots in front wing; abdomen banded with longitudinal light and dark stripes; southeastern and central U.S.

Hexagenia limbata Serville (Fig. 69)—18-30 mm; generally pale yellow; thorax reddish brown; outer margin of front wing with dark border, anterior margin purplish; abdomen yellow with brown markings; most of North America.

GENERAL REFERENCES

Berner, L. 1950. The Mayflies of Florida. Univ. Fla. Studies, Biol. Sci. Ser. no. 4. 267 pp.

Burks, B.D. 1953. The Mayflies, or Ephemeroptera, of Illinois. Ill. Nat. Hist. Surv. Bull. 26:1-216.

Day, W.C. 1956. Ephemeroptera. In *Aquatic Insects of California,* R.L. Usinger (ed.). Univ. California Press, Berkeley. Pp. 79-105.

Edmunds, G.F., Jr. 1959. Ephemeroptera. In *Freshwater Biology,* W.T. Edmondson (ed.). Wiley, N.Y. Pp. 908-916.

———. 1972. Biogeography and Evolution of the Ephemeroptera. Ann. Rev. Ent. 17:21-42.

Edmunds, G.F.; Jensen, S.L.; and Berner, L. 1976. *The Mayflies of North and Central America.* Univ. Minnesota Press, Minneapolis. 330 pp.

Leonard, J.W., and Leonard, F.A. 1962. *Mayflies of Michigan Trout Streams.* Cranbrook Institute of Science, Bloomfield Hills, MI. 139 pp.

McCafferty, W.P. 1975. The Burrowing Mayflies (Ephemeroptera: Ephemeroidea) of the United States. Trans. Amer. Ent. Soc. 101:447-504.

Needham, J.G.; Traver, J.R.; and Hsu, Y.-C. 1935. *The Biology of Mayflies, with a Systematic Account of North American Species.* Comstock Publ., Ithaca, N.Y. 759 pp.

Pennak, R.W. 1953. *Fresh-Water Invertebrates of the United States.* Ronald Press, N.Y. 769 pp.

ORDER ODONATA[3]
Dragonflies and Damselflies

Adult Odonata are moderate- to large-size predaceous insects with big compound eyes, four abundantly veined wings, and a long, slender abdomen. Many are brightly colored and most are found flying or resting on plants near ponds and rivers during the day. Nymphs are aquatic predators with a hemimetabolous mode of development.

The order Odonata is divided into dragonflies (Suborder Anisoptera) (Fig. 73A) and damselflies (Suborder Zygoptera) (Fig. 80A). Adult dragonflies are stouter and more robust than damselflies, their flight is stronger, and

3. Odonata: *odon,* tooth (refers to the toothed mandibles).

they rest less frequently on plants. The heavily veined hind wings of adult dragonflies are broader at the base than the front wings and both wings do not narrow at the base (Fig. 70A). The compound eyes meet dorsally (Fig. 73A) or are no more than one diameter apart (Fig. 78A). In contrast, the hind wings of damselflies are similar to the front wings and both narrow at the base (Fig. 70B). The compound eyes are more than one diameter apart. The Odonata are unable to fold their wings completely. Dragonfly wings project outward at right angles to the thorax (Fig. 73A) whereas damselflies hold their wings together or slightly divergent above the body when resting (Fig. 80A). The wing position is a good field mark for separating damselflies from dragonflies.

Eyesight in both suborders is extremely good due to the many thousands of single eye units that form each compound eye. The antennae are very short and bristlelike. The legs of Odonata are not adapted for walking but are used for clinging to plants, climbing, and grasping prey. Adults have chewing mouthparts and prey on small flying insects. Odonata generally live 2-3 weeks. Overwintering occurs in the egg or nymphal stages in northern latitudes.

Adult dragonflies occasionally are found swarming during the day or at twilight and some nomadic tropical species follow rainbelts. Males will defend an area (territory) against other males of the same species. If a female enters the area the male may attempt to mate with her in a posture unique to dragonflies and damselflies. The male bends his abdomen to place his sperm capsule on a ventral copulatory organ located at the base of his second abdominal segment (Fig. 71B). He grasps the prothorax or the top of the female's head using appendages at the tip of his abdomen. She brings the tip of her abdomen forward to touch his copulatory organ and receives the capsule. Much of the time the female's abdomen is directed backward and during their tandem flight copulation occurs. Females of some species deposit eggs on the water surface by dipping their abdomens or dropping the eggs. In other species eggs are placed within or on submergent or emergent plant tissue.

Odonata nymphs (naiads) are aquatic and are generally found climbing on submerged vegetation, resting on a silty or sandy bottom, or burrowing in sand or mud of streams and ponds. The presence or absence of external respiratory gills is the simplest criterion used to identify nymphs of the two suborders. Damselflies have three visible, leaflike gill plates at the tip of the abdomen (Fig. 80B) whereas dragonfly gills are concealed in the rectal chamber of the abdomen (Fig. 73C). The intake and expulsion of water from the rectum for gas exchange by dragonfly nymphs also acts as a jet propulsion for swimming. Damselflies swim by undulating their bodies and gills. Nymphs prey on aquatic invertebrates including mosquitoes and snails, tadpoles, and small fishes. The labium of their chewing mouthparts is modified into a large, folded, prehensile (grasping) organ that is rapidly thrust forward to grasp the food with movable hooks and draw it back to the mouth (Figs. 73D; 82C). Nymphs may require from 36 days to six years to mature. They leave the water just prior to the last molt and the cast skins of emerged adults remain on banks or plants.

Damselflies are usually less difficult to catch than dragonflies because they fly more slowly and perch on vegetation more often. It may be necessary to become familiar with the repetitious flight path of large, rapid-flying dragonflies in order to catch these insects. The individuals are intercepted by swinging a net from behind the specimen or partially concealing oneself in a bush to reduce the insect's eyesight advantage. It may be necessary to wade streams and ponds to capture those that land on reeds offshore or patrol over the water. Adults are placed in envelopes immediately after being killed and their wings are folded above the body. Although colors tend to fade, rapid drying helps retain most of the colors. Specimens may be dried in the sun, an oven, or under a lamp. Color preservation is also improved by placing individuals in paper triangles and submerging the triangles into a sealed jar of acetone for 24 hours. The specimens are allowed to air dry (caution: acetone is flammable) and are then placed in permanent cellophane envelopes with a labeled card inserted behind each insect.

If specimens are pinned, the pin may be inserted sideways through the thorax at the base of the wings with the insect's left side uppermost. Several days of drying in an envelope prior to pinning may be needed to insure that the wings stay folded over the body. An alternate display method is to spread the wings on a spreading board after inserting the pin through the dorsum of the thorax. Adults that are newly emerged will not preserve well if dried because they are soft, not fully colored, and have wrinkled wings. They should be preserved in alcohol if kept at all.

Nymphs may be collected with a net in mud and bottom debris of streams and ponds, and among submerged vegetation such as reeds and sedges. An apron net works especially well to probe, gather, and sift debris. Nymphs are preserved in 70-80% alcohol.

Species: North America, 413; world, 5,000.
Families: North America, 11.

KEY TO FAMILIES OF ODONATA

Adult males may be recognized by the three (Anisoptera) or four (Zygoptera) forceplike caudal appendages and a copulatory organ on the ventral side of the second abdominal segment (Fig. 71B). The female has two caudal appendages and the apex of the abdomen is swollen in those families with ovipositors (Petaluridae, Aeshnidae, all Zygoptera).

1a	**Hind wings wider at base than front wings (Fig. 70A); wings extended horizontally when at rest (dragonflies, Suborder Anisoptera) 2**
1b	**Front and hind wings similar in size and shape and narrowed at base (Fig. 70B); wings held together or slightly divergent above and nearly parallel to body (Fig. 80A) (damselflies, Suborder Zygoptera) 8**
2a(1a)	**Triangles of front and hind wings similar in shape and about equidistant from arculus (Figs. 73B; 74A) 3**
2b	**Triangles of front and hind wings dissimilar in shape and triangle in front wing farther from arculus than triangle in hind wing (Figs. 70A; 76B; 77A). 6**

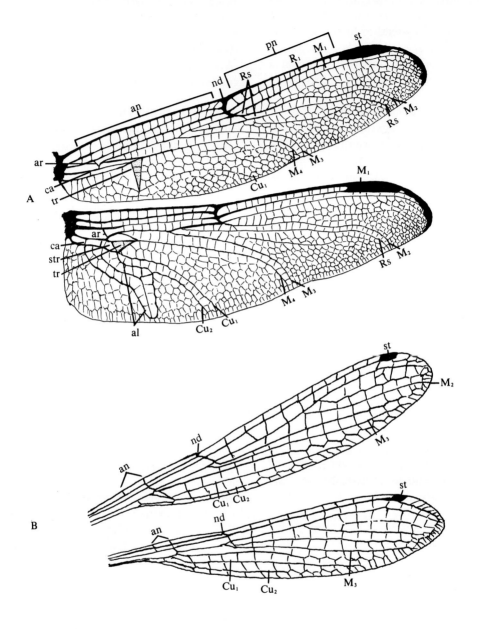

Figure 70 A, dragonfly wings (Libellulidae); B, damselfly wings (Coenagrionidae). al, anal loop; an, antenodal crossveins; ar, arculus; ca, cubitoanal crossvein; nd, nodus; pn, postnodal crossveins; st, stigma; str, subtriangle; tr, triangle.

3a(2a)	Brace vein at inner end of stigma (Fig. 72A)................ 4		7a(6b)	Anal loop elongated and divided lengthwise by a vein into two rows of cells (Fig. 77A); base of triangle in hind wing opposite arculus (Fig. 77A)...... (p. 77) *Corduliidae*
3b	No brace vein at inner end of stigma..... (p. 76) *Cordulegastridae*			
			7b	Anal loop rounded and not divided into two long rows of cells (Fig. 76B); base of triangle in hind wing distal to arculus (Fig. 76B)................ .... (p. 76) *Macromiidae*
4a(3a)	Compound eyes meet for much of length on dorsal side of head (Figs. 71A; 73A)................ .... (p. 74) *Aeshnidae*			
4b	Compound eyes separated on dorsal side of head 5		8a(1b)	Ten or more antenodal crossveins in front wing (Fig. 79A); wings not narrowed to a stalk at base and may have red or black markings (Fig. 79A) ... (p. 79) *Calopterygidae*
5a(4b)	Central lobe of labium notched (Fig. 72B); stigma more than 8 mm long.......... (p. 74) *Petaluridae*			
5b	Central lobe of labium not notched (Fig. 74C); stigma less than 8 mm long .... (p. 75) *Gomphidae*		8b	Two antenodal crossveins in front wing (Fig. 70B); wings narrowed to a stalk at base and clear, faintly tinged with brown or occasionally smoky (Fig. 70B).................... 9
6a(2a)	Anal loop of hind wing usually foot-shaped and with well-developed "toe" (Fig. 70A); hind margin of compound eyes straight or with a very small lobe (Fig. 78B); male without small lateral lobe on each side of second abdominal segment .. .. (p. 78) *Libellulidae*		9a(8b)	M_3 vein originates nearer arculus than nodus (Fig. 82A); wings at rest usually divergent over body .... (p. 81) *Lestidae*
			9b	M_3 vein originates nearer nodus than arculus (Fig. 70B); wings at rest usually held together over body (Fig. 80A).................. 10
6b	Anal loop of hind wing rounded (Fig. 76B), or if elongated into a foot shape there is little development of "toe" (Fig. 77A); hind margins of compound eyes slightly lobed (Fig. 77B); males with small lateral lobe on each side of second abdominal segment 7		10a(1b)	Cu_1 and Cu_2 veins well developed (Fig. 70B); widely distributed .... (p. 80) *Coenagrionidae*
			10b	Cu_1 vein short, forming the border of 3-4 cells distal to arculus, and Cu_2 vein rudimentary or absent (Fig. 81); southern Texas................. .... (p. 81) *Protoneuridae*

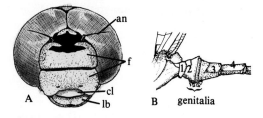

Figure 71   A, head of a darner (Aeshnidae); B, male genitalia on second abdominal segment. an, antenna; cl, clypeus; fr, frons; lb, labrum.

SUBORDER ANISOPTERA

Family Petaluridae—Graybacks

There are only two North American species in this family. The triangles in the wings of adults are about equidistant from the arculus (Fig. 73B), there is a brace vein behind the proximal end of the stigma (Fig. 72A), and the front margin of the central lobe of the labium is notched (Fig. 72B). *Tachopteryx thoreyi* (Hagen) occurs in the eastern U.S. and Texas, flying in sunny openings of woods or landing on tree trunks or stones. The thorax is gray-brown and the body is about 75 mm long. Nymphs of this species live in bogs or small pools and have thick, 7-segmented antennae. The rare *Tanypteryx hageni* (Selys) occurs in high elevations of the Pacific Northwest. The thorax is black with yellow spots and the metathorax bears ventrally a round, hairy tubercle. Nymphs live in wet moss.

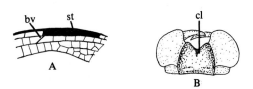

Figure 72   Graybacks (Petaluridae). A, part of front wing; B, labium of adult (ventral view). bv, brace vein; cl, central lobe; st, stigma.

Family Aeshnidae—Darners

Adults are typically large (50-116 mm), fast-flying individuals that occasionally wander some distance from water. The triangles in the front and hind wings are about equidistant from the arculus (Fig. 73B) and the enlarged compound eyes meet along the top of the head (Figs. 71A; 73A). Their names "darning needles" or "darners" may have come from the slim appearance produced when these species are flying high.

Nymphs climb on submerged plants and debris in ponds or slow-moving streams. They have smooth, elongated bodies, long legs, and the mentum of the labium is nearly flat (Fig. 73D). The antennae have 6-7 bristlelike, slender segments.

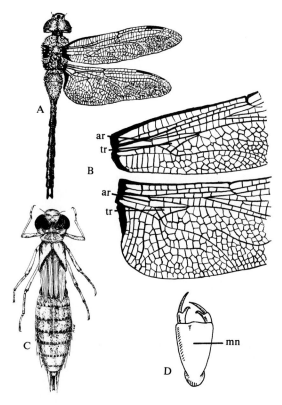

Figure 73   Darners (Aeshnidae). A, green darner, *Anax junius;* B, darner wings; C, *A. junius* nymph; D, labium of a nymph. ar, arculus; mn, mentum; tr, triangle.

Common Species

Aeshna constricta Say—68-72 mm; greenish face without black lines; black "T" spot on frons (Fig. 71A) with stalk directed dorsally and only slightly widened at its base; thorax brown with pale yellow stripes; abdomen brown with blue spots; throughout U.S. except southernmost states, southern Canada (marshes).

Aeshna umbrosa Walker—68-78 mm; yellow or brown markings on posterior of head; ventral side of abdominal segments 4, 5, and 6 has row of pale spots; primarily eastern and central North America, also northwestern U.S. (streams).

Anax junius (Drury) (Figs. 73A, C). Green darner—68-80 mm; green thorax and bluish abdomen; frons has a dark spot surrounded anteriorly by 2 semicircles; throughout U.S. and southern Canada (ponds).

Family Gomphidae—Clubtails

Adults, 31-90 mm long, are dark-colored insects with yellow or greenish stripes on their bodies. Their eyes are widely separated on the head, the front margin of the central lobe of the labium has no notch (Fig. 74C) and the triangles of the wings are equally distant from the arculus (Fig. 74A). The family name is derived from the swollen, terminal abdominal segments of many species in the genus *Gomphus* (Fig. 74D). Adults hover and soar much less than other dragonflies, more often spending their time resting on flat objects or darting from place to place. Some species are commonly found around large, muddy streams while others frequent cleaner streams and ponds. The female has no ovipositor and deposits eggs by dipping her abdomen in the water.

Nymphs burrow in the bottom sediment of streams and ponds. They have a depressed and often wedge-shaped head, and the labium is flat. The antennae are 4-segmented and the tarsi have a 2-2-3 segmentation. The tip of the abdomen of many species is narrowed and elongated (Fig. 74E).

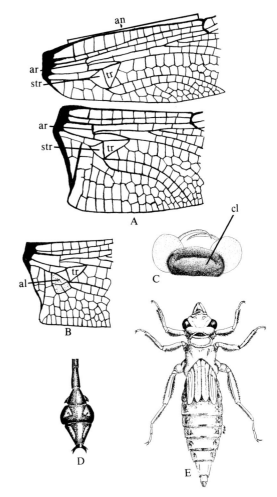

Figure 74 Clubtails (Gomphidae). A, *Gomphus* spp., wing section; B, *Ophiogomphus* spp. hind wing section; C, labium of an adult (ventral view); D, "clubtail" of some *Gomphus* spp.; E, nymph. al, anal loop; an, antenodal crossveins; ar, arculus; cl, central lobe; str, subtriangle; tr, triangle.

Common Species

Gomphus exilis Selys—39-48 mm; greenish yellow with pale brown stripes; middorsal thoracic stripe widens downward into a broad triangle; hind wing with 8 antenodal crossveins; blackish tibiae; dorsum of 8th abdominal segment more black than yellow; hamule yellow (♂); eastern U.S., southern Canada (ponds and streams).

Gomphus externus Hagen—52-59 mm; yellowish green with brown-striped thorax; hind wing without anal loop or crossveins in triangle and subtriangle (Fig. 74A); dark middorsal thoracic stripe; wide yellow band on dorsum of 9th abdominal segment; eastern U.S. south to FL and TX, southern Canada (large streams).

Hagenius brevistylus Selys—73-90 mm; primarily black; sides of thorax yellow with black stripes; crossveins in triangles but not in subtriangles; eastern U.S. south to FL and TX, southern Canada (adults perch on vegetation or banks of streams).

Ophiogomphus rupinsulensis (Walsh)—45-54 mm; greenish with brown shoulder stripes on thorax; no brown middorsal stripe on thorax; 3 cells in the semicircular anal loop of hind wing (Fig. 74B); tibiae yellow on outer side; eastern U.S., southern Canada (rapid streams).

Family Cordulegastridae—Biddies

Adults in this small family are brownish black with yellow markings and are 45-88 mm long. The triangles of the front and hind wings are about equally distant from the arculus (Fig. 73B) and there is no brace vein (Fig. 72A) at the proximal end of the stigma. Adults fly over woodland streams. All eight species are in the genus *Cordulegaster*.

The hairy nymphs bury themselves up to their eyes in the sand or muck bottoms of woodland streams. The labium is spoon-shaped and the lateral lobes are deeply cut forming large, irregular-shaped teeth (Fig. 75).

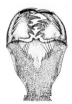

Figure 75 Biddies (Cordulegastridae). Labium of a nymph.

Common Species

Cordulegaster dorsalis Hagen—70-85 mm; similar to *C. maculata* but top of frons not darker than its anterior surface; 2 cubitoanal crossveins in front wing; yellow middorsal abdominal spots; U.S. and Canada west of Rocky Mts.

Cordulegaster maculata Selys—64-76 mm; eyes meeting at one median point only; occiput brown to black and flattened; top of frons much darker than more anterior surface; lateral spots on abdomen; eastern U.S. south to FL, southeastern Canada.

Family Macromiidae—Belted Skimmers and River Skimmers

Adults in the two genera of this family are large (56-91 mm), blackish dragonflies with yellow *(Macromia)* or brownish *(Didymops)* markings. The anal loop of the hind wing is rounded and lacks a bisecting vein (Fig. 76B). *Macromia* adults typically fly over rivers and slow streams and *Didymops* adults are usually found around boggy lakes.

Nymphs sprawl on the bottom of shallow lakes and slow-moving sections of large streams. They have a spoon-shaped labium and a large pyramidal horn on the head in front of the eyes (Fig. 76C).

Common Species

Didymops transversa (Say)—56-60 mm; light brown; nodus of front wing ca. halfway between wing base and apex; basal antenodal cells of both wings tinged with brown; eastern U.S. south to FL and TX, southeastern Canada.

Macromia illinoiensis Walsh—65-76 mm; yellowish brown; 2 pointed cones on top of head; vertex of head all black; nodus of front wing beyond middle of wing (Fig. 76B); auricles black (♂); eastern U.S. and southeastern Canada.

Macromia magnifica McLachlan (Fig. 76A)—69-74 mm; bright yellow and brown; head cones and nodus (Fig. 76B) as in *M. illinoiensis;* vertex of head broadly marked with yellow; 2 separate and lateral yellow spots on 3rd abdominal segment; no yellow on sides of 1st abdominal segment; Far West, southwestern Canada.

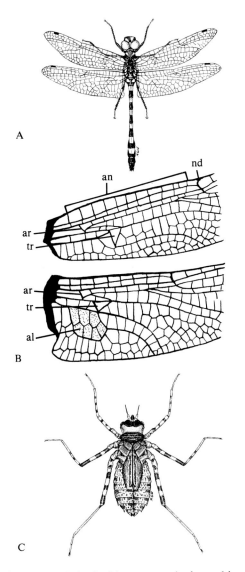

Figure 76 Belted skimmers and river skimmers (Macromiidae). A, *Macromia magnifica*; B, *Macromia* spp. wings; C, *M. magnifica* nymph. al, anal loop; an, antenodal crossveins; ar, arculus; nd, nodus; tr, triangle.

Family Corduliidae— Greeneyed Skimmers

Adults are 28-78 mm long and usually black or metallic in color. The anal loop of the hind wing is elongated into a "foot" which is bisected into 2 even rows (Fig. 77A), but there is no "toe" development as with libellulids (Fig. 70A). The base of the triangle of the hind wing is opposite the arculus (Fig. 77A) and not toward the wing tip as with macromiids (Fig. 76B). The eyes of most species are bright green when the specimen is alive.

Nymphs sprawl on silt bottoms or climb over debris and vegetation of ponds, swamps or bogs. Nymphs have a spoon-shaped labium and are very similar to nymphs of Libellulidae (Fig. 78C) but the surface of the labial palpi of most Corduliidae nymphs is scalloped into rounded scoops that are one-fourth to one-half as long as they are wide. The labial palpi of libellulid nymphs have very shallow indentations that are one-tenth to one-sixth as long as they are wide.

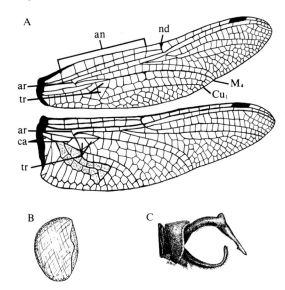

Figure 77 Greeneyed skimmers (Corduliidae). A, *Somatochlora tenebrosa* wings; B, lobed hind margin of adult eye; C, elbowed caudal appendage of male *S. tenebrosa*. an, antenodal crossveins; ar, arculus; ca, cubitoanal crossvein; nd, nodus; tr, triangle.

Common Species

Cordulia shurtleffi Scudder—43-50 mm; orange-brown markings on face; bronze-green thorax; abdomen shiny black; venation like *E. cynosura* except outer side of front wing triangle nearly straight and not convex (Fig. 77A); ventral caudal appendage forked twice at tip (♂); northern U.S. and southern Canada from Pacific to Atlantic coasts.

Epitheca cynosura (Say)—38-43 mm; brownish black with yellow face and yellow-spotted abdomen; veins M_4 and Cu_1 converge to margin in front wing (Fig. 77A); no spots at wing nodus or stigma; triangle of front wing 2-celled (Fig. 77A) and outerside convex; 1 cubitoanal crossvein in hind wing; hind wing base has brown extending to include 1st antenodal crossvein; eastern U.S. south to FL and TX, southeastern Canada.

Epitheca princeps (Hagen)—58-76 mm; head and thorax yellowish brown or olive and abdomen black on dorsum; triangle of hind wing 2-celled; large brown wing spots (at base, stigma, and usually nodus) covering 1/4 or less of wing area; eastern U.S. southwest to TX, southeastern Canada.

Somatochlora tenebrosa (Say)—48-64 mm; dark brown with metallic green and bronze reflections; orange labrum; body markings pale and dull colored; veins M_4 and Cu_1 converge to margin in front wing (Fig. 77A); clear hind wing has 2 cubitoanal crossveins (Fig. 77a); tibia outer face black; 2nd abdominal segment with 1 large lateral spot anterior to auricle and caudal appendages distinctly elbowed (♂) (Fig. 77C); eastern U.S., southeastern Canada (occasionally twilight-flying).

Family Libellulidae— Common Skimmers

This family is one of the largest in the order and contains the most common dragonflies. Adults are brightly colored, 20-75 mm long, and frequently perch on vegetation. Most species occur around ponds, lakes, and marshes; a few frequent quieter sections of streams. The triangle in the front wing is farther from the arculus than the triangle in the hind wing (Fig. 70A). The hind wing has a foot-shaped anal loop (Fig. 70A). Most dragonflies with colored spots or bands on their wings will be species in this family, one of the exceptions being the genus *Epitheca* in the family Corduliidae. Reddish orange species around ponds in the autumn often belong to the genus *Sympetrum*.

Nymphs are somewhat stocky and sprawl on mud or crawl on trash in quiet, shallow water. The labium is mask- or spoon-shaped (Fig. 78D) and covers the face up to the eyes. The labial palpi have shallow indentations that are one-tenth to one-sixth as long as they are wide. The metasternum has no tubercle near its posterior margin. The abdomen of some species contains middorsal projections which may be spines or stubs but are not sickle-shaped.

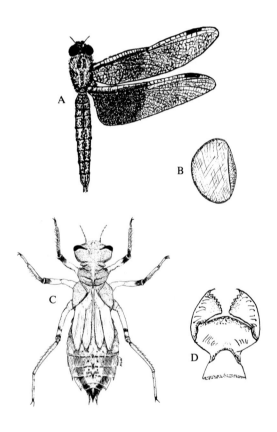

Figure 78 Common skimmers (Libellulidae). A, *Libellula luctuosa;* B, straight hind margin of adult eye; C, *Libellula* sp. nymph; D, labium of a nymph.

Common Species

Celithemis elisa (Hagen)—29-34 mm; yellowish or reddish brown; M_2 veins evenly curved but not wavy; 2-4 cells in triangle of front wing (Fig. 70A); anal loop midrib nearly straight or bent at "ankle" (Fig. 70A); wings with large brown spots near nodus and a roundish spot just beyond nodus; wing tips with dark edges beyond stigma; east of Rocky Mts. and south to FL, southeastern Canada.

Leucorrhinia intacta (Hagen)—29-33 mm; brownish black; upper face white; labrum yellow; labium black; M_2 veins curved but not wavy; midrib of anal loop angular; 2 crossveins under stigma of wing; stigma twice as long as wide (♂); 2 pale yellow spots on dorsum of 7th abdominal segment; northern U.S. and southern Canada from Pacific to Atlantic coasts (bogs, farm ponds).

Libellula luctuosa Burmeister (Fig. 78A). The widow—42-50 mm; dark brown with yellow and black stripes on abdomen; basal 3rd of both wings completely covered by blackish band and bordered distally by white (♂); M_2 veins wavy (Fig. 70A); 3 or more cells in front wing triangle (Fig. 70A); arculus very close to 2nd antenodal crossvein (Fig. 70A); Cu_1 vein in hind wing arises from posterior angle of triangle and not its side (Fig. 70A); throughout U.S. except Northwest, southeastern Canada.

Libellula pulchella Drury. Ten spot—52-57 mm; brown with yellow and black markings; venation like *L. luctuosa;* wings spotted with brownish black including tip beyond stigma; 10 white spots between brown on wings (♂); throughout U.S. and southern Canada.

Pachydiplax longipennis (Burmeister)—28-45 mm; white face with dorsum of head metallic blue; thorax brown above and yellow-green with lateral brown stripes on sides; abdomen black with yellow parallel lines on dorsum; wings clear or smoky in part, either no crossvein below stigma or 1 crossvein under distal end leaving a long cell below stigma; throughout U.S. and southern Canada.

Plathemis lydia (Drury)—42-48 mm; yellow face and brown thorax with dark areas; yellow and white markings on body; wing venation like *Libellula* spp. (Fig. 70A) except arculus is midway between 1st and 2nd antenodal crossvein; middle crossband of wings uniformly dark brown (♂) or wing tips brown (♀); rows of lateral and oblique yellow spots on abdominal segments 3-8 (♀); dorsum of abdomen white (♂); throughout U.S. and southern Canada (often fly with *L. pulchella*).

Sympetrum rubicundulum (Say)—33-34 mm; yellow face; thorax reddish brown in front and the remainder olive color; red (♂) or yellowish (♀) abdomen with lateral band of black triangles; M_2 veins and anal loop like *L. intacta;* 1 crossvein almost centered under stigma; tibiae outer face black; dorsal caudal appendages with prominent tooth (♂) and ventral (subgenital) plate deeply forked (♀); hamuli forked for 1/3 of length (♂); northern U.S. and southern Canada from Pacific to Atlantic coasts.

SUBORDER ZYGOPTERA

Family Calopterygidae—Broadwinged Damselflies

Many of these relatively large damselflies have striking metallic bodies, dark wings, or bright spots of color on their wings. The wings of adults gradually narrow at their bases (Fig. 79A) but they are not stalked as in other damselfly families. There are 10 or more antenodal crossveins (Fig. 79A). Adults fly or rest along streams.

The slender nymphs prefer streams or woodland pools and cling with their long legs to stems and roots of submerged vegetation. The first antennal segment is equal to or longer than the remaining six and the mentum has a deep, open, median cleft (Fig. 79C).

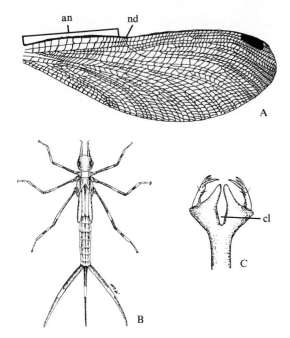

crossveins, and M_3 arises nearer the nodus than the arculus (Fig. 70B). When an adult lands, it typically holds its body horizontally and its wings together over the body (Fig. 80A). Adults frequent slow streams, ponds and marshes. Males are usually more colorful than females and often have a different color pattern. Males of the subspecies *Argia fumipennis violacea* (Hagen) are a striking violet color and common in northern U.S. and southeastern Canada.

Nymphs (Fig. 80B) live in a great variety of aquatic habitats. The first antennal segment is much shorter than the other segments combined. The mentum is flat, gradually narrowed toward its base (Fig. 80C), and when the labium is folded the base of the mentum reaches to or slightly beyond the prothoracic coxae. The gills are somewhat oval and usually pointed at the tips.

Figure 79 Broadwinged damselflies (Calopterygidae). A, front wing; B, *Calopteryx aequabilis* nymph; C, labium of a nymph. an, antenodal crossveins; cl, cleft; nd, nodus.

Common Species

Calopteryx aequabilis (Say) (Fig. 79B)—44-52 mm; body color like *C. maculata;* outer 1/2 of hind wing and 1/3 of front wing black; northern U.S. (primarily northeastern), southern Canada.

Calopteryx maculata (Beauvois)—40-42 mm; body metallic green with blue-black cast; wings black without a stigma (♂) or dark gray with white stigma (♀); eastern U.S., northern U.S. from MN to CA, southern Canada.

Hetaerina americana (Fabricius). American or common ruby spot—42-44 mm; body reddish to greenbronze or dark brown; red (♂) or red-amber (♀) spot in basal 1/3 or 1/4 of wings; throughout U.S. (except Northwest), southeastern Canada.

Family Coenagrionidae— Narrowwinged Damselflies

Most damselfly species belong to this family and range from 25-45 mm in length. The wings are stalked and contain only two antenodal

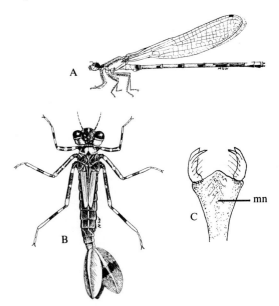

Figure 80 Narrowwinged damselflies (Coenagrionidae). A, adult; B, *Argia* sp. nymph; C, labium of a nymph. mn, mentum.

Common Species

Argia sedula (Hagen) (Fig. 80A)—29-33 mm; body tan (♀) or dorsum of abdominal segments 4-6 mostly black with a wide black middorsal stripe on mesothorax (♂); tibia with spines much longer than intervals between spines; ventral caudal appendages forked (♂); eastern U.S. south to FL and west to southern CA.

Enallagma civile (Hagen). Blue damselfly—31-38 mm; pale blue (♂) or greenish blue (♀) with black markings covering at least the distal 1/4 of abdominal segments 2-7 (♂) or black almost continuous on dorsum of abdomen (♀); occiput pale; stigma similar in front and hind wings (♂); tibiae with spines ca. as long as interval between spines; spine on 8th abdominal segment (♀); dorsal caudal appendages forked with the upper branch larger than lower and fork filled with a pale nodule (♂); throughout U.S. and southern Canada.

Ischnura verticalis (Say)—23-33 mm; dark colored with green stripes on thorax and tip of abdomen blue (♂); young ♀ brownish orange with black markings, older ♀ bluish green or bluish gray with faint dark markings and outer face of hind tibia black; tibia spination and occiput like *E. civile* in both sexes; eastern U.S. and southeastern Canada.

Family Protoneuridae

Two species of damselflies in this family occur in the U.S. (southern Texas). These insects are 32-37 mm long, reddish brown, and occur along streams. *Neoneura aaroni* Calvert is common and *Protoneura cava* Calvert is occasionally collected.

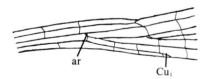

Figure 81 Protoneuridae wing section. ar, arculus.

Family Lestidae—Spreadwinged Damselflies

The stalked wings of adults in this family contain two antenodal crossveins and M_3 arises nearer the arculus than the nodus (Fig. 82A). In contrast to species of Coenagrionidae, their bodies are usually held vertically at rest and the wings are partly spread. Adults frequent the margins of ponds and marshes, occasionally wandering from these habitats. Lestidae contains the largest damselfly in the U.S., *Archilestes grandis* (Rambur), a yellow to greenish brown fall species ranging up to 62 mm in length.

Nymphs live in ponds or streams with abundant emergent vegetation and boggy margins. The first antennal segment is much shorter than all the others combined. The labium is spoon-shaped, the basal half of the mentum is greatly narrowed (Fig. 82C), and when the labium is folded the base of the mentum reaches to or beyond the mesothoracic coxae (Fig. 82D). The gills are nearly parallel-sided and usually rounded at the tips.

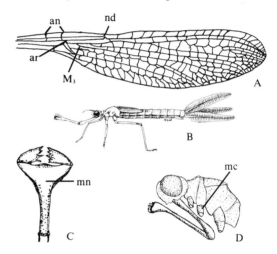

Figure 82 Spreadwinged damselflies (Lestidae). A, front wing section; B, *Lestes dryas* nymph; C, labium of a nymph; D, labium-head-thorax relationship of a nymph. an, antenodal crossveins; ar, arculus; mc, mesothoracic coxa; mn, mentum; nd, nodus.

Common Species

Lestes dryas Kirby (Fig. 82B)—38-40 mm; metallic green; sides of body yellow; M_2 vein originates several cells beyond nodus (Fig. 82A); ventral caudal appendages reach beyond midpoint of dorsal appendages and their tips are expanded (♂); abdomen 26-28 mm long (♂); northern U.S. and southern Canada from Pacific to Atlantic coasts (temporary ponds, marshes).

Lestes unguiculatus Hagen—36-38 mm; head and thorax brown; occiput yellow (♀); abdomen greenish with pale yellow sides; M_2 vein and length of ventral caudal appendages like *L. dryas;* ventral caudal appendages not expanded distally and their tips are "S"-shaped and directed outwards; northern U.S. and southern Canada from Pacific to Atlantic coasts (ponds, marshes).

GENERAL REFERENCES

Corbet, P.S. 1963. *A Biology of Dragonflies.* Quadrangle Books, Chicago. 247 pp.

Gloyd, L.K., and Wright, M. 1959. Odonata. In *Fresh-Water Biology,* W.T. Edmondson (ed.). Wiley, N.Y. Pp. 917-940.

Needham, J.G., and Westfall, M.J., Jr. 1955. *A Manual of the Dragonflies of North America (Anisoptera).* Univ. California Press, Berkeley. 615 pp.

Smith, R.F., and Pritchard, A.E. 1956. Odonata. In *Aquatic Insects of California,* R.L. Usinger (ed.). Univ. California Press, Berkeley. Pp. 106-153.

Walker, E.M. 1953. *The Odonata of Canada and Alaska.* Vol. 1. *General, the Zygoptera—Damselflies.* Univ. Toronto Press, Toronto. 292 pp.

———. 1958. *The Odonata of Canada and Alaska.* Vol. 2. *The Anisoptera—Four Families.* Univ. Toronto Press, Toronto. 318 pp.

Walker, E.M., and Corbet, P.S. 1975. *The Odonata of Canada and Alaska.* Vol. 3. *The Anisoptera—Three Families.* Univ. Toronto Press, Toronto. 307 pp.

ORDER PLECOPTERA[4]
Stoneflies

Stoneflies are small- to medium-sized (4-60 mm long), flattened insects that live in and around lakes and streams. Most nymphs occur under stones in well aerated streams. The Plecoptera inhabit primarily the cooler, temperate parts of the world.

Adults of most species are drab colored and have four membranous wings that at rest are held flat over the abdomen (Fig. 83A). The front wings are narrower than the hind wings (Fig. 84A) and the expanded anal lobes of the hind wings are often folded fanlike at rest. The wings of males of several species are reduced or absent. The long antennae are slender and many-segmented. The cerci are often long and prominent. Stoneflies have chewing mouthparts and the nymphs undergo hemimetabolous development. Adults of many species emerge and reproduce during the fall and coldest months of the winter. These "winter" stoneflies generally fly during the day, feed on blue-green algae and plant foliage, and probably live for several weeks. Adults of most species emerge during the late spring and summer; many are nocturnal and/or do not feed. The families Capniidae, Leuctridae, and Taeniopterygidae have been classified as subfamilies of Nemouridae in the past by some taxonomists.

Females deposit several egg masses, which together may total over 1,000 eggs, by flying over water or occasionally by crawling up to the water. Nymphs have flattened and elongated bodies with two long cerci and most also have tufts of branched respiratory gills on the sides of the thorax and around the bases of the legs (Fig. 83B). Mayfly nymphs (Ephemeroptera) are similar in general appearance but all except a few genera have three caudal filaments and the gills are leaflike and located on the sides of the abdomen. Stonefly nymphs feed on algae, diatoms, mosses, and immature aquatic invertebrates including mayflies and midges. Some nymphs are known to molt 12-36 times and require one to three years to mature. Full-grown nymphs leave the water, cling to shoreline vegetation and debris, and molt into the adult stage. Stonefly nymphs form an important portion of the diet of stream fishes.

---

4. Plecoptera: *pleco,* twisted or braided; *ptera,* winged (refers to the folded posterior region of the resting hind wing).

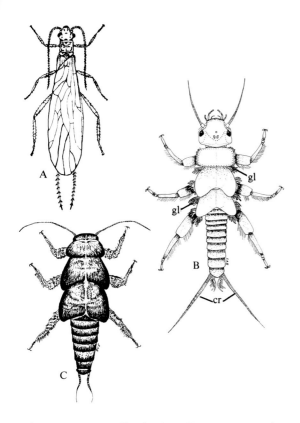

Figure 83 Stoneflies. A, *Allocapnia pygmaea* (Capniidae); B, nymph; C, *Yoraperla brevis* nymph. cr, cerci; gl, gills.

Adults may be collected from stones, logs, and vegetation along water courses, ponds, and lakes. Large numbers of stoneflies often occur in hilly and mountainous areas which have swift streams with rocky bottoms. In late spring and summer, sweeping vegetation (especially willows) along streams may be the most productive collecting method. Some species hide under the loose bark of logs extending into the water and others are attracted to lights at night. From late fall to early spring, winter stoneflies are often attracted to concrete bridges over streams; bridges generally are good sites to collect stoneflies throughout the year. Some species rest on fence posts and snow on the warmer days of late winter. All adults are preserved in 70% alcohol.

Nymphs occur primarily under stones in cool, unpolluted streams. Some species frequent rocky shores of cold lakes, cracks of submerged logs, and debris that accumulates around stones, branches, and water diversion grills. Individuals may be handpicked from their habitat or recovered from a screen downstream after their habitat has been purposely disturbed. Nymphs are preserved in 70% alcohol.

Species: North America, 470; world, 1,750. Families: North America, 9.

KEY TO FAMILIES OF PLECOPTERA

The tip of the abdomen of males has a complex of dorsal structures including a slender, single or forked projection that may curve upward or be recurved (Fig. 86A). Females lack these dorsal structures.

1a	Anal areas of front wings with 2 or more rows of crossveins (Fig. 87A); body usually over 25 mm long . . . . (p. 86) *Pteronarcyidae*	
1b	Anal areas of front wings lack crossveins (Fig. 84A) or with 1 row; size variable 2	
2a(1b)	Cerci equal to or shorter than greatest width of pronotum 3	
2b	Cerci longer than greatest width of pronotum. 6	
3a(2a)	2 ocelli on dorsum of head between compound eyes; cockroachlike form (Fig. 83C) . . . . . (p. 86) *Peltoperlidae*	

Order Plecoptera 83

| 3b | 3 ocelli on dorsum of head between compound eyes; typical body form .... 4 |

| 4a(3b) | Second tarsal segment about as long as each of the other tarsal segments; cerci 1- to 6-segmented; primarily winter and spring .... (p. 85) *Taeniopterygidae* |

| 4b | Second tarsal segment much shorter than other segments; cerci short and 1-segmented; primarily spring and early summer .... 5 |

| 5a(4b) | Front wings flat at rest, outer third with 4 veins forming large "X" (Fig. 84B). ...... (p. 85) *Nemouridae* |

| 5b | Front wings bent down around sides of abdomen, outer third without veins forming an "X" .... (p. 86) *Leuctridae* |

| 6a(2b) | First tarsal segment about as long as third; primarily winter and spring... .... (p. 85) *Capniidae* |

| 6b | First tarsal segment much shorter than third; primarily spring and summer .... 7 |

| 7a(6b) | Remnants of branched or filamentous gills on sides or ventral surface of thorax...... (p. 88) *Perlidae* |

| 7b | Gill remnants usually absent, if present they are unbranched and fingerlike .... 8 |

| 8a(7b) | Pronotum nearly oval with front corners broadly rounded (Fig. 85A); anal lobe of hind wing small and usually with 3 or fewer veins reaching wing margin behind vein 1A (Fig. 84C) ... (p. 87) *Chloroperlidae* |

| 8b | Pronotum rectangular with corners narrowly rounded (Fig. 85B); anal lobe of hind wing well developed and with 5 or more veins reaching the wing margin behind vein 1A....... .... (p. 87) *Perlodidae* |

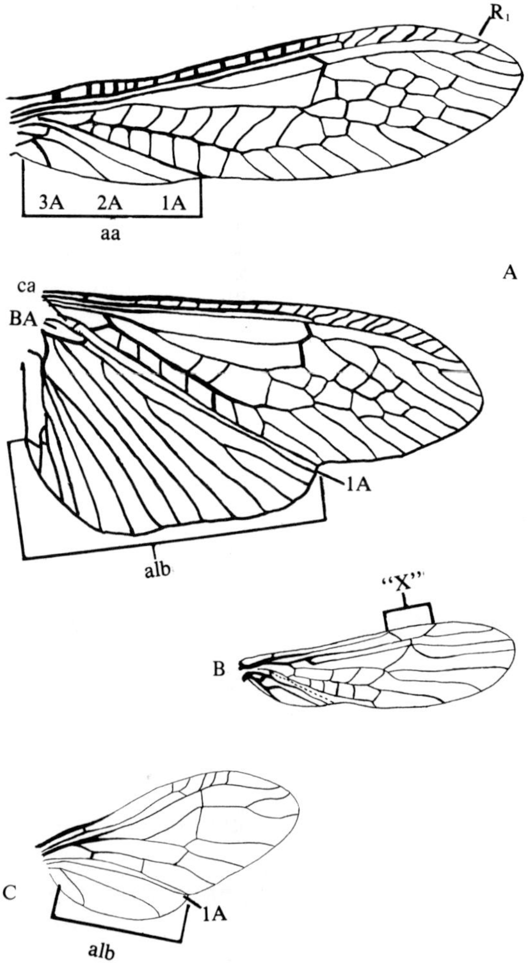

Figure 84 Stonefly wings. A, Perlidae; B, front wing of Nemouridae; C, hind wing of *Alloperla* spp. (Chloroperlidae). aa, anal area; alb, anal lobe; BA, basal anal cell; ca, cubitoanal crossvein.

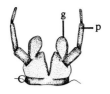

Figure 85 A, head and pronotum of Chloroperlidae; B, head and pronotum of Perlodidae; C, paraglossae and glossae of labium of nymphs. g, glossae; p, paraglossae.

Family Nemouridae

Adults are brown or black, 15 mm or less in length, and many emerge from April to June. This family is similar to the Leuctridae except that the front wings are flat when at rest rather than bent down around the abdomen. The hind wing pads of nymphs diverge strongly from the axis of the body. Nymphs frequent small streams with sandy bottoms.

Common Species

Malenka californica (Claassen)—8-9 mm to wing tips; brown; veins 1A and 2A of front wing do not unite at wing margins; ventral cervical gills with 5 + branches that join before reaching gill base; 9th sternite with lobe (♂); spinelike process at base of cercus directed inward and much shorter than cercus (♂); Rocky Mts. and westward (Apr.-Dec., primarily late summer-fall; small creeks).

Zapada cinctipes (Banks)—9-13 mm to wing tips; blackish brown with banded legs and wings; 2 pairs of ventral cervical gills; veins 1A and 2A of front wings do not unite at wing margins (♂); 9th sternite with lobe (♂); western North America north of AZ (Feb.-Oct.).

Family Taeniopterygidae

Adults are dark brown to black, mostly under 15 mm long, and have very long antennae. The second tarsal segment of nymphs is as long as the first.

Common Species

Strophopteryx fasciata (Burmeister)—10-15 mm to wing tips; dark brown; 3 ocelli; cerci yellowish with 5-6 segments; subgenital plate of 9th abdominal sternite extends beyond tip of abdomen and turns up at right angle (♂); eastern U.S., southeastern Canada (Mar.-April).

Taenionema pacificum (Banks). Pacific salmonfly—12-15 mm to wing tips; brown to blackish; yellowish or reddish markings on head, thorax, and tip of abdomen; 3 ocelli; pigmented band near middle of wings; 5- to 6-segmented cerci; western U.S. south to AZ.

Taeniopteryx maura (Pictet)—8-12 mm to wing tips; brown to blackish; often a light middorsal line on abdomen; 3 ocelli; hind femora with a strong tooth; cerci with 1 (♂) or 6-7 (♀) segments; eastern U.S., southern Canada (Jan.-March).

Taeniopteryx nivalis Fitch—11-17 mm to wing tips; dark brown; 3 ocelli; hind femora without a strong tooth; cerci with 1 (♂) or 8-10 (♀) segments; IL eastward (Feb.-April).

Family Capniidae

These blackish stoneflies are generally less than 12 mm long, emerge from November to June, and may be very abundant crawling on snow or resting on concrete bridges over streams. The second tarsal segment is much shorter than the other segments and the long cerci have at least four segments. The wings of some males are short or rudimentary. Nymphs have three ocelli, hind wing pads that are nearly parallel to the body, and the dorsal and ventral halves of abdominal segments 1-9 are divided laterally by a membranous fold. Species in the genus *Capnia* occur in the western U.S. and western Canada; the genus is characterized by the anal areas of the wings reaching not more than half the length of the wings.

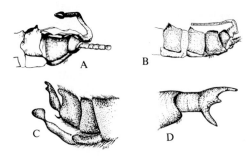

Figure 86 Tip of abdomen of male stoneflies. Capniidae: A, *Allocapnia pygmaea;* B, *Capnia gracilaria.* Leuctridae: C, *Paraleuctra occidentalis;* D, *P. sara* cercus.

Common Species

Allocapnia pygmaea (Burmeister) (Fig. 83A)—3-8 mm to wing tips; dark brown to black; anal areas of wings as long as entire wing; 7th tergite without a tubercle; wide middorsal pale stripe from base of abdomen to posterior margin of 8th segment; ventral surface of abdominal segments 7-8 dark and fused near middle without a pale transverse stripe separating segments; dorsal process at tip of abdomen recurved and split into an upper and lower branch (♂) (Fig. 86A); eastern U.S. (Dec.-April).

Capnia gracilaria Claassen—5 mm to wing tips; blackish; R_1 of front wing bent upward at origin; gently curved S-shaped and slender projection on 10th tergite is round but not oval in cross section and ca. 10 times longer than wide (♂) (Fig. 86B); northwestern U.S. east to MT and UT, southwestern Canada (Jan.-May).

Isocapnia grandis (Banks)—14-16 mm to wing tips; black; R_1 of front wings not bent upward at origin; cerci > 11-segmented; slender projection of 10th tergite is S-shaped in profile (♂); northwestern U.S. (March-June).

Family Leuctridae

These stoneflies are brown or black and generally 10 mm or less in length. The front wings at rest are bent down over the sides of the abdomen. Nymphs resemble those of Capniidae except only abdominal segments one to seven, or fewer, are divided into upper and lower halves by a membranous lateral fold. Most of the species are western in distribution and adults appear in spring and summer.

Common Species

Paraleuctra occidentalis (Banks)—6-9 mm to wing tips; brown to black; < 6 veins in anal area of hind wing; ventral lobe on 9th abdominal sternite (♂); tip of abdomen with a ventral, upcurved, and stalked process that lacks a membranous apical bulb (♂) (Fig. 86C); northwestern U.S., southwestern Canada (Feb.-Aug.).

Paraleuctra sara (Claassen)—6-9 mm to wing tips; dark brown to black; wings blackish; < 6 veins in anal area of hind wing; ventral lobe on 9th abdominal sternite (♂); cercus forked with upper prong longer than lower prong and a small tooth on inner margin (♂) (Fig. 86D); eastern U.S. (March-Aug.).

Family Pteronarcyidae— Giant Stoneflies

This family contains the largest stoneflies—usually over 25 mm long. Nymphs have branched gills on the ventral side of the first two abdominal segments.

Common Species

Pteronarcys californica Newport (Fig. 87A). California salmonfly—32-46 mm to wing tips; dark grayish brown with faint reddish shades; paler underneath; abdominal gills only on 1st 2 segments; 9th abdominal sternite truncated (♂); western U.S., southwestern Canada (April-Aug.; nocturnal).

Pteronarcys dorsata (Say). Giant stonefly—40-60 mm to wing tips; dark brown; paler underneath; abdominal gills only on 1st 2 segments; 9th abdominal sternite extended backward beyond tip of abdomen (♂); eastern U.S. northwest to MT and southwest to KS, central and southeastern Canada (May-July; nocturnal).

Family Peltoperlidae— Roachlike Stoneflies

This small family contains five genera in North America and most species are western and northern in distribution. Adults emerge from late spring to summer. The nymphs are somewhat cockroachlike in form (Fig. 83C), have two ocelli, and the ventral surface of the thorax is covered by large, overlapping, shieldlike plates. *Yoraperla brevis* (Banks) (Fig. 83C) (9-13 mm to wing tips) is common in northwestern U.S. and southwestern Canada. *Peltoperla arcuata* Needham (14-18 mm to wing tips) and *Viehoperla zipha* Frison (12 mm to wing tips) occur in the eastern U.S.

Family Perlodidae

Adults of one of the major genera, *Isoperla,* are diurnal and commonly have yellowish or greenish bodies as well as greenish wings. Species in other genera are brown or black in color. Adult males of most perlodid genera are identified by the presence of a lobe on the seventh sternite but those of the genus *Arcynopteryx* lack a lobe. Perlodid nymphs have the paraglossae extended beyond the glossae (Fig. 85C), thoracic gills are single or double but not profusely branched, hind wing pads protrude at an angle from the body's axis, and the body is pigmented in a distinct pattern.

Common Species

Isoperla bilineata (Say) (Fig. 87B)—10-14 mm to wing tips; yellowish with brown spot in ocellar triangle of head; brown pronotum with wide, yellow, median stripe; wings greenish and transparent with brown veins; lobe on posterior margin of 9th sternite (♂); northern 1/2 of U.S. west to CO, southern 1/2 of Canada west to Saskatchewan (April-June).

Isoperla patricia Frison—10 mm to wing tips; pale yellowish brown with darker head and pronotal patterns; posterior 1/2 of mesothorax and metathorax dark brown; anterior margins of wings pale yellowish; lobe on posterior margin of 8th sternite (♂); patches of short, stout hairs on posterior margin of 9th tergite; often reddish cast to some abdominal segments; northwestern U.S. east to SD and CO, southwestern Canada (April-Aug.).

Family Chloroperlidae— Green Stoneflies

Adults are 6-15 mm long, greenish or yellowish in color, and emerge in the spring. The paraglossae of nymphs extend beyond the glossae (Fig. 85C), and the thoracic gills are single or double but not profusely branched. The hind wing pads of nymphs are nearly parallel to the body and the brown body lacks a distinct pattern. Adults in the common genus *Alloperla* have an anal lobe in the hind wing that is small (Fig. 84C) or absent.

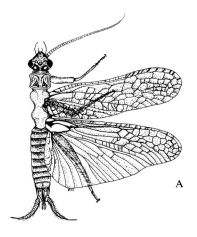

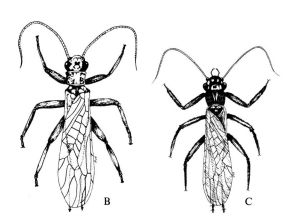

Figure 87 A, California salmonfly, *Pteronarcys californica* (Pteronarcyidae); B, *Isoperla bilineata* (Perlodidae); C, *Neoperla clymene* (Perlidae).

Common Species

Alloperla atlantica Baumann—10-12 mm to wing tips; light pastel green; without fingerlike process at base of cerci (♂); abdominal segments covered with numerous small hairs; eastern U.S. and eastern Canada (May-Aug.).

Hastaperla brevis (Banks)—6-9 mm to wing tips; yellowish green to bright green; hind wing without folded anal area; northern 1/2 of U.S., southeast to GA, southern Canada (May-July).

Suwallia pallidula (Banks)—7-10 mm to wing tips; lemon to whitish yellow; anal area of hind wing reduced or absent; basal segment of cerci not elongated but with fingerlike process at base (♂); northwestern U.S., WY, UT, southwestern Canada (May-Oct.).

Family Perlidae— Common Stoneflies

This family is the largest and its members are the most commonly collected. Adults are generally 20-40 mm long, fly in the spring and summer, and do not feed. The paraglossae of nymphs extend beyond the glossae (Fig. 85C) and the thoracic gills are profusely branched. Adult males of *Acroneuria,* a genus with many common species, are recognized by the absence of hooks on the lateral angles of the tenth abdominal tergite and the presence of a disclike structure on the posterior half of the ninth sternum.

Common Species

Hesperoperla pacifica (Banks)—32-37 mm to wing tips; dark brown; head yellowish; wings tinted brown; no hooks on lateral angles of 10th tergite (♂); anal gills present; northwestern U.S. east to MT and CO, southwestern Canada (April-Aug.).

Neoperla clymene (Newman) (Fig. 87C)—10-18 mm to wing tips; brownish yellow; 2 ocelli on black spot; ocelli closer to each other than to compound eyes; head wider than long; eastern North America west to NM (June-Aug.).

Perlesta placida (Hagen)—9-14 mm to wing tips; brown to blackish; head and anterior wing margin yellow; thorax with yellow middorsal stripe; 3 ocelli; eastern North America west to UT and WY (May-Sept.).

Phasganophora capitata (Pictet)—16-24 mm to wing tips; brown; abdomen and anterior edges of wings yellowish; cubitoanal crossvein (ca) occurs before apex of basal anal cell (BA) (Fig. 84A); eastern U.S. west to the Dakotas, southeastern Canada (May-July).

GENERAL REFERENCES

Claassen, P.W. 1931. Plecoptera Nymphs of America (North of Mexico). Thomas Say Foundation Publ. 3. 199 pp.

Frison, T.H. 1935. The Stoneflies, or Plecoptera, of Illinois. Ill. Nat. His. Surv. Bull. 20:281-471.

Gaufin, A.R.; Nebeker, A.V.; and Sessions, J. 1966. The Stoneflies (Plecoptera) of Utah. Univ. Utah Biol. Ser. 14:9-89.

Gaufin, A.R.; Ricker, W.E.; Miner, M.; Milam, P.; and Hays, R.A. 1972. The Stoneflies (Plecoptera) of Montana. Trans. Amer. Ent. Soc. 98:1-161.

Hitchcock, S.W. 1974. The Plecoptera or Stoneflies of Connecticut. Guide to the Insects of Conn. VII. Bulletin 107. 262 pp.

Hynes, H.B.N. 1976. Biology of Plecoptera. Ann. Rev. Ent. 21:135-153.

Jewett, S.G., Jr. 1956. Plecoptera. In *Aquatic Insects of California*, R.L. Usinger (ed.). Univ. Calif. Press, Berkeley. Pp. 155-181.

———. 1959. The Stoneflies (Plecoptera) of the Pacific Northwest. Ore. State Monog. No. 3. 95 pp.

Needham, J.G., and Claassen, P.W. 1925. A Monograph of the Plecoptera or Stoneflies of America North of Mexico. Thomas Say Foundation Publ. 2. 397 pp.

Ricker, W.E. 1959. Plecoptera. In *Fresh-Water Biology*, W.T. Edmondson (ed.). Wiley, N.Y. Pp. 941-957.

Ross, H.H., and Ricker, W.E. 1971. The Classification, Evolution, and Dispersal of the Winter Stonefly Genus *Allocapnia*. Ill. Biol. Monog. 45. 166 pp.

Surdick, R.F., and Kim, K.C. 1976. Stoneflies (Plecoptera) of Pennsylvania. A Synopsis. Penn. St. Univ. Coll. of Agric. Bull. 808. 73 pp.

ORDER ORTHOPTERA[5]
Grasshoppers, Katydids and Crickets

Most Orthoptera are medium to large insects with enlarged hind legs adapted for jumping (Fig. 88A). The front wings (tegmina) are typically thickened and usually narrow, and the hind wings are membranous and broad (Fig. 88A). Although some species such as the locusts can fly hundreds of miles, flying ability is often poorly developed and many species are short-winged or wingless. Immature and adult orthopterans have chewing mouthparts and feed primarily on vegetation although there is a tendency toward cannibalism under stressful conditions. The female's ovipositor is well developed in most families and it may be as long as her body (Fig. 95D). Adults produce sounds by rubbing various body parts together. The songs are produced primarily by males to attract females and are also associated with courtship displays. The singing Orthoptera usually possess auditory organs (tympana) on their bodies or front legs. Nymphs undergo a hemimetabolous type of development.

Opinions of taxonomists differ as to the classification of orthopterans and their allies. Some authorities include cockroaches, mantids and walkingsticks as families in the order Orthoptera; others give these groups ordinal status as is done in this text. Certain taxonomists upgrade the subfamilies of katydids and crickets to family levels and have redesignated or changed the names of orthopteroid groups ranging from subfamilies to orders. Some of the alternative subfamily names for grasshoppers are given in parentheses next to the names used in this text.

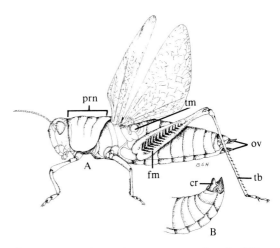

Figure 88 A, General structure of a female differential grasshopper, *Melanoplus differentialis;* B, tip of male's abdomen; cr, cercus; fm, femur; ov, ovipositor; prn, pronotum; tb, tibia; tm, tympanic organ.

SUBORDER CAELIFERA
Grasshoppers, Pygmy Mole Crickets, Pygmy Grasshoppers or Grouse Locusts

The antennae of these orthopterans are short (less than 30 segments), the segments are filiform or flattened, the hind femora are enlarged, and the tarsi contain three or less segments (Fig. 88A). A tympanic (auditory) organ, if present, is found at the base of the abdomen (Fig. 88A) on each side and consists of a vibrating, thin, cuticular membrane. The ovipositor is short (Fig. 88A). Body colors of species living in dry fields or in arid climates tend to simulate the background color of the habitat. Many tropical species found in lush vegetation are very brightly colored. A few species are able to change color in less than an hour and others over a period of days.

5. Orthoptera: *ortho,* straight; *ptera,* wings.

Most grasshoppers stridulate by moving various types of serrations (pegs, bristles) on the inner surface of the femur across raised veins of the front wing. In the Romaleinae subfamily part of the hind wing is rubbed against the underside of the tegmen to produce sound. The bandwinged grasshoppers make a crackling or buzzing sound in flight by rapidly opening and closing their hind wings like a fan (crepitation).

Most grasshoppers oviposit in soil containing sparse to heavy vegetation. The female manipulates the appendages that form her ovipositor and extends her abdomen to dig a hole. She releases several dozen eggs while simultaneously discharging a frothy, mucilaginous substance that surrounds the eggs. The material dries and forms a hard, cylindrical egg pod. The eggs of most species overwinter and hatch in spring and summer but numerous species overwinter as nymphs or adults. Nymphs of most species molt 5-7 times and take 40-60 days to mature. The adult life span varies from 1-3 months. Grasshoppers generally eat a wide variety of plant foods although some are restricted to grasses, seeds, or a few plant species. Under stressful conditions such as starvation, lack of water, etc., grasshoppers will eat dead or dying insects including their own kind. A few species release irritating fluids or disagreeable odors as a defense mechanism, and when handled most species regurgitate a brown fluid from the digestive system.

Certain species of grasshoppers that have rapid population increases and migrate enmasse are called locusts. The change in behavior and morphology of locusts from a solitary (isolated) phase to a gregarious (crowded) phase is termed phase polymorphism. It is due to hormonal changes induced primarily from crowding but modified by temperature, humidity and food supply. *Locusta migratoria* (Linnaeus), the migratory locust, and *Schistocerca gregaria* (Forskål), the desert locust, are two destructive species found in Africa and western Asia that have been recorded since biblical times. In the U.S., *Melanoplus spretus* (Walsh), the Rocky Mountain grasshopper, migrated in hordes throughout the Great Plains during the late 1800's and caused great damage to crops. Although the species became extinct in the early 1900's, large numbers of preserved specimens have been found in certain glaciers of Montana.

Grasshoppers may be collected by stalking and netting them or beating and sweeping them from vegetation. After pinning, the left wings should be spread at right angles to the body to expose their venation and colors. Rapid drying helps preserve colors and with large grasshoppers and katydids it may be desirable to slit the first three or four abdominal segments on the ventral side and remove the internal contents with forceps. The body cavity is then swabbed out with a small wad of cotton and another piece of cotton inserted loosely to replace the contents. Dehydration of the stuffed specimen in a freezer for a few weeks will help preserve colors. Nymphs and soft bodied adults are preserved in 70% alcohol.

SUBORDER ENSIFERA
Katydids (Includes Longhorned Grasshoppers), Cave and Camel Crickets, Crickets, Mole Crickets

The antennae of these insects are long (more than 30 segments), the segments are slender, the hind femora are enlarged for jumping, and the tarsi are 3- or 4-segmented (Fig. 98). The tympana, if present, are small membranes found at the base of the front tibiae (Fig. 95B). The ovipositor is relatively long and either cylindrical (Fig. 98A) or sword-shaped (Fig. 95D). Many species are color camouflaged to blend in with the green vegetation and some simulate the shape of leaves. Species stridulate by raising both front wings and rubbing the edge of one front wing over a filelike ridge on the ventral side of the other.

Female ensiferans use their ovipositors to insert eggs into soil (many crickets, cave and camel crickets) or plant material (most katydids and tree crickets). The eggs may be laid in rows or clumped but are not encased in a pod as with grasshoppers. Nymphs molt 5-9 times. Food varies with the species and includes leaves, flowers, seeds, roots, and other insects.

Specimens on vegetation may be collected by general sweeping or stalking individuals. Burrowing species such as mole crickets may be forced to the surface by inserting a stem (they grasp it) or pouring water down the burrow. Nocturnal singing species on vegetation are obtained by stalking with a light and ground-dwelling species such as crickets may be collected overnight by baiting a pitfall trap with molasses. Rapid drying helps preserve colors (see previous section on grasshopper preservation). Nymphs and soft-bodied adults are preserved in 70% alcohol.

Species: North America, 1,200; world, > 15,000. Families: North America, 10.

KEY TO FAMILIES OF ORTHOPTERA

Females are recognized by the presence of an ovipositor (Figs. 88A; 95D; 96F; 98A). Males typically have an enlarged 9th abdominal sternum that contains the genitalia and it may be rounded on the end (grasshoppers) (Fig. 88B) or bear a pair of styles (katydids and crickets) (Figs. 95B; 98B).

1a	Front legs greatly enlarged and modified for digging (Figs. 90; 99); tarsi 1- to 3-segmented 2	
1b	Front legs not greatly enlarged and modified for digging, if slightly enlarged the tarsi are 4-segmented; tarsi 2- to 4-segmented 3	
2a(1a)	Tarsi 3-segmented; hind femora not greatly enlarged (Fig. 99); body pubescent . . . . . (p. 100) *Gryllotalpidae*	
2b	Front and middle tarsi 2-segmented, hind tarsi 1-segmented or absent; hind femora greatly enlarged for jumping (Fig. 90); body not pubescent . . . . (p. 92) *Tridactylidae*	
3a(1b)	Antennae usually short (Fig. 88A), usually half or less length of body and with less than 30 segments; tympana (auditory organs), if present, on sides of first abdominal segment (Fig. 88A); ovipositor short and stout (Fig. 88A). 4	
3b	Antennae as long as or longer than body (Fig. 95) and with more than 30 segments; tympana, if present, form narrow openings on front tibiae (Fig. 95B); ovipositor long, and straight or curved, and flat or cylindrical (Figs. 95D; 98A). 7	
4a(3a)	Pronotum extended and tapered backward over abdomen (Fig. 91); front and middle tarsi 2-segmented. . . . . (p. 93) *Tetrigidae*	
4b	Pronotum not extended backward over abdomen; all tarsi 3-segmented 5	
5a(4b)	Antennae shorter than front femora; wingless; brushy deserts of southwestern U.S. . . (p. 93) *Eumastacidae*	
5b	Antennae longer than front femora; commonly with wings; most are widely distributed. 6	

6a(5b) Wings and tympana absent; antennae long, longer than body in male; cylindrical stridulatory ridge on side of third abdominal tergite of males; deserts of southwestern U.S. .... (p. 93) *Tanaoceridae*

6b Wings and tympana usually present; antennae not long; males without stridulatory ridge; widely distributed .... (p. 94) *Acrididae*

7a(3b) Tarsi 3-segmented; ovipositor usually slender and cylindrical (Fig. 98A)... .... (p. 99) *Gryllidae*

7b Middle (and usually all) tarsi 4-segmented; ovipositor flat and sword-shaped (Fig. 95D) 8

8a(7b) Usually wingless, if wings present they have 8 or more major longitudinal veins and males lack stridulatory structures on ventral surface at base of front wings; front tibiae without auditory organs; usually gray or brown .... (p. 98) *Gryllacrididae*

8b Usually with wings of variable size, less than 8 major longitudinal veins and males with stridulatory structures on front wings (Fig. 89); front tibiae with auditory organs; color generally green 9

9a(8b) Hind tibiae long, slender, and dorsal side with many spines in each of 2 rows; hind femora usually extend beyond tip of abdomen; widely distributed ...... (p. 96) *Tettigoniidae*

9b Hind tibiae short, stout, and dorsal side with 8 or fewer spines in each of 2 rows; hind femora extend to tip of abdomen; northwestern U.S. and southwestern Canada............ .... (p. 98) *Prophalangopsidae*

SUBORDER CAELIFERA

Family Tridactylidae—Pygmy Mole Crickets or Pygmy Sand Crickets

Adults are small (less than 10 mm long) insects that leap vigorously when approached. Their front legs are enlarged for digging, the front and middle tarsi are 2-segmented, and the hind femora are greatly enlarged for jumping (Fig. 90). Tridactylids inhabit minute burrows in damp, sparsely vegetated sandy margins of ponds, ditches, streams, and sand bars. These insects are able to swim using plates on their hind tibiae. The Tridactylidae are not crickets but because they are similar to the Gryllotalpidae, the mole crickets, the name continues to be used.

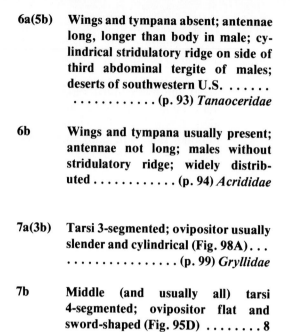

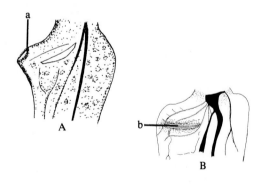

Figure 89 Stridulatory or song-producing structures on the front wings of katydids and crickets. A, scraper at base of right wing; B, ventral view of file at base of left wing. a, scraper; b, file.

Figure 90 Minute pygmy mole cricket, *Ellipes minuta* (Tridactylidae).

Common Species

Ellipes minuta Scudder (Fig. 90). Minute pygmy mole cricket—4-5 mm; black or blackish brown with pale pronotal margins and other light markings on body; no hind tarsi; eastern U.S. (rare in New England) southwest to CA.

Neotridactylus apicialis (Say). Larger pygmy mole cricket—6-10 mm; black or dark brown; body usually unspotted; tegmina yellowish; hind tarsi 1-segmented; eastern U.S. southwest to CA.

Family Tetrigidae—Pygmy Grasshoppers or Grouse Locusts

Members of the family are 6-13 mm in length and possess a distinctive, elongated, tapered pronotum that extends over most of the abdomen (Fig. 91). The tegmina are reduced to small flaps (rarely absent) and may be exposed or hidden under the pronotum. The hind wings are usually well developed and functional. Tetrigids occupy a wide variety of habitats but are more common in damp areas and stream borders. When disturbed tetrigids occasionally may be seen swimming or temporarily clinging to an underwater stem. Adults overwinter and are more commonly seen in the spring and early summer. Tetrigids are distributed throughout most of North America, being more abundant in the southeastern U.S.

Common Species

Tettigidea lateralis (Say). Sedge or blacksided pygmy grasshopper—8-13 mm; black, gray, dark or light brown; occasionally dorsum of pronotum cream or brown; 20-22 antennal segments; anterior margin of pronotum bowed and not sharply pointed; head not pointed from a dorsal view; eastern U.S. to NB and AZ, southeastern Canada.

Tetrix arenosa Burmeister (Fig. 91). Sanded or obscure pygmy grasshopper—6-9 mm; tan, brown, gray or blackish gray; 12-14 antennal segments; lateral view: pronotal outline slightly to moderately concave, head outline strongly slanted, and lateral margin broadly notched at level of ocelli; eastern U.S. to NB and TX, southeastern Canada.

Tetrix subulata (Linnaeus). Slender or awlshaped pygmy grasshopper—7-11 mm; gray to black; occasional middorsal stripe; 12-14 antennal segments; lateral view: head outline anterior to compound eyes is truncated and not concave; middle femora ca. 4 times as long as broad; slender species; western, northcentral and northeastern U.S., all but northeastern Canada.

Family Eumastacidae—Monkey Grasshoppers

The small (8-22 mm) adults of this family are agile insects with pointed heads and antennae shorter than the front femora. The North American species are wingless. These uncommon grasshoppers occur on scrub in hot deserts, chaparral of hillsides, and timbered mountain areas of California, Nevada, Utah, and Arizona.

Family Tanaoceridae—Tanaocerids

Adults are wingless, gray to blackish insects that are 8-26 mm long. The antennae are long and those of males exceed the body length. The two species occurring in the U.S. are nocturnal and only *Tanaocerus koebelei* Bruner is likely to be collected. Adults occur in the spring in the southern deserts of California, Nevada, and Utah.

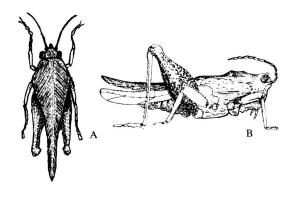

Figure 91 Sanded or obscure pygmy grasshopper, *Tetrix arenosa* (Tetrigidae).

Family Acrididae—Grasshoppers

This very common family encompasses the medium to large grasshoppers with short antennae, short, stout ovipositors and auditory organs on the sides of the first abdominal segment (Fig. 88A). Nymphs with developing wings may be separated from short-winged adults by noting that the developing hind wings of nymphs overlap the front wings. Grasshoppers are common and cosmopolitan in distribution, known throughout North America from subtropical and desert to arctic zones and found up to about 4,300 m in elevation. The highest populations of acridids occur in the semiarid grasslands of the Great Plains and the Southwest.

Romalea microptera (Palisot de Beauvois). Eastern lubber grasshopper—43-90 mm; dull yellow or tawny with numerous blackish brown markings to nearly all black; reddish stripes on tegmina; hind wings reddish; southeastern U.S. (commonly sold for classroom use).

Subfamily Gomphocerinae (= Acridinae of other authors)—Slantfaced Grasshoppers; slanting face, no prosternal spine.

Aulocara elliotti (Thomas). Bigheaded grasshopper—20-27 mm; grayish brown; pale middorsal stripe; hind tibiae blue with pale ring at bases; tegmina nearly reaching tip of abdomen; west of Mississippi River, southwestern Canada.

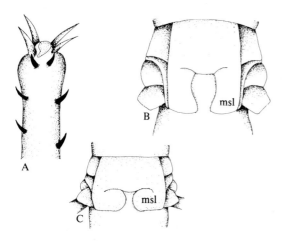

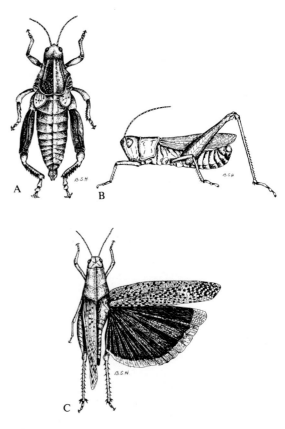

Figure 92 A, spines at tip of Romaleinae tibia; B, mesosternal lobes of Cyrtacanthacridinae; C, mesosternal lobes of Melanoplinae. msl, mesosternal lobe.

Common Species

Subfamily Romaleinae—Lubber Grasshoppers; inner and outer small spines at tip of tibiae (Fig. 92A), spine or low tubercle on prosternum between front legs.

Brachystola magna (Girard) (Fig. 93A). Lubber grasshopper—38-79 mm; brown or grayish green; thick bodied; oval tegmina very short and pink with black dots; ND to southwestern U.S.

Figure 93 Grasshoppers (Acrididae). A, lubber grasshopper, *Brachystola magna;* B, meadow grasshopper, *Chorthippus curtipennis;* C, Carolina grasshopper, *Dissosteria carolina.*

Chorthippus curtipennis (Harris) (Fig. 93B). Meadow grasshopper—13-24 mm; light brown with green markings; black bar on each side of pronotum; pronotum with lateral ridges entire length; tegmina lengths vary from short to slightly exceeding tip of abdomen; tibiae with outer rows of 12+ spines; yellowish sternum; throughout most of U.S., southern Canada (meadows, wet areas).

Orphulina speciosa (Scudder). Pasture grasshopper—13-21 mm; brown or green body and tegmina; lateral ridges of pronotum slightly incurved but the shortest distance between them is > 2/3 greatest distance; tegmina usually do not exceed tips of hind femora; hind tibiae with 12-15 spines on outer margin; North America east of Rocky Mts.

Subfamily Oedipodinae—Bandwinged Grasshoppers; hind wings usually colored, median longitudinal ridge on pronotum.

Arphia pseudonietana (Thomas). Redwinged grasshopper—22-33 mm; black or brown but not yellowish body; body speckled; pronotum with high median ridge; hind wings orange-red or rarely yellow; crackles in flight; west of Mississippi River, northcentral U.S., southern Canada.

Camnula pellucida (Scudder). Clearwinged grasshopper—17-25 mm; light brown to blackish; lateral ridges of pronotum distinct and uninterrupted along most of length; tegmina with dark, rounded spots and yellowish blotches; straw-colored stripe occurs along overlapping margins of folded tegmina; most of U.S. except southeast, southern Canada.

Chortophaga viridifasciata (De Geer). Greenstriped grasshopper—17-32 mm; usually brown (♂) or green with brownish streak on tegmen (♀); abdomen reddish brown; hind wing base faintly yellowish; crackles in flight; primarily east of Rocky Mts., south to FL, southern Canada (early spring, often damp meadows).

Dissosteira carolina (Linnaeus) (Fig. 93C). Carolina grasshopper—24-40 mm; gray, grayish or reddish brown; small dark spots on pronotum and tegmina; pronotum with high median ridge; central area of wing black and margin pale yellow; crackles in flight; throughout U.S., southern Canada (common on roads, dusty paths).

Subfamily Cyrtacanthacridinae (=Catantopinae, Melanoplinae of other authors)—Bird Grasshoppers; mesosternal lobes angular (Fig. 92B), tubercle or spine on prosternum between front legs.

Schistocerca americana (Drury). American or American bird grasshopper—39-55 mm; reddish brown; yellow middorsal stripe; 3 light bands on side of pronotum; dark blotches on tegmina; hind tibiae red; southern half of U.S., occasional fall flights to southern Canada.

Schistocerca emarginata (Scudder). Prairie or spotted bird grasshopper—23-50 mm; reddish brown to olive green; yellow middorsal stripe dominant in western forms but faint in eastern forms; yellow spots on posterior area of pronotum; tegmina brownish to greenish; hind tibiae brown, pink or blackish; southwestern forms sometimes with green body and red hind tibiae; Rocky Mts. eastward except LA to SC, southcentral Canada.

Figure 94 Subgenital plates of males. A, *Melanoplus femurrubrum;* B, *M. sanguinipes.*

Subfamily Melanoplinae (=Cyrtacanthacridinae of other authors)—mesosternal lobes rounded (Fig. 92C), tubercle or spine on prosternum between front legs.

Melanoplus bivittatus (Say). Twostriped grasshopper—23-40 mm; olive or yellowish brown to dark brown; head, pronotum and tegmina with prominent light stripe on each side; hind femora blackish on upper half; throughout U.S., southern half of Canada.

Melanoplus differentialis (Thomas) (Fig. 88). Differential grasshopper—28-44 mm; dark brownish green; pronotum and tegmina not striped; hind femora yellow with prominent black chevron markings arranged in herringbone fashion; hind tibiae yellow; throughout U.S., southern Canada.

Melanoplus femurrubrum (De Geer). Redlegged grasshopper—17-27 mm; dark to light brown, reddish brown, or occasionally yellowish green; horizontal black bar on each side of pronotum; hind femora and tibiae usually reddish; subgenital plate broadly notched along top margin (♂) (Fig. 94A); throughout U.S., southern Canada.

Melanoplus sanguinipes (Fabricius). Migratory grasshopper—17-27 mm; body, pronotum, and hind femora and tibiae colors similar to *M. femurrubrum;* tegmina relatively narrow; several transverse bands on hind femora; minute notch in center of upper margin of subgenital plate (♂) (Fig. 94B); throughout most of North America.

SUBORDER ENSIFERA

Family Tettigoniidae— Katydids (Includes Longhorned Grasshoppers)

The medium to large size insects in this family have very long, slender antennae often extending far beyond the tip of the abdomen, 4-segmented tarsi, and the females possess sword-shaped ovipositors (Figs. 95D; 96F). Most species are green or greenish brown and are difficult to locate in vegetation. Some winged forms are common in dense foliage of trees and shrubs (true, bush, and roundheaded katydids) and sing primarily at night. Others, sometimes collectively called longhorned grasshoppers, sing at night while on low vegetation (coneheaded katydids) or both day and night in wet, grassy meadows and margins of streams and ponds (meadow katydids). The wingless, brown-to-black shieldbacked katydids occur in woods or fields and may sing both day and night. The subfamilies that follow are the more common ones and are given family status by some taxonomists.

Common Species
Subfamily Phaneropterinae—Bush and Roundheaded Katydids, or False Katydids; prosternal spines absent.

Amblycorypha oblongifolia (De Geer). Oblong-winged katydid—21-25 mm; bright green (rarely pink); abdomen and usually front and middle femora yellowish or brownish green; tegmina oblong-elliptical and > 30 mm long; hind femora 3/4 length of tegmina; tegmina with brown basal patch much larger than dorsal surface of pronotum; east of Rocky Mts. (song is "zzzzzz-zik-zik" repeated every few seconds and heard from tall bushes at night).

Microcentrum rhombifolium (Saussure) (Fig. 95A). Broadwinged katydid—25-30 mm; color like *S. furcata;* margins of pronotum with blunt spine in middle; hind wings longer than tegmina; tegmina broad and wider at middle than at apex; hind femora less than 2/3 length of tegmina; ovipositor short and bent abruptly upward (♀); throughout U.S., less common in northern states (songs are 20-40 loud ticks at 3-10 ticks per second and becoming louder and faster as the series progresses, or 2-3 loud lisps within 5 seconds; from trees at night).

Scudderia furcata Brunner (Fig. 95B). Forktailed bush katydid—15-22 mm; dark green; head, pronotum, and sternum greenish yellow; hind wings longer than tegmina; tegmina narrow and only

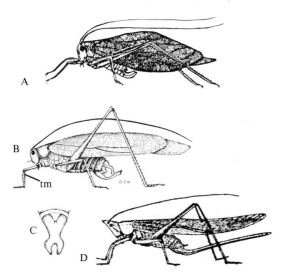

Figure 95 Katydids (Tettigoniidae). A, broadwinged katydid, *Microcentrum rhombifolium;* B, forktailed bush katydid, *Scudderia furcata;* C, forked process at tip of abdomen of male *S. furcata;* D, swordbearing conehead, *Neoconocephalus ensiger,* tm, tympanum.

slightly broader at middle than at apex; middle femora without ventral spines; deeply forked and downcurved dorso-caudal process with lobes of fork rounded above (♂) (Fig. 95C); throughout U.S., southern Canada (song is slow, soft lisps; from bushes and weeds day and night).

Subfamily Pseudophyllinae—True Katydids; prosternal spines, pronotum with 2 transverse grooves.

Pterophylla camellifolia (Fabricius). Northern true katydid—25-34 mm; large, robust; dark to pale green; tegmina as long as or longer than hind wings; tegmina broadly oval and convex; eastern U.S. southwest to TX (song is loud, harsh 2- to 3-part phrase delivered ca. 1 phrase per second; from trees at night).

Subfamily Copiphorinae—Coneheaded Katydids; apex of head conical and extending beyond first antennal segment, also known as coneheaded grasshoppers.

Neoconocephalus ensiger (Harris) (Fig. 95D). Swordbearing conehead—24-30 mm; green or occasionally brown; head produced forward as a distinct cone that is notched beneath; cone much longer than wide and narrows rapidly to apex; lower surface of cone with only sides and tip black; east of Rocky Mts., southeastern Canada (song is rapid lisps, e.g., "sip-sip-sip"; from fields and roadsides at night).

Subfamily Conocephalinae—Meadow Katydids; apex of head terminates in a rounded tubercle not extending beyond first antennal segment, front tibiae without dorsal spines, also known as meadow grasshoppers.

Conocephalus brevipennis (Scudder). Shortwinged meadow katydid—11-14 mm; pale reddish brown or brownish green; pronotum with green sides and dark brown middorsal stripe margined with yellow lines; head and prosternal spines like *C. fasciatus;* tegmina and wings shorter than abdomen; cerci stout with a tooth and flattened inner side (♂) (Fig. 96A); east of Rocky Mts., southeastern Canada (song is very soft buzz less than 5 seconds long with 2-4 ticks following, e.g., "bzzzzzz-zip-zip-zip-zip-bzzzzzz"; near ground in low vegetation of fields day and night).

Conocephalus fasciatus (De Geer). Slender meadow katydid—12-15 mm; face and sides of pronotum and abdomen green; tegmina and rest of body brown; top of head produced forward into rounded tubercle with concave sides; hindwings 2-3 mm longer than tegmina; tegmina longer than abdomen but occasionally shorter in western states; no spines on dorsal surface of front tibiae; prosternum with 2 short spines; cerci slender with stout tooth near middle of inner margin (♂) (Fig. 96B); abdomen with broad and dark middorsal stripe; ovipositor shorter than hind femur and upper margin straight (♀); throughout most of U.S., southern Canada (song is very soft buzz 10-30 seconds long with 10-25 ticks following; near ground in damp meadows day and night).

Orchelimum gladiator (Bruner). Gladiator meadow katydid—19-23 mm; pale green; pronotal band; head tubercle and prosternal spines like *O. vulgare;* ovipositor almost 2/3 length of hind femora (♀); tooth of cercus as long as apical 1/2 of shaft (♂) (Fig. 96C); northern 1/2 of U.S. from Pacific to Atlantic coasts (song is like *O. vulgare* but not louder at end; primarily marshy areas day and night).

Orchelimum nigripes Scudder. Blacklegged meadow katydid—18-21 mm; blue-green or reddish brown; pronotal band; head tubercle and prosternal spine like *O. vulgare;* tibiae and antennae blackish; dorsal surface of cerci with curved ridge (♂) (Fig. 96D); east of Rocky Mts. (song is 2-6 buzzes per 5 seconds prefaced by 2-3 rapid ticks; on tall plants of ponds and marshes).

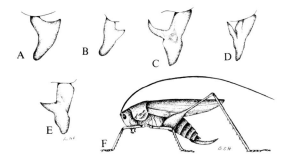

Figure 96 Katydids (Tettigoniidae). A-E, cerci of males: A, shortwinged meadow katydid, *Conocephalus brevipennis;* B, slender meadow katydid, *C. fasciatus;* C, gladiator meadow katydid, *Orchelimum gladiator;* D, blacklegged meadow katydid, *O. nigripes;* E and F, common meadow katydid, *O. vulgare.*

Orchelimum vulgare Harris (Fig. 96F). Common meadow katydid—18-23 mm; green or pale reddish brown; green face; pronotum with reddish brown band often bordered with darker lines; head tubercle like *C. fasciatus;* prosternum with long slender spines; ovipositor less than 1/2 length of hind femora and upper margin distinctly curved (♀); tooth of cercus shorter than apical 1/2 of shaft (♂) (Fig. 96E); east of Rocky Mts., southeastern Canada (song is loud 5 second buzz increasing in volume toward the end and followed by > 4 ticks; weedy fields, occasionally marshes, day and night).

Subfamily Decticinae—Shieldbacked Katydids; pronotum extends back to abdomen; also known as shieldbacked grasshoppers.

Anabrus simplex Haldeman. Mormon cricket—24-40 mm; brown, black, green, or yellowish; pronotum without a sharp, well-developed longitudinal ridge; pronotum shiny and > 1/3 body length; tegmina shorter than pronotum (♂); tip of subgenital plate straight across and without a notch (♀); west of Mississippi River (song is loud, sharp, high-pitched and rapid "zwee-zwee-zwee").

Atlanticus testaceus (Scudder) (Fig. 97A). Shortlegged shieldbearer—18-25 mm; robust; grayish or dark brown; sides of pronotum and tegmina blackish; yellow line on side of pronotum; pronotum shieldlike and extends back to abdomen; tegmina shorter than pronotum but > 1/2 its length (♂); eastern U.S. south to KY (song is group of soft buzzes lasting 1/2 second to several minutes; on ground in open woodlands day and night).

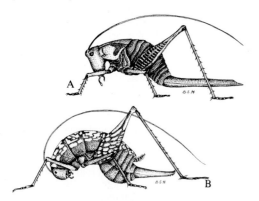

Figure 97 A, shortlegged shieldbearer, *Atlanticus testaceus* (Tettigoniidae); B, spotted camel cricket, *Ceuthophilus maculatus* (Gryllacrididae).

Family Gryllacrididae—Cave and Camel Crickets

Most U.S. members of this family are wingless, brown or gray nocturnal insects. They occur in moist habitats such as under logs and stones, in caves and hollow trees, and in burrows created by other animals. The hind femora are stout, the tibiae are long and spined, and some species have a high, arched back (Fig. 97B). *Ceuthophilus* is the common genus. The camel crickets may be caught by placing molasses in pitfall traps near wooded areas. Some authorities place most of the species in an alternate family, Rhaphidophoridae, and others place the Jerusalem crickets in a separate family, Stenopelmatidae.

Common Species

Ceuthophilus maculatus Harris (Fig. 97B). Spotted camel cricket—14-19 mm; brown with pale median line and yellow spots on dorsal surface; spines of hind tibiae distinctly unequal in length; 9th tergite deeply bowed inward at middle (♂); east of Rocky Mts. south to AR, southeastern Canada (dry woods).

Stenopelmatus fuscus Haldeman. Jerusalem cricket—30-50 mm; brown; shiny and smooth; jaws, head and abdomen large; head and pronotum without dark markings; hind femora heavily spined and do not extend beyond apex of abdomen; abdominal segments banded black; west of Mississippi River, southwestern Canada (in soil, under rocks, slow and clumsy).

Family Prophalangopsidae—Primitive Katydids

Two species of this family occur in North America and both inhabit mountains of northwestern U.S. and southwestern Canada. The more common *Cyphoderris monstrosa* Uhler is brown with light markings, robust, and 21-25 mm long. Males occur on branches and trunks of trees at night and their songs are high-pitched, slightly pulsating trills lasting two seconds or less.

Family Gryllidae—Crickets

Crickets are small to medium size insects with long antennae, 3-segmented tarsi, and cylindrical ovipositors (Fig. 98A). The tegmina are bent down abruptly at the sides of the body. Like the Tettigoniidae, crickets possess auditory organs on the front tibiae and the males stridulate by rubbing the front tegmina together. The snowy tree cricket chirps so that the number of chirps in 15 seconds + 40 gives the approximate air temperature (°F). The Chinese and Japanese have caged crickets and kept them as household songsters. The Chinese have promoted cricket fights for centuries by rearing and selecting pugnacious male field crickets for contests involving thousands of dollars.

Common Species

Subfamily Gryllinae—House and Field Crickets; ocelli arranged in a triangle, adults usually > 14 mm long, numerous *Gryllus* spp. occur in the U.S. but their morphological variation often makes identification difficult.

Acheta domesticus (Linnaeus). House cricket—15-17 mm; pale yellowish brown or straw color; reddish brown bars on head; throughout U.S. and southern Canada (song is 1 chirp per second or slower; in or near buildings day and night).

Gryllus pennsylvanicus Burmeister (Fig. 98A). Fall field cricket—14-30 mm; dark brown or black; hind tibiae at least 3/4 length of hind femora and with 5-9 spines on each side; eastern U.S. south to northern GA, southwest to OK, northwest to Rocky Mts., southern Canada (song is > 1 chirp per second, loud, after late July; lawns, fields, day or night).

Gryllus veletis (Alexander and Bigelow). Spring field cricket—Same description and range as *G. pennsylvanicus*; song is in spring and summer prior to late July.

Subfamily Nemobiinae—Ground Crickets; have ocelli, adults usually < 12 mm long.

Allonemobius allardi (Alexander and Thomas). Allard's ground cricket—7-12 mm; reddish brown to black; dorsal head stripes faint; head and anterior edge of pronotum narrower than posterior edge; Rocky Mts. and eastward, southern Canada (song is slow trill but not a buzz; moist stream banks to dry sand, especially lawns and pastures, day and night).

Allonemobius fasciatus (De Geer) (Fig. 98B). Striped ground cricket—7-11 mm; dark reddish brown to black; top of head with distinct dark and light longitudinal stripes; head and anterior edge of pronotum as broad as posterior edge of pronotum; Rocky Mts. and eastward, southern Canada (song is soft, buzzing chirp, < 1 chirp per second; primarily in moist areas, also fields and lawns, day and night).

Subfamily Oecanthinae—Tree Crickets; no ocelli, long head, pale greenish.

Oecanthus fultoni T.J. Walker (Fig. 98C). Snowy tree cricket—12-15 mm; whitish or pale green; head and 1st basal segment of antennae yellowish; 1 black

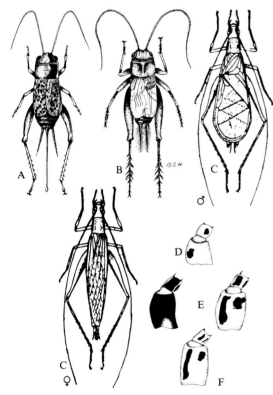

Figure 98 Crickets (Gryllidae). A, fall field cricket, *Gryllus pennsylvanicus;* B, striped ground cricket, *Allonemobius fasciatus;* C, snowy tree cricket, *Oecanthus fultoni.* D-F, markings on first two antennal segments of *Oecanthus* spp.: D, *O. fultoni;* E, *O. nigricornis;* F, *O. quadripunctatus.*

spot on each of 1st 2 antennal segments (Fig. 98D); tegmina equal to or slightly shorter than hind wings; tegmina and hind wings broad, paddlelike and lie flat on back (♂), or narrow with edges wrapped over back (♀); most of U.S., southern Canada (song is "treet-treet-treet" delivered so that number/15 seconds + 40 approximates air temperature (°F); bushes and trees at night).

Oecanthus nigricornis F. Walker. Blackhorned tree cricket—12-14 mm; greenish yellow; head and pronotum usually black or latter with 1-3 longitudinal black stripes; 1st and 2nd basal segments of antennae have 2 dark marks which may run together (Fig. 98E); tegmina and hind wings like *O. fultoni*; distal part of legs and sternum of abdomen blackish; primarily eastern U.S., also Pacific coast, southern Canada (song is continuous chirps without break; heavy vegetation [not trees] of fields and damp areas, day and night).

Oecanthus quadripunctatus Beutenmuller. Four-spotted tree cricket—11-14 mm; pale green or straw color; pronotum not black and without black stripes; outside stripes on 1st and 2nd antennal segments usually paler than inside marks and sometimes rounded or absent (Fig. 98F); tegmina like *O. fultoni*; throughout U.S., southern Canada (song is continuous chirps without break; grasses and weeds of dry fields [not trees], day and night).

Family Gryllotalpidae— Mole Crickets

These brown, pubescent crickets are usually more than 25 mm long and have relatively short antennae. The front legs are very broad and modified for burrowing (Fig. 99). Mole crickets resemble the Tridactylidae in general appearance but have 3- instead of 1-segmented tarsi and their hind legs are not enlarged for jumping as are tridactylids. Mole crickets are nocturnal and many burrow up to 20 cm into moist sand and mud under logs and along the margins of ponds and streams. They feed primarily on plant roots. The genus *Scapteriscus,* characterized by having two "fingers" on the front tibiae, contains several species that cause damage to crops and grasses growing in moist, sandy soils of the southeastern states.

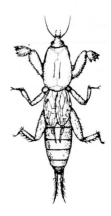

Figure 99 Northern mole cricket, *Neocurtilla hexadactyla* (Gryllotalpidae).

Common Species

Neocurtilla hexadactyla (Perty) (Fig. 99). Northern mole cricket—21-30 mm; brown; pointed head; fine short hairs; front tibiae with 4 "fingers" that curve outward; apices of hind tibiae with 8 spines; eastern and midwestern North America (song is loud, low-pitched chirps delivered 1-3 per second with regularity; marshes at night).

Scapteriscus vicinus Scudder—26-29 mm; pale brownish yellow; pronotum with irregular blotch but not definite spots; front tibiae with 2 "fingers," distance between "fingers" at base < 1/2 width of each "finger"; southeastern U.S. (lawns, golf courses, cultivated fields). *S. acletus* Rehn and Hebard similar but "finger" distance > 1/2 basal width; also along edges of water.

GENERAL REFERENCES

Blatchley, W.S. 1920. *Orthoptera of Northeastern America*. Nature Publ. Co., Indianapolis. 785 pp.

Helfer, J.R. 1963. *How to Know the Grasshoppers, Cockroaches, and Their Allies*. Wm. C. Brown Company Publishers, Dubuque, Iowa. 353 pp.

Hubbell, T.H. 1936. A Monographic Revision of the Genus *Ceuthophilus* (Orthoptera, Gryllacrididae, Rhaphidophorinae). Univ. Fla. Pub. Biol. Sci. Ser. 2:1-551.

Rehn, J.A.G., and Grant, H.J., Jr. 1961. A Monograph of the Orthoptera of North America (North of Mexico), vol. 1. Acridoidea in part, covering the Tetrigidae, Eumastacidae, Tanaoceridae, and Romaleinae of the Acrididae. Acad. Nat. Sci. Phil. Monog. no. 12.

Rentz, D.C., and Burchim, J.D. 1968. Revisionary Studies of Nearctic Decticinae. Mem. Pac. Coast Ent. Soc. 3:1-173.

Uvarov, B.P. 1966,1977. *Grasshoppers and Locusts. A Handbook of General Acridology. Vol. 1. Anatomy, Physiology and Development, Phase Polymorphism, Introduction to Taxonomy.* Cambridge Univ. Press, Cambridge. 481 pp. (1966). Vol. 2. *Behaviour, Ecology Biogeography, Population Dynamics.* Centre for Overseas Pest Research, London. 613 pp. (1977).

There are numerous references on Orthoptera and allied orders of individual states, regions, and Canadian provinces. Examples of these references in abbreviated form follow.

Alexander, G. Univ. Colo. Studies Ser. D, vol. 1:129-164 (1941:Colorado).

Ball, E.D.; Tinkham, E.R.; Flock, R.; and Vorheis, C.T. Ariz. Agr. Expt. Sta. Tech. Bull. 93:257-373 (1942:Arizona).

Brooks, A.R. Can. Ent. Suppl. 9 to vol. 90 (1958:Alberta, Saskatchewan, Manitoba).

Cantrall, I.J. Mich. Ent. 1:299-346 (1968: Michigan).

Coppock, S., Jr. Okla. St. Univ. Expt. Sta. Processed Ser. P-399 (1962:Oklahoma).

Dakin, M.E., Jr., and Hays, K.L. Auburn Univ. Agr. Expt. Sta. Bull. 404 (1970:Alabama).

Froeschner, R.C. Iowa St. Coll. J. Sci. 29:163-354 (1954:Iowa).

Hebard, M. Proc. Acad. Nat. Sci. Phil. 77:33-155 (1925:South Dakota); ibid. 80:211-306 (1928: Montana); ibid. 81:303-425 (1929:Colorado); ibid. 82:377-403 (1931:Alberta); ibid. 83: 119-227 (1931:Kansas); Univ. Minn. Agr. Expt. Sta. Tech. Bull. no. 85 (1932:Minnesota); Ill. Nat. Hist. Surv. Bull. 20:125-279 (1934:Illinois); N.D. Agr. Coll. Expt. Sta. Tech. Bull. 284 (1936:North Dakota); Ent. News 48:219-225, 274-280; 49:33-38, 97-103, 155-159 (1937-1938: Pennsylvania); Okla. Agr. Expt. Sta. Tech. Bull. 5:1-31 (1938: Oklahoma); Trans. Amer. Ent. Soc. 68: 239-311 (1943:Texas).

Hewitt, G.B., and Barr, W.F. Univ. Idaho Agr. Expt. Sta. Res. Bull. 72 (1967:Idaho).

LaRivers, I. Amer. Midl. Nat. 39:652-720 (1948:Nevada).

Pfadt, R.E. Univ. Wy. Agr. Expt. Sta. Mimeo Circular 210 (1965:Wyoming).

Rehn, J.A.G., and Hebard, M. Proc. Acad. Nat. Sci. Phil. 68:87-314 (1916:southeastern U.S.).

Stein, J.L., and McCafferty, W.P. Purdue Univ. Agr. Expt. Sta. Res. Bull. 921 (1975:Indiana).

Strohecker, H.F.; Middlekauff, W.W.; and Rentz, D.C. Bull. Calif. Insect Surv. 10:1-177 (1968:California).

Tinkham, E.R. Amer. Midl. Nat. 40:521-663 (1948:Texas and Midwest).

Vickery, V.R.; Johnstone, D.E.; and McE. Kevan, D.K. Lyman Ent. Mus. Res. Lab. Mem. 1:1-204 (1974:Quebec, Atlantic provinces).

Vickery, V.R., and McE. Kevan, D.K. Proc. Ent. Soc. Ont. 97:13-68 (1967:Ontario).

ORDER PHASMIDA[6]
Walkingsticks

Walkingsticks are usually large (12-178 mm long), slow-moving, sticklike insects that are wingless in the U.S. with the exception of one short-winged species in southern Florida. The antennae are long and slender, the prothorax is short, and the mesothorax and metathorax are usually very elongated (Fig. 100). The legs often are quite long and slender and can be partially regenerated if lost. The Phasmida, predominantly tropical insects, are more common in the southern U.S. The wings and legs of some tropical species assume a broad, leaflike appearance and these species are called leaf insects (Family Phylliidae). Some tropical walkingsticks reach a length of 325 mm; several are parthenogenic, and others are able to change color in a matter of hours or days. Phasmida are classified by some as a suborder of Orthoptera and others prefer to name the order Phasmatoptera, Phasmatodea or Cheleutoptera.

Phasmids have chewing mouthparts and feed on plants. *Diapheromera femorata* (Say), the most common walkingstick in the U.S., periodically defoliates oaks and other hardwood trees in local areas. Eggs of walkingsticks are usually dropped to the ground singly (some adhere them to leaves or dig holes) and the winter is passed in this stage. There is one generation per year although eggs may not hatch until the second spring after they are laid.

6. Phasmida: *phasma*, apparition, phantom.

Nymphs have a hemimetabolous type of development. Phasmids occur in trees, shrubs and low vegetation. They are captured by beating branches and sweeping vegetation, and are preserved by pinning except for very young nymphs which are placed in 70% alcohol.

Species: North America, 27; world, 2,025. Families: North America, 1.

Figure 100 Walkingstick, *Diapheromera femorata* (Phasmatidae).

Family Phasmatidae— Walkingsticks

This family's description is the same as for the order. Females have a slightly enlarged eighth abdominal sternite. In California and Arizona a genus *(Timema)* of small walkingsticks occurs locally that is characterized by a broad body less than 32 mm long and 3-segmented tarsi. The genus is sometimes classified as a separate family, Timemidae.

Common Species

Anisomorpha buprestoides (Stoll). Twostriped walkingstick—39-77 mm; brownish yellow or dark brown; 3 conspicuous black stripes; mesonotum less than 3 times length of pronotum; SC to FL and MS.

Diapheromera femorata (Say) (Fig. 100). Walkingstick or northern walkingstick—65-101 mm; brown, green, or gray; head slightly longer than wide; hind femora usually spined ventrally; middle femora swollen and usually banded (σ); eastern U.S. south to FL, northwest to ND and southwest to AZ, southeastern Canada.

Parabacillus hesperus Hebard. Western shorthorned walkingstick—63-90 mm; light to dark brown; antennae shorter than front femora; last abdominal segment as long as wide (σ); southwestern U.S., OR.

Pseudosermyle straminea (Scudder). Gray walkingstick—40-75 mm; grayish brown or greenish; middle femora without spines; thorax with dorsolongitudinal ridges and fine spines (φ); tip of cerci 3-pronged (σ); southwestern U.S.

GENERAL REFERENCES
Blatchley, W.S. 1920. *Orthoptera of Northeastern America*. Nature Publ. Co., N.Y. 785 pp.
Hebard, M. 1943. The Dermaptera and Orthopterous Families Blattidae, Mantidae and Phasmidae of Texas. Trans. Amer. Ent. Soc. 68:239-311.
Helfer, J.R. 1963. *How to Know the Grasshoppers, Cockroaches, and Their Allies*. Wm. C. Brown Company Publishers, Dubuque, Iowa. 353 pp.
Hodson, A.C. 1972. Distribution and Abundance of the Northern Walkingstick, *Diapheromera femorata*. Ann. Ent. Soc. Amer. 65:876-882.

ORDER DICTYOPTERA[7]
Cockroaches and Mantids

These medium- to large-sized insects have an enlarged pronotum, legs modified for running, and five tarsal segments. The front wings (tegmina) are thickened and the large hind wings are membranous and folded. Most dictyopterans fly poorly and some are short-winged or wingless. They lack auditory organs and usually stridulatory organs. Dictyopterans have chewing mouthparts and the nymphs have a hemimetabolous type of development. Most species occur in the subtropics and some, such as the German and oriental cockroaches, have been accidentally introduced into the U.S. The European and Chinese mantids have also been

7. Dictyoptera: *dicty*, net; *ptera*, wings.

introduced and purposely spread in the U.S. to be used as predators of some insect pests. Dictyoptera is classified as a suborder of Orthoptera by some specialists and others consider cockroaches and mantids as separate orders (Blattodea and Mantodea, respectively).

SUBORDER BLATTARIA
Cockroaches

Cockroaches are flattened insects with long, slender antennae that ceaselessly sweep side to side, and a head that is concealed under a shieldlike pronotum (Fig. 102A). Individuals run rapidly using spiny legs and the integument is smooth and waxy. Cockroaches are usually shades of brown to black although some tropical species are brightly colored. These insects are usually nocturnal and frequent dark, humid habitats during the day. The common cockroaches occur in heated buildings that simulate warm, damp, subtropical conditions. In the northern states a few native species occur outdoors under the bark of dead trees whereas numerous native species occur outdoors in the humid, southern portion of the U.S. Cockroaches are one of the most ancient groups of living insects.

Females produce 16-40 eggs in double rows which are glued together into a leathery capsule (ootheca) (Fig. 102). The ootheca may be deposited in sheltered areas several days after it is formed but two to three months before hatching (American and oriental cockroaches) or it may be carried partially extruded from the female's abdomen until hatching time (German cockroach) (Fig. 102C). In the family Blaberidae eggs hatch within the female's abdomen. Maturation of nymphs is slow, taking from three months (German cockroach) up to one year or more.

Cockroaches are omnivorous although they prefer plant materials. The household species contaminate food, leave an unpleasant odor, and are generally nuisances. They play only a limited role in transmitting human diseases.

Domestic cockroaches frequent kitchens, food storage areas, and damp, warm, dark areas of buildings such as basements and bathrooms. They occur outdoors under stones and bark and on vegetation in the southern states. Some are attracted to lights at night. Pitfall traps baited with fruit, molasses, bran, or carrion may be used for collecting specimens. Laying a long, thin trail of oatmeal at night along paths cleared through vegetation and inspecting the trails hourly has also been a successful collecting method. Adults and older nymphs are pinned whereas young nymphs are preserved in 70% alcohol.

SUBORDER MANTODEA
Mantids or Praying Mantids[8]

Mantids are medium to large, rather slow-moving green or brown predaceous insects. They usually have a greatly elongated prothorax and spined front legs modified to grasp and hold prey (Fig. 103A). The front legs are usually held upraised in a "praying" position. Mantids are one of the few insects able to turn their heads. Some tropical species are mimics of flowers, leaves, and other insects, a condition which lures prey or protects mantids against predators. Although mantids feed on other insects, their population is too low and their feeding too general to control significantly specific insect pests in a garden. Cannibalism may occur and after mating the female of some species may seize and devour the male.

Mantid egg masses (oothecae) are attached to branches or other objects (Fig. 103B) and those of most species have a distinctive

---

8. Mantid: prophet or soothsayer.

shape and size. Each mass consists of rows of eggs which are surrounded by a thick, frothy, quick-drying liquid that provides a protective coating. A female may produce three to six oothecae, each containing 10-400 eggs. Small pinholes seen in an ootheca are the exit holes of wasps and flies parasitic on mantid eggs. There is one mantid generation per year in the northern states and the winter is spent in the egg stage; there may be two generations in the southern states.

Mantids may be collected by sweeping vegetation in fields and beating shrubs. Occasionally specimens fly to lights at night. Adults and larger nymphs are pinned and smaller nymphs are placed in 70% alcohol.

Cockroach species: North America, 60; world, 3,500. Families: North America, 5.
Mantid species: North America, 20; world, 2,000. Families: world, 1.

KEY TO COMMON FAMILIES OF DICTYOPTERA

Both sexes have jointed cerci but male dictyopterans also have a pair of shorter, unjointed styli near the apex of the subgenital plate.

1a	Prothorax much longer than mesothorax; large front legs with spines for grasping prey (Fig. 103A) .... (p. 106) *Mantidae*	
1b	Prothorax wide and not greatly lengthened; front legs not modified for grasping and similar to middle legs (Fig. 102A) 2	
2a(1b)	Middle and hind femora with many spines on ventroposterior margins 3	
2b	Middle and hind femora without spines on ventroposterior margins 5	
3a(2a)	Front femora with a row of stout spines on ventroposterior margins and shorter and more slender spines distally (Fig. 101B); 2-3 apical spines on front femora (genus *Parcoblatta*) .... (p. 106) *Blattellidae* (in part)	
3b	Front femora with a row of spines on ventroposterior margins that decrease gradually in size and length distally, or spines are nearly equal in length (Fig. 101A); apical spine number on front femora variable .. 4	

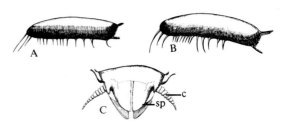

Figure 101 Cockroaches. A, front femur of Blattidae; B, front femur of Blattellidae (genus *Parcoblatta*); C, subgenital plate of Blattidae (female). c, cercus; sp, subgenital plate.

4a(3b)	Subgenital plate divided longitudinally (female) (Fig. 101C); slender styli are straight and similar (male); adults 18 mm or longer..... .... (p. 105) *Blattidae*
4b	Subgenital plate not divided longitudinally (female); styli often unequal in size or asymmetrical (male); adults usually less than 18 mm long (genus *Blattella, Supella*)..................... .... (p. 106) *Blattellidae* (in part)

5a(2b)	Tip of abdomen covered by extended seventh dorsal and sixth ventral sclerites; wingless; body nearly parallel-sided (colonies in rotting logs). . . . . . . (p. 105) *Cryptocercidae*
5b	Tip of abdomen not covered by sclerites, usually winged, body usually oval 6
6a(5b)	Anal area of hind wings folded fanlike at rest; length over 16 mm or color pale green . . . . (p. 106) *Blaberidae*
6b	Anal area of hind wings flat and not folded fanlike at rest; length usually less than 16 mm; never green . . . . (p. 106) *Polyphagidae*

SUBORDER BLATTARIA
Cockroaches

Family Cryptocercidae
The only species in the U.S. is *Cryptocercus punctulatus* Scudder. It is 23-29 mm long, shiny reddish brown, wingless, and the dorsum is finely pitted. This species inhabits decaying logs (especially oak) and occurs in hills and mountains from New York to Georgia and Washington to California.

Family Blattidae
These moderately large cockroaches have numerous spines on the ventral margins of the front femora that gradually decrease in size distally Fig. 101A).

Common Species
Blatta orientalis Linnaeus (Fig. 102B). Oriental cockroach—18-24 mm; dark brown to black; shiny; tegmina leave 2 or more segments of abdomen exposed (♂) or are reduced to small oval pads (♀); throughout U.S., southern Canada (indoors, outdoors in warm climates).

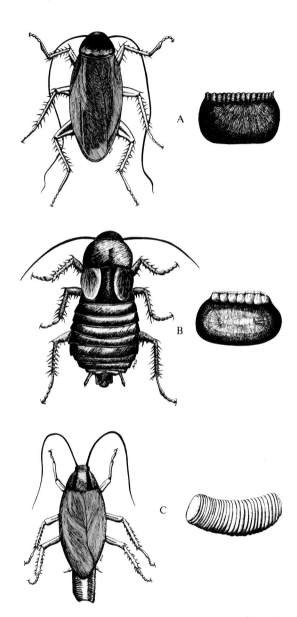

Figure 102 Cockroaches and egg cases (oothecae). A, American cockroach, *Periplaneta americana* (Blattidae); B, oriental cockroach, *Blatta orientalis* (Blattidae); C, German cockroach, *Blattella germanica* (Blattellidae).

Periplaneta americana (Linnaeus) (Fig. 102A). American cockroach—27-37 mm; reddish brown; pronotum yellowish with 2 faint brown blotches; wings well developed; last abdominal segment deeply notched between cerci (Fig. 101C); throughout most of U.S., southern Canada (indoors).

Periplaneta fuliginosa (Serville). Smokybrown cockroach—24-33 mm; dark brown to black; shiny; tegmina longer than abdomen; Gulf coast states (outdoors).

Family Polyphagidae

All but one species of these small (less than 25 mm long) cockroaches occur in the Southwest; the exception is a species found in Florida. Females are wingless in the genus *Arenivaga*; winged males frequently come to lights. Some species burrow in sand dunes and leave molelike ridges.

Family Blattellidae

The numerous species in this family are generally small (12 mm long or less). The wood cockroaches (genus *Parcoblatta*) are the most common outdoor species, occurring under forest litter, loose bark of logs, trash, and signs on trees.

Common Species

Blattella germanica (Linnaeus) (Fig. 102C). German cockroach—10-13 mm; pale brownish yellow; 2 longitudinal stripes on pronotum; throughout North America (indoors).

Parcoblatta pennsylvanica (De Geer). Pennsylvania wood cockroach—16-25 mm (♂); 12-18 mm (♀); brown; sides of pronotum yellowish; tegmina with yellowish transparent margins 1/3-3/4 of their length; 1st 2 abdominal segments concave dorsally at base and concavity overhung by 2 ridges with many hairs on undersurface of ridges (♂); tegmina exceed abdomen (♂) or cover ca. 2/3 of abdomen (♀); eastern U.S. southwest to TX, southcentral Canada (outdoors).

Supella longipalpa (Fabricius). Brownbanded cockroach—10-12 mm; shiny yellowish brown, pronotum and tegmina with large brown spots; southern U.S., locally in more northern states and southern Canada (indoors).

Family Blaberidae

The species in this family are usually uncommon and occur outdoors primarily in the southern states. *Blaberus craniifer* Burmeister, the death's-head cockroach, is the largest cockroach in the U.S. (40-59 mm long). It has a dark brown, facelike marking on a yellowish white pronotum and occurs in southern Florida. The large (38-53 mm long), pale brown Madeira cockroach, *Leucophaea maderae* (Fabricius), is occasionally found indoors in the eastern U.S. *Pycnoscelus surinamensis* (Linnaeus), the Surinam cockroach, is a southern species that sometimes is a pest in greenhouses; the basal fourth of the tegmina has many small pits. *Panchlora nivea* (Linnaeus), the Cuban cockroach, is a pale green species found in southern Texas.

SUBORDER MANTODEA
Mantids or Praying Mantids

Family Mantidae—Mantids

This is the only mantid family and is described earlier under the suborder Mantodea. Species in the genus *Yersiniops* occur in the southwestern deserts and have conical, pointed eyes, short front legs, and run rapidly.

Common Species

Litaneutra minor (Scudder). Minor ground mantid—25-32 mm; reddish or yellowish brown to dark brown; pronotum slightly longer than front coxae; tegmina cover 1/3 of abdomen or less (♀); west of Mississippi River, southwestern Canada.

Mantis religiosa Linnaeus (Fig. 103A). European mantid—47-65 mm; green or brown; large black-ringed spot near base of front coxae on inside surface; most of eastern U.S., southeastern Canada.

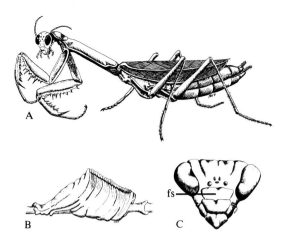

Figure 103 Mantids. A, European mantid, *Mantis religiosa;* B, egg case (ootheca) of the Chinese mantid, *Tenodera aridifolia sinensis;* C, facial shield of the Carolina mantid, *Stagmomantis carolina.* fs, facial shield.

Stagmomantis carolina (Linnaeus). Carolina mantid—48-57 mm; green or grayish brown with green tarsi; width of facial shield more than twice that of height (Fig. 103C); tegmina shorter than abdomen and with black spot at middle near anterior margin (♀); abdomen greatly widened at middle (♀); southern U.S. north to MD, IN, UT, and CA.

Tenodera aridifolia sinensis Saussure (Fig. 103B). Chinese mantid—77-104 mm; green, sometimes brown (♂); width of facial shield less than twice that of height; tegmina brown with green margins; hind wings heavily mottled; northeastern U.S. south to VA, northcentral U.S., CA.

GENERAL REFERENCES

Blatchley, W.S. 1920. *Orthoptera of Northeastern America.* Nature Publ. Co., Indianapolis. 785 pp.

Cornwell, P.B. 1968. *The Cockroach,* vol. 1. Hutchinson, London. 391 pp.

Gurney, A.B. 1951. Praying Mantids of the United States: Native and Introduced. Smiths. Inst. Rept. 1950:339-362.

Guthrie, D.M., and Tindall, A.R. 1968. *The Biology of the Cockroach.* E. Arnold Ltd., London. 408 pp.

Hebard, M. 1943. The Dermaptera and Orthopterous Families Blattidae, Mantidae and Phasmidae of Texas. Trans. Amer. Ent. Soc. 68:239-311.

Helfer, J.R. 1963. *How to Know the Grasshoppers, Cockroaches, and Their Allies.* Wm. C. Brown Company Publishers, Dubuque, Iowa. 353 pp.

McKittrick, F.A. 1964. Evolutionary Studies of Cockroaches. Cornell Univ. Agric. Expt. Sta. Mem. 389. 197 pp.

Roth, L.M., and Willis, E.R. 1960. The Biotic Associations of Cockroaches. Smiths. Misc. Coll. 141. 470 pp.

ORDER GRYLLOBLATTODEA[9]
Rock Crawlers or Icebugs

Adults are yellowish brown to gray wingless insects, 15-30 mm long, with the eyes small or vestigial (Fig. 104). The antennae are moderately long and the legs have 5-segmented tarsi and are adapted for running. The long cerci have five (Siberian species) or nine segments (American species) and the ovipositor is sword-shaped. Some specialists prefer the ordinal name Notoptera and others classify the group as a family of Orthoptera.

Grylloblattids occur in the mountains of northwestern U.S. beneath stones, in and beneath rotting logs and debris, along the edges of or on snow fields, in ice caves, and in other low-temperature habitats. These omnivorous insects have chewing mouthparts, are generally nocturnal, and are relatively uncommon. Nymphs have a hemimetabolous type of development. Specimens should be preserved in 70% alcohol.

Species: North America, 11; world, 17. Families: world, 1.

---

9. Grylloblattodea: *gryll,* cricket; *blatta,* cockroach.

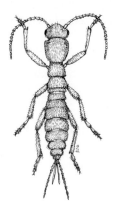

Figure 104 Rocky Mountain rock crawler, *Grylloblatta campodeiformis* (Grylloblattidae).

Family Grylloblattidae

Common Species

Grylloblatta campodeiformis E.M. Walker (Fig. 104). Rocky Mountain rock crawler—16-30 mm; amber yellow; usually <30 antennal segments; apical 1/2 of last abdominal segment (10th tergite) nearly symmetrical from dorsal view (♂); MT and WY, Alberta and British Columbia.

GENERAL REFERENCES

Gurney, A.B. 1948. The Taxonomy and Distribution of the Grylloblattidae. Proc. Ent. Soc. Wash. 50:86-102.

Helfer, J.R. 1963. *How to Know the Grasshoppers, Cockroaches, and Their Allies.* Wm. C. Brown Company Publishers, Dubuque, Iowa. 353 pp.

Imms, A.D.; Richards, O.W.; Davies, R.G. 1977. *Imms' General Textbook of Entomology.* 10th ed. Vol. 2, 910 pp. Halsted Press, N.Y.

Kamp, J.W. 1963. Descriptions of Two New Species of Grylloblattidae and the Adult of *Grylloblatta barberi,* with an Interpretation of Geographic Distribution. Ann. Ent. Soc. Amer. 56:53-68.

ORDER DERMAPTERA[10]
Earwigs[11]

Earwigs are small- to medium-sized (4-26 mm long), elongated insects with a pair of forceps-like appendages, the cerci, at the tip of the abdomen (Fig. 106). The sclerotized front wings (tegmina) are very short, veinless, and meet in a straight line down the back. The rounded hind wings are folded beneath the front wings and usually only the tips are exposed. Tegmina and hind wings are absent on a few species. Earwigs have chewing mouthparts and nymphs undergo a hemimetabolous type of metamorphosis. These insects resemble rove beetles (Staphylinidae) but the latter lack the forcepslike cerci.

Earwigs are nocturnal and hide during the day under bark, stones, cracks, compost, and in other concealed areas. They are more common in the southern and coastal areas of the U.S. Earwigs are omnivorous but a few species become pests by feeding on flower petals and young plants in the garden and greenhouse, and by entering houses. Some highly specialized species in the tropics are ectoparasites of bats and rodents. The cerci of earwigs can be maneuvered to pinch and several species squirt a foul-smelling fluid as a means of defense. Eggs and newly hatched young are guarded by the female, an unusual trait among insects. There is one generation per year and nymphs mature fairly rapidly.

Earwigs may be collected during the day from flowers, under bark, stones, and leaves, and other protected places. Pitfall traps baited

10. Dermaptera: *derma,* skin; *ptera,* wings.

11. The name earwig may be derived either from an old English tale that these insects crawl into the ears of sleeping persons (rarely recorded) or from the earlike shape of the hind wings of some species.

with ripe fruit or rose petals may trap them at night, and some are attracted to lights. Adults and older nymphs are pinned and young nymphs are placed in 70% alcohol.

Species: North America, 18; world, > 1,100. Families: North America, 6.

KEY TO COMMON FAMILIES OF DERMAPTERA

Males have 9 and females 7 distinct abdominal tergites. The cerci of males are greatly bowed (Figs. 106A, D) and those of females are nearly straight.

1a	Second tarsal segment cylindrical and not extended beneath third segment (Fig. 105A)................. 2
1b	Second tarsal segment lobed and extends beneath third segment (Fig. 105B).................. 4
2a(1a)	Antennae 10- to 16-segmented, segments 4-6 together are longer than the first segment... (p. 109) *Labiidae*
2b	Antennae 14- to 24-segmented, segments 4-6 together are usually shorter than first segment........ 3
3a(2b)	Wings present; cerci of males symmetrical ... (p. 109) *Labiduridae*
3b	Wingless; right cerci of males curved more than left (Fig. 106C) .... (p. 110) *Carcinophoridae*
4a(1b)	Lobe of second tarsal segment narrow and with a ventral brush of long hairs (Fig. 105C); 12-segmented antennae; California .... (p. 110) *Chelisochidae*
4b	Lobe of second tarsal segment broad and with very short ventral hairs; 12- to 16-segmented antennae; widely distributed .... (p. 110) *Forficulidae*

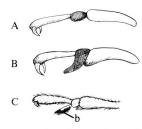

Figure 105 Tarsi of earwigs. A, cylindrical; B, lobed; C, brush of hairs present. b, brush.

Family Labiidae

Species in this family are small (4-7 mm long) and antennal segments 4-6 together are longer than the first segment.

Common Species

Labia minor (Linnaeus) (Fig. 106A). Little earwig—4-5 mm; head nearly black; last abdominal segment and forceps reddish brown; body clothed with yellowish pubescence; sternum of last abdominal segment has backward projecting median tooth (♂); throughout U.S., southern Canada (flies to lights).

Marava pulchella (Serville). Handsome earwig—6-7 mm; shiny reddish or dark brown; 4th antennal segment shorter than 3rd; hind wings often lacking; southern U.S.

Family Labiduridae

Antennal segments 4-6 together are shorter than the first segment and all species are winged.

Common Species

Labidura riparia (Pallas) (Fig. 106B). Shore earwig—18-26 mm; light brown; southern CA to NC (under debris along seashore and rivers, flies to lights).

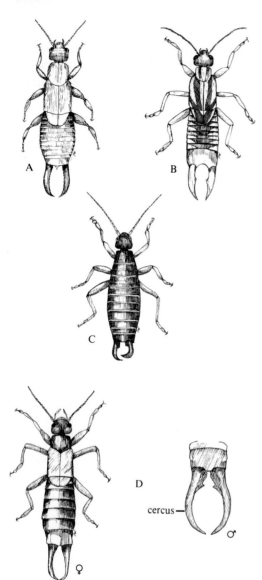

Family Carcinophoridae

The best known species in this family are wingless and the cerci of males are asymmetrical. This family has been classified as part of Labiduridae or named Psalididae by some authors.

Common Species

Anisolabis maritima Géné (Fig. 106C). Seaside earwig—16-25 mm; shiny brownish black; legs and ventral surface lighter; 24-segmented antennae; Atlantic, Gulf and Pacific coasts (under seaweed, boards, and other seashore debris).

Euborellia annulipes (Lucas). Ringlegged earwig—9-11 mm; shiny brownish black; 15-16 segmented antennae; legs and ventral surface yellowish; dark rings at bases of tibiae and middle of femora; South Atlantic, Gulf, and Pacific coasts.

Family Forficulidae

In this family the second tarsal segment is lobed and extends beneath the third segment.

Common Species

Doru aculeatum (Scudder). Spinetailed earwig—7-11 mm; dark brown; tegmina and edges of pronotum yellow; 12-segmented antennae; short tooth near tip of each cercus; eastern U.S. south to FL, northwest to NB, Ontario.

Forficula auricularia Linnaeus (Fig. 106D). European earwig—10-17 mm; blackish brown with tegmina lighter; 14-15 antennal segments; throughout U.S., southwest and southeast Canada.

Family Chelisochidae

The extended and narrow second tarsal segment has a brush of long hairs ventrally. The only species in the U.S., *Chelisoches morio* (Fabricius), the black earwig, is black, 16-20 mm long, and occurs in California.

Figure 106 Earwigs. A, little earwig, *Labia minor* (Labiidae); B, shore earwig, *Labidura riparia* (Labiduridae); C, seaside earwig, *Anisolabis maritima* (Carcinophoridae); D, European earwig, *Forficula auricularia* (Forficulidae).

GENERAL REFERENCES
Blatchley, W.S. 1920. *Orthoptera of Northeastern America*. Nature Publ. Co., Indianapolis. 784 pp.

Helfer, J.R. 1963. *How to Know the Grasshoppers, Cockroaches, and Their Allies.* Wm. C. Brown Company Publishers, Dubuque, Iowa. 353 pp.

Imms, A.D.; Richards, D.W.; Davies, R.G. 1977. *Imms' General Textbook of Entomology.* 10th ed. Vol. 2, 910 pp. Halsted Press, N.Y.

Langston, R.L., and Powell, J.A. 1975. The Earwigs of California (Order Dermaptera). Bull. Calif. Ins. Surv. 20. 25 pp.

ORDER EMBIOPTERA[12]
Webspinners

The Embioptera are small (4-7 mm long), slender, usually brownish insects that may have wings (males) or be wingless (some males and all females). The four nearly equal wings have alternating brown and pale longitudinal bands (Fig. 107A). The antennae are slender and 16- to 32-segmented. The tarsi are 3-segmented and the unique basal segments of the front tarsi are greatly enlarged for producing silk from hollow hairs issuing on the basal and midsegments (Fig. 107B). The hind femora are also thickened. These insects are rapid runners, often running backwards. The bodies of males are flattened but those of females and immatures are more cylindrical. The cerci are 2-segmented although the left cercus of some males is 1-segmented. Webspinners have chewing mouthparts and the nymphs have a hemimetabolous type of development.

The Embioptera, which are primarily tropical insects, occur in the southern part of the U.S. but are not often noticed. They live in colonies and exhibit limited maternal care for eggs and young. Silk galleries are spun under stones and bark, in debris, cracks in the soil and bark, and among grass roots, lichens, mosses, and epiphytic plants. Webspinners feed on dead plant material as well as lichens and mosses found around their galleries.

Webspinners may be collected in the habitats mentioned above, and are more numerous during the rainy season. Winged males of some species are attracted to lights at night. All stages are preserved in 70% alcohol.

Species: North America, 11; world, 200. Families: North America, 3.

KEY TO FAMILIES OF EMBIOPTERA

1a   All instars with 2 ventral papillae on basal segment of hind tarsus (Fig. 107C); Oregon, California, Arizona, Texas (*Haploembia solierii* [Rambur]) ... (p. 112) *Oligotomidae*

1b   All instars with 1 ventral papilla on basal segment of hind tarsus 2

2a(1b)   Males without apical teeth on mandibles; left cercus of males 1-segmented; Arkansas to So. California .... *Anisembiidae*

2b   Males with 2-3 apical teeth on mandibles; left cercus of males 2-segmented 3

3a(2b)   Males with wing vein R_{4+5} forked; coastal southeastern U.S. to Texas .. .. *Teratembiidae*

3b   Males with wing vein R_{4+5} unforked (Fig. 107A); southeastern and southwestern U.S. (genus *Oligotoma*) .... (p. 112) *Oligotomidae*

12. Embioptera: *embio*, lively; *ptera*, wings.

Family Oligotomidae

There are three introduced species in the continental U.S. The two described below are the most commonly collected, partly because winged males are attracted to lights.

Common Species

Oligotoma nigra (Hagen)—8-10 mm; dark brown with legs and pigmented patches of wings lighter brown; left side of 10th abdominal tergite extended posteriorly as a slender process 5 times as long as broad (♂); tip of 9th sternite hooklike and on a vertical axis (♂); southwestern U.S.

Oligotoma saundersii (Westwood) (Fig. 107A)—9-13 mm; light brown with darker head (♂); chocolate brown (♀); left side of 10th abdominal tergite extended posteriorly as a spatulate process twice as long as broad (♂) (Fig. 107D); tip of 9th sternite sickle-shaped and extended horizontally beneath left side of genitalia (♂) (Fig. 107D); southeastern U.S.

GENERAL REFERENCES
Ross, E.S. 1940. A Revision of the Embioptera of North America. Ann. Ent. Soc. Amer. 33:629-676.
———. 1944. A Revision of the Embioptera, or Webspinners, of the New World. Proc. U.S. Natl. Mus. 94:401-504.
———. 1970. Biosystematics of the Embioptera. Ann. Rev. Ent. 15:157-172.

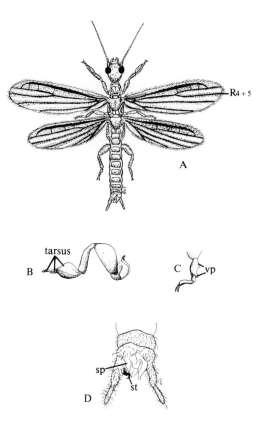

Figure 107 Webspinners. A, *Oligotoma saundersii* (Oligotomidae); B, front leg of a webspinner; C, ventral papillae on basal segment of hind tarsus; D, tip of abdomen of male *O. saundersii*. sp, spatulate process of 10th tergite; st, 9th sternite; vp, ventral papillae.

ORDER ISOPTERA[13]
Termites[14]

Termites are small- to medium-sized, soft-bodied insects that feed on wood and other forms of dead plant material with their chewing mouthparts. They live in colonies consisting of three common social castes—reproductives and their worker and soldier offspring—which vary in form, function and numbers as the result of pheromone and hormone production. The workers are immatures and/or sterile adults that are wingless and usually pale and eyeless (Fig. 110C). This group maintains the nest and cares for the royal pair, their young and the soldiers. The soldier caste generally consists of sterile, wingless adults with enlarged, heavily

13. Isoptera: *iso,* equal; *ptera,* wings.
14. Termites: *termit,* a wood worm.

sclerotized heads and mandibles (Figs. 109A, C, D; 110D, E). These individuals defend the colony against intruders, usually ants, by cutting and slashing with their mandibles; soldiers of a few species block holes in the nest with their pluglike heads. Still other species have soldiers (nasutes) with a long snout (nasus) through which they spray a sticky, toxic liquid that effectively repels or entangles invaders.

The reproductive caste consists of pigmented males and females with compound eyes and four wings of equal size (Fig. 110A). Diurnal or nocturnal swarms of reproductives disperse by flying short distances from the nest during certain seasons—spring, fall or monsoon—and after landing they typically break off their wings near the base, leaving only stubs (Fig. 110B). The female, followed by a male, seeks out a suitable nesting site in wood or soil. The pair mate periodically and remain together in the nest as king and queen of the new colony for the duration of their lives. The queen will lay thousands of eggs during her life span of a few to many years and a large colony may consist of several hundred thousand individuals. A nest may contain short-winged supplementary reproductives which also produce eggs. Termite immatures have a hemimetabolous type of metamorphosis.

Termite and ant kings and queens are similar in appearance and swarming behavior. Since termite swarms may indicate an existing pest problem, it is useful to be able to differentiate between termites and ants. All four wings of termites are similar in size; the hind wings of ants are smaller than the front. The abdominal-thoracic connection is broad in termites but constricted in ants and the latter have one or two nodes arising between the thorax and abdomen. The antennae of termites are straight and those of ants are elbowed.

Termites are most abundant in tropical rain forests. Some live in underground galleries in moist soil and in wood beneath or on the surface of the ground. Certain species build earthen tubes from soil to wood above the ground surface which conserve moisture and provide predator protection. Others live above ground in wooden buildings, posts, stumps, and trees. Various species in the tropics inhabit mounds up to 9 m high which they construct of soil, excrement, and saliva.

Termites feed primarily on cellulose-containing materials such as sound and decaying wood, paper, and fungi; dead termites, cast skins, and fecal material from other termites are also ingested. Cellulose cannot be utilized by the more primitive termites without first being digested by symbiotic protozoans that live in their digestive tracts. Termites in the family Termitidae (which contains about 75% of all termites) have no protozoans but rely on bacteria and their own enzymes to digest cellulose. In some areas of the eastern and southern U.S., in the Southwest, and on the Pacific coast, termites cause heavy damage to the wood foundations and frameworks of buildings as well as to furniture and books. However, over 90% of the species are not pests and play an important role in the breakdown and recycling of dead plant material.

Termites can be collected from dead branches and logs in contact with soil, stumps, fences, and infested wood in buildings. They are best preserved in 70-95% alcohol.

Species: North America, ca. 39; world, 2,200.
Families: North America, 4.

KEY TO FAMILIES OF ISOPTERA

A. Winged Adults

Female reproductives differ from males by having an enlarged seventh sternum which overlies the remaining terminal sterna. A wing stub (scale) remains after the termite has broken off most of the wing along a line of weakness, the humeral suture (Fig. 110B).

1a	Fontanelle usually present (Fig. 108); anterior part of wings with 2 prominent, thick veins 2
1b	Fontanelle absent; anterior part of wings with 3 or more prominent, thick veins 3
2a(1a)	Stub of front wing longer than pronotum (Fig. 110B); cerci 2-segmented; widely distributed .... (p. 115) *Rhinotermitidae*
2b	Stub of front wing shorter than pronotum; cerci 1- or 2-segmented; southwestern U.S. .... (p. 116) *Termitidae*
3a(1b)	Ocelli present; antennae usually with less than 22 segments; shafts of tibiae lack heavy spines; southeastern Florida and western U.S. .... (p. 115) *Kalotermitidae*
3b	Ocelli absent; antennae with at least 22 segments; shafts of tibiae with heavy spines; Far West and southwestern U.S. .... (p. 115) *Hodotermitidae*

Figure 108   Termite head. f, fontanelle.

B. Soldiers

1a	Head with fontanelle (Fig. 108) ... 2
1b	Head without fontanelle. 3
2a(1a)	Pronotum flat, almost as wide as head, and without anterior lobes; mandibles not obviously toothed; widely distributed. .... (p. 115) *Rhinotermitidae*
2b	Pronotum saddle-shaped, much narrower than head, and with anterior lobes; each mandible with an obvious tooth; southwestern U.S. .... (p. 116) *Termitidae*
3a(1b)	Cerci long and 3- to 8-segmented; antennae at least 22-segmented; Far West and southwestern U.S.. .... (p. 115) *Hodotermitidae*
3b	Cerci short and 2-segmented; antennae usually 10- to 19-segmented; southeastern Florida and western U.S. .... (p. 115) *Kalotermitidae*

Family Kalotermitidae

These insects lack a fontanelle, the reproductives have ocelli, and the soldiers have 2-segmented cerci. They do not make earthen tubes and usually dwell in wood, leaving sand-like fecal pellets in abandoned chambers or outside their tunnels. Some species live in the hot desert areas of the Southwest and members of this family generally require less moisture than termites of other families.

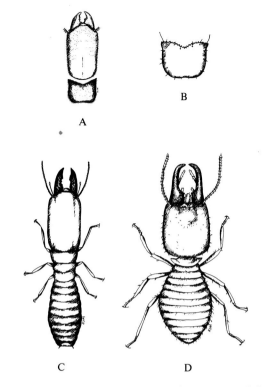

Figure 109   Termites. A, head and pronotum of the western drywood termite (soldier), *Incisitermes minor* (Kalotermitidae); B, pronotum of *I. snyderi* (soldier); C, Pacific dampwood termite (soldier), *Zootermopsis angusticollis* (Hodotermitidae); D, Wheeler's desert termite (soldier), *Amitermes wheeleri* (Termitidae).

Common Species

Cryptotermes brevis (Walker). Roughheaded powderpost termite—winged forms: 10-11 mm to wing tips, uniform light to medium brown, and faint, light colored median vein usually unites with dark "subcostal" veins near middle of wings; soldiers: head pluglike, short and roughened, truncated and concave anteriorly, mandibles short; Gulf coast states (in dry wood; flies at night, May-Oct.).

Incisitermes minor (Hagen). Western drywood termite—winged forms: 11-12 mm to wing tips, head and thorax chestnut brown, abdomen blackish, dark wings; soldiers: > 1 marginal tooth on mandibles, anterior margins of pronotum angular and only moderately cut (Fig. 109A); Pacific coast east to UT and AZ (flies in summer [AZ] and fall [CA]; in dry, sound wood).

Incisitermes snyderi (Light)—winged forms: < 9.5 mm to wing tips; yellowish or yellowish brown; soldiers: > 7 mm long, 12-13 antennal segments, 3rd antennal segment nearly as long as 4th and 5th combined, anterior margin pronotum with moderately deep indentation at midline and minutely toothed (Fig. 109B); southeastern U.S. (flies at night primarily in spring, attracted to lights).

Paraneotermes simplicicornis (Banks)—winged forms: 11-12 mm to wing tips, dark brown, wings clear, dark colored median vein parallels "subcostal" veins to wing tips; soldiers: reddish brown head, mandibles short and thick with narrow tips, antennae 14-segmented, femora slightly swollen; immatures: pronotum narrower than head, spotted and translucent abdomen; southwestern U.S. (in stumps, roots, other wood on or in ground).

Family Hodotermitidae

Three species in the genus *Zootermopsis* occur in the U.S. Adults are large (18 mm or more in length), lack a fontanelle, and the tibiae are spined in all castes.

Common Species

Zootermopsis angusticollis (Hagen) (Fig. 109C). Pacific dampwood termite—winged forms: 23-29 mm to wing tips, yellowish brown to chestnut, wings gray or pale brown; soldiers: head longer than broad; Pacific coast and inland mts. of Northwest (especially in damp Douglas fir logs).

Zootermopsis nevadensis (Hagen)—winged forms: 18-23 mm to wing tips, chocolate brown, wings dark gray; soldiers: head longer than broad, parallel-sided; Pacific coast and mts. of Northwest east to MT and NV (in dead tree trunks and logs; higher and drier areas than *Z. angusticollis*).

Family Rhinotermitidae

Adults of this widespread family are small, the winged forms of the common species are nearly black, and the wingless forms are very pale. All members have a fontanelle on the front of the head (Fig. 108). The insects must maintain contact with the soil as their source of moisture and may construct extensive earthen tubes to reach wood above the ground. *Reticulitermes flavipes* (Kollar), the most common termite in the eastern U.S. and the only species occurring in the Northeast, is the most destructive species in North America.

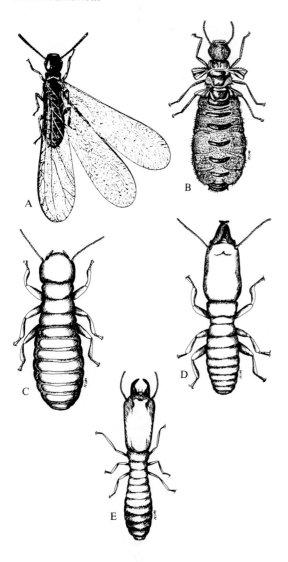

Figure 110 Termites (Rhinotermitidae). A-D, eastern subterranean termite, *Reticulitermes flavipes*: A and B, reproductive female with wings and after wings are removed; C, worker; D, soldier; E, western subterranean termite (soldier), *R. hesperus*.

Common Species

Reticulitermes flavipes (Kollar) (Figs. 110A-D). Eastern subterranean termite—winged forms: ca. 10 mm to wing tips, dark brown to black with translucent brownish wings, ocelli > own diameter from compound eyes; workers and soldiers ivory-white; soldiers: mandibles elongate S-shaped, length of head 1 1/2 times width; eastern and southern U.S. west to WI and OK, southeastern Canada west to Ontario.

Reticulitermes hesperus Banks (Fig. 110E). Western subterranean termite—winged forms: 9 mm to wing tips, brownish black with light gray wings, tibiae slightly darkened but not black; soldiers: white with long, narrow, yellow heads twice as long as wide, mandibles elongate S-shaped; Pacific coast east to ID and NV (usually moist habitats).

Reticulitermes tibialis Banks—winged forms: 9-10 mm to wing tips, black, tibiae blackish; soldiers: mandibles elongate S-shaped, short and broad yellow heads; inland valleys, deserts and mountains of western U.S., east to somewhat beyond Mississippi River (more open, drier habitats than above 2 spp.).

Reticulitermes virginicus (Banks)—winged forms: 8 mm to wing tips, dark brown or black, ocelli < own diameter from compound eyes, wings colorless; soldiers: mandibles elongate S-shaped, maximum width of rodlike plate in midventral portion of head is 2 1/2 times minimum width; eastern U.S. from NY south to FL, west to southern IL and eastern TX.

Family Termitidae

Although this is the largest world family of termites, only about 13 species occur in the U.S. and they are limited to the arid areas of the Southwest. The insects have a fontanelle, the front wing stubs of the reproductives are shorter than the pronotum, and the pronotum of the soldiers is saddle-shaped.

Common Species

Amitermes wheeleri (Desneux) (Fig. 109D). Wheeler's desert termite—winged forms: 10-12 mm to wing tips, body dark brown above, lighter below, head <1 mm wide, fontanelle long and narrow; soldiers: pale yellowish head with gray abdomen, short mandibles slender and strongly curved with a sharp, inward-projecting, marginal tooth at base; southern CA to TX (wood in or on ground, in cow chips).

Gnathamitermes perplexus (Banks)—winged forms: 13-16 mm to wing tips, body dark brown above, lighter below, fontanelle nearly circular; soldiers: long, straight mandibles slightly incurved at tip and each with pointed tooth in basal 1/2; southern CA to TX (under rocks, construct earthen tubes and shelters over wood, vegetation, and cow chips).

GENERAL REFERENCES

Ebeling, W. 1968. Termites: Identification, Biology, and Control of Termites Attacking Buildings. Univ. Calif. Agric. Expt. Sta. Ext. Serv. Man. 38. 74 pp.

Krishna, K., and Weesner, F.M. (eds.). 1969-1970. *Biology of Termites*. Academic Press, N.Y. vol. 1 (1969), 598 pp., vol. 2 (1970), 643 pp.

Lee, K.E., and Wood, T.G. 1971. *Termites and Soils*. Academic Press, N.Y. 251 pp.

Snyder, T.E. 1949. Catalogue of the Termites (Isoptera) of the World. Smiths. Misc. Coll. 112:1-490.

Weesner, F.M. 1965. *The Termites of the United States—A Handbook*. Nat. Pest Control Assn., Elizabeth, N.J. 70 pp.

Wilson, E.O. 1971. *The Insect Societies*. Belknap Press of Harvard Univ. Press, Cambridge, Mass. 548 pp.

ORDER ZORAPTERA[15]
Zorapterans

The Zoraptera are uncommon, minute insects (1.5-3.0 mm long) that have either four wings or are wingless (Fig. 111). The wings have only a few veins and the hind pair are smaller than the front. Like termites, adults break off their wings and leave only stubs attached to the body. The dark-colored winged forms have compound eyes and three ocelli; the pale wingless forms lack both characteristics. The threadlike antennae are 9-segmented, the tarsi are 2-segmented, and the short, unsegmented cerci terminate in a long bristle. Zorapterans have chewing mouthparts and a hemimetabolous type of development.

The Zoraptera are primarily tropical, living in colonies under the bark of dead trees and in decaying wood. These insects feed on fungi and small, dead arthropods. The order contains only one family, Zorotypidae, a single genus, *Zorotypus*, and, in the U.S., only two species. *Z. hubbardi* Caudell (Fig. 111) ranges from Pennsylvania south to Florida and west to Texas and Iowa. This species occurs under slabs of wood buried in piles of old sawdust, under loose bark, and in fallen, decaying logs. *Z. snyderi* Caudell occurs in Florida.

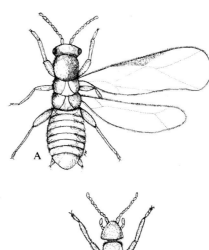

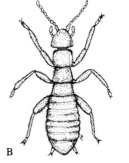

Figure 111 Zorapterans. *Zorotypus hubbardi* (Zorotypidae). A, winged; B, wingless.

15. Zoraptera: *zor*, pure; *aptera*, wingless (originally all were thought to be wingless).

Zorapterans are collected by using a Berlese funnel or searching suitable habitats. They are preserved in 70% alcohol or mounted on microscope slides.

Species: North America, 2; world, 24. Families: world, 1.

GENERAL REFERENCES

Gurney, A.B. 1938. A Synopsis of the Order Zoraptera, with Notes on the Biology of *Zorotypus hubbardi* Caudell. Proc. Ent. Soc. Wash. 40:57-87.

Riegel, G.T. 1963. The Distribution of *Zorotypus hubbardi* (Zoraptera). Ann. Ent. Soc. Amer. 56:744-747.

ORDER PSOCOPTERA[16]
Psocids

The Psocoptera are small (usually less than 6 mm long), winged or wingless insects that occur both indoors and outdoors. The four membranous wings are usually held rooflike over the abdomen; the wings of some species are very short or the hind wings are absent. The psocids are soft-bodied, the antennae are moderately long, and the head is inflated anteriorly giving it a slight bulging appearance. These insects have chewing mouthparts and nymphs have a hemimetabolous type of development. The classification of families of Psocoptera varies among taxonomists; alternate family names are shown in parentheses next to several family names used in this text. Older texts may refer to the order as Corrodentia.

Although the psocids are sometimes called "lice" none of them is parasitic. They feed on fungi, cereals, paste, pollen, dead insect fragments, and other organic matter. The small number of species occurring indoors are primarily wingless and are called booklice because they are often found around books and paper. Indoor species also occur on straw, flour, dried insect and plant collections, and in cracks of dusty window sills and shelves. Most psocids are winged and occur outdoors on bark and living or dead foliage of trees and shrubs, under bark or rocks, in bird nests, and on lichens and algae; they are called barklice. Some species live together under silken webs and often spin them on tree trunks and branches.

Psocids living indoors and outdoors are collected by inspecting the habitats described earlier. Vegetation may also be swept, dead leaves gathered and inspected, and branches beaten to dislodge the insects. All specimens are preserved in 80-90% alcohol or mounted on slides.

Species: North America, 270; world, 2,200. Families: North America, 11.

KEY TO COMMON FAMILIES OF PSOCOPTERA

1a	Tarsi 2-segmented	2
1b	Tarsi 3-segmented	4
2a(1a)	One-segmented labial palpi broad and triangular (Fig. 113A) . . . . (p. 121) *Polypsocidae*	
2b	One-segmented labial palpi short and often oval.	3

16. Psocoptera: *psoco,* rub small; (refers to chewing food into powder); *ptera,* wings.

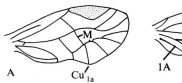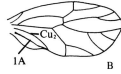

Figure 112 Front wings of psocids. A, Psocidae; B, Psyllipsocidae.

3a(2b) Cu₁ₐ in front wing joined to M (Fig. 112A); if short-winged then head has glandular setae................ (p. 121) *Psocidae*

3b Cu₁ₐ in front wing usually not fused with M or often absent; head without glandular setae................ (p. 121) *Pseudocaeciliidae*

4a(1b) Antennae usually with 15-17 segments and segments 7-17 are faintly ringed; body flat; hind femora flat and broad............ (p. 120) *Liposcelidae*

4b Antennae with more than 20 segments and segments are not ringed; body and femora usually not flat or broad.................. 5

5a(4b) Head long and anterior part is perpendicular (Fig. 113B); Cu₂ and 1A veins in front wing converge to a point at wing margin (Fig. 112B) .... (p. 120) *Psyllipsocidae*

5b Head short and anterior part is slanted; Cu₂ and 1A veins end separately at front wing margin....... 6

6a(5b) Tarsal claws with preapical tooth (Fig. 113C); body and wings covered with scales.................... .... (p. 119) *Lepidopsocidae*

6b Tarsal claws without preapical tooth; body and wings without scales..... .... (p. 119) *Trogiidae*

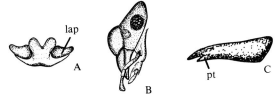

Figure 113 A, labial palpi (Polypsocidae); B, lateral view of head (Psyllipsocidae); C, tarsal claw with preapical tooth (Lepidopsocidae). lap, labial palpus; pt, preapical tooth.

Family Trogiidae (= Atropidae)
Members of this family either have short wings or are wingless.

Common Species

Trogium pulsatorium (Linnaeus). Larger pale booklouse—1.5-2.0 mm; pale yellowish white; 3 distinct thoracic segments; front wings reduced to small convex scales; throughout North America (in houses, libraries, barns, granaries; makes ticking noise; often with other booklice [Liposcelidae]).

Family Lepidopsocidae
Members of this family have scales on their bodies and wings, and the wings are generally pointed apically.

Common Species

Echmepteryx hageni Packard—3 mm; head pale with dark brown markings and fine hairs; thorax brown with fine hairs; wing scales form dark patches and bands; long hairs on wing apex; abdomen pale yellowish with brown markings; eastern U.S. south to FL, west to TX and IA, southeastern Canada (on tree trunks, stone bluffs).

Family Psyllipsocidae

Common Species

Psyllipsocus ramburii Selys (Fig. 114A)—2 mm; gray; 3rd antennal segment shorter than head; apical segment of maxillary palpus tapers at apex and not broad or slanted; short- and long-winged forms; eastern U.S. (in buildings, cellars, caves).

Family Liposcelidae (= Troctidae)

Although the name *Liposcelis divinatorius* (Müller) is still used in textbooks for the booklouse (=cereal psocid), most taxonomists are unable to state to what species this name really refers. Eventually the name may be dropped from the literature. Members in the genus *Liposcelis* are about 1 mm long and wingless, the head and hind femora are enlarged, and only two thoracic segments are distinct (Fig. 114B). Many occur throughout North America in buildings (e.g., shelves, cracks, warm and dusty areas, books, cereal boxes, and dry museum specimens).

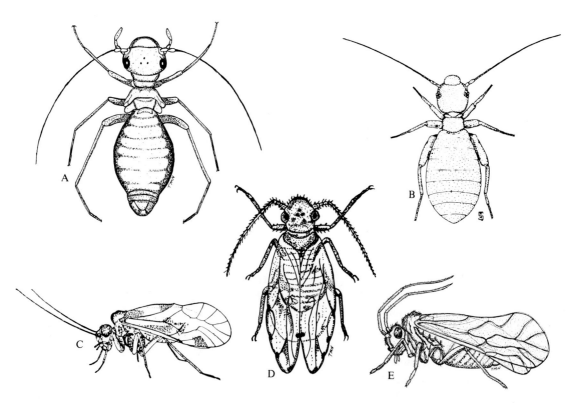

Figure 114 Psocids. A, *Psyllipsocus ramburii* (Psyllipsocidae); B, *Liposcelis* sp. (Liposcelidae); C, *Hyalopsocus striatus* (Psocidae); D, *Ectopsocus meridionalis* (Pseudocaeciliidae); E, *Lachesilla nubilis* (Pseudocaeciliidae).

Family Psocidae

Wings are almost always well developed in the 56 species of this family. The genus *Psocus* has no hairs on the wing veins and the base of the subcosta is distinct.

Common Species

Hyalopsocus striatus (Walker) (Fig. 114C)—4-5 mm; brown; front wings clear except for brown spot in stigma and another on posterior margin about midway that may connect across wing; eyes large and black; eastern 1/2 U.S. (comes to lights).

Metylophorus novaescotiae (Walker)—3.5-5.5 mm; brown to blackish; 3 pairs faint to distinct dark spots on head; front wings with faint to distinct spots or bands; eastern U.S. south to FL, southeastern Canada (on tree trunks or dead limbs, especially pine; often with *P. leidyi*).

Psocus leidyi Aaron—3.5-5.0 mm; ivory white with sparse dark markings; whitish hairs; large blackish spot on clypeus; small to large dark spot where M and Cu veins join; wings clear except for opaque stigma; most of U.S., southern Canada (on trunks of deciduous trees).

Family Pseudocaeciliidae
(= Ectopsocidae, Peripsocidae, Lachesillidae)

This is the largest family (about 84 spp.) and most species occur outdoors.

Common Species

Ectopsocus meridionalis Ribaga (Fig. 114D)—2-3 mm; grayish brown; tan spots on top of head; eyes bluish black; front wings with 10 small and scattered spots at end of veins; Gulf coast states north to IL and CT (on fungus on leaves of trees, shrubs; greenhouses).

Lachesilla nubilis (Aaron) (Fig. 114E)—1.5 mm; head and thorax brownish; abdomen purplish gray and striped; brownish spots at end of wing veins; most of U.S. except Northwest, southcentral Canada (on trees, shrubs, dried corn stalks, grain).

Lachesilla pedicularia (Linnaeus). Cosmopolitan grain psocid—1.3-1.6 mm; head and thorax golden to dark brown; abdomen dull white with dorsal grayish stripes; wings clear with dark veins; wing length 0.1-1.9 mm; throughout North America (in buildings at windows, on furniture, cereals, straw; outdoors on dead and living leaves and branches).

Peripsocus quadrifasciatus Harris—1.5-2.0 mm; head and thorax brown to dark brown; 2 yellowish brown dots on top of head; wings grayish brown and divided into 4 sections by 3 opaque bands; abdomen dull white with gray rings; throughout U.S., southern Canada (on dead or living tree leaves and trunks).

Family Polypsocidae
(= Caeciliidae)

Common Species

Caecilius aurantiacus Hagen—2.0-2.5 mm; dull white to bright yellow with brownish markings; area between ocelli dark brown or black; dorsum of thorax darker than sides; distal 1/2 of veins in front wing darker than basal portion; most of U.S., southern Canada (on leaves of deciduous and coniferous trees).

Polypsocus corruptus (Hagen)—2.5-3.0 mm; brown head and thorax; shining bronze wings with clear band at apex; purplish gray abdomen; throughout U.S. (on dead leaves of deciduous trees, tree trunks).

> GENERAL REFERENCES
> Chapman, P.J. 1930. Corrodentia of the United States of America. I. Suborder Isotecnomera. J. N.Y. Ent. Soc. 38:219-290, 319-403.
> Gurney, A.B. 1950. Corrodentia. In *Pest Control Technology, Entomological Section*. National Pest Control Assn., New York. Pp. 129-163.
> Mockford, E.L. 1951. The Psocoptera of Indiana. Proc. Ind. Acad. Sci. 60:192-204.
> Mockford, E.L., and Gurney, A.B. 1956. A Review of the Psocids, or Book-Lice and Bark-Lice, of Texas (Psocoptera). J. Wash. Acad. Sci. 46:353-368.

ORDER MALLOPHAGA[17]
Chewing Lice

The Mallophaga are small (0.5-11.0 mm long), flat-bodied, wingless insects that are external parasites of birds and, less frequently, mammals. Mallophaga differ from the sucking lice (Anoplura) by having mouthparts modified for chewing, a head that is as wide as or wider than the prothorax (Fig. 115), and one or two small tarsal claws. Anoplura have sucking mouthparts, a head that is narrower than the prothorax, and one large tarsal claw. Mallophaga and Anoplura are classified by some taxonomists as suborders of a single order, Phthiraptera.

Females cement up to 100 eggs on the hair or feathers of hosts. Nymphs, appearing very similar to adults, have a hemimetabolous type of development and remain on the hosts. Chewing lice feed on pieces of feathers, hair, skin, and cell secretions, and often occur only in specific areas of the body. A few species consume blood, some feed partly on mites infesting the host, and others live in the primary quills of birds. Chewing lice are unable to survive more than a few days after the host dies. Movement from one host to another occurs when two hosts come in contact. Most species are confined to only one or a few host species and they do not attack humans. Chewing lice may become major pests of poultry and other domestic animals by causing severe skin irritation which weakens the hosts. There are more species (11) of Mallophaga on chickens than any other known fowl.

Mallophaga are obtained by examining domestic or wild birds and mammals. Hosts collected in the field should be placed in a tightly closed bag for later examination and any lice that leave the host will remain in the bag. Forceps or a comb can be used to remove the lice or the host can be shaken over paper to dislodge them. A small brush dipped in alcohol can be used to stun and pick up live lice. Lice are preserved in 70% alcohol or mounted on slides.

Species: North America, 700; world, 2,900.
Families: North America, 8.

KEY TO FAMILIES OF MALLOPHAGA

1a Antennae clubbed or otherwise enlarged apically, usually concealed in grooves on underside of head, and normally 4-segmented; maxillary palpi present. 2

1b Antennae filiform, not concealed in grooves, and 3- to 5-segmented; maxillary palpi absent. 7

2a(1a) Middle and hind legs with 1 tarsal claw or none; on guinea pigs. . . . . (p. 123) *Gyropidae*

2b Middle and hind legs with 2 tarsal claws; primarily on birds 3

3a(2b) Antennae capitate and 5-segmented; on dogs. . . . . (p. 123) *Boopidae*

3b Antennae clavate and 4-segmented. 4

4a(3b) Metathorax resembles the inverted prothorax in size and shape (Fig. 116C); on guinea pigs. . . . . (p. 124) *Trimenoponidae*

4b Metathorax and prothorax normal. 5

17. Mallophaga; *mallo*, wool; *phaga*, eat.

5a(4b)	Head broadly triangular and expanded behind eyes (Fig. 115B)... .... (p. 124) *Menoponidae*	
5b	Head not broadly triangular or expanded behind eyes.......... 6	
6a(5b)	Each side of head with a prominent swelling in front of the eyes at the base of the antennae (Fig. 116A) ... . (p. 123) *Laemobothriidae*	
6b	Sides of head without swellings (Fig. 116B) ...... (p. 124) *Ricinidae*	
7a(1b)	Tarsi with 2 claws; antennae 5-segmented; on birds .... (p. 124) *Philopteridae*	
7b	Tarsi with 1 claw; antennae usually 3-segmented; on mammals....... .... (p. 125) *Trichodectidae*	

Family Gyropidae— Rodent Chewing Lice

Only two species occur in the U.S. and both are introduced species found on guinea pigs.

Common Species

Gliricola porcelli (Schrank)—1.1-1.5 mm; pale yellow, head and abdomen darker; transverse dark marks on dorsum of abdomen; head longer than wide; maxillary palpi 2-segmented; 2nd tarsal segment of middle leg not clawlike; throughout North America.

Gyropus ovalis Burmeister (Fig. 115A). Oval guineapig louse—1.0-1.5 mm; pale yellowish brown; oval; head wider than long with large projecting posterior lobes; maxillary palpi 4-segmented; 2nd tarsal segment of middle and sometimes hind leg enlarged and clawlike; dorsal setae of abdomen 1/4 or less length of segments on which they occur; throughout North America.

Family Boopidae

The only species in North America is *Heterodoxus spiniger* (Enderlein), the dog large body louse, which occurs on dogs and coyotes.

Family Laemobothriidae

Members of this small family are found on birds of prey and water birds. The species are 6.5-11.0 mm long and represent the largest known bird lice. *Laemobothrion glutinans* Nitzsch occurs on turkey and black vultures, *L. maximum* (Scopoli) parasitizes hawks and osprey, and *L. tinnunculi* (Linnaeus) is found on falcons and small hawks. *L. atrum* (Nitzsch) parasitizes the American coot and *L. chloropodis* (Schrank) occurs on the common gallinule.

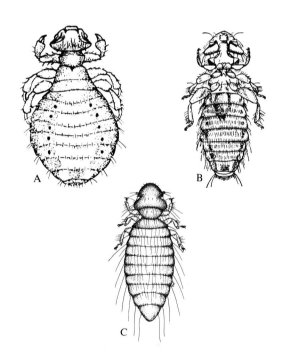

Figure 115 Chewing lice. A, oval guineapig louse, *Gyropus ovalis* (Gyropidae); B, chicken body louse, *Menacanthus stramineus* (Menoponidae); C, shaft louse, *Menopon gallinae* (Menoponidae).

Family Menoponidae—Bird Body Lice

This is a common family that parasitizes birds.

Common Species

Holomenopon leucoxanthum (Burmeister)—1.5-2.0 mm; yellowish brown; prosternum with highly serrated margin; fringe of setae around ventral portion of anus includes 4 atypical thickened setae each separated by 2 typical setae; North America (on ducks, geese).

Menacanthus stramineus (Nitzsch) (Fig. 115B). Chicken body louse—2.0-3.5 mm; pale yellow to brownish; forehead with ventral spinelike process; short hairs on dorsum of meso- and metathorax; dorsum of abdominal segments 3-7 with 2 transverse rows of long, fine hairs; throughout North America (on skin of chickens, turkeys, other fowl raised near chickens).

Menopon gallinae (Linnaeus) (Fig. 115C). Shaft louse—1.0-1.5 mm; pale yellow; forehead without ventral spinelike processes; throughout North America (on feathers of chickens, ducks, pigeons, turkeys).

Trinoton querquedulae (Linnaeus). Large duck louse—4.0-4.5 mm; dorsal posterior surface of head and prothorax without stout spinelike setae but side of face with these setae; throughout North America (on ducks).

Family Ricinidae

Species in this small family are thought to be blood feeders and most are parasitic on songbirds (passerines) and hummingbirds. Most species of songbirds probably harbor members of the genus *Ricinus*. The genus is characterized by membranous, lobelike extensions of the labrum that arise ventrally and laterally (Fig. 116B), and 4-segmented antennae that sit inside two capsules on the ventral side of the head.

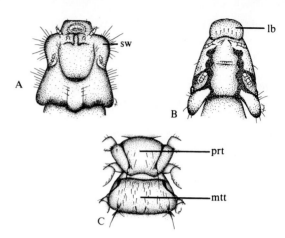

Figure 116 A, head of a laemobothriid (Laemobothriidae); B, head of a ricinid (Ricinidae); C, metathorax and prothorax of *Trimenopon hispidum* (Trimenoponidae). lb, labrum; mtt, metathorax; prt, prothorax; sw, swelling.

Family Trimenoponidae

The only species in North America, *Trimenopon hispidum* (Burmeister), occurs on guinea pigs.

Family Philopteridae—Feather Chewing Lice

This is the largest family in the order and most kinds of water and land birds are hosts. Species in the genera *Philopterus* and *Quadraceps* occur commonly on most birds.

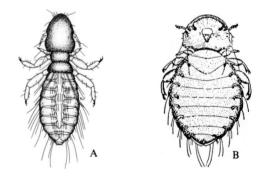

Figure 117 Feather chewing lice (Philopteridae). A, chicken head louse, *Cuclotogaster heterographus;* B, chicken fluff louse, *Goniocotes gallinae*.

Common Species

Anatoecus dentatus (Scopoli)—1 mm; dark red; throughout North America (on ducks, geese, coots).

Columbicola columbae (Linnaeus). Slender pigeon louse—0.8-1.8 mm; yellowish white, very elongated body; throughout North America (on pigeons).

Cuclotogaster heterographus (Nitzsch) (Fig. 117A). Chicken head louse—2.5-2.8 mm; dark gray; head longer than wide; 4 patches of long hairs on posterior margin of metathorax; throughout North America (on heads of chickens, pheasants reared near chickens).

Goniocotes gallinae (De Geer) (Fig. 117B). Chicken fluff louse—1 mm; head wider than long; 2 long hairs each side of head behind antennal region; throughout North America (on fluff around vent [anus] of chickens).

Lipeurus caponis (Linnaeus). Wing louse—3 mm; grayish; slender; head longer than wide and greatest width in front of antennae; 1st antennal segment with an appendage; throughout North America (on large wing feathers of chickens).

Family Trichodectidae— Mammal Chewing Lice

Members of this family are parasitic on mammals. The genus *Geomydoecus* contains small species that parasitize only pocket gophers. Members of the genus are stout, the anterior end of the head is broadly notched, and their range includes all but northeastern U.S. Two species of *Tricholipeurus* infest deer.

Common Species

Bovicola bovis (Linnaeus) (Fig. 118A). Cattle chewing louse—2.0-2.5 mm; yellowish white with reddish head; abdomen with 8 dark crossbands; throughout North America (on cattle; most common in winter).

Bovicola equi (Denny). Horse chewing louse— 2.0-2.5 mm; head and thorax brown; abdomen yellowish with dark crossbands; throughout North America (on horses, donkeys, mules; around neck and base of tail).

Bovicola ovis (Schrank) (Fig. 118B). Sheep chewing louse—1.0-1.5 mm; yellowish brown with reddish head; head rounded in front and ca. as wide as long; transverse row of hairs across dorsum of each abdominal segment; throughout North America (on sheep; most common in winter).

Felicola subrostratus (Burmeister). Cat louse—1.0-1.5 mm; yellowish brown; stout; pointed head; distinct bands on body; abdominal segments without plates on sides; throughout North America (on cats).

Trichodectes canis (De Geer) (Fig. 118C). Dog chewing louse—2 mm; clear yellow; dull lateral bands; abdominal segments with plates on sides; throughout North America (on dogs, coyotes, wolves).

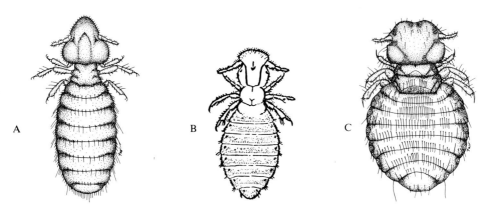

Figure 118 Mammal chewing lice (Trichodectidae). A, cattlz chewing louse, *Bovicola bovis;* B, sheep chewing lousz, *Bovicola ovis;* C, dog chewing louse, *Trichodectes canis.*

GENERAL REFERENCES

Clay, T. 1969. A Key to the Genera of the Menoponidae (Amblycera: Mallophaga: Insecta). Bull. Brit. Mus. (Natur. Hist.), Ent. 24:1-26.

Emerson, K.C. 1956. Mallophaga (Chewing Lice) Occurring on the Domestic Chicken. J. Kan. Ent. Soc. 29:63-79.

———. 1972. *Checklist of the Mallophaga of North America (North of Mexico). Part 1. Suborder Ischnocera.* 200 pp. Part 2. *Suborder Amblycera.* 118 pp. Part 3. *Mammal Host List.* 28 pp. Part 4. *Bird Host List.* 216 pp. Deseret Test Center, Proving Ground, Dugway, Utah.

Flynn, R.J. 1973. *Parasites of Laboratory Animals.* Iowa St. Univ. Press, Ames. 884 pp.

Nelson, B.C. 1972. A Revision of the New World Species of Ricinus (Mallophaga) Occurring on Passeriformes (Aves). Univ. Calif. Pub. Ent. 68:1-175.

ORDER ANOPLURA[18]
Sucking Lice

The Anoplura are small (2-5 mm long), wingless, blood-feeding insects that are external parasites of mammals. The body is flat and the head is narrower than the prothorax (Fig. 119B). Each 1-segmented tarsus has a single claw that fits against a thumblike projection at the end of the tibia (Fig. 119A) enabling the insect to grasp the hairs of the host. The piercing-sucking mouthparts are retracted into the head when the lice are not feeding. Nymphs have a hemimetabolous type of development. Anoplura and Mallophaga are sometimes classified as suborders of the order Phthiraptera.

The female human body louse, *Pediculus humanus humanus* Linnaeus, lays about 300 eggs, at the rate of 8-12 eggs daily, on clothing next to the skin. The eggs take 6-14 days to hatch and nymphs mature in 8-16 days. Sucking lice are restricted to either one or a few hosts. Most species of mammals are parasitized by either Anoplura, Mallophaga, or both. Anoplurans cause blood loss and severe irritation of livestock, and irritate skin and transmit diseases to humans.

Sucking lice are collected from mammals in the manner described for Mallophaga and preserved in 70% alcohol or mounted on microscope slides.

Species: North America, 70; world, 490. Families: North America, 5.

KEY TO FAMILIES OF ANOPLURA

1a Many short, stout spines on body; some with scales on body; on marine mammals .... (p. 128) *Echinophthiriidae*

1b Spines and hairs on body; no scales on body; on land mammals 2

2a(1b) Head with eyes or eye tubercles; on man and other primates .... (p. 127) *Pediculidae*

2b Head without eyes; on mammals except man and other primates 3

3a(2b) Dorsolateral areas of abdominal segments lack plates or tubercles (Figs. 120B, C); dorsum of abdomen smooth ..... (p. 128) *Linognathidae*

3b Dorsolateral areas of at least 1 abdominal segment set off from central area in the form of a plate (Figs. 120A, D) or tubercle; dorsum

---

18. Anoplura: *anopl*, unarmed; *ura*, tail.

of abdomen finely wrinkled or smooth..................... 4

4a(3b) Dorsolateral abdominal plates with freely projecting apical margin (Fig. 120D); dorsum of abdomen not wrinkled; primarily on rodents..... .... (p. 128) *Hoplopleuridae*

4b Dorsolateral abdominal plates without freely projecting apical margins (Fig. 120A); dorsum of abdomen finely wrinkled; primarily on cattle, horses, and sheep.......... .... (p. 127) *Haematopinidae*

Family Haematopinidae— Wrinkled Sucking Lice

These species occur on domestic animals.

Common Species

Haematopinus asini (Linnaeus). Horse sucking louse—2.5-3.5 mm; head, thorax, and sides of abdomen brown; dorsum of abdomen yellowish; throughout North America (on horses, donkeys).

Haematopinus eurysternus Denny. Shortnosed cattle louse—2.0-3.2 mm; yellowish gray; head bluntly rounded and nearly as wide as long; prominent protuberance behind antennae on each side of head; 2 black blotches on last abdominal segment; throughout North America (on cattle).

Haematopinus suis (Linnaeus) (Fig. 120A). Hog louse—5-6 mm; bluish gray or gray brown; reddish brown thorax; head long and narrow; abdomen very wide; throughout North America (on hogs).

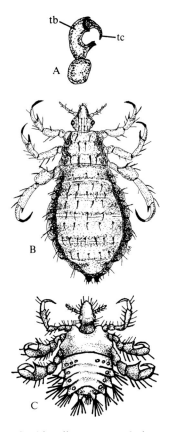

Figure 119 Sucking lice. A, tarsal claw and tibia; B, body louse, *Pediculus humanus humanus* (Pediculidae); C, crab louse, *Pthirus pubis* (Pediculidae). tb, tibia; tc, tarsal claw.

Family Pediculidae—Human Lice

The head and body lice, usually considered as two subspecies of a single species, and the crab louse occur on humans. The body louse ("cootie") spends most of its time on underclothes and may feed on the host while attached to the host's clothing; it is transferred from clothing and bedding. The body louse is the major vector of epidemic typhus, a serious disease caused by a rickettsial organism. Infection occurs from scratching the louse's feces or crushing the louse into the skin, but the disease is not transmitted by the bite of the louse. Relapsing fever is a bacterial disease also transmitted by the body louse. The head louse is often common on school children because it is easily disseminated by contact regardless of sanitary conditions. This subspecies occurs on the head; its eggs are glued to the hair and may be transferred by combs, brushes, and hats. There is rarely any disease problem with the head louse.

Common Species

Pediculus humanus capitis De Geer. Head louse—2.0-3.5 mm; grayish or resembles host's hair color; throughout North America (on head).

Pediculus humanus humanus Linnaeus (Fig. 119B). Body louse—2-4 mm; pale gray or brownish; throughout North America (on clothing).

Pthirus pubis (Linnaeus) (Fig. 119C). Crab louse—1.5-2.0 mm; grayish white with reddish legs; broadly oval and crab-shaped; tarsal claws of middle and hind legs much larger than front claws; throughout North America (in pubic or other hairy regions).

Family Echinophthiriidae

Members of this family infest seals, sea lions, and walruses.

Family Linognathidae— Smooth Sucking Lice

Common Species

Linognathus ovillus (Neumann)—2.0-2.5 mm; yellowish brown; smooth dorsum; prothorax with single seta; large thoracic spiracle; throughout North America (on sheep).

Linognathus setosus (Olfers) (Fig. 120B). Dog sucking louse—1.5-2.0 mm; yellowish or pinkish; throughout North America (on dogs).

Linognathus vituli (Linnaeus) (Fig. 120C). Longnosed cattle louse—2 mm; brown or bluish; anterior part of head very elongated; slender body; throughout North America (on cattle, especially calves).

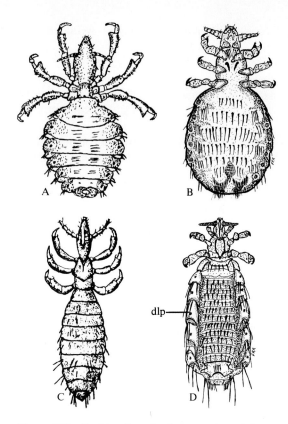

Figure 120 Sucking lice. A, shortnosed cattle louse, *Haematopinus suis* (Haematopinidae); B, dog sucking louse, *Linognathus setosus* (Linognathidae); C, longnosed cattle louse, *L. vituli* (Linognathidae); D, *Hoplopleura acanthopus* (Hoplopleuridae). dlp, dorsolateral plate.

Family Hoplopleuridae— Smallmammal Sucking Lice

Members of the genus *Hoplopleura* are confined to rodents including rats, mice, squirrels, and flying squirrels. Members of the genus usually have large, rectangular, dorsolateral plates on each abdominal segment. *Hoplopleura hesperomydis* (Osborn) is a small louse (0.8-1.2 mm long) common on the whitefooted mouse and other *Peromyscus* species of mice. *H. acanthopus* (Burmeister) (Fig. 120D) occurs on various native voles, mice, and occasionally laboratory mice. Several members of the genus *Neohaematopinus* occur on squirrels; males of the genus are generally characterized by the modified second abdominal tergite which has a group of setae arranged in an asterlike formation at each end of the tergite's broadly notched posterior margin.

GENERAL REFERENCES

Ferris, G.F. 1951. The Sucking Lice. Pacific Coast Ent. Soc. Mem. 1. 320 pp.

Hopkins, G.H.E. 1949. The Host-associations of the Lice of Mammals. Proc. Zool. Soc. London 119:387-604.

James, M.T., and Harwood, R.F. 1969. *Herms's Medical Entomology,* 6th ed. Macmillan, N.Y. 484 pp.

Stojanovich, C.J. and Pratt, H.D. 1965. Key to Anoplura of North America. U.S. Dept. Health, Education, and Welfare, Public Health Service. CDC, Atlanta, Georgia. 24 pp.

ORDER THYSANOPTERA[19]
Thrips[20]

The Thysanoptera in the U.S. are minute (most are 0.5-5.0 mm long), elongated insects that may be wingless or may have four narrow wings fringed with long hairs (Fig. 123). The antennae are short and the mouthparts form a short, conical proboscis used for sucking liquids. The 1- or 2-segmented tarsi are bladderlike at the tips. Thrips often crawl with the abdomen curved up over the back. Parthenogenesis occurs in this order and males of some species are unknown. Metamorphosis is a modification of the hemimetabolous type: there are two larval stages followed by two to three inactive pupal stages.

Females in the four families of the suborder Terebrantia have a short ovipositor and the last abdominal segment is rounded (Figs. 121B, C; 123B, C). The sawlike ovipositor, except in species where it is greatly reduced, is used to slit living plant tissue for egg insertion. Females of the suborder Tubulifera lack an ovipositor and the last abdominal segment is tubular (Figs. 121A; 123A). Eggs are glued onto leaf or bark surfaces or pushed between flower parts and under bark. Thrips have one to several generations per year and adults or immatures overwinter in debris. Most thrips feed on flowers, leaves, buds, and fruit, and many species are serious pests of crops. A few feed on fungal spores or are predaceous on mites and small insects. Some thrips transmit plant viruses.

Thrips may be collected by sweeping vegetation and placing all the contents in a killing jar for later sorting. Flowers, especially dandelions, daisies, and clover, may be shaken, pulled apart over a piece of white paper, or placed in a bag and inspected later. Other methods include placing fungi, debris, and flower heads of grasses in a Berlese funnel, inspecting the inner surface of bark, and shaking thrips from branches into a beating umbrella. Specimens should be placed in 60% alcohol and then mounted on microscope slides. To reduce brittleness thrips may be killed and left for a few weeks in the following mixture: 60% alcohol (10 parts), glycerin (1 part), and acetic acid (1 part). The specimens should then be transferred to 80% alcohol for storage or mounting on slides following the usual procedures.

Species: North America, > 600; world, 5,000.
Families: North America, 5.

---

19. Thysanoptera: *thysano,* fringe; *ptera,* wings.
20. Thrips: wood louse or wood worm.

KEY TO FAMILIES OF THYSANOPTERA

1a Last abdominal segment tubular in both sexes (Figs. 121A; 123A); female without a ventral ovipositor; front wings (if present) without veins or with a short median vein that does not reach wing tip (suborder Tubulifera) ...... (p. 130) *Phlaeothripidae*

1b Last abdominal segment broadly rounded (males) or conical (females) (Fig. 121B); female usually with sawlike ovipositor (Figs. 121B, C); front wings (if present) with 1 or 2 longitudinal veins that reach the wing tip (suborder Terebrantia) 2

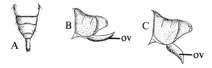

Figure 121 Thrips. A, tubular tip of abdomen (Phlaeothripidae); B, conical tip of abdomen with upturned ovipositor (Aeolothripidae); C, ovipositor downturned (Thripidae); ov, ovipositor.

2a(1b) Antennae 9-segmented; ovipositor upturned (Fig. 121B) or straight; front wings broad and rounded at tips. .... (p. 131) *Aeolothripidae*

2b Antennae 6- to 9-segmented; ovipositor downturned (Fig. 121C); front wings narrow and usually pointed at tips 3

3a(2b) Sense organs near the apex of the third and fourth antennal segments are slender and simple or forked (Fig. 122A). ...... (p. 131) *Thripidae*

3b Sense organs of the third and fourth antennal segments form flat areas at apex of each segment (Figs. 122B, C), or peglike cones at the apex. 4

4a(3b) Antennae 8-segmented; sense organs at apices of third and fourth antennal segments are flat (Fig. 122B) .... (p. 132) *Merothripidae*

4b Antennae 9-segmented; sense organs of third and fourth antennal segments are small and flat ringlike areas that encircle apex of each segment (Fig. 122C), or they are short and blunt cones at apex .... (p. 132) *Heterothripidae*

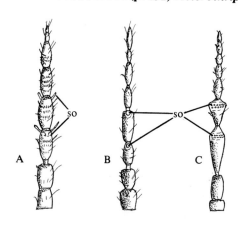

Figure 122 Sense organs on third and fourth antennal segments. A, Thripidae; B, Merothripidae; C, Heterothripidae. so, sense organ.

SUBORDER TUBULIFERA

Family Phlaeothripidae

Thrips in this family are typically dark colored and the tip of the abdomen is tubular. Species in leaf litter are often yellow or yellow-brown.

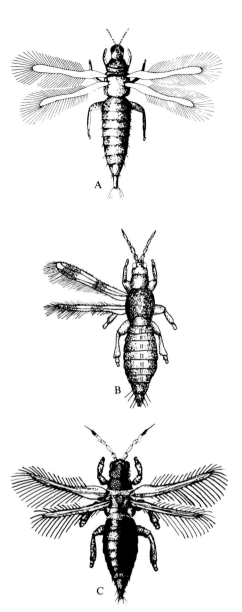

Figure 123 Thrips. A, Mullein thrips, *Neoheegeria verbasci* (Phlaeothripidae); B, banded thrips, *Aeolothrips fasciatus* (Aeolothripidae); C, greenhouse thrips *Heliothrips haemorrhoidalis* (Thripidae).

Common Species

Neoheegeria leucanthemi (Schrank)—2 mm; dark brown or black; stalk of 7th antennal segment about as narrow as 6th; most of North America (in daisy and yarrow flowers).

Neoheegeria verbasci (Osborn) (Fig. 123A). Mullein thrips—2.0-2.5 mm; black; head long; antennal segments 3-6 yellow; prothoracic setae long; most of North America (on common mullein).

SUBORDER TEREBRANTIA

Family Aeolothripidae

The dark-colored adults have relatively broad front wings with two longitudinal veins and several crossveins. The wings are often banded or mottled and the ovipositor is straight or upturned. Many species are predaceous.

Common Species

Aeolothrips fasciatus (Linnaeus) (Fig. 123B). Banded thrips—1.6 mm; yellowish to dark brown; 2 dark crossbands on front wing; 6th antennal segment shorter than 1/2 length of 5th segment (♀); larvae yellow with orange abdomen; most of North America (red clover flower heads, grasses, other plants).

Family Thripidae

This family is the largest in the order and contains most of the species of economic importance. The wings are relatively narrow, the ovipositor turns downward, and the antennae are 6- to 9-segmented.

Common Species

Anaphothrips obscurus (Müller). Grass thrips—1.5 mm; yellow with dark blotches; 9-segmented antennae; antennal segments 3 and 4 brownish; pronotum without well-developed setae; last 2 abdominal segments yellow; North America (grasses—especially bluegrass—and corn).

Order Thysanoptera

Frankliniella occidentalis (Pergande). Western flower thrips—1.2-1.3 mm; clear lemon yellow or yellowish brown; 8-segmented antennae; front wing with 22-27 bristles on anterior vein and 17-20 on posterior vein; western U.S. (cotton, truck crops, grass, alfalfa, fruit, melons).

Frankliniella tritici (Fitch). Flower thrips—1.2-1.3 mm; yellow and orange; antennae: 8-segmented, 4th segment with forked sense cone and 3rd segment with middle ring; anterior angles of prothorax each with long, strong bristle; larvae white to yellow; east of Rocky Mts. (in flowers).

Heliothrips haemorrhoidalis (Bouché) (Fig. 123C). Greenhouse thrips—1.0-1.2 mm; dark brown or black; 8-segmented antennae; ridged body has a checked or netlike surface; hind coxae widely separated; yellowish white larvae; injures plants in greenhouses throughout U.S. and southern Canada, outdoors in southern U.S.

Taeniothrips inconsequens (Uzel). Pear thrips—1.2-1.3 mm; brown or black; wings grayish; 8-segmented antennae; front tarsus with apical tooth; row of small comblike setae on 8th abdominal tergite; most of North America, especially east and west coasts (injures buds and fruits of pears, plums, cherries, other fruit and shade trees).

Taeniothrips simplex (Morison). Gladiolus thrips—1.5 mm; brown or black; basal 1/3 of wings and 3rd antennal segment white; throughout U.S. (injures gladiolus and iris).

Thrips tabaci Lindeman. Onion thrips—1.0-1.2 mm; pale yellowish to dark brown; antennae 7-segmented; head wider than long; posterior longitudinal vein of front wing with 13-17 regularly spaced bristles; larvae white, green, or yellow with red eyes; throughout U.S. (injures onions, beans, tobacco, cabbage and other crops; in greenhouses; may swarm and bite).

Family Merothripidae

Members of this family have enlarged front and hind femora and two longitudinal sutures on the pronotum. *Merothrips morgani* Hood, the only common species, occurs in eastern U.S. under bark, in fungi, and in debris. It is pale brownish yellow, 1 mm long, and the antennae are 8-segmented.

Family Heterothripidae

Species of the genus *Heterothrips* occur in flowers (azalea, wild rose, jack-in-the-pulpit), buds of wild grapes, and trees (oak, buckeye, willow). *Oligothrips oreios* has short, blunt cones on the apex of the third and fourth antennal segments; this species occurs in blossoms of madrone and manzanita in California and Oregon.

GENERAL REFERENCES
Bailey, S.F. 1957. The Thrips of California, Part 1: Suborder Terebrantia. Bull. Calif. Insect Surv. 4: 143-220.
Cederholm, L. 1963. Ecological Studies on Thysanoptera. Opusc. Ent. Suppl. 22. 214 pp.
Cott, H.E. 1956. *Systematics of the Suborder Tubulifera (Thysanoptera) in California.* Univ. Calif. Press, Berkeley. 216 pp.
Lewis, T. 1973. *Thrips, Their Biology, Ecology and Economic Importance.* Academic Press, N.Y. 349 pp.
Stannard, L.J., Jr. 1957. The Phylogeny and Classification of the North American Genera of the Suborder Tubulifera (Thysanoptera). Ill. Biol. Monog. No. 25. 200 pp.
———. 1968. The Thrips, or Thysanoptera, of Illinois. Ill. Nat. Hist. Surv. Bull. 29: 215-552.

ORDER HEMIPTERA[21]
True Bugs, Cicadas, Leafhoppers, Planthoppers, Aphids, Scale Insects, and Others

The Hemiptera are a large group of highly specialized and diversified insects that are characterized by having piercing-sucking mouthparts and usually four wings. The species range in length from less than 1 mm to over 100 mm. They may be soft- or hard-bodied, winged or wingless, brightly or somberly colored, bisexual or parthenogenic, and viviparous or oviparous. Many species have glands for secreting odors, waxes, or scalelike coverings. Nymphs have a hemimetabolous type of development except for the semi-holometabolous maturation of whiteflies and male scales.

The Hemiptera are divided into the suborders Heteroptera (true bugs) and Homoptera (cicadas, aphids, etc.). Many taxonomists classify these suborders as the orders Hemiptera (=Heteroptera) and Homoptera and there are also numerous differences of opinions on the limits of families and subfamilies. This text includes a key to suborders and a key to families of each suborder. Collecting and preserving techniques are described in the introduction to each suborder.

Heteroptera species: North America, 4,500; world, 23,000. Families: North America, 44.

Homoptera species: North America, 7,000; world, 33,000. Families: North America, 31.

KEY TO SUBORDERS

1a   Beak arises ventrally toward front of head (Fig. 125); basal portion of front wings hardened and outer portion membranous (Fig. 124), or if outer portion not membranous then entire wing is sculptured and lacelike (Figs. 136A-C); wings at rest held flat over body ..... (p. 133) *Heteroptera*

1b   Beak arises ventrally toward posterior part of head (Figs. 144A, C) or appears to arise between front coxae; front wings either uniformly membranous (Figs. 147; 153) or slightly hardened (Fig. 149); wings at rest are held rooflike or vertically over body ..... (p. 156) *Homoptera*

SUBORDER HETEROPTERA[22] — True Bugs

The basal portions of a true bug's front wings are usually thickened and colored whereas the membranous overlapping tips are colored or transparent (Figs. 124; 126). These types of front wings with their different sections are called hemelytra (singular: hemelytron). The hind wings are entirely membranous and hidden under the flattened front wings. The scutellum is usually exposed and triangular. The slender 3- to 4-segmented beak or rostrum arises from the front of the head and curves backward to extend along the ventral side of the body (Fig. 125). The beak is used to feed on plant sap, seeds, fungi, fruit juices, or the blood of insects and other animals including humans. Various species are plant pests and a few transmit animal diseases, but the great majority are of no direct economic consequence to humans. The predaceous species exert some control over certain insect populations. Scent glands on the side of the thorax emit char-

---

21. Hemiptera: *hemi*, half; *ptera*, wings (refers to the partly thickened bases and membranous tips of the front wings of many families).
22. Heteroptera: *hetero*, different; *ptera*, wings.

acteristic odors probably for self-protection (Fig. 128A). Although most species are terrestrial, some common groups are aquatic or semiaquatic.

The Heteroptera are obtained by standard collecting methods such as sweeping and beating vegetation, looking underneath bark, collecting and placing debris in a Berlese funnel, or operating a light at night. An aspirator is commonly used for small individuals. Adults are pinned through the right front wing although the stink bugs and other large bugs may be pinned through the scutellum. Many collectors prefer to mount all but the large specimens on points. Nymphs are preserved in 70% alcohol.

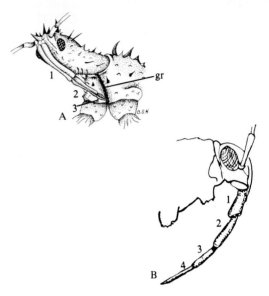

Figure 125 Beak (proboscis) of true bugs. A, 3-segmented (Reduviidae); B, 4-segmented. gr, groove.

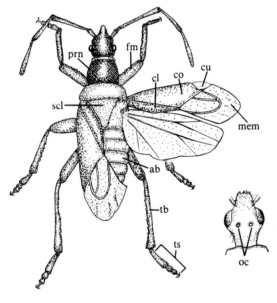

Figure 124 General structure of a true bug (Hemiptera: suborder Heteroptera). ab, abdomen; cl, clavus; co, corium; cu, cuneus; fm, femur; mem, membrane; oc, ocelli; prn, pronotum; scl, scutellum; tb, tibia; ts, tarsus.

KEY TO COMMON FAMILIES OF HETEROPTERA

Accurate use of this key requires the recognition of thickened front legs and the correct interpretation of the front wing structure. Thickened front legs (Figs. 128B, C) are often used as raptorial appendages and consist of expanded femora that bear large spines on the lower margin. The tibiae (often spined) may be pressed against the lower femoral surface. The thickened, basal portion of the front wing (hemelytron) is separated by a suture into two areas, the clavus and the corium (Figs. 124; 126). The corium may have a narrow strip along its anterior margin called the embolium (Fig. 126A) which is sometimes difficult to discern. The cuneus is a distinct area of the corium (separated by a suture) along the anterior margin and adjacent to the wing membrane (Figs. 124; 126B). The membranous area of the front wing usually contains veins.

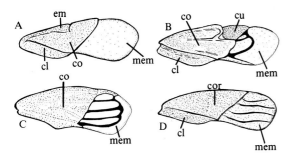

Figure 126 Front wings (hemelytra) of true bugs. A, Anthocoridae; B, Miridae; C, Saldidae; D, Lygaeidae. cl, clavus; co, corium; cu, cuneus; em, embolium; mem, membrane.

1a	Antennae shorter than the head, frequently hidden in grooves beneath the eyes and not visible from above; primarily aquatic	2
1b	Antennae as long as the head or longer, not usually hidden in grooves, visible from above; terrestrial or on water's surface.	7
2a(1a)	Ocelli present (Fig. 124); shoreline habitat . . . . . (p. 138) *Gelastocoridae*	
2b	Ocelli absent; aquatic habitat although may fly to lights	3
3a(2b)	Hind tarsi with claws	4
3b	Hind tarsi without claws.	6
4a(3a)	Membrane of front wing without veins; body generally less than 15 mm long. . . . . (p. 139) *Naucoridae*	
4b	Membrane of front wing with veins; length usually 20 mm or more.	5
5a(4b)	Abdomen with very short filaments at tip; hind legs flattened and distinctly fringed . . . . (p. 139) *Belostomatidae*	
5b	Abdomen with very long and slender breathing filaments extending from tip (Fig. 130A); hind legs long and cylindrical, not fringed. . . . . (p. 140) *Nepidae*	
6a(3b)	Body flattened dorsally and with fine transverse lines (Fig. 130B); posterior margin of head overlaps anterior margin of pronotum; front tarsi scoop-shaped (Fig. 130C). . . . . (p. 140) *Corixidae*	
6b	Body convex dorsally and dorsal side usually lighter than ventral side; posterior margin of head inserts into pronotum; front tarsi not scoop-shaped . . . . . . (p. 141) *Notonectidae*	
7a(1b)	Tarsal claws arise well before tip of last tarsal segment and often nearer base (Fig. 127A).	8
7b	Tarsal claws arise at tip of last tarsal segment (Fig. 127B)	10
8a(7a)	Middle legs attach to body about midway between front and hind legs . . . . . . (p. 143) *Veliidae* (in part)	

Figure 127 Tarsal claws. A, arise before tip; B, arise at tip; C, pads (arolia) under claws. pd, pad; tc, tarsal claw.

8b	Middle legs attach closer to hind legs than front legs 9	14a(12b)	Ocelli present 15
		14b	Ocelli absent................ 26
9a(8b)	Hind femora reach much past the tip of abdomen ...... (p. 141) *Gerridae*	15a(14a)	Middle and hind tarsi 1- or 2-segmented..................... .... (p. 143) *Hebridae* (in part)
9b	Hind femora reach little if at all beyond tip of abdomen .... (p. 143) *Veliidae* (in part)	15b	Middle and hind tarsi 3-segmented 16
10a(7b)	Head equal to or longer than thorax (Fig. 132B); body and legs extremely elongate and slender; aquatic or semiaquatic .... (p. 144) *Hydrometridae*	16a(15b)	Front wings with a cuneus or at least membrane separated from corium (Fig. 126A)... (p.144) *Anthocoridae*
		16b	Front wings without a cuneus 17
10b	Head shorter than thorax; body and legs variable in shape, if unusually slender then insect is terrestrial; chiefly terrestrial 11	17a(16b)	Beak apparently 3-segmented (Fig. 125A)................. 18
		17b	Beak 4-segmented (Fig. 125B).... 21
11a(10b)	Antennae 4-segmented 12	18a(17a)	Front legs usually thickened and/or spined (Figs. 128B, C); beak fits into midventral groove on prosternum (Fig. 125A)................. 19
11b	Antennae 5-segmented 32		
12a(11a)	Front wings with numerous tiny closed cells giving a lacelike or netlike sculptured appearance (Figs. 136A-C); usually 5 mm or less in length................... 13	18b	Front legs not thickened or spined; no midventral groove on prosternum 20
12b	Front wings not as above; size variable 14	19a(18a)	Fourth antennal segment stouter than third; front femora very thickened (Fig. 128B); front tarsi concealed in groove of tibiae .... (p. 146) *Phymatidae*
13a(12a)	Ocelli present; membrane of front wing without closed cells .... (p. 148) *Piesmatidae*		
13b	Ocelli absent; membrane of front wing with closed cells (Figs. 136A, B) .... (p. 147) *Tingidae*	19b	Fourth antennal segment not enlarged; front femora usually slightly enlarged (Fig. 128C); front tarsi

	usually not concealed in groove of tibiae .. (p. 145) *Reduviidae* (in part)
20a(18b)	Front wing membrane with 4 to 5 large closed cells (Fig. 126C); always winged .... (p. 144) *Saldidae*
20b	Front wing membrane without closed cells and sometimes absent; often wingless...... (p. 143) *Mesoveliidae*
21a(17b)	Front legs thickened; no pads under tarsal claws .... (p. 145) *Nabidae* (in part)
21b	Front legs usually not thickened; pads present under tarsal claws (Fig. 127C)................. 22
22a(21b)	Body and legs very elongate and slender (Fig. 139A); first antennal segment long and enlarged at tip, last segment spindle-shaped; femora broadest at tip.... (p. 151) *Berytidae*
22b	Body shape variable; antennae and femora not as above 23
23a(22b)	Front wing membrane with only 4 to 5 veins (Fig. 126D) .... (p. 150) *Lygaeidae* (in part)
23b	Front wing membrane with 6 or more veins..................... 24
24a(23b)	Lips around scent gland openings on side of thorax between middle and hind coxae; greatly reduced or absent; generally pale-colored and less than 10 mm long (1 common species is black with reddish lines) .... (p. 153) *Rhopalidae*
24b	Large and distinct lips around scent gland openings on side of thorax between middle and hind coxae (Fig. 128A)................. 25
25a(24b)	Distance between compound eyes greater than length of anterior margin of scutellum; width and length of head nearly equal to that of pronotum (Fig. 141C) .... (p. 153) *Alydidae*
25b	Distance between compound eyes less than length of anterior margin of scutellum; width and length of head distinctly less than that of pronotum .... (p. 152) *Coreidae*
26a(14b)	Groove in prosternum for reception of short 3-segmented beak (Fig. 125A).................... .... (p. 145) *Reduviidae* (in part)

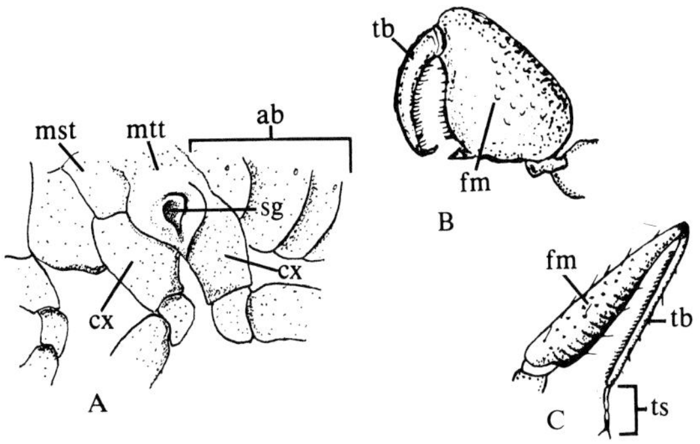

Figure 128 A, thoracic scent gland opening; B and C, raptorial front legs. ab, abdomen; cx, coxa; fm, femur; mst, mesothorax; mtt, metathorax; sg, scent gland; tb, tibia; ts, tarsus.

Order Hemiptera

26b	Groove in prosternum absent 27		32b	Color and size variable, sometimes shining black; wings normal in length; front legs not thickened; stout and shield-shaped, do not resemble ants.................. 33
27a(26b)	Wings absent; beak 3-segmented .... (p. 144) *Cimicidae*			
27b	Wings usually present and well developed; beak 4-segmented (Fig. 125B)................. 28		33a(32b)	Tibiae with hairs but without rows of stout spines; rarely shining black; generally more than 8 mm long; often shield-shaped .... (p. 154) *Pentatomidae*
28a(27b)	Cuneus present on front wings (Fig. 126B); 1 or 2 closed cells in wing membrane........ (p. 148) *Miridae*		33b	Tibiae with 2 or more rows of stout black spines (Fig. 143); usually shining black; generally 8 mm long or less; rarely shield-shaped 34
28b	Cuneus absent on front wing; wing membrane not as above 29			
29a(28b)	Tarsi 2-segmented; body very flattened........ (p. 148) *Aradidae*		34a(33b)	Scutellum very large, broadly rounded posteriorly and almost reaching tip of abdomen (Fig. 143A) .... (p. 155) *Thyreocoridae*
29b	Tarsi 3-segmented; body not especially flat 30		34b	Scutellum moderate in size, triangular, and not near tip of abdomen (Fig. 143B) .... (p. 156) *Cydnidae*
30a(29b)	Lips of scent gland opening on side of metathorax well developed (Fig. 128A)................... .... (p. 150) *Lygaeidae* (in part)			

30b Lips of scent gland opening very small or opening absent 31

Family Gelastocoridae—
Toad Bugs

These broadly oval insects with a slightly bumpy dorsum and bulging eyes superficially resemble small toads. Toad bugs occur along the sandy and muddy margins of ponds and streams where they prey by sometimes leaping on insects and mites.

31a(30b) Pronotum with a distinct margin on sides .... (p. 152) *Pyrrhocoridae*

31b Pronotum rounded on sides and without a margin .. (p. 151) *Largidae*

Common Species

Gelastocoris oculatus (Fabricius) (Fig. 129A). Toad bug—7.5-9.0 mm; usually dull brownish yellow with blackish mottling, but may vary from yellow to blackish; legs with dark and light bands, 2 claws on front tarsi; most of U.S., rare in some northern states.

32a(11b) Shining black; wings sometimes very short (Fig. 133C); front legs thickened; slender, resemble ants; 5-7 mm long ..... (p. 145) *Nabidae* (in part)

Nerthra martini Todd—7-9 mm; brown or reddish brown; wings separate and not fused; front tibia appears to have a claw at tip; AZ, CA, NV.

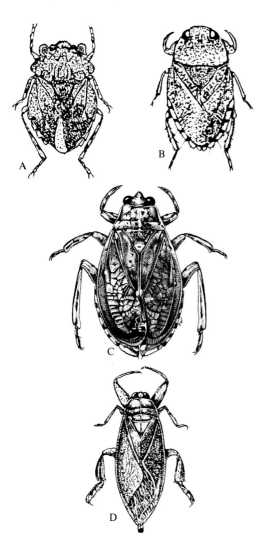

Figure 129 A, toad bug, *Gelastocoris oculatus* (Gelastocoridae); B, a creeping water bug, *Pelocoris femoratus* (Naucoridae); C and D, giant water bugs (Belostomatidae): C, *Abedus indentatus*; D, *Lethocerus griseus*.

Family Naucoridae— Creeping Water Bugs

Insects in this small family are generally 9-13 mm long, broadly oval, and yellowish brown in color. They have large raptorial front legs and lack veins in the membrane of the front wings. Naucorids occur in dense vegetation and debris of ponds and streams. Small aquatic arthropods serve as food; naucorids also can bite humans when handled. The two common genera are *Pelocoris* (anterior margin of the pronotum is nearly straight) and *Ambrysus* (anterior margin of the pronotum is very concave).

Common Species

Ambrysus mormon Montandon—9-12 mm; yellowish brown; head and pronotum with longitudinal dark markings; front wings dark brown, often mottled yellow; embolium flaplike; western U.S.

Pelocoris femoratus (Palisot de Beauvois) (Fig. 129B)—9-12 mm; greenish yellow or yellowish brown; posterior part pronotum sometimes lighter; eastern U.S. south to FL, west to Dakotas.

Family Belostomatidae— Giant Water Bugs

Members of this small family are 12-65 mm long, oval, and very flattened. The raptorial front legs are distinctly enlarged. Belostomatids are common in ponds, lakes, and quiet areas of streams. Lights often attract them at night. These bugs prey on aquatic insects, tadpoles, snails, and small fish, and will bite if handled carelessly. Females of some species deposit eggs on the backs of males.

Common Species

Three genera in U.S.: *Lethocerus* (1st beak segment wider than long and shorter than 2nd segment); *Belostoma* (1st beak segment longer than wide and = 2nd segment length, front wing membrane wider than clavus); *Abedus* (beak like *Belostoma*, front wing membrane small and usually narrower than clavus).

Abedus indentatus (Haldeman) (Fig. 129C)—27-37 mm; brown; broadly oval and distinctly tapering toward front; long hairs on abdominal sternites; CA.

Belostoma flumineum Say—22 mm; light brown, legs darker; margins on underside of abdominal segments 2-5 covered with hairs; most of U.S., chiefly eastern and central. *B. lutarium* (Stål), similar, head projects forward, eastern 1/2 U.S., chiefly southeastern. *B. bakeri* Montandon, like *B. flumineum,* West and Southwest.

Lethocerus americanus (Leidy). Giant water bug—50-60 mm; brown, front femora with 2 grooves on front margin; 3 dark brown bands on middle and hind femora; northern U.S., southern Canada. *L. uhleri* (Montandon), similar, 40-50 mm, mesosternum with tiny spines rather than hairs, eastern U.S. west to Great Plains.

Lethocerus griseus (Say) (Fig. 129D)—45-65 mm; dull yellowish brown; front femora not grooved; middle and hind femora with bands; eastern U.S. west to Great Plains.

Family Nepidae—Waterscorpions

The 12 North American species of waterscorpions are quickly recognized by the unusually long pair of appendages that protrude from the tip of the abdomen. When held together the appendages form a breathing tube. These insects cling to or crawl on aquatic vegetation in shallow areas of ponds and streams. The breathing tube is periodically thrust upward to reach the overlying air. The front legs are distinctly raptorial. Insects or small fish are seized for food. The most common species (genus *Ranatra*) are very elongate and slender. The eastern *Nepa apiculata* Uhler is the only broad North American species in the family and resembles a small belostomatid but lacks the flattened middle or hind legs.

Common Species

Ranatra fusca Palisot de Beauvois (Fig. 130A)—35-42 mm; dark reddish brown, gray brown, or dull yellow; prosternum with wide shallow groove; front femora narrowed toward middle and with a tooth near tip; most of U.S. *R. nigra* Herrich-Schäffer, similar, pale yellowish or dull brown, no tooth near tip of front femora, eastern U.S. west to MN and TX. *R. brevicollis* Montandon, 25-26 mm, brown, CA.

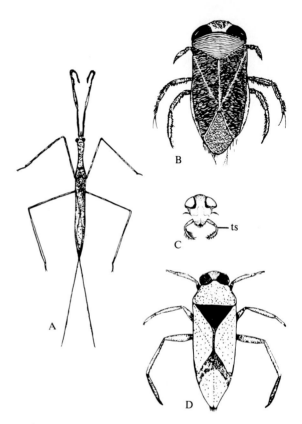

Figure 130 A, a waterscorpion, *Ranatra fusca* (Nepidae); B, a water boatman, *Hesperocorixa interrupta* (Corixidae); C, front legs of a water boatman; D, a backswimmer, *Notonecta undulata* (Notonectidae). ts, tarsus.

Family Corixidae—Water Boatmen

The abundant water boatmen are generally less than 13 mm long, the broad head and eyes overlap the front of the prothorax, and the front tarsi are curved like a scoop (Fig. 130C). Dark, fine lines often cross the body. The unusual beak is very short and broad; it is used to feed on algae, larvae, and particulate matter scraped from surfaces and strained by the front legs. Water boatmen do not bite humans. Corixids are most common along margins of lakes and ponds although a few species frequent streams and brackish water. These insects are sometimes abundant at lights. To breathe underwater an air bubble is trapped at the water's surface and the silvery bubble is held around most of the body and under the wings.

Common Species
Hesperocorixa atopodonta Hungerford—8.0-9.5 mm; reddish brown; pronotum with 9 dark crossbands; corium with yellow band on posterior margin; northern 1/2 U.S., southeastern Canada.

Hesperocorixa interrupta (Say) (Fig. 130B)—9-11 mm; eye width greater than distance between eyes; pronotum with 8-10 dark crossbands; dorsal surface finely roughened, wing membrane shiny; eastern U.S. west to AR and NB, eastern Canada.

Hesperocorixa laevigata (Uhler)—10-11 mm; dark brown; pronotum with irregular crossbands; front wings with distinct netlike pattern; most of U.S. except Gulf coast states, chiefly Far West.

Sigara alternata (Say)—5.5-7.0 mm; brown or reddish brown; pronotum with 8-9 narrow crossbands; corium darkest along membrane margin; eastern U.S. west to Great Plains, rarely in Far West.

Trichocorixa calva (Say)—5 mm; dark brown; hind margin of eye longer than distance between eyes; pronotum with 8-9 crossbands; front wings with distinct network of dark markings; eastern 1/2 U.S. southwest to AZ.

Family Notonectidae— Backswimmers
These aquatic bugs are 16 mm or less in length and swim or rest with the ventral side up. Air stores are held on the ventral side. When resting at the water surface, the body usually is dipped in the water at an angle with the head downward. Backswimmers resemble the Corixidae but the dorsal side of backswimmers is more convex, keel-shaped, and often light-colored, and the front legs are not scooplike. Most species occur in lakes and ponds but some are also found in slow-moving streams. Backswimmers are predators and feed on insects, other arthropods, and sometimes tadpoles and minnows; they also can bite people. The most conspicuous species are in the genus *Notonecta;* they have 4-segmented antennae and are relatively large.

Common Species
Buenoa margaritacea Torre Bueno—6.0-7.5 mm; color variable, dull white to dark brown; antennae 3-segmented; hemelytra with depression or pit just beyond tip of scutellum; 3rd segment of beak with lateral projections = to or < length of segment; most of U.S. except extreme northern and southern parts.

Notonecta insulata Kirby—12.5-16.0 mm; resembles large *N. undulata;* head and anterior 1/2 pronotum pale greenish yellow; posterior 1/2 pronotum and scutellum black; wings dull yellow-white or gray, black crossband near middle, sometimes traces of red or orange on corium; northeastern U.S. south to MD, west to Great Plains. *N. kirbi* Hungerford, similar, darker, more brown and dull red markings, TX to Dakotas and westward.

Notonecta irrorata Uhler—12.5-14.0 mm; dark brown; front wings dark, greatly mottled with yellow and brown spots; eastern U.S. southwest to AZ.

Notonecta undulata Say (Fig. 130D)—10-12 mm; dull yellowish white usually with black markings; scutellum usually blackish; greatest distance between eyes > length of head; most of U.S. (most common species east of Rocky Mts. except in South). *N. indica* Linnaeus, similar, often black crossband on wings, greatest distance between eyes < length of head, southern 1/3 U.S.

Family Gerridae—Water Striders
Water striders are familiar, long-legged insects that dart over the surface of water in ponds and streams. The dark, slender body is typically clothed with waterproof hairs and these hairs

on the underside of the tarsi enable the insect to walk on the water's surface. Some species have both winged and wingless forms. Species in the genus *Halobates* occur on the surface of the subtropical areas of oceans and are the only genuine oceanic insects. Water striders feed on small, living or dead insects on the water's surface.

Common Species

Large common spp. are in genus *Gerris* (eyes with concave inner margin; pronotum dull and clothed with short, velvety hairs). Most of common smaller spp. are in genus *Metrobates* (inner margin of eyes round; 1st antennal segment longer than other 3 combined) or *Trepobates* (Fig. 131A) (inner margin of eyes round; 1st antennal segment shorter than other 3 combined, 3rd segment with fine hairs but not bristles).

Gerris marginatus Say—8-9 mm; dark brown; anterior corners of pronotum lack a pale stripe; throughout U.S.

Gerris remigis Say (Fig. 131B)—11.5-16.0 mm; dark brown to blackish; pronotum with yellowish or orange middorsal stripe; winged or wingless; eastern (streams) and western (ponds, lakes) North America (most common sp.).

Limnogonus hesione (Kirkaldy)—7-8 mm; shiny black; light area at base of head and anterior part of pronotum; lateral and posterior margins of pronotum pale; light area at base of head and anterior part of pronotum; eastern U.S. west to NB and TX.

Metrobates hesperius Uhler—3-5 mm; broad body; velvety black; tip of head orange; pronotum very small; mesonotum with 3 dull gray stripes; bases of legs yellow; eastern U.S. southwest to KS.

Trepobates pictus (Herrich-Schäffer)—3.5-5.0 mm; black and yellow stripes and patches; head yellow with middorsal black patch; eastern U.S. west to IL.

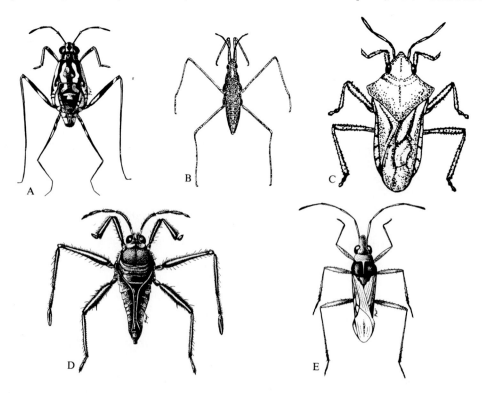

Figure 131 A and B, water striders (Gerridae): A, *Trepobates.* sp.; B, *Gerris remigis*. C and D, broad-shouldered or smaller water striders (Veliidae): C, *Microvelia pulchella;* D, *Rhagovelia obesa*. E, a water treader (Mesoveliidae), *Mesovelia mulsanti*.

Family Veliidae— Broadshouldered Water Striders or Smaller Water Striders

The Veliidae are small (1.5-5.5 mm long), usually wingless, dark-colored bugs that may have silvery markings. The thoracic area of many species is broad. These bugs live on the surface of water (and occasionally shorelines) feeding on small insects. Species in the genus *Rhagovelia*, sometimes called riffle bugs, occur in groups in the riffle areas of streams. *Velia* and *Microvelia* species frequent ponds and placid areas of streams.

Common Species

Microvelia pulchella Westwood (Fig. 131C)—1.5-2.0 mm; velvety dark brown; front of pronotum, middle of abdomen and base of coxae yellowish; often wingless; most of U.S. (associated with duckweed on ponds).

Rhagovelia obesa Uhler (Fig. 131D)—3-4 mm; black; basal 1/3 of 1st antennal segment cream-colored; often wingless; pronotum with 2 orange spots near anterior margin; mesonotum exposed (wingless ♂) or covered by posterior margin of pronotum (winged ♂); sides of abdomen wrap up and over abdomen to nearly meet at midline (♀); eastern U.S. to MN, southeastern Canada. *R. distincta* Champion, similar, > 4 mm, western U.S.

Family Mesoveliidae— Water Treaders

These small insects (2-4 mm long) occur on the surface of ponds and lakes with dense vegetation. The clavus of the front wing is membranous and the wing membrane is veinless. Ocelli occur in the winged forms. Mesoveliids prey on small insects.

Common Species

Mesovelia mulsanti White (Fig. 131E)—3.5-4.0 mm; green or yellowish green; often wingless; winged forms: posterior part pronotum dark brown with pale yellow middorsal stripe, front wings whitish and often with brown areas on sides and at tip; most of U.S.

Family Hebridae— Velvet Water Bugs

These pubescent bugs are less than 3 mm long and live on the surface of shallow ponds near shorelines filled with dense vegetation. A few species live on damp shorelines. Hebrids feed on small insects.

Common Species

Merragata hebroides White (Fig. 132A)—1.5-2.0 mm; black; head, pronotum and antennae reddish brown; antennae 4-segmented; legs yellowish; large white areas on front wing including 4 white spots on the large membrane; most of U.S.

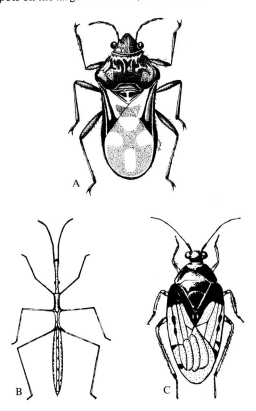

Figure 132 A, a velvet water bug (Hebridae), *Merragata hebroides;* B, a water measurer or marsh treader (Hydrometridae), *Hydrometra martini;* C, a shore bug (Saldidae), *Saldula pallipes.*

Family Hydrometridae— Water Measurers or Marsh Treaders

Hydrometrids are very slender insects with a long head and threadlike legs. Wings are usually absent. This group resembles small walkingsticks (Phasmatidae). Water measurers live on the surface of ponds with dense vegetation where they feed on small insects and crustaceans. All seven species in North America are in the genus *Hydrometra*.

Common Species

Hydrometra martini Kirkaldy (Fig. 132B)—8-11 mm; brown, often with blue tinges; wings whitish, veins dark; eastern North America, occasionally western states.

Family Saldidae—Shore Bugs

Shore bugs are oval, flattened bugs that are commonly brown and white or black and white. The front wing membrane has four or five long, closed cells. Saldids occur on sandy shorelines, salt marshes, and mud flats, and on rocks in springs, streams, and bogs. They are predators of other insects.

Common Species

Ioscytus politus (Uhler)—6-7 mm; blackish; 1st 2 antennal segments and corium of front wing red; rest of antennae and clavus black; western U.S.

Pentacora ligata (Say)—5-6 mm; black with yellow spots dorsally; legs yellowish with black bands; eastern 1/2 U.S. (on rocks in streams).

Saldula pallipes (Fabricius) (Fig. 132C)—5-6 mm; head, pronotum, scutellum and basal 1/4 front wing black; rest of wing yellowish or pale white; some forms with more black on wings; most of U.S.

Family Anthocoridae— Minute Pirate Bugs

Anthocorids are 2-5 mm long, flattened, and often black with white markings. The front wing has a distinct cuneus and the membrane has few or no veins and lacks closed cells. Most of these bugs are predators of insects. Many occur on flowers (they are sometimes called flower bugs) and others live under bark or frequent needles and leaves of trees, animal nests, and other similar habitats.

Common Species

Lyctocoris campestris (Fabricius)—3-4 mm; dull red or reddish brown; front wing membrane clouded brown, 1 distinct vein; most of North America (in granaries, houses, straw, animal nests).

Orius insidiosus (Say) (Fig. 133A)—2 mm; dark brown or black; wings pale yellow except for dark brown cuneus and area around it; eastern 1/2 North America. *O. tristicolor* (White), minute pirate bug, similar, clavus entirely dark, western U.S.

Family Cimicidae—Bed Bugs

Bed bugs are 6 mm or less in length, flattened, and nearly wingless. They are often reddish brown in color. These insects are bloodsucking external parasites of birds and mammals. The bed bug, *Cimex lectularius* Linnaeus (Fig. 133B), feeds primarily on humans and may become a pest in sleeping areas. It is nocturnal but does not necessarily feed nightly. During the day it hides in cracks in walls, mattresses, etc. Numerous species attack bats and birds. Nests of cliff swallows, purple martins, woodpeckers, raptors, chickens, and other birds contain various kinds of bed bugs.

Family Nabidae—Damsel Bugs

Damsel bugs are 3-12 mm long and usually dull brown, slender insects with a 4-segmented beak. There is a series of elongated cells around the margin of the front wing membrane. The wings of some shiny black species may be very small and these nabids resemble ants. Nabids are predators of other insects.

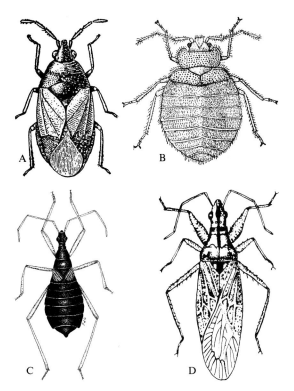

Figure 133 A, a minute pirate bug, *Orius insidiosus* (Anthocoridae); B, bed bug, *Cimex lectularius* (Cimicidae); C and D, damsel bugs (Nabidae): C, *Nabicula subcoleoptrata;* D, common damsel bug, *Nabis americoferus.*

Common Species

Nabicula subcoleoptrata (Kirby) (Fig. 133C)—7-9 mm; shiny black; minute wings, usually fully winged form rare; resembles ant; antennae, beak, legs, and margins of abdomen yellow; eastern 1/2 U.S. south to KS and NJ, southern Canada.

Nabis americoferus Carayon (Fig. 133D). Common damsel bug—6-9 mm; dull gray or gray-yellow; head and pronotum with dark stripe; throughout U.S. *N. alternatus* Parshley, western damsel bug, similar, wings speckled brown, reddish brown spots on hind tibiae, west of Mississippi R.

Family Reduviidae— Assassin Bugs

Members of this family are small to large bugs that range from stout to extremely slender and elongate. The important identification character is the longitudinal groove on the prosternum that receives the 3-segmented beak (Fig. 125A). The first segment of the beak forms a rounded, downward curve helping to give the head an elongated appearance. The abdomen of some species is wider than the wings and flattened dorsally. Assassin bugs are predators of other insects and many will bite humans. Several southwestern species in the genus *Triatoma* transmit trypanosome (protozoan) diseases to rodents and man (Chagas disease in humans).

Common Species

Apiomerus crassipes (Fabricius) (Fig. 134A)—12-19 mm; black; pronotum, scutellum, and abdomen broad and with dark red fringe; head, thorax, and legs with dense hairs; most of North America (especially West).

Arilus cristatus (Linnaeus). Wheel bug—28-36 mm; dark brown with thick grayish pubescence; pronotum with semi-circular crest containing 8-12 stout tubercles; NY to IL and south to FL.

Emesaya brevipennis (Say) (Fig. 134B)—33-37 mm; extremely slender and long-legged; grayish or silvery covering of fine hairs; front tarsi 3-segmented; most of U.S. (in barns, old buildings).

Melanolestes picipes (Herrich-Schäffer)—15-20 mm; black; apical 1/4 of tibiae irregularly dilated; front wings often very short (♀); eastern 1/2 U.S., southwest to CA. *M. abdominalis* (Herrich-Schäffer), similar, red abdomen.

Rasahus biguttatus (Say)—16-20 mm; dark reddish brown; most of wing dark yellow with large yellow spot in middle of black membrane; southern U.S. west to CA, sometimes Midwest. *R. thoracicus* Stål, similar, amber, amber spot in middle of black wing membrane, Far West.

Reduvius personatus (Linnaeus) (Fig. 134C). Masked hunter—17-23 mm; dark brown to blackish; prominent vertical hairs; eastern U.S. west to KS (often at lights, in houses; immatures thickly covered with dust).

Repipta taurus (Fabricius) (Fig. 134D)—11-13 mm; slender; orange-red or yellow; 2 spines on head; posterior 1/2 pronotum with 4 longitudinal stripes, 2 long black spines; PA to FL, west to TX.

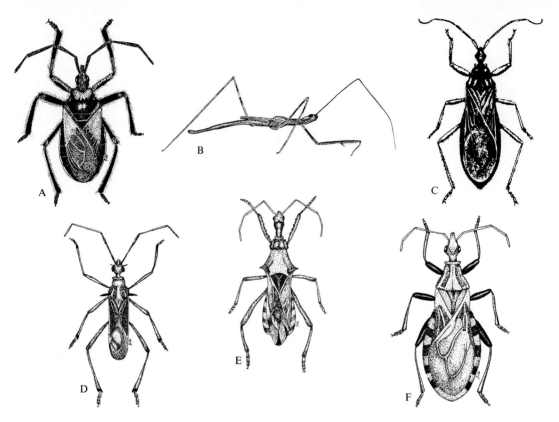

Figure 134 Assassin bugs (Reduviidae). A, *Apiomerus crassipes;* B, *Emesaya brevipennis;* C, masked hunter, *Reduvius personatus;* D, *Repipta taurus;* E, spined assassin bug, *Sinea diadema;* F, bloodsucking conenose, *Triatoma sanguisuga.*

Sinea diadema (Fabricius) (Fig. 134E). Spined assassin bug—12-14 mm; dark brown to dull reddish brown; pronotum with prominent sharp spines; each abdominal segment with a lateral spot; most of U.S. *S. spinipes* (Fabricius), similar, pronotum with tubercles but not spines, east of Rocky Mts.

Triatoma sanguisuga (LeConte) (Fig. 134F). Bloodsucking conenose—16-21 mm; mixed reddish and dark brown to blackish; corium with reddish yellow area near base and apex; lateral areas of broad abdomen with alternating red and black patches; FL to TX, north to NJ, IL, KS (often indoors; severe bite). *T. protracta* (Uhler), western bloodsucking conenose, smaller, black, UT, AZ, CA.

Family Phymatidae— Ambush Bugs

These stout-bodied bugs are 13 mm or less in length, the raptorial front femora are greatly enlarged, and the abdomen distinctly widens posteriorly. Ambush bugs are commonly yellow or greenish yellow with brown or black markings and individuals often wait on similarly colored flowers (e.g., goldenrod) for an insect to land. The bug then quickly seizes the prey (typically a fly, bee, or wasp) and sucks out its blood. Phymatids do not bite humans.

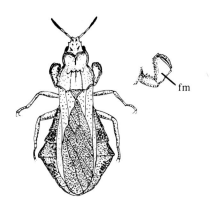

Figure 135 An ambush bug, *Phymata pennsylvanica* (Phymatidae). fm, femur.

Common Species

Phymata fasciata (Gray)—9-12 mm; pale yellow to greenish yellow; scattered dark markings; wide brown band across middle of abdomen; most of U.S.

Phymata pennsylvanica Handlirsch (Fig. 135)— 8.5-9.5 mm; yellow with brown or black areas, sometimes greenish when alive; front wing membrane brown; 4th antennal segment longer than 2nd and 3rd combined (♂); 4th abdominal segment abruptly expanded (♂); eastern U.S. *P. americana* Melin, similar, 4th antennal segment shorter than 2nd and 3rd segments combined (♂), abdomen evenly expanded to 4th segment (♂), throughout U.S.

Family Tingidae—Lace Bugs

Lace bugs are small (usually less than 6 mm long) insects with the front wings and expanded thorax sculptured into a lacelike pattern of cells. Most adults are grayish white and the nymphs are blackish. Lace bugs are plant feeders and often found on leaves of trees and shrubs; they tend to be host specific. One species is used in control of a weed in certain parts of the world. The largest and most common genus in North America is *Corythucha* which is characterized by a prominent pronotal hood that extends over the head and greatly expanded, lacy, pronotal margins that bear small, sharp spines (Fig. 136A).

Common Species

Corythucha ciliata (Say) (Fig. 136A). Sycamore lace bug—4 mm; uniformly milky white except for dark mark on thorax and center of each wing; east of Rocky Mts. (on sycamore, ash, hickory, mulberry).

Corythucha marmorata (Uhler). Chrysanthemum lace bug—3.0-3.5 mm; opaque white; front wings with many faint brown spots forming 2 crossbands near tips; most of U.S., southern Canada (on chrysanthemum, aster, goldenrod, other composits).

Corythucha pruni Osborn and Drake—4 mm; whitish with brownish black crossbar near front wing apex; pronotal hood low; eastern 1/2 U.S., UT, OR (on wild cherry).

Melanorhopala clavata (Stål) (Fig. 136B)—4-7 mm; brown; enlarged tip of 3rd and all of 4th antennal segment dark brown, antenna nearly or = to length of body; primarily northern U.S. west to WY, southern Canada (in weedy fields, on goldenrod).

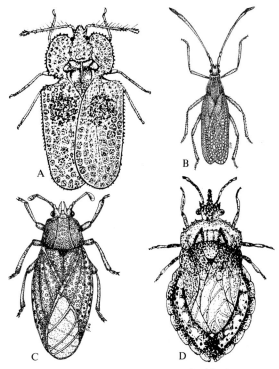

Figure 136 A and B, lace bugs (Tingidae): A, sycamore lace bug, *Corythucha ciliata;* B, *Melanorhopala clavata.* C, a piesmatid or ash-gray leaf bug, *Piesma cinerea* (Piesmatidae); D, a flat bug, *Aradus acutus* (Aradidae).

Family Piesmatidae— Piesmatids or Ash-Gray Leaf Bugs

Members of this family are small (about 2-4 mm long) and the cell network in the front wings resembles that of certain lace bugs (Tingidae). The pronotum has five longitudinal ridges. These insects feed on leaves of trees and weeds. All 7 species are in the genus *Piesma*.

Common Species

Piesma cinerea (Say) (Fig. 136C)—2.5-3.5 mm; dull gray or yellow; 3rd antennal segment length = to 1st and 2nd combined; most of U.S. (on pigweed, other weeds).

Family Aradidae—Flat Bugs

These insects are generally 3-11 mm long, very flat, and brown to black in color. There are no ocelli and the wings do not cover the entire abdomen. Flat bugs occur on or under loose bark of dead trees and feed on the liquids of fungi and decaying wood (rarely living wood).

Common Species

Aradus acutus Say (Fig. 136D)—7-10 mm; blackish brown; 2nd antennal segment clubbed; white spot on side of each abdominal segment; most of U.S. (on oak).

Aradus robustus Uhler—5-7 mm; brown to gray; head and pronotum same length; 2nd antennal segment thick and covered with short stout spines; eastern U.S. (on red and black oak).

Neuroctenus simplex (Uhler)—4.5-5.0 mm; dark reddish brown or blackish; tip of 4th antennal segment yellow; tarsi yellow; front wing membrane whitish; narrow transverse ridge behind anterior margin of abdominal segments 4-6; eastern U.S. (on oak, beech, weeds).

Family Miridae—Plant Bugs

These very common bugs are generally less than 10 mm long and the front wing has a cuneus and one or two closed cells in the membrane (Fig. 126B). Ocelli are absent. Most species feed on plant juices and many are serious pests (e.g., lygus bugs in the genus *Lygus*). Some species are predaceous on small insects. Fleahoppers (genus *Halticus*) are jumping mirids with enlarged hind femora and short, evenly thickened front wings (Fig. 137A). Numerous wingless species are remarkable mimics of ants. This family contains about 1,700 species in North America and is the largest in the suborder Heteroptera. Keys to genera are given in Slater and Baranowski (1978).

Common Species

Adelphocoris rapidus (Say). Rapid plant bug—7-8 mm; dark brown; 2 dark spots on posterior 1/2 of pronotum; sides of front wing dull yellow-white; eastern U.S. *A. lineolatus* (Goeze), alfalfa plant bug, 8-9 mm, green or greenish yellow, faint clouded areas on front wings, on alfalfa and sweet clover, throughout U.S.

Halticus bractatus (Say) (Fig. 137A). Garden fleahopper—1.5-2.5 mm; black; 1st antennal segment and part of 2nd and 3rd often pale; front wings of ♀ short, thickened, with or without membrane, with silvery flat hairs; front wings of ♂ normal length; hind legs enlarged; jumps; eastern 1/2 U.S. (often on white clover in lawns).

Ilnacora malina (Uhler)—5.5 mm; black; base of pronotum and front wings deep green; eastern and central U.S. (on goldenrod in damp areas).

Leptopterna dolabrata (Linnaeus) (Fig. 137B). Meadow plant bug—7-9 mm; pale greenish white with 2 broad longitudinal bands on pronotum (♂, ♀) or reddish brown to orangish yellow (♂); wings often short (♀); northern 1/2 U.S.

Lygus lineolaris (Palisot de Beauvois) (Fig. 137C). Tarnished plant bug—5-6 mm; yellowish brown with some dull reddish, brown, and black mottling; Y-shaped yellow marking on scutellum; eastern 1/2 U.S.

Neurocolpus nubilus Say (Fig. 137D). Clouded plant bug—6.5-7.0 mm; bright reddish tan to yellow brown; 1st antennal segment with flattened hairs, 2nd antennal segment slightly swollen at base; front wings with tiny yellow dots; eastern 1/2 North America.

Plagiognathus politus Uhler—3.5-4.0 mm; black with whitish hairs; tibiae and tips of femora pale; later in summer not as dark, yellowish shade on thorax; eastern U.S. (on ragweed, goldenrod, other weeds).

Poecilocapsus lineatus (Fabricius) (Fig. 137E). Fourlined plant bug—7.0-7.5 mm; green or yellowish; head may be yellowish red; 4 broad, black longitudinal stripes on pronotum and front wings; eastern 1/2 U.S., southern Canada.

Pseudatomoscelis seriatus (Reuter). Cotton fleahopper—3-4 mm; pale yellowish green; covered with brownish black specks and minute black hairs; South and Southwest (on cotton, other plants).

Slaterocoris stygicus (Say)—4.5 mm; shiny black; broadly oval; eastern 1/2 North America, MT, CA (on goldenrod).

Stenodema vicinum (Provancher) (Fig. 137F)—7-8 mm; dull yellow or straw color with greenish tones; antennae reddish, 1st antennal segment thick and hairy; pronotum with 2-3 dark longitudinal stripes; most of U.S.

Stenotus binotatus (Fabricius)—6 mm; yellowish orange; pronotum and front wings with 2 broad black stripes sometimes changing to spots on posterior 1/2 pronotum; northern and central U.S.

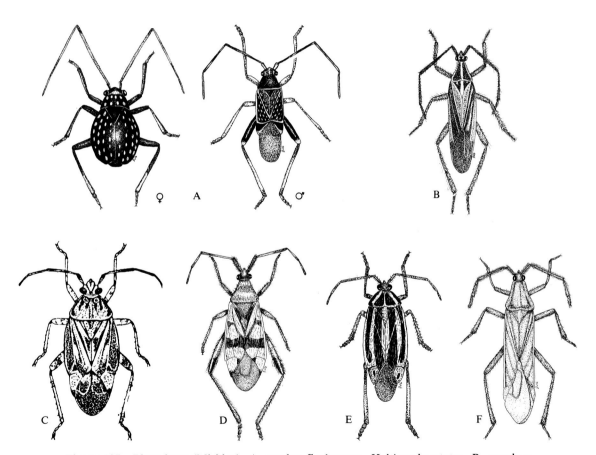

Figure 137 Plant bugs (Miridae). A, garden fleahopper, *Halticus bractatus* B, meadow plant bug, *Leptopterna dolabrata;* C, tarnished plant bug, *Lygus lineolaris;* D, clouded plant bug, *Neurocolpus nubilus;* E, fourlined plant bug, *Poecilocapsus lineatus;* F, *Stenodema vicinum.*

Family Lygaeidae—Seed Bugs

Seed bugs are brownish or sometimes brightly colored insects that vary from 2-18 mm in length. They are recognized by the 4-segmented beak and antennae, the presence of ocelli, and four or five distinct veins in the membrane of the front wing. Some species resemble the Miridae but the latter have a cuneus and no ocelli. Lygaeids may resemble certain species of Alydidae, Coreidae, or Rhopalidae, but the latter species have more than five veins in the front wing membrane. Most lygaeids feed on mature seeds, typically injecting saliva and sucking out the dissolved contents of the seed. Some species are sap feeders (e.g., chinch bug) or predators (e.g., *Geocoris* species prey on other insects). Various species are agricultural pests. Seed bugs, with over 250 species in North America, are the second largest family in the suborder Heteroptera.

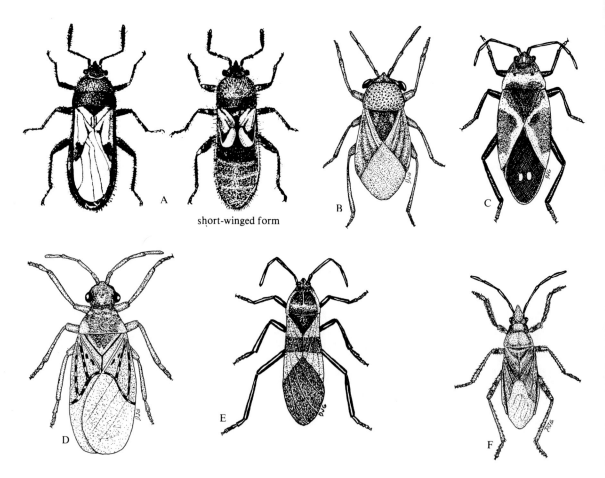

Figure 138 Seed bugs (Lygaeidae). A, chinch bug, *Blissus leucopterus;* B, *Geocoris punctipes;* C, small milkweed bug, *Lygaeus kalmii;* D, *Nysius niger;* E, large milkweed bug, *Oncopeltus fasciatus;* F, *Ortholomus scolopax.*

Common Species

Blissus leucopterus (Say) (Fig. 138A). Chinch bug—3-4 mm; dull black and white with thick gray pubescence; winged and wingless forms; front wings white with large black spot on corium; Midwest (in corn and grain fields).

Geocoris punctipes (Say) (Fig. 138B)—3-4 mm; grayish or dull yellowish; top of head smooth and shiny, with a middorsal grove; scutellum with 2 pale stripes; southern 2/3 U.S. *G. bullatus* (Say), large bigeyed bug, similar, top of head roughened, most of U.S. *G. pallens* Stål, western bigeyed bug, pale yellowish green, small black spots on head and thorax, scutellum black and without a middorsal ridge, western 1/2 North America.

Ligyrocoris diffusus Uhler—5-7 mm; head and front of pronotum gray or black; front wings dull reddish brown; posterior part of pronotum with 4 faint darker stripes; pale spot near inner corner of corium; most of U.S. except Southeast (on composit flowers).

Lygaeus kalmii Stål (Fig. 138C). Small milkweed bug—11-13 mm; red and black; red spot at base of head; irregular red "X" formed by wings; front wing membrane with 2 white spots or none (subspecies in eastern 1/2 U.S.) or 1 large white spot (subspecies in western 1/2 U.S.); on milkweed.

Melanopleurus belfragei Stål—10 mm; head, pronotum, scutellum and front wing membrane black; clavus and corium orange-red; TX to CA.

Myodocha serripes Olivier—8.0-9.5 mm; head shiny black, extended posteriorly into a distinct long neck; pronotum gray; wings dark brown, margin pale; legs pale yellow, dark bands on femora; eastern 1/2 U.S., southeastern Canada.

Neacoryphus bicrucis (Say)—7.0-9.5 mm; body black above; posterior part of pronotum and corium bright red; southern U.S.

Nysius niger Baker (Fig. 138D)—3-4 mm; dull yellow-gray; 2 irregular longitudinal rows of black spots on corium; membrane clear; most of North America.

Oncopeltus fasciatus (Dallas) (Fig. 138E). Large milkweed bug—10-18 mm; dark brown to black; top of head, sides of pronotum, and 2 broad crossbands on wings are orange; east of Rocky Mts. (on milkweed).

Ortholomus scolopax (Say) (Fig. 138F)—5-6 mm; dull gray-brown; tip of corium dull reddish; short silvery or gray hairs on body; most of U.S., southern Canada.

Paromius longulus (Dallas)—6-7 mm; pale tan; head projecting prominently; southern U.S.

Family Berytidae—Stilt Bugs

These slender bugs are 3-10 mm long and have unusually long legs and antennae. The first antennal segment is much longer than the others. Stilt bugs resemble the shape of Hydrometridae and certain Reduviidae; however, stilt bugs do not occur on water or have thickened front legs, and the beak is 4-segmented. Berytids feed on plants (a few may be partly predaceous on other insects) and occur on dense vegetation or on the ground.

Common Species

Jalysus spinosus (Say) (Fig. 139A). Spined stilt bug—6-9 mm; dull reddish brown or yellowish brown; scutellum with angled spine; eastern 1/2 U.S.

Family Largidae—Largid Bugs

These medium- to large-sized bugs are often red and black. They have a 4-segmented beak, no ocelli, and seven or eight branched veins plus two large cells in the front wing membrane. Some have short wings and resemble ants. Largids are plant feeders and generally southern in distribution.

Common Species

Largus succinctus (Linnaeus) (Fig. 139B)—13-17 mm; black or dark yellowish brown with short gray hairs; orange-red or orange-yellow pronotal, wing, and abdominal margins, and also base of femora; southern U.S. west to CO and AZ, north to NY.

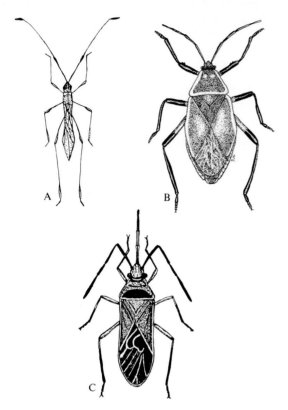

femora red; posterior area of pronotum with dark brown patch; front wings dark brown with dull yellow margins; FL and southeastern U.S.

Family Coreidae—Coreid Bugs

Members of this family are 7-40 mm long, dark-colored, and elongate but thick-bodied insects. The front wing membrane contains many veins. Scent gland openings are present on the thorax between the middle and hind coxae (Fig. 128A). Because some species have dilated, leaflike areas on the antennae and/or the hind legs, the family name leaffooted bugs has been used. They are also known as the squash bug family. Most species feed on plants and some are agricultural pests.

Figure 139 A, spined stilt bug, *Jalysus spinosus* (Berytidae); B, a largid bug, *Largus succinctus* (Largidae); C, cotton stainer, *Dysdercus suturellus* (Pyrrhocoridae).

Family Pyrrhocoridae— Red Bugs and Stainers

Pyrrhocorids are medium- to large-sized bugs that are often marked with red, yellow, brown, and white. Many pyrrhocorids resemble the largid bugs but the former's body lacks hairs, the beak extends to the abdomen, and the sixth sternite of the female is not split. All seven North American species are in the genus *Dysdercus* and occur in the southernmost states.

Common Species

Dysdercus suturellus (Herrich-Schäffer) (Fig. 139C). Cotton stainer—11-17 mm; head, base of 1st antennal segment, front of pronotum, scutellum and

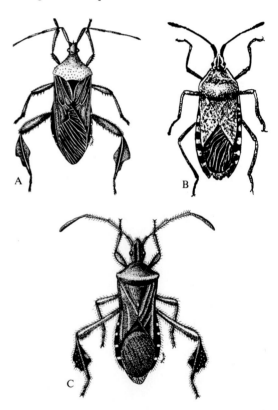

Figure 140 Coreid bugs (Coreidae). A, *Acanthocephala femorata;* B, squash bug, *Anasa tristis;* C, *Leptoglossus oppositus.*

Common Species

Acanthocephala femorata (Fabricius) (Fig. 140A)—25-28 mm; reddish brown or yellowish brown; thick yellowish hairs; small tubercles on pronotum; hind femora greatly thickened and curved (♂), leaflike expansions of hind tibiae nearly reach tip of tibiae; southern U.S.

Acanthocephala terminalis (Dallas)—20-22 mm; brown; 1st 3 antennal segments dark brown or dark reddish brown in contrast to orange-yellow 4th segment; leaflike expansions of hind tibiae reach about 2/3 length of tibiae; eastern 1/2 U.S.

Anasa tristis (De Geer) (Fig. 140B). Squash bug—13-18 mm; dull brownish yellow; body with dark gray punctures; head with 3 yellow stripes; sides of each abdominal segment partly yellow; most of North America (on squash, cucumber, pumpkin).

Euthochtha galeator (Fabricius)—14.5-17.0 mm; dull brown; stout-bodied like *A. tristis;* 4th antennal segment dark brown; antennalike tubercles on body with a spine on their outer sides; hind femora with large teeth or spines ventrally; eastern U.S.

Leptoglossus oppositus (Say) (Fig. 140C)—18-20 mm; brown without distinct white bands on wings; leaflike expansion on hind tibiae has indentation on outer margin; eastern U.S. northwest to MN, southwest to AZ. *L. occidentalis* Heidemann, similar, hind tibia expansion elongate and with indentations, western 2/3 U.S.

Leptoglossus phyllopus (Linnaeus). Leaffooted bug—18-20 mm; brown with distinct white crossband on front wing; leaflike expansion on hind tibiae; southeastern U.S. north to NY and MO. *L. clypealis* Heidemann, similar, spine projects upward at tip of head, western 2/3 U.S.

Family Rhopalidae—Rhopalid Bugs

Rhopalids resemble small species of Coreidae but either have very small lips around the scent gland openings or lack openings. The ocelli are distinctly raised. Rhopalids are sometimes called scentless plant bugs and are classified as Corizidae in some texts.

Common Species

Arhyssus lateralis (Say)—5-7 mm; color variable; reddish brown to dull yellow; head and pronotum red, pale yellowish brown, or combination of both; abdomen commonly yellow; most of U.S., southern Canada.

Leptocoris trivittatus (Say) (Fig. 141A). Boxelder bug—11-14 mm; black; dark reddish pink stripe down center of pronotum and edges of pronotum and corium; most of U.S. (on boxelders, congregates in masses in fall).

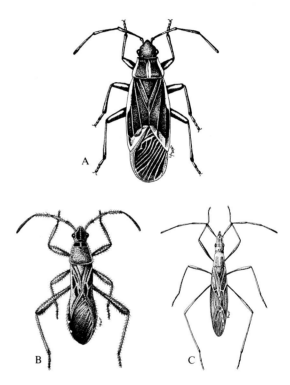

Figure 141 A, boxelder bug, *Leptocoris trivittatus* (Rhopalidae); B and C, alydid bugs (Alydidae): B, *Alydus eurinus;* C, *Protenor belfragei.*

Family Alydidae—Alydid Bugs

These yellowish brown or black bugs resemble the Coreidae but have broad, prominent heads and more slender bodies. They often produce strong odors from thoracic scent glands. Aly-

Order Hemiptera 153

dids are sometimes called broadheaded bugs and are classified as Coriscidae in some texts. Nymphs of some species (genus *Alydus*) resemble ants; a few adults mimic wasps. Alydids are plant feeders and occur on foliage and flowers of weeds including grasses.

Common Species

Alydus eurinus (Say) (Fig. 141B)—11-15 mm; blackish; sides of each abdominal segment with small yellow spot; throughout U.S. *A. pilosulus* Herrich-Schäffer, similar, light brown, pronotal margins white, anterior corners of pronotum sharp, throughout U.S. but chiefly southern.

Protenor belfragei Haglund (Fig. 141C)—12-15 mm; shiny brown; slender and parallel sided; antennae and often beak and legs yellow-red; body with dark punctures; northern U.S. from MD to Rocky Mts.

Family Pentatomidae—Stink Bugs

Stink bugs are common, broad, shield-shaped bugs with 5-segmented antennae. The large scutellum is distinct and in the subfamily Scutellerinae (shield bugs) it extends to the tip of the abdomen (Fig. 142D). The shield bugs are considered a separate family, Scutelleridae, in some texts. Most stink bugs are plant feeders, some are predators, and others feed on both plants and animals. Pentatomids produce odors from thoracic glands for self-defense.

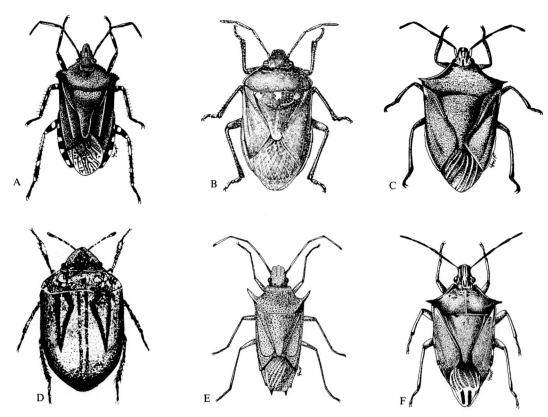

Figure 142 Stink bugs (Pentatomidae). A, rough stink bug, *Brochymena quadripustulata;* B, Say stink bug, *Chlorochroa sayi;* C, onespot stink bug, *Euschistus variolarius;* D, *Homaemus bijugis;* E, *Oebalus pugnax;* F, spined soldier bug, *Podisus maculiventris.*

Common Species

Acrosternum hilare (Say). Green stink bug—12.0-17.5 mm; bright green; elongate body; head and pronotum with narrow yellow, orange or reddish border; lateral pronotal margins narrow anteriorly in straight line rather than curving; most of U.S.

Brochymena quadripustulata (Fabricius) (Fig. 142A). Rough stink bug—14-18 mm; grayish or dark reddish brown; body marked with dark and light areas and resembles tree bark; anterior corners of pronotum bluntly pointed; throughout U.S. *B. sulcata* Van Duzee, similar, southwestern U.S., CA (in orchards).

Chlorochroa sayi Stål (Fig. 142B)—Say stink bug—15-17 mm; bright green; white spots or tiny raised points on dorsum, 3 white or pale orange spots across base of scutellum; western 2/3 U.S. *C. ligata* (Say), Conchuela, similar, often dark olive green, lacks 3 white or pale orange spots at base of scutellum.

Cosmopepla bimaculata (Thomas)—5-7 mm; shiny black; yellowish or bright red stripes along lateral and anterior margins of pronotum and middorsum of pronotum (anterior 1/2); tip of scutellum with red dot on each side; most of U.S.

Euschistus variolarius (Palisot de Beauvois) (Fig. 142C). Onespot stink bug—11-14 mm; hind tibiae with distinct groove for complete length; edges of anterior abdominal sternities without spots; large black spot on genital capsule at tip of abdomen (♂); chiefly northern 1/2 U.S. *E. servus* (Say), brown stink bug, similar, abdomen with black spots on ventral side near margins but lacks row of midventral black spots, most of U.S., especially South.

Holcostethus limbolarius Stål—7-8 mm; dark grayish yellow; dense dark punctures; lateral margin of pronotum, basal 1/2 of lateral margin of corium, and tip of scutellum are whitish; lateral margin of black abdomen uniformly pale yellow; most of U.S. *H. abbreviatus* Uhler, similar, alternating light and dark patches on lateral margins of abdomen, western 1/2 U.S.

Homaemus bijugis Uhler (Fig. 142D)—6-8 mm; convex scutellum nearly reaches tip of abdomen; light to dark yellowish tan; head black, lateral margins reddish tan; black punctures form diverging and parallel stripes on dorsum; west of Mississippi R., IL. *H. parvulus* Germar, similar, 4-6 mm, paler yellow, southern 1/2 U.S.

Mormidea lugens (Fabricius)—6-7 mm; dull bronze; scutellum with whitish margins; east of Rocky Mts.

Murgantia histrionica (Hahn). Harlequin bug—10-11 mm; variegated orange and black markings; chiefly southern U.S. (on cabbage, other crucifers).

Oebalus pugnax (Fabricius) (Fig. 142E)—9-11 mm; dull yellow; elongate; distinct pronotal spines project forward; east of Rocky Mts., chiefly southern (on rice, wheat, sorghum, grasses).

Podisus maculiventris (Say) (Fig. 142F). Spined soldier bug—6-8 mm; brown; often pink toward tip of corium; prothorax pointed laterally; base of abdomen with spine projecting between hind coxae; 2 dark spots near tip of front wing membranes; most of U.S. (predaceous).

Thyanta accerra McAtee. Redshouldered stink bug—9-10 mm; pale green; pronotum with anterior corners pointed and lateral margins narrowly edged with dull red; east of Rocky Mts.

Family Thyreocoridae— Negro Bugs

These shiny black, broadly oval bugs are generally 3-6 mm long. The huge scutellum covers the abdomen giving the appearance of small beetles. The front tibiae are not flattened for digging as are those of the similar burrower bugs (Cydnidae). Negro bugs occur in weedy fields. The name Corimelaenidae is used for the family in some texts.

Common Species

Common genera: *Galgupha* (front wing with groove along lateral margin of corium; corium is black) and *Corimelaena* (no groove; corium with a colored lateral margin).

Corimelaena pulicaria (Germar). Negro bug—2.5-3.5 mm; black; lateral margin of corium with pale stripe that widens basally, tip of corium rounded; most of U.S., southern Canada.

Galgupha atra Amyot and Serville (Fig. 143A)—4.5-6.0 mm; shiny black; scutellum evenly and broadly rounded posteriorly; most of U.S.

Family Cydnidae—Burrower Bugs

Members of this family are usually 7 mm or less in length, black or reddish brown, and resemble small stink bugs (Pentatomidae). Cydnids differ from stink bugs by having strongly spined tibiae. These bugs burrow into soil by using their front legs and occur around plant roots and under boards and rocks. Burrower bugs are most likely to be encountered at lights.

Common Species

Pangaeus bilineatus (Say) (Fig. 143B)—6-8 mm; black or reddish black; pronotum with groove paralleling anterior margin to form a collar; corium and clavus with minute cracks; eastern U.S. southwest to CA.

Sehirus cinctus (Palisot de Beauvois)—4-7 mm; shiny black, bluish black or dark reddish brown; lateral margins of body with white line; throughout U.S., southern Canada (on mints, nettles).

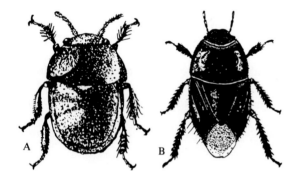

Figure 143 A, a negro bug, *Galgupha atra* (Thyreocoridae); B, a burrower bug, *Pangaeus bilineatus* (Cydnidae).

SUBORDER HOMOPTERA[23]

The front wings of the Homoptera are uniformly clear and membranous (Figs. 147; 153) or colored and slightly thickened (Fig. 149). They are commonly held at an angle (rooflike) over the body when the insect is resting. The ventral beak originates from the posterior part of the head (Figs. 144A, C). Homopterans feed by sucking plant sap. Numerous species injure crops by feeding and many transmit plant diseases. A few species are used as sources of dyes, shellac, and waxes. The Homoptera are divided into two groups (suborders of some authors): the Auchenorrhyncha (active fliers and jumpers) includes the cicadas, treehoppers, leafhoppers, froghoppers or spittlebugs, and about 11 families of planthoppers; the Sternorrhyncha (most of which are relatively inactive) include the psyllids or jumping plant lice, whiteflies, three or four aphid families, and 15 or 16 families of scale insects in the U.S.

Most homopterans are collected by sweeping vegetation. Trees and shrubs may be shaken or beaten followed by sweeping underneath the plant. The entire contents of the net may be killed and sorted later or individual specimens may be collected from the net with an aspirator. Less active or inactive forms are trapped in a killing jar or scraped or picked off the host. Cicadas in trees may be collected by using a longhandled net or dislodged to a lower branch by using a long pole or throwing a rock.

Homoptera are generally pinned or mounted on points. Soft-bodied immatures, aphids, and sometimes psyllids and whiteflies are preserved in 80% alcohol or mounted on slides. Host plants with scales may be collected, dried, and pinned, or the insects may be scraped off and placed in alcohol. Male scales are obtained primarily by rearing immatures. To identify most scale insects to families and species, females must be mounted on micro-

23. Homoptera: *homo*, same or uniform; *ptera*, wings.

scope slides after clearing in potassium hydroxide and staining.

KEY TO FAMILIES OF SUBORDER HOMOPTERA

This key is modified from that in Borror, DeLong, and Triplehorn (1976). Scale insects are keyed only to superfamily; the paper by Howell and Williams (1976) should be consulted for a key to the families.

1a	Antennae very short and bristlelike; tarsi 3-segmented	2
1b	Antennae usually long and filiform; tarsi 1- or 2-segmented	16
2a(1a)	Antennae arising from front of head between eyes (Fig. 144B) or in front of eyes (Fig. 144A)	3

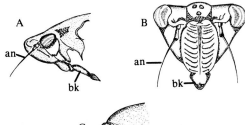

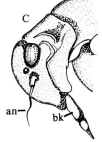

Figure 144 Beak and antennae positions in suborder Homoptera. A, antennae arise in front of eyes; B, antennae arise between eyes; C, antennae arise beneath eyes. an, antenna; bk, beak.

2b	Antennae arising from sides of head beneath eyes (Fig. 144C).	6
3a(2a)	Three ocelli present (Fig. 144B); front femora much larger than middle femora; membranous front wings; large insects. . . . . (p. 160) *Cicadidae*	
3b	Two (rarely 3) ocelli present; front and middle femora similar in size; front wings sometimes thickened; small insects	4
4a(3b)	Pronotum extends backward over abdomen and may form spines or other processes (Fig. 148). (p. 161) *Membracidae*	
4b	Pronotum does not extend backward over abdomen (Fig. 149).	5
5a(4b)	Hind tibiae with 1 or more rows of small spines down its entire length (Fig. 145A). . . . (p. 163) *Cicadellidae*	
5b	Hind tibiae with 1 or 2 large spines and usually several at apex (Fig. 145B). . . . . (p. 162) *Cercopidae*	
6a(2b)	Hind tibiae with a large, movable spur at apex (Fig. 145C) . . . . (p. 164) *Delphacidae*	
6b	Hind tibiae without large, movable spur at apex	7

Order Hemiptera

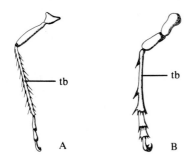

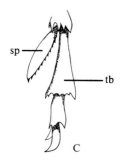

Figure 145 Hind legs. A, rows of spines on tibia of a leafhopper (Cicadellidae); B, single or grouped spines on tibia of a spittlebug (Cercopidae); C, large spur on apex of delphacid tibia (Delphacidae). sp, spur; tb, tibia.

7a(6b) Anal area of hind wings with a network of many crossveins (Fig. 150B) ..... (p. 164) *Fulgoridae*

7b Anal area of hind wings without a network of crossveins........... 8

8a(7b) Hind tarsi with 2 spines on apex of second segment, apex rounded 9

8b Hind tarsi with a row of spines on apex of second segment, apex notched or truncated 12

9a(8a) Front wings longer than body, at rest held almost vertically at sides of body, triangular anal area (clavus) at least partly granular, many crossveins in costal area (Fig. 151D); often mothlike in form... (p. 166) *Flatidae*

9b Front wings variable in length and position, triangular anal area (clavus) not granular, usually without numerous crossveins in costal area; form variable 10

10a(9b) Apex of front wings more densely veined and set off from rest of wing by crossveins; Florida to Mississippi .... *Tropiduchidae*

10b Apex of front wings not as described above; widely distributed 11

11a(10b) Front wings broadly oval, longer than body, at rest held almost vertically at sides of body (Fig. 151C), and costal area net-veined; somewhat mothlike in form and usually green .... (p. 166) *Acanaloniidae*

11b Front wings variable in size and shape, often shorter than abdomen, and costal area not net-veined; hind tibiae with spines on sides and apex; often stout and frequently beetle-like .... (p. 165) *Issidae*

12a(8b)	Terminal segment of beak not more than 1 1/2 times as long as wide; often slender or fragile forms with bright coloration . . (p. 164) *Derbidae*	16b	Tarsi 1-segmented and with 1 claw; female wingless, often legless, and body covered with hard shell or waxy secretions (Fig. 155); male with 1 pair of wings and no beak (Fig. 155E) . . . . (p. 170) *Superfamily Coccoidea*
12b	Terminal segment of beak at least twice as long as wide; form variable 13	17a(16a)	Antennae 5- to 10-segmented (usually 10); front wings often thicker than hind wings; jumping insects . . . . (p. 166) *Psyllidae*
13a(12b)	Front wings overlap at apex (Fig. 150C); body somewhat flattened . . . . (p. 165) *Achilidae*		
13b	Front wings do not overlap at apex; body not flattened 14	17b	Antennae 3- to 7-segmented; wings clear or whitish; not jumping insects 18
14a(13b)	Head strongly elongated at front (Fig. 151B), or front with 2-3 raised ridges; no median ocellus . . . . (p. 165) *Dictyopharidae*	18a(17b)	Wings opaque whitish and covered with a whitish powder; hind wings nearly as large as front wings; no cornicles . . . . (p. 167) *Aleyrodidae*
14b	Head not strongly elongated at front although sometimes slightly extended, front of head without elevated ridges or only 1; median ocellus usually present 15	18b	Wings clear and without whitish powder; hind wings much smaller than front wings; cornicles usually present (Fig. 153) 19
		19a(18b)	Front wings with 4-5 (rarely 6) veins below darkened stigma, veins reach wing margin (Fig. 153); cornicles usually present (Fig. 153); antennae generally 6-segmented. 20
15a(14b)	Abdominal terga 6-8 are chevron-shaped, sometimes sunk below rest of terga, and with wax-secreting pores; western U.S. (genus *Oeclidius*) . . . . *Kinnaridae*		
15b	Abdominal terga 6-8 are rectangular; widely distributed . . (p. 165) *Cixiidae*	19b	Front wings with 3 veins below stigma, veins reach wing margin; cornicles absent; antennae 3-5 segmented 21
16a(1b)	Tarsi 2-segmented and with 2 claws; 4 wings or none. 17	20a(19a)	Cornicles typically present and conspicuous (Fig. 153); M vein in front wing branched (Fig. 153); no waxy or woolly secretions. . . . . (p. 167) *Aphididae*

20b   Cornicles minute or absent; M vein in front wing not branched (Fig. 146A); many wax glands and waxy or woolly secretions .... (p. 168) *Pemphigidae*

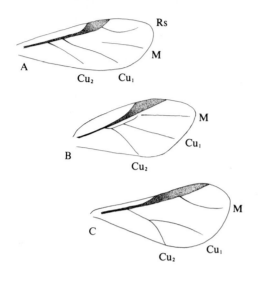

Figure 146   Front wings. A, woolly and gallmaking aphids (Pemphigidae); B, pine and spruce aphids (Adelgidae); C, phylloxerans (Phylloxeridae).

21a(19b)   Wings held rooflike at rest; veins Cu_1 and Cu_2 in front wing do not join at bases (Fig. 146B); wingless females covered with threads and masses of wax; on conifers. . (p. 169) *Adelgidae*

21b   Wings held horizontally at rest; veins Cu_1 and Cu_2 in front wing join in a common stalk at bases (Fig. 146C); wingless females not covered with threads and masses of wax, at most covered with powder .... (p. 170) *Phylloxeridae*

Family Cicadidae—Cicadas

Cicadas are medium to large (16-50 mm long), robust insects with four membranous wings. The large beak is easily seen and the antennae are short and bristlelike. Cicadas are usually dull or dark with lighter colored markings. Males of each species produce loud, sometimes shrill, characteristic buzzing sounds generally from popping in and out riblike bands associated with a large resonating air sac; the bands and sac are located in each of two ventral or lateral cavities at the base of the abdomen. Each cavity also contains an auditory organ (tympanum) and may be partly covered by a plate. Some species stridulate or make clicking noises with their wings.

    Cicadas are primarily arboreal in the eastern U.S.; many frequent forbs and grasses in the southwestern states where fewer trees grow. Females deposit eggs in twigs of trees and shrubs by slitting the bark and wood, usually causing the twig tips to die. The nymphs drop to the ground, enter the soil, and feed on plant roots. The hemimetabolous development takes 4-17 years depending on the species. The mature nymphs dig out of the soil, climb onto an object such as a tree, and molt leaving prominent skins attached to the tree.

    The dogday group of cicadas are large, blackish species that occur from July to September. The periodical cicadas of the eastern U.S. (genus *Magicicada*) occur from May to early July, have reddish eyes and wing veins, and 13- or 17-year life cycles. There are numerous broods of each of the three 13-year or three 17-year *Magicicada* species. The broods emerge in different years, have different geographical ranges, and may consist of one to three species of each life-cycle type. The large broods emerge only every few years. The name "locusts" has been commonly misused for these cicadas because of the enormous numbers that periodically emerge (up to 20,000-40,000 under a tree), leading to confusion with the biblical locusts or migratory grasshoppers (Orthoptera).

Figure 147 Linne's cicada, *Tibicen linnei* (Cicadidae).

Common Species

Magicicada septendecim (Linnaeus). Periodical cicada or Linnaeus' 17-year cicada—27-33 mm; black and reddish; eyes reddish and sides of thorax reddish; ventral part of abdomen with reddish brown or yellow markings; eastern U.S. west to NB and southwest to TX (song is 1-3 second buzz that drops in pitch at end and sounds like "Phaaaroah"; begins at dawn in May and June).

Magicicada tredecim Walsh and Riley. Riley's 13-year cicada—like *M. septendecim* but range is southern U.S. north to IA, west to OK.

Okanagana bella Davis—24 mm; shiny blue-black with dark orange markings; western U.S., southwestern Canada.

Platypedia areolata (Uhler). Orchard cicada—19-20 mm; dark bronze; amber and yellow markings; greenish yellow crossband between pro- and mesothorax; western U.S., southwest Canada (common in orchards).

Platypedia minor Uhler. Minor cicada—17 mm; like *P. areolata* but lacks pale band on thorax; western U.S. (song is brief, sharp wing clicks repeated 4-5 times; in trees on grassy slopes and along creeks).

Tibicen canicularis (Harris). Dogday cicada—32 mm; green and black; anterior part of pronotum (collar) green or brown; wing margins greenish; wing length usually < 40 mm; eastern and midwestern U.S. (song is even, whining buzz without vibrato and lasting 1 minute or less).

Tibicen linnei (Smith and Grossbeck) (Fig. 147). Linne's cicada—like *T. canicularis* but wing length usually > 40 mm; eastern and midwestern U.S. (song a shrill 15-25 second vibrato; from trees in afternoon).

Tibicen pruinosa (Say)—43-47 mm; green and black; wings and legs greenish; mid- and southwestern U.S. (song is "za-wie"; often with *T. linnei*).

Family Membracidae— Treehoppers

Treehoppers are leaping insects usually less than 12 mm long, and are characterized by the large pronotum that projects over the head and extends backward over the abdomen. The insects may appear humpbacked, thornlike, or have various spines and horns as part of the pronotum. Treehoppers feed primarily on stems of trees and shrubs although some occur and feed on weeds, grasses, and crops. Nymphs have a less-expanded pronotum and often branched spines on the dorsum of the abdomen. Many nymphs and some adults are tended by ants for the sweet anal secretion treehoppers produce.

Common Species

Campylenchia latipes Say (Fig. 148A)—6-8 mm; reddish brown; without spots; anterior pronotal horn with prominent ventral keel and 2 lateral ridges; pronotum hairy and densely punctured; throughout North America (on goldenrod, alfalfa, clover, grasses, others).

Enchenopa binotata (Say). Twomarked treehopper—5-6 mm; dark brown; 2 yellow spots on mid-dorsal line of pronotum; anterior horn projects up and forward, is slightly expanded at tip, and has 3 ridges along length; eastern U.S.

Platycotis vittata (Fabricius)—10 mm; dull olive green, blue-green or bronze; stripes or red-orange dots; fine punctures; middle horn long, short, or absent; 2 short lateral horns; most of U.S. (on oaks).

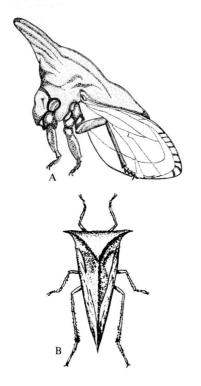

Figure 148 Treehoppers (Membracidae). A, *Campylenchia latipes;* B, buffalo treehopper, *Stictocephala bubalus.*

Stictocephala bubalus (Fabricius) (Fig. 148B). Buffalo treehopper—9-10 mm; bright green or yellowish; no colored bands; sternum yellowish; pronotum with 2 stout side horns projecting outward; front wings not covered by pronotum; throughout North America (weeds, legumes, trees [especially elm, apple]).

Stictocephala diceros (Say)—9-10 mm; dark brown; brown spots; 2 sharp lateral horns; posterior pronotal process with 2 white bands; front wings not covered by pronotum; eastern U.S. (elderberry, shrubs of wet areas).

Stictocephala festina (Say)—5 mm; bright green; no pronotal horns; most of U.S., southern Canada (on alfalfa, other forage, weeds, fruit trees).

Stictocephala inermis (Fabricius)—7-9 mm; uniformly green or reddish; no pronotal horns; tip of femora black; throughout U.S. (especially on grasses, alfalfa, clover).

Family Cercopidae—Spittlebugs or Froghoppers

Adults are small (less than 13 mm), very common hopping insects that are shades of gray or brown. Spittlebugs are distinguished from the similar leafhoppers by the one or two stout spines on the hind tibiae rather than the rows of small spines (Fig. 145B). Spittlebugs feed on weeds, grasses, shrubs, and a few trees. The name is derived from the inactive nymphs that surround themselves with a conspicuous mass of white spittlelike froth which provides a moist habitat.

Common Species

Aphrophora permutata Uhler—8-10 mm; mottled brown; eyes twice as wide as long; space between ocellus and posterior margin of head = or > diameter of ocellus; western U.S. east to MI (on dandelion, dock, lettuce, others).

Aphrophora quadrinotata Say—7.0-8.5 mm; pale; red ocelli; head and pronotum with slightly raised middorsal line; anterior margin of front wing with 2 large, clear spots ringed with dark brown; North America east of Rocky Mts.

Aphrophora saratogensis (Fitch). Saratoga spittlebug—9-11 mm; brown; light middorsal stripe on head and pronotum; front wings with tan and silvery mottled crossbands; North America east of Rocky Mts. (on pines, especially red pines in Great Lakes states).

Clastoptera proteus Fitch. Dogwood spittlebug—4-5 mm; black with 2 yellow crossbands on head and 1 on thorax and each front wing; scutellum yellow; throughout U.S., southern Canada.

Lepyronia quadrangularis (Say) (Fig. 149A). Diamondbacked spittlebug—6-8 mm; brownish; black underneath; densely clothed with fine hairs; brown diamond pattern formed on front wings when together; blackish, curved marking at tip of front wing; most of North America.

Philaenus spumarius (Linnaeus). Meadow spittlebug—5-6 mm; straw colored to dark brown; 2 white or yellow spots on costal margin of front wing; throughout North America (on clovers, alfalfa, weeds).

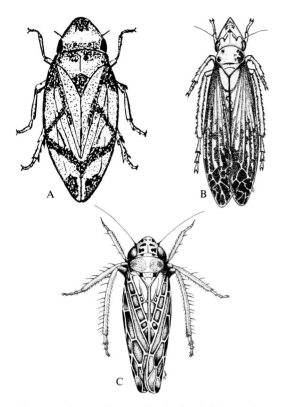

Figure 149 A, diamondbacked spittlebug, *Lepyronia quadrangularis* (Cercopidae); B and C, leafhoppers (Cicadellidae): B, watercress sharpshooter, *Draeculacephala mollipes;* C, painted leafhopper, *Endria inimica.*

Family Cicadellidae—Leafhoppers

This very large family (nearly 3,000 North American species) consists of small, jumping insects chiefly under 13 mm in length. Leafhoppers are similar to froghoppers but have one or more rows of small spines on the hind tibiae. Many are brightly colored and some produce weak sounds. Cicadellids occur on most types of plants and in most habitats. Some are limited to one host, whereas others are found on many plant species. Many are plant pests due to feeding on plant sap, destroying cells, plugging vessels, and transmitting viruses. Some species migrate to southern states in the fall and northern states in the spring.

Common Species

Aceratagallia sanguinolenta (Provancher). Clover leafhopper—2.5-3.5 mm; light brown or gray with dark markings; dark spot above each ocellus; longitudinal light brown stripe each side of pale middorsal line; front wings pale brown with white on posterior area; North America east of Rocky Mts. (on legumes, other crops).

Circulifer tenellus (Baker). Beet leafhopper—3.0-3.5 mm; greenish yellow to dark brown; front wings with dark markings or broken bands; western U.S. east to IL, southwestern Canada (on beets, sugar beets, tomatoes, others).

Draeculacephala mollipes (Say) (Fig. 149B). Watercress sharpshooter—6-8 mm; yellow and green; head extended and pointed anteriorly in dorsal and lateral views; throughout U.S., southern Canada.

Empoasca fabae (Harris). Potato leafhopper—3 mm; pale to irridescent green; 6 or 8 white spots on anterior margin of pronotum; small white "H" marking on scutellum; front wings with crossveins only near tip and without a broad cell that wraps around apex; 1 apical cell in hind wing; North America east of Rocky Mts. (on potato, alfalfa, clover, beans, apple).

Endria inimica (Say) (Fig. 149C). Painted leafhopper—4 mm; grayish yellow; 2 black spots on top of head, anterior margin of pronotum, and base of scutellum; most of North America (on lawns, meadows).

Exitianus exitiosus (Uhler). Gray lawn leafhopper—3.5-5.0 mm; color variable; sometimes brownish; dorsum of head grayish white and sometimes tinged orange-yellow; red ocelli with large, round, black spots on margins; irregular black spots on pronotum; clear front wings with dark veins; long, broad cell wraps around wing tip; throughout U.S., southern Canada (on grasses, wheat, legumes).

Graphocephala coccinea (Förster)—8-9 mm; top of head yellow with black band on anterior margin; pronotum red; front wings red with longitudinal green stripes; eastern U.S. southwest to TX, southeastern Canada (on ornamental shrubs, e.g., *Forsythia*).

Idiocerus pallidus Fitch—6.0-6.5 mm; white with green tinge; without markings; eyes reddish brown; throughout U.S., southern Canada (one of dozens of *Idiocerus* spp. on willows, poplar, aspen, cottonwood).

Paraphlepsius irroratus (Say)—5.5-6.0 mm; dark brown; heavily covered with minute spots; top of head 1/4 longer at middle than next to eyes; head at least width of pronotum; most of North America.

Family Delphacidae— Delphacid Planthoppers

This is the largest planthopper family. Its members are small, fairly common, and characterized by a large apical spur on the hind tibia.

Common Species

Stobaera tricarinata (Say) (Fig. 150A)—4 mm; face and front wings banded with whitish and dark markings; mesonotum pale or brownish yellow; legs white and black; apical spur on hind tibia thick and convex on sides with teeth on hind margin; most of U.S., southern Canada.

Family Derbidae— Derbid Planthoppers

The terminal segment of the beak is about as long as wide in this family. Some species resemble small moths or caddisflies and the wings of many are broader toward the tips. *Otiocerus* spp. have branched antennae and elongated heads that are bluntly but not sharply angled.

Common Species

Apache degeerii (Kirby) (Fig. 151A)—11 mm to wing tips; pink; veins of front wings dark pink; lower margins of head yellow and crossed with black lines; head sharply angled; southern and midwestern U.S. (on deciduous trees).

Family Fulgoridae— Fulgorid Planthoppers

Some of the largest planthoppers in the U.S. (up to 25 mm wingspan) occur in this family which is recognized by the many crossveins in the anal areas of the hind wings.

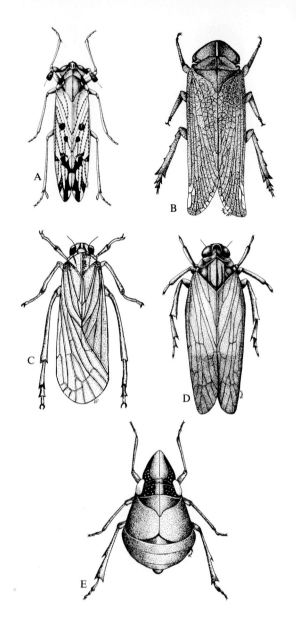

Figure 150 Planthoppers. A, *Stobaera tricarinata* (Delphacidae); B, *Cyrpoptus belfragei* (Fulgoridae); C, *Synecdoche impunctata* (Achilidae); D, *Oliarus humilis* (Cixiidae); E, *Bruchomorpha oculata* (Issidae).

Common Species

Cyrpoptus belfragei Stål (Fig. 150B)—8-9 mm; grayish brown with reddish tinge; head broad, flattened, and rounded anteriorly; wings with small blackish patches; smoky bar near tips of wings and clear areas along anterior margins; southern U.S. north to OH.

Family Dictyopharidae— Dictyophard Planthoppers

Members in the genus *Scolops* have the head produced forward into a long, slender, beaklike structure (Fig. 151B). The head is not produced forward in the genus *Phylloscelis;* instead there are two to three raised ridges on the head and the front femora are broad and flattened.

Common Species

Scolops pallidus Uhler—6-7 mm; yellowish and gray; southwestern U.S.

Scolops sulcipes (Say) (Fig. 151B)—6 mm; brownish gray; front wings with many netlike crossveins; eastern U.S. (meadows).

Family Achilidae— Achilid Planthoppers

The apices of the front wings overlap in species of this family. In the genus *Catonia,* the lateral compartments of the pronotum are shorter than the eyes.

Common Species

Synecdoche impunctata (Fitch) (Fig. 150C)—5 mm; face with black spots separated by white crossband; front wings uniformly brownish yellow; eastern U.S. southwest to OK, southeastern Canada.

Family Cixiidae— Cixiid Planthoppers

The head is not usually extended forward in this family. The genus *Cixius* has two to three spines on the hind tibiae and three longitudinal ridges on the mesonotum.

Common Species

Oliarus humilis (Say) (Fig. 150D)—5 mm; dark brown; hind tibiae with 2-3 conspicuous spines; mesonotum with 5 longitudinal ridges; apical 1/3 of front wings with smoky or black patch; eastern U.S., southeastern Canada.

Family Issidae— Issid Planthoppers

Most species are dark-colored and somewhat broad; some have a weevillike snout and short wings.

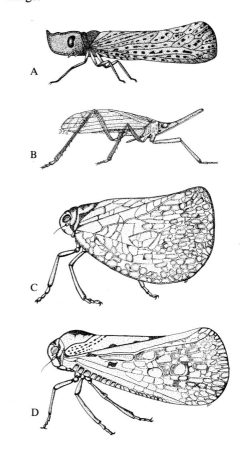

Figure 151 Planthoppers. A, *Apache degeerii* (Derbidae); B, *Scolops sulcipes* (Dictyopharidae); C, *Acanalonia bivittata* (Acanaloniidae); D, *Metcalfa pruinosa* (Flatidae).

Order Hemiptera

Common Species
Bruchomorpha oculata Newman (Fig. 150E)—3 mm; black; distinct, short snout; front wings either reach base of abdomen (short wing form) or extend slightly beyond tip of abdomen; eastern North America.

Family Acanaloniidae—
Acanaloniid Planthoppers
These greenish insects have broadly oval front wings held almost vertically at rest. Acanaloniidae are similar to the Flatidae but lack granules in the triangular anal area (clavus) of the front wing.

Common Species
Acanalonia bivittata (Say) (Fig. 151C)—7 mm to wing tips; green, yellowish green, or brownish yellow; brown head is broad, rounded, and not conical; 2 longitudinal dark stripes from eyes to end of front wings; eastern North America.

Family Flatidae—
Flatid Planthoppers
The front wings of members of this family have a uniform row of crossveins along the anterior margins and granules in the triangular anal area (clavus).

Common Species
Anormenis septentrionalis (Spinola)—9-11 mm; dull green; front wings without dots; front wings truncate and not rounded at tip; eastern U.S. west to Rocky Mts.

Metcalfa pruinosa (Say) (Fig. 151D)—8-9 mm; dark gray above; underneath paler; whitish powder on body; front wings with dark dots; eastern North America.

Family Psyllidae—
Jumping Plant Lice or Psyllids
These small (2-5 mm) jumping insects resemble tiny cicadas although many psyllids have long antennae. The front wings are usually membranous but may be thickened and the four wings are held rooflike over the body. Many nymphs produce strands and masses of white wax; nymphs in the genus *Pachypsylla* cause numerous galls to form on hackberry leaves. A few species are pests of crops.

Common Species
Paratrioza cockerelli (Sulc). Potato psyllid, tomato psyllid—1.4 mm; black; white crossband on 1st abdominal segment; white "Y" marking on last segment; western U.S. except WA and OR, southwestern Canada (on potato, tomato, wild Solanaceae).

Psylla floccosa Patch. Cottony alder psyllid—2-4 mm; pale yellowish green or brownish; nymphs yellow and green; northern 1/2 of U.S. (cottony wax masses on alder in spring and early summer; late summer masses are woolly alder aphids [Pemphigidae]; brown nymphs with dark abdominal crossbands under waxy masses in CA and NV are *P. alni americana* Crawford).

Psylla pyricola Förster (Fig. 152A). Pear psylla—1.5-4.0 mm to wing tips; light orange, reddish brown or brown to black; longitudinal stripes (usually 4) on mesothorax; black spot at tip of anal area on front wing; cross stripes on abdomen; eastern U.S., Far West, southwestern Canada.

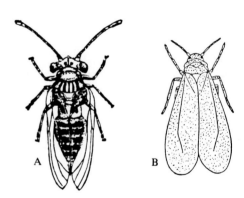

Figure 152 A, pear psylla, *Psylla pyricola* (Psyllidae); B, greenhouse whitefly, *Trialeurodes vaporariorum* (Aleyrodidae).

Family Aleyrodidae—Whiteflies

Adult whiteflies are small (1-3 mm), winged insects that are covered with white, powdery wax and resemble tiny moths. Larvae (nymphs) have a modified type of hemimetabolous development: (1) the first instar is active; (2) the following instars are legless, produce waxy secretions over their bodies, and the wings develop internally; and (3) the next to last instar (pupa) is inactive. Whiteflies are primarily subtropical and tropical. In the U.S. several species are important greenhouse and citrus pests. Besides feeding on sap, whiteflies secrete a sugary anal material (honeydew) on plants that attracts a sooty fungus.

Common Species

Dialeurodes citri (Ashmead). Citrus whitefly—2 mm; whitish powder on body and wings; wings without darkened area in middle; FL and other Gulf coast states, CA (on citrus, ornamentals).

Trialeurodes vaporariorum (Westwood) (Fig. 152B). Greenhouse whitefly—1.5 mm; whitish powder on body and wings; nymphs pale green, flat and oval, long filaments extend from bodies; throughout North American (in greenhouses), southern 1/2 of U.S. (outdoors on fruit, vegetables, flowers).

Family Aphididae— Aphids or Plantlice

Aphids are small (1-6 mm), winged or wingless, pear-shaped insects that are found clustered on stems or leaves of plants. Most species have a pair of tubes (cornicles) near the posterior end of the abdomen (Fig. 153). When an aphid is disturbed the cornicles secrete alarm pheromones which stimulate defensive behavior (e.g., dropping from the plant) of nearby aphids. Aphids produce a sugary anal secretion, honeydew, often seen as a shiny coating on leaves. Ants and other insects feed on honeydew and ants often tend aphids. Many species are serious plant pests: they weaken a plant by sap removal and may cause it to wilt, they cause leaves to curl and turn yellow, and some transmit virus diseases of plants.

Aphids form a large family whose species are known for their variation (polymorphism) in shape, size, and winged condition, and their complex life cycle. Typically, small numbers of eggs overwinter and hatch in the spring into wingless females. These females reproduce parthenogenetically (without fertilization) and give birth to young rather than laying eggs. Several generations of wingless females may occur until winged females are produced which migrate to another species of host plant to continue reproducing. Late in the season winged forms return to the original species of food plant, a generation of males and females is produced, mating occurs, and eggs are deposited for overwintering.

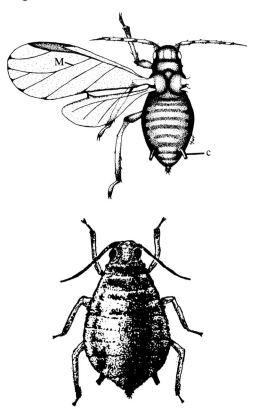

Figure 153 Winged and wingless bean aphids, *Aphis fabae* (Aphididae). c, cornicle.

Common Species

Acyrthosiphon pisum (Harris). Pea aphid—2.0-3.5 mm; pale green; tips of tibiae, tarsi, and cornicles blackish; winged forms with yellowish brown head and thorax; throughout North America (on peas, alfalfa, clover, other legumes).

Aphis fabae Scopoli (Fig. 153). Bean aphid—1.5-2.5 mm; dark olive green to black; tips of tibiae, abdomen, cornicles, and all of tarsi black; caudal appendage bushy; throughout North America (on beans, beets, dock).

Aphis gossypii Glover. Cotton aphid, melon aphid—1-3 mm; tan, brown, green, or black; head tubercles between antennae not prominent; several dark dashes on dorsum of abdomen; cornicles black; throughout U.S., southern Canada (on melons, cucumber, cotton, citrus, vegetables, ornamentals, pigweed, dock, other weeds).

Brevicoryne brassicae (Linnaeus). Cabbage aphid—1.5-2.5 mm; grayish green; striped; dusty appearing from wax covering; throughout North America (on cabbage, cauliflower, broccoli).

Hyalopterus pruni (Geoffroy). Mealy plum aphid—1-2 mm; green; covered with whitish powder; 3 longitudinal darker green lines; winged form with head and thorax brown; throughout U.S., southern Canada (on plum, other stone fruits).

Macrosiphoniella sanborni (Gillette). Chrysanthemum aphid—1.5-2.0 mm; amber brown to black; shiny; cornicles black and most of length with scale-like markings; most of U.S., southern Canada (on chrysanthemums, in greenhouses).

Macrosiphum avenae (Fabricius). English grain aphid—1.5-2.5 mm; grass-green; sometimes yellow or pink; head with brown marking; darker dorsal blotch on abdomen of wingless form; darker median line in winged form; throughout North America (on grasses, cereals).

Macrosiphum rosae (Linnaeus). Rose aphid—2.0-3.5 mm; green or pinkish; winged form with black antennae, head, thorax and cornicles; throughout U.S., southern Canada (on roses).

Myzus persicae (Sulzer). Green peach aphid—1.5-2.5 mm; wingless form pale to dark green with stripes in summer and pinkish in fall; winged form dark brown with dark dorsal patch on yellowish abdomen; 2 prominent, converging head tubercles between antennae in all forms; throughout U.S., southern Canada (on fruit, vegetables, shrubs, flowers).

Rhopalosiphum maidis (Fitch). Corn leaf aphid—1.3-2.5 mm; bluish green; slender; throughout U.S., southern Canada (on corn, sorghums).

Schizaphis graminum (Rondani). Greenbug—1.2-2.0 mm; pale green; wingless form with dark green median stripe; winged form with brownish yellow head and black cornicles; cornicles narrower than hind tibiae; no reddish blotches at bases of cornicles; throughout North America, chiefly in South and Midwest (wheat, other grains, grasses, alfalfa).

Therioaphis maculata (Buckton). Spotted alfalfa aphid—1.5-2.0 mm; pale yellow; 6+ rows of black spots on dorsum; smoky areas along wing veins; most of U.S. except New England.

Family Pemphigidae (= Eriosomatidae)—Woolly and Gallmaking Aphids

The cornicles are small or absent in this family and most members have wax glands that produce large amounts of woolly or waxy material that covers part or all of the body. Mouthparts are absent in the males and females of the bisexual generation. The family is classified as a subfamily (Pemphiginae) of the aphids (Aphididae) by some taxonomists.

Most species alternate between host plants and galls are produced on trees and shrubs by many species. The vagabond aphid gall of poplar and cottonwoods, caused by *Mordwilkoja vagabunda* (Walsh), appears as an irregular, folded and twisted leaf (Fig. 154B). The leaf petiole gall of poplar is an oval gall with a transverse opening on the leaf stem; it is caused by *Pemphigus populitransversus* Riley (Fig. 154C).

Common Species

Eriosoma lanigerum (Hausmann) (Fig. 154A). Woolly apple aphid—0.5-2.5 mm; reddish or purplish; bluish white woolly masses on body; throughout North America (on roots and bark of apple, pear, hawthorn).

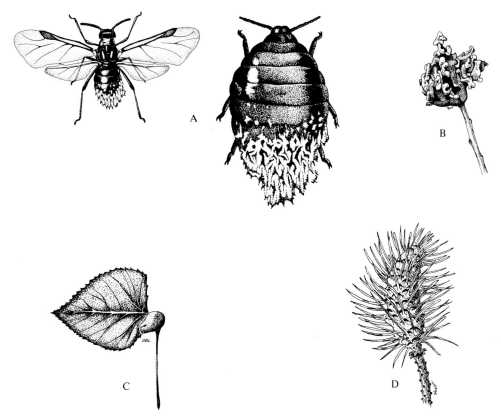

Figure 154 A-C, woolly and gallmaking aphids (Pemphigidae): A, woolly apple aphid, *Eriosoma lanigerum;* B, vagabond aphid gall of poplar and cottonwood caused by *Mordwilkoja vagabunda;* C, leaf petiole gall of poplar caused by *Pemphigus populitransversus.* D, gall on spruce caused by eastern spruce gall aphid, *Adelges abietis* (Adelgidae).

Prociphilus tessellatus (Fitch). Woolly alder aphid—0.5-2.5 mm; woolly masses cover body; no cornicles; east of Rocky Mts. (on alders and maples from middle summer to early fall; early summer masses are cottony alder psyllid [Psyllidae]).

Family Adelgidae (= Chermidae) Pine and Spruce Aphids

The winged or wingless members of this family lack cornicles, veins Cu_1 and Cu_2 are separated at their bases in the front wing (Fig. 146B), and wingless females are often covered with waxy filaments. These species live and feed only on needles, twigs, or in galls of conifers. Generally a different kind of conifer acts as an alternate host during part of the year.

Common Species

Adelges abietis (Linnaeus) (Fig. 154D). Eastern spruce gall aphid—1.5-2.0 mm; brownish yellow; northeastern and northcentral U.S., southeastern Canada (forms large, brown or gray pineapple-shaped galls primarily on twigs of Norway and white spruce).

Adelges cooleyi (Gillette). Cooley spruce gall aphid—1.5 mm; reddish brown; white, woolly wax; similar to *A. abietis;* northern U.S. from Pacific to Atlantic coasts, range of white spruce in Canada (forms green or reddish brown galls on blue, Engelman, Stika, and white spruces; wax on Douglas fir).

Adelges piceae (Ratzeburg). Balsam woolly aphid—1 mm; black; convex; covered with white, curly wax threads; northeastern U.S., Appalachian Mts., OR, WA, British Columbia, southeastern Canada (on firs but not Douglas firs).

Pineus strobi (Hartig). Pine bark aphid—1-2 mm; dark brown; covered with white, fluffy wax; forms spots and patches on trunks and limbs; most of U.S., southern Canada (on Austrian, Scotch, and white pines).

Family Phylloxeridae—Phylloxerans

Phylloxerans are tiny, winged or wingless insects that have 3-segmented antennae, lack cornicles, and have the bases of veins Cu_1 and Cu_2 joined in a common stalk (Fig. 146C). The wings are held horizontally; wingless forms sometimes have a whitish powder on them but never waxy threads. The grape phylloxera, *Daktulosphaira vitifoliae* (Fitch), has been a serious pest of European grapes grown in the western U.S. but causes little damage to native grapes in the eastern states. The tiny (0.3-1.2 mm), wingless, yellowish or greenish brown phylloxerans form galls about half the size of a pea on leaves of wild and sometimes cultivated grapes in the eastern U.S. The Pacific coast populations cause root galls and eventually kill the European grapes. Grafting European grapes to resistant rootstocks of native grapes has alleviated most of the problem caused by the grape phylloxera.

Most of the leaf and leaf stem galls on hickory trees are produced by species in the genus *Phylloxera*. The disk-shaped to globular galls are recognized by the presence of an opening on the upper or lower surface.

SUPERFAMILY COCCOIDEA—Scale Insects

The families in this group consist of small (0.5-8.0 mm), winged and wingless insects that are often unlike the other Homoptera in appearance. Adult females are always wingless, very sluggish or fixed in position, and usually are covered by a hard scale or waxy secretion, or have a tough integument. Some females lack legs and have reduced antennae, and often the abdominal segmentation is indistinct. Adult males have no beak, one pair of well-developed wings (rarely wingless) on the mesothorax, a pair of filaments (halteres) on the metathorax, and sometimes an elongate process at the tip of the abdomen (Fig. 155E). The absence of mouthparts and the presence of the abdominal process help separate males from the similarly appearing gnats (Diptera).

The active, first instar nymphs (crawlers) have legs and antennae and disperse themselves on the host plant. After the first molt the legs and antennae are generally lost and most species become sedentary. A waxy or scalelike covering may be secreted as the nymph develops toward the adult stage. Adult females reproduce under their permanent scale or wax covering. The wings of males develop in the pupalike resting stage of the last instar; adult males are tiny, active, flying insects.

Family Margarodidae—Margarodid Scales or Giant Coccids and Ground Pearls

Females are often large and brightly colored. Yellowish green, beadlike wax cysts are formed by members of the genus *Margarodes* and are called ground pearls; they occur on roots of many plants in the warm areas of the U.S. *M. laingi* Jakubski and *M. meridionalis* Morrison are pests of lawns in the southern U.S.

Common Species

Icerya purchasi Maskell. Cottonycushion scale—10-15 mm with egg sac (♀); reddish brown; large, white, fluted cottony egg sac; SC and Gulf coast states to CA (on citrus, other trees).

Family Ortheziidae—Ensign Scales

Members in this family live on plant leaves, stems, and roots. Females are usually covered with white, waxy plates and at times carry a long egg sac that protrudes posteriorly from the abdomen. The egg sac of all other scale insects is attached to a substrate.

Common Species
Orthezia insignis Browne (Fig. 155A). Greenhouse orthezia—8-9 mm to tip of wax egg sac; dull green; white and longitudinally ribbed waxy egg sac extends > 2 times body length; throughout North America (ornamental plants in greenhouses).

Family Kerriidae—Lac Scales
Most species are subtropical and tropical. The Indian lac insect of India and southeast Asia, *Laccifer lacca* (Kerr), produces wax or lac used in making shellac and varnish. Six species (genus *Tachardiella*) occur in southwestern U.S. on desert plants; the lac of some species is highly pigmented.

Family Coccidae—Soft Scales
Females are oval and usually convex (but sometimes flattened) with a hard, smooth cuticle, or the cuticle is covered with wax. Some species are turtle-shaped (Fig. 155B). Certain species are plant pests and several Asian species produce wax used in candles and medicines.

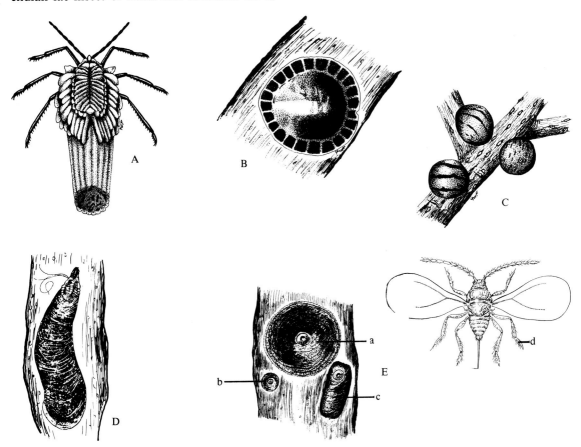

Figure 155 Scale insects. A, greenhouse orthezia, *Orthezia insignis* (Ortheziidae); B, terrapin scale, *Lecanium nigrofasciatum* (Coccidae); C, a gall-like scale, *Kermes* sp. (Kermesidae); D, oystershell scale, *Lepidosaphes ulmi* (Diaspididae); E, San Jose scale, *Quadraspidiotus perniciosus* (Diaspididae). a, mature female; b, immature female; c, immature male; d, mature male.

Order Hemiptera

Common Species

Ceroplastes ceriferus (Anderson)—5-20 mm (♀); spherical, black body with hornlike posterior tube; covered with thick, white, doughlike wax; throughout U.S. (in greenhouses), southern 1/3 U.S. (subtropical ornamentals, citrus).

Coccus hesperidum Linnaeus. Brown soft scale—3-5 mm (♀); flat, oval body; brown or marbled brown, sometimes pale yellow or greenish; throughout U.S. (in greenhouses), southern 1/2 U.S. (outdoors).

Lecanium corni Bouché. European fruit lecanium—2.5-5.0 mm (♀); brown when mature; oval; very convex; throughout North America.

Lecanium nigrofasciatum Pergande (Fig. 155B). Terrapin scale—2.5-3.5 mm (♀); reddish to dark brown; very convex; squarish areas around slightly crinkled margin outlined by black lines giving turtle-shell appearance; eastern U.S., southwest to NM.

Pulvinaria innumerabilis (Rathvon). Cottony maple scale—6 mm (♀); brown or reddish brown; secretes large, white, cottony mass of wax which elevates posterior part of body at angle; most of North America (on maple, other deciduous trees).

Saissetia coffeae (Walker). Hemispherical scale—3 mm (♀); shiny brown; very convex; elliptical; smooth; throughout U.S. (on ferns and other plants in greenhouses and homes, citrus).

Family Aclerdidae

Most members of this family feed near the bases of perennial grasses and sedges. All species from North America are in the genus *Aclerda*.

Family Kermesidae—
Gall-like Scales

Members of the genus *Kermes* (Fig. 155C) resemble tiny spherical galls or berries and are found almost exclusively on oak leaves and twigs.

Family Asterolecaniidae—
Pit Scales

Some species in this small family produce gall-like pits in the bark of their hosts (commonly oak) and others occur on leaves. Most species are in the genus *Asterolecanium;* mature females are covered with a pale green, glassy secretion. *A. puteanum* Russell, which occurs on holly and southern buckthorn primarily in the Atlantic coast states, is slightly convex and sometimes has wax threads at the margins.

Family Dactylopiidae—
Cochineal Insects

Females have reddish, rotund bodies covered with a white, waxy filamentous secretion. The cochineal insect, *Dactylopius coccus* Costa, occurs on prickly pear cactus in Mexico and a few other countries (not the U.S.) and still retains some value for its red cochineal dye. *D. opuntiae* (Cockerell), the cactus mealybug, and *D. confusus* (Cockerell) occur on *Opuntia* and other cacti in the southern third of the U.S. and are occasionally encountered in the warm deserts of Colorado, Utah, and Wyoming.

Family Diaspididae—
Armored Scales

The bodies of immature males and females and adult females are covered with a scale composed of cast skins and a waxy secretion. Diaspididae is the largest family of scale insects (over 300 species in the U.S.) and contains numerous, important pests of deciduous fruit and shade trees, citrus trees, and ornamental shrubs.

Common Species

Aonidiella aurantii (Maskell). California red scale—scale 1.7-2.0 mm (♀); red body gives transparent scale a reddish color; nearly circular with 1st cast skin appearing nipplelike near center; CA to southeastern U.S. and FL (chiefly on citrus; nearly identical to more yellowish yellow scale, *A. citrinia* [Coquillett]).

Aspidiotus nerii Bouché. Oleander scale—scale 1-2 mm (♀); light gray; circular; flat; yellowish brown cast skin near center; throughout U.S. (on English ivy, oleander, olive, citrus, subtropical ornamentals).

Aulacaspis rosae (Bouché). Rose scale—scale 1.5-2.0 mm (♀); white or grayish white, circular, flat, body reddish (♀); scale white, narrow, 3 ridges (♂); throughout U.S., southern Canada (on roses, bush fruits).

Chionaspis furfura (Fitch). Scurfy scale—scale 2.5-3.0 mm (♀); scale 0.7-0.9 mm (♂); grayish white; oystershell-shaped and thin (♀) or white, elongate-oval with 3 ridges (♂); throughout U.S. (apples, pears, other trees).

Chionaspis pinifoliae (Fitch). Pine needle scale—scale 2.5-3.0 mm (♀); scale 1 mm (♂); snowy white; oystershell-shaped elongate-oval; throughout North America (on needles of pine, other conifers).

Lepidosaphes ulmi (Linnaeus) (Fig. 155D). Oystershell scale—scale 2.0-3.5 mm (♀); scale 1.0-1.5 mm (♂); gray or brownish, elongate oystershell-shaped; very convex; throughout U.S., southern Canada.

Pseudaulacaspis pentagona (Targioni-Tozzetti). White peach scale—1.7-2.8 mm (♀); opaque white; nearly circular (♀) or elongate and without ridges (♂); convex; cast skins near margin, 1st skin often projecting; Atlantic coast states, southern 1/2 U.S. (on peaches, other fruit and ornamental trees, lilac).

Quadraspidiotus perniciosus (Comstock) (Fig. 155E). San Jose scale—scale 1.5-2.0 mm (♀); scale 0.8-1.0 mm (♂); gray; nearly circular with 1st cast skin appearing nipplelike at center (♀) or oval with skin at one end (♂); bodies bright yellow; throughout U.S., southern Canada.

Unaspis euonymi (Comstock). Euonymus scale—scale 2.0 mm (♀); scale 1.5 mm (♂); dark grayish brown to blackish, yellowish cast skin attached, broadly pear-shaped, convex (♀); white, 3 ridges (♂); most of U.S., southern Canada (on euonymus [chiefly in greenhouses]).

Family Pseudococcidae—Mealybugs

Most members of this relatively large family (over 300 species in the U.S.) have a whitish covering of mealy or powdery wax. The margin is often adorned with several waxy filaments. The female's body is elongate-oval and the legs are normally well developed. These insects range from the tropics to the arctic circle but are predominantly in warm climates; some are major pests in greenhouses and on citrus. The tamarisk manna scale of the Middle East secretes copious amounts of honeydew to form layers with the tree leaves. The edible sugary product is believed to have produced the manna in the biblical account of the Children of Israel.

Common Species

Planococcus citri (Risso) (Fig. 156A). Citrus mealybug—3 mm (♀); white; median stripe darker; anal filaments at most 1/4 body length; citrus regions (on citrus, avocados, cotton, ornamentals), frequently in greenhouses.

Pseudococcus longispinus (Targioni-Tozzetti) (Fig. 156B). Longtailed mealybug—3 mm (♀); white; body margin with 16 or 17 long, waxy filaments, the 2 filaments at tip of abdomen as long as or longer than body; throughout North America (greenhouses, citrus, other subtropical plants).

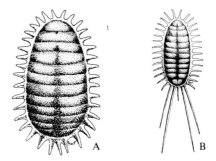

Figure 156 Mealybugs (Pseudococcidae). A, citrus mealybug, *Planococcus citri;* B, longtailed mealybug, *Pseudococcus longispinus.*

Order Hemiptera

Pseudococcus maritimus (Ehrhorn). Grape mealybug—3-5 mm (♀); white; powdery wax covering is very thin; lateral wax filaments short and slender; slender anal filaments 1/4-1/2 body length; throughout U.S. (in greenhouses), southern 1/3 U.S. straying north to MI (grapes, citrus, peaches, pears, walnuts, ornamentals).

Family Eriococcidae

These insects are similar to Pseudococcidae but the body usually is bare or only lightly covered with wax.

Common Species

Gossyparia spuria (Modeer). European elm scale—6-10 mm (♀); reddish brown; oval body surrounded by white cottony fringe; most of U.S. (on leaves and limbs of elms).

GENERAL REFERENCES ON HETEROPTERA

References to smaller families appear in Slater and Baranowski (1978).

Blatchley, W.S. 1926. *Heteroptera or True Bugs of Eastern North America, with Especial Reference to the Faunas of Indiana and Florida.* Nature Publ. Co., Indianapolis. 1116 pp.

Britton, W.E. *et al.* 1923. The Hemiptera or Sucking Insects of Connecticut. Guide to the Insects of Connecticut, Part IV. Conn. St. Geol. and Nat. Hist. Surv. Bull. No. 34. 807 pp.

DeCoursey, R.M. 1971. Keys to the Families and Subfamilies of the Nymphs of North American Hemiptera-Heteroptera. Proc. Ent. Soc. Wash. 73:413-428.

Drake, C.J., and Ruhoff, F.A. 1965. Lacebugs of the World: A Catalogue (Hemiptera: Tingidae). Bull. U.S. Natl. Mus. 243. 634 pp.

Froeschner, R.C. 1941-1961. Contributions to a Synopsis of the Hemiptera of Missouri. Pts. I-V. Amer. Midl. Nat. 26:122-146; 27:591-609; 31:638-683; 42:123-188; 67:208-240.

Herring, J.L., and Ashlock, P.D. 1971. A Key to the Nymphs of the Families of Hemiptera (Heteroptera) of America North of Mexico. Fla. Ent. 54:207-212.

Hungerford, H.B. 1920. The Biology and Ecology of Aquatic and Semiaquatic Hemiptera. Kans. Univ. Sci. Bull. 11:1-328.

———. 1948. The Corixidae of the Western Hemisphere (Hemiptera). Univ. Kan. Sci. Bull. 32:1-827.

———. 1959. Hemiptera. In *Fresh-Water Biology,* W.T. Edmundson (ed.). Wiley, N.Y. Pp. 958-972.

Knight, H.H. 1941. The Plant Bugs or Miridae of Illinois. Ill. Nat. Hist. Surv. Bull. 22:1-234.

———. 1968. Taxonomic Review: Miridae of the Nevada Test Site and the Western United States. Brigham Young Univ. Sci. Bull., Biol. Ser. 9:1-282.

Parshley, H.M. 1925. *A Bibliography of the North American Hemiptera-Heteroptera.* Smith College, Northampton, Mass. 252 pp.

Pennak, R.W. 1953. *Fresh-Water Invertebrates of the United States.* Ronald Press, N.Y. 769 pp.

Readio, P.A. 1927. Studies on the Biology of the Reduviidae of America North of Mexico. Univ. Kan. Sci. Bull. 17:1-291.

Slater, J.A. 1964. *A Catalogue of the Lygaeidae of the World.* Vols. I and II. Univ. Connecticut, Storrs, Conn. 1,668 pp.

Slater, J.A. and Baranowski, R.M. 1978. *How to Know the True Bugs (Hemiptera: Heteroptera).* Wm. C. Brown Company Publishers, Dubuque, Iowa.

Usinger, R.L. 1956. Aquatic Hemiptera. In *Aquatic Insects of California,* R.L. Usinger (ed.). Univ. California Press, Berkeley. Pp. 182-228.

Van Duzee, E.P. 1917. Catalogue of the Hemiptera of America North of Mexico, Excepting the Aphididae, Coccidae, and Aleurodidae. Univ. California Pubs. Tech. Bull. Ent. 2. 902 pp.

GENERAL REFERENCES ON HOMOPTERA

Annand, P.N. 1928. A Contribution Toward a Monograph of the Adelginae (Phylloxeridae) of North America. Stanford Univ. Pub. Biol. Sci. 6:1-146.

Beirne, B.P. 1956. Leafhoppers (Homoptera: Cicadellidae) of Canada and Alaska. Can. Ent. Suppl. 2:1-180.

Britton, W.E. *et al.* 1923. The Hemiptera or Sucking Insects of Connecticut. Conn. St. Geol. and Nat. Hist. Surv. Bull. 34. 807 pp.

Crawford, D.L. 1914. Monograph of the Jumping Plant Lice or Psyllidae of the New World. U.S. Natl. Mus. Bull. 85. 186 pp.

DeLong, D.M. 1948. The Leafhoppers, or Cicadellidae, of Illinois (Eurymelinae-Balcluthinae). Ill. Nat. Hist. Surv. Bull. 24:91-376.

Doering, K. 1930. Synopsis of North American Cercopidae. J. Kan. Ent. Soc. 3:53-64, 81-108.

Ferris, G.F. 1937-1955. *Atlas of the Scale Insects of North America.* Vol. 1-7. Stanford Univ. Press, Stanford, Calif.

Hottes, F.C., and Frison, T.H. 1931. The Plant Lice, or Aphididae, of Illinois. Ill. Nat. Hist. Surv. Bull. 19:121-447.

Howell, J.O., and Williams, M.L. 1976. An Annotated Key to the Families of Scale Insects (Homoptera:Coccoidea) of America, North of Mexico, Based on Characteristics of the Adult Female. Ann. Ent. Soc. Amer. 69:181-189.

Kennedy, J.S., and Stroyan, H.L.G. 1959. Biology of Aphids. Ann. Rev. Ent. 4:139-160.

MacGillivray, A.D. 1921. *The Coccidae.* Scarab, Urbana, Ill. 502 pp.

McKenzie, H.L. 1967. *The Mealybugs of California.* Univ. California Press, Berkeley. 525 pp.

Metcalf, Z.P. 1923. Fulgoridae of Eastern North America. J. Elisha Mitchell Sci. Soc. 38:139-230.

———. 1945. *A Bibliography of the Homoptera (Auchenorrhyncha).* Dept. Zool. & Ent., No. Carolina St. Coll., Raleigh, N.C. Vol. 1-2.

———. 1954-1963. *General Catalogue of the Homoptera.* No. Carolina St. Coll., Raleigh, N.C. Fasc. IV. Fulgoroidea. Parts 11-18. Fasc. VII. Cercopoidea. Parts 1-4. Fasc. VIII. Cicadoidea. Parts 1-2.

———. 1962-1967. *General Catalogue of the Homoptera.* USDA, ARS. Fasc. VI. Cicadelloidea. Parts 1-17.

Moore, T.E. 1966. The Cicadas of Michigan (Homoptera:Cicadidae). Pap. Mich. Acad. Sci. 51:75-96.

Morrison, H., and Renk, A.V. 1957. A Selected Bibliography of the Coccoidea. USDA Agric. Res. Serv. Misc. Pub. No. 734. 222 pp. First supplement (Morrison, H., and E. Morrison; 1965; Ibid. 987). Second supplement (Russell, L.M.; M. Kosztarab; and M.P. Kosztarab; 1970; Ibid. 1281).

Oman, P.W. 1949. The Nearctic Leafhoppers (Homoptera:Cicadellidae), a Generic Classification and Check List. Ent. Soc. Wash. Mem. No. 3. 253 pp.

Osborn, H. 1938. The Fulgoridae of Ohio. Ohio Biol. Surv. Bull. 35:283-349.

———. 1940. The Membracidae of Ohio. Ohio Biol. Surv. Bull. 37:51-101.

Palmer, M.A. 1952. *Aphids of the Rocky Mountain Region.* Thomas Say Foundation Vol. 5. Hirschfeld Press, Denver. 452 pp.

Smith, C.F. 1972. Bibliography of the Aphididae of the World. N.C. Agric. Expt. Sta. Tech. Bull. 216. 717 pp.

Wade, V. 1966. General Catalogue of the Homoptera. Species Index of the Membracoidea and Fossil Homoptera (Homoptera:Auchenorrhyncha). A Supplement to Fascicle 1—Membracidae of the General Catalogue of the Hemiptera. N.C. Agric. Expt. Sta. Pap. No. 2160. 40 pp.

Williams, M.L., and Kosztarab, M. 1972. Morphology and Systematics of the Coccidae of Virginia, with Notes on Their Biology (Homoptera:Coccoidea). Va. Polytech. Inst. and St. Univ. Res. Div. Bull. 74. 215 pp.

ORDER COLEOPTERA[24]
Beetles

Beetles comprise the largest order of insects with over 300,000 species forming about 40 percent of the known insects. Beetles have thickened front wings (elytra) that are soft, leathery, or hard and brittle. The elytra usually meet in a straight line along the middle of the back and cover most or all of the longer, folded, membranous hind wings (Fig. 157A). Some beetles have short elytra and/or hind wings or are wingless. Most beetles use only the hind wings for flight; the elytra are raised to allow for the hind wing movement. The pronotum is generally distinct whereas the mesonotum and the metanotum are hidden beneath the elytra. The mouthparts of adults and most larvae are adapted for chewing. Some predatory larvae are able to suck body fluids of prey. Larvae have a holometabolous type of development.

The appearance, life history, and habitats of beetles are so diverse that only general statements can be made. Their sizes vary from less than 1 mm to over 200 mm in length and up to 75 mm in width (e.g., tropical rhinoceros and goliath beetles). Body shapes vary from round to very slender, and flattened to stout or cylindrical. Some adults have long snouts, horns, or tubercles on the head, extremely long antennae, greatly enlarged mandibles, or other diverse features. The body and elytral surface may be smooth or rough, brightly colored or dull.

24. Coleoptera: *coleo,* sheath; *ptera,* wings.

Many beetles produce sounds by rubbing body parts together (e.g., femur of the hind leg against the margin of the elytron; head rubbed against the front margin of the prothorax) or by expelling air from the spiracles. Some members of the Lampyridae, Phengodidae, Drilidae, and Elateridae emit light (biolumenescence) from special organs. Certain beetles in the families Staphylinidae, Pselaphidae, Scydmaenidae, and others live in the nests of ants and termites.

Beetles are primarily plant feeders and scavengers, although a significant number are predators of insects and other small invertebrates and a few are parasites. The plant-feeding habits (fruit, seed, root, stem, and wood borers; leaf feeders) of adults and larvae have made many beetles serious pests of agricultural crops and forests. Some beetles also feed on museum specimens, clothing, and stored food. A few species transmit bacterial and fungal diseases to plants. Life cycles vary from several weeks in warm climates to four or more years in cool climates. Wood- and root-boring beetles usually have longer life cycles than leaf-feeding species. Eggs are usually deposited in an area suitable for larval development. The larvae, sometimes called grubs, typically molt three to five times. The immobile pupa is soft, pale, and resembles the adult.

Beetles may be collected from almost any habitat by sweeping, beating, sifting, using Berlese funnels and light traps, etc. Pit traps baited with carrion such as old meat, a dead mouse, or a fish head will attract scavenger beetles. Molasses or brown sugar syrup placed in a pit trap or in a container suspended from a low tree branch will draw in many kinds of beetles. Wood-boring beetles are attracted to a screen-covered container of turpentine suspended in a tree. Adults are pinned through the base of the right elytron. Small beetles are mounted on paper points. Larvae are preserved in 70% alcohol. Degreasing large members of the families Carabidae and Scarabaeidae is sometimes practiced to remove surface grease that forms after pinning. These specimens are immersed in ether, xylene, or benzene, and the solvent renewed until it is no longer discolored.

Species: North America, 28,000; world, >300,000. Families: North America, about 113.

KEY TO COMMON FAMILIES OF COLEOPTERA

Familiarity with the external morphology of beetles (Fig. 157) is important in identification. For example, determining that the hind coxae divide the posterior margin of the first abdominal sternite (Fig. 157B) will help separate a very common family, Carabidae (and also the suborder Adephaga), from the remaining beetles (Fig. 158). The difference between open and closed coxal cavities must be recognized (Fig. 164). The number, shape, and size of tarsal segments is important; a 5-5-4 tarsal formula means that the front and middle tarsi have five segments and the hind tarsus has four. If the tarsus appears to be 4-segmented and the third segment is bilobed ("U"-shaped), the tarsus usually is 5-segmented with the extra tiny segment hidden inside the third segment (Fig. 162D).

1a   Head rarely extended into a snout; antennae variable.............2

1b   Front of head usually extended into a snout (Fig. 213A) or broad and short muzzle (Fig. 211B); antennae arise along snout, often clubbed and elbowed (Fig. 160G), occasionally filiform or moniliform (Figs. 160A, B); if no snout or muzzle then: body is less than 9 mm long, cylindrical, antennae elbowed and usually with a large club, and front tibia with row of teeth or its tip extended into a stout spur (Fig. 212)...........74

176   Subclass Pterygota

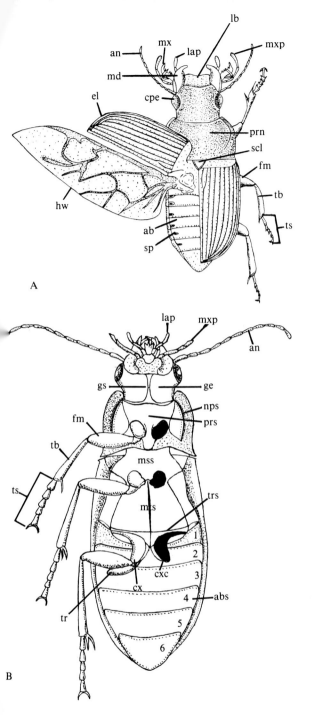

(suborder Adephaga). ab, abdomen; abs, abdominal sternite; an, antenna; cpe, compound eye; cx, coxa; cxc, coxal cavity; el, elytra (front wing) with grooves; fm, femur; ge, gena; gs, gular suture; hw, hind wing; lap, labial palpus; lb, labrum; md, mandible; mss, mesosternum; mts, metasternum; mx, maxilla; mxp, maxillary palpus; nps, notopleural suture; prn, pronotum; prs, prosternum; scl, scutellum; sp, spiracle; tb, tibia; tr, trochanter; trs, transverse suture; ts, tarsus.

Figure 157 General structure of a ground beetle, *Harpalus caliginosus* (Carabidae). A, dorsal view; B, ventral view showing the posterior margin of the first abdominal sternite divided by the hind coxae

2a(1a) Hind coxae flattened and widened into large plates that cover most of abdomen (Fig. 169B); aquatic; oval; 5 mm long or less . . . . (p. 191) *Haliplidae*

2b Hind coxae not greatly expanded; habitat, size, and shape variable 3

3a(2b) Posterior margin of first abdominal sternite divided by hind coxae (Fig. 157B); antennae almost always filiform (Fig. 160A); tarsi usually 5-5-5; hind trochanters large and offset toward center of body (Fig. 157B) (Suborder Adephaga) 4

3b Posterior margin of first abdominal sternite not divided by hind coxae and extending across abdomen without interruption (Fig. 158); antennae and tarsi variable; hind trochanters small and not offset (Fig. 158) (Suborder Polyphaga) . . . 8

4a(3a) Hind legs flattened and fringed with long hairs (Fig. 170C); aquatic, also found on land (some fly to lights) . . 5

4b Hind legs generally not flattened or fringed with long hairs; usually terrestrial 6

Order Coleoptera 177

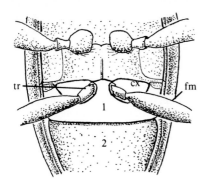

Figure 158 Ventral view of a beetle in the suborder Polyphaga showing the undivided posterior margin of the first abdominal sternite. 1, first abdominal sternite; cx, coxa; fm, femur; tr, trochanter.

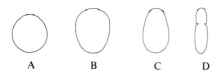

Figure 159 General body shapes. A, spherical; B, broadly oval; C, oval; D, elongate.

5a(4a)	Compound eyes divided into upper and lower halves (Fig. 171B); black and oval; antennae very short; swim in circles on water's surface .... (p. 192) *Gyrinidae*	
5b	Compound eyes not divided; color brown, black, or greenish, often with pale yellow margins on elytra and pronotum; oval; antennae long and filiform; do not swim in circles on water's surface................. .... (p. 191) *Dytiscidae*	
6a(4b)	No transverse suture anterior to hind coxae; 3 or more longitudinal grooves on pronotum (Fig. 166); 5.5-7.5 mm long................. .... (p. 188) *Rhysodidae*	
6b	Transverse suture anterior to hind coxae (Fig. 157B); longitudinal grooves on pronotum present or absent; size variable 7	
7a(6b)	Head including eyes as wide or wider than pronotum (Fig. 167); mandibles long and sickle-shaped; clypeus enlarged laterally beyond bases of antennae; elytra not grooved or punctured longitudinally (Subfamily Cicindelinae) .... (p. 188) Carabidae (in part)	
7b	Head including eyes generally narrower than pronotum (Fig. 168); mandibles usually not long and sickle-shaped; clypeus not enlarged as above; elytra often grooved or punctured longitudinally (Fig. 168B) .... (p. 188) *Carabidae* (in part)	
8a(3b)	Tarsal segmentation 5-5-5, 4-4-4, 3-4-4, or 3-3-3................. 9	
8b	Tarsal segmentation 5-5-4....... 49	
9a(8a)	Tarsal segmentation actually or apparently 4-4-4, 3-4-4, or 3-3-3 .. 10	
9b	Tarsal segmentation 5-5-5....... 11	
10a(9a)	Tarsal segmentation actually or apparently 4-4-4 or 3-4-4........ 62	
10b	Tarsal segmenation 3-3-3 72	
11a(9b)	Antennae lamellate (Fig. 160D) or flabellate (Fig. 160E) 12	

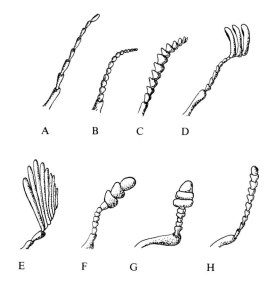

Figure 160 Antennae. A, filiform; B, moniliform; C, serrate; D, lamellate; E, flabellate; F, 3-segmented club; G, elbowed and clubbed; H, clavate.

11b	Antennae not lamellate or flabellate 18
12a(11a)	Antennae flabellate 13
12b	Antennae lamellate. 14
13a(12a)	Stout-bodied; hairy on ventral side; front tibiae wide, flat, and scalloped or toothed; western U.S. . . . . (p. 198) *Scarabaeidae* (in part)
13b	Not as above; first 4 tarsal segments with paired lobes beneath (Fig. 162A), hairy projection between tarsal claws; widely distributed . . . . (p. 203)*Rhipiceridae*
14a(12b)	Last 3-4 plates of lamellate antennae flattened and can be closed into a ball (Fig. 160D); tibiae often flattened and toothed or scalloped . . . . (p. 198) *Scarabaeidae* (in part)
14b	Last 3-4 plates of lamellate antennae rounded or flattened but cannot be closed into a ball (Fig. 160D) 15
15a(14b)	Last tarsal segment nearly as long as other tarsal segments combined (Fig. 162B) 16
15b	Last tarsal segment similar in length to other segments 17
16a(15a)	Length 1-8 mm; elytra cover entire abdomen; silvery, gray, brown or blackish; aquatic or shoreline beetles . . . . (p. 202)*Dryopidae* (in part)
16b	Length 15-35 mm; elytra leave about 3 abdominal segments exposed (Fig. 173A); black or orange and black; terrestrial beetles associated with carrion . . . . (p. 193) *Silphidae* (in part)
17a(15b)	Pronotum with deep median groove; elytra with conspicuous longitudinal grooves (Fig. 176C); head with short horn on top that curves forward; mandibles not unusually long; shining black; 30-40 mm long . . . . (p. 197) *Passalidae*
17b	Pronotum without median groove; elytra generally smooth; head usually without horn; mandibles long, half the body length or longer in some males; brown or black; 8-40 mm long . . . . (p. 197) *Lucanidae*

Order Coleoptera

18a(11b) Maxillary palpi long, slender, and equaling or exceeding length of antennae (Fig. 172A); hind legs curved and flattened for swimming (Fig. 172A); antennae clubbed . . . . (p. 193) *Hydrophilidae*

18b Maxillary palpi much shorter than antennae; hind legs not modified for swimming; antennae variable 19

19a(18b) Elytra usually cover less than half the length of abdomen (Fig. 174); elytra occasionally pubescent. . . . . (p. 194) *Staphylinidae*

19b Elytra cover more than half or all of abdomen; elytral pubescence present or absent 20

20a(19b) Elytra leave 1-2 abdominal segments exposed; tip of abdomen sharply pointed (Fig. 173C); elytra shining . . . . (p. 194) *Scaphidiidae*

20b Not as above. 21

21a(20b) Abdomen with 7-8 ventral segments visible. 22

21b Abdomen with less than 7 ventral segments visible 24

22a(21a) Middle coxae separated from each other; elytra usually with netlike sculpturing and posterior half noticeably wider (Fig. 184); often brightly colored . . . . (p. 207) *Lycidae*

22b Middle coxae touching or nearly so; elytra usually parallel-sided, sometimes posterior half slightly wider; color variable. 23

23a(22b) Head generally concealed by pronotum and not visible from above (Fig. 183A); episternum of metathorax with nearly straight inner margin (Fig. 161A); many with yellowish band on ventral side of abdomen . . . . (p. 206) *Lampyridae*

23b Head slightly or not concealed by pronotum and visible from above; episternum of metathorax with "S" curved inner margin (Fig. 161B); no yellowish ventral abdominal band . . . . (p. 206) *Cantharidae*

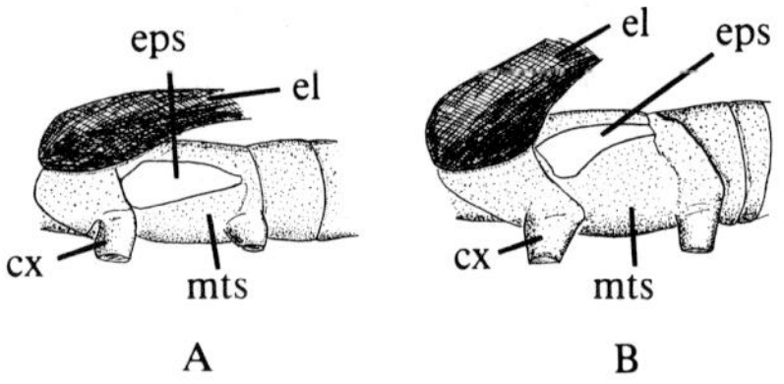

Figure 161 Lateroventral views of the metathorax to compare the inner margin of the episterna. A, firefly (Lampyridae); B, soldier beetle (Cantharidae). cx, coxa; el, elytron; eps, episternum; mts, metasternum.

24a(21b) Abdomen with 6 ventral segments visible. 25

24b Abdomen with 5 ventral segments visible. 27

25a(24a) Flattened, oval or broadly oval, generally blackish beetles (Fig. 173B); antennae with 3- or 4-segmented pubescent club; usually

	over 10 mm long; elytra sparsely or not pubescent. . . . . (p. 193) *Silphidae* (in part)	28b	Not with above combination of features 29
25b	Usually cylindrical; antennae variable; length usually 10 mm or less; elytra pubescent 26	29a(28b)	Over half or all of head concealed from above by pronotum 30
		29b	Over half or all of head visible from above 37
26a(25b)	Middle coxae conical (Fig. 163A) and prominent; hind coxae prominent and extend ventrally below abdomen; elytra much broader posteriorly producing a stout, wedge-shaped appearance (Fig. 190). . . . . (p. 211) *Melyridae*	30a(29a)	Antennae long and usually at least half the body length, filiform, with or without basal process on some outer segments (Fig. 180A) 31
		30b	Antennae short and usually less than half the body length, shape variable, no basal processes. 32
26b	Middle coxae rounded (Fig. 163B) and not prominent; hind coxae somewhat flattened and not prominently extending below abdomen; elytra sometimes slightly broader posteriorly but generally with an elongated and narrowed appearance . . . . (p. 210) *Cleridae* (in part)	31a(30a)	Antennae with basal process on segments 4-10 (Fig. 180A) . . . . (p. 202) *Ptilodactylidae*
		31b	Antennae without basal processes . . . . (p. 208) *Ptinidae*
27a(24b)	Antennae both elbowed and clubbed (Fig. 160G); elytra short and squared-off at ends leaving 2 abdominal segments exposed (Fig. 175B) . . . . (p. 196) *Histeridae*	32a(30b)	Body elongate, elongate-oval, or cylindrical (Fig. 159D) 33
		32b	Body broadly oval and convex to nearly spherical (Figs. 159A, B) . . 36
27b	Antennae, elytra, and abdominal segments not in above combination; if antennae are clubbed they are not also elbowed. 28	33a(32a)	Antennae very short, clavate or lamellate with most segments broader than long (Fig. 180B). . . . . (p. 202) *Dryopidae* (in part)
		33b	Antennae not as above 34
28a(27b)	Dense covering of white, gray, or yellowish hairs on light brown or dark orange body; 3-segmented antennal club; second and third tarsal segments lobed; body elongate with elytra parallel-sided . . . . (p. 217) *Byturidae*		

Order Coleoptera

34a(33b) Pronotum and elytra sculptured with curved ridges giving a slightly wrinkled appearance (Fig. 180C); rows of punctures on elytra . . . . (p. 203)*Elmidae*

34b Pronotum and elytra not highly sculptured or wrinkled 35

35a(34b) Pronotum often covers and encloses all of head from above; at least last 3 segments of antennae lengthened, often expanded and pectinate or serrate; hind coxae with grooves for receiving femora (Fig. 163D) . . . . (p. 208) *Anobiidae* (in part)

35b Pronotum does not enclose but usually covers all of head from above; antennae not lengthened or expanded distally; pronotum with tubercles, hooks, or rasplike teeth near front. . . . . (p. 209) *Bostrichidae*

36a(32b) At least last 3 segments of antennae lengthened, often expanded and pectinate or serrate; fourth tarsal segment not bilobed, tarsi decreasing in size distally; hind coxae with grooves for holding femora (Fig. 163D); head enclosed and concealed by pronotum; body oval. . . . . (p. 208) *Anobiidae* (in part)

36b Last 3 segments of antennae not lengthened, antennae weakly serrate; fourth tarsal segment bilobed and larger than third; no grooves in hind coxae; head often concealed by shelflike anterior expansion of pronotum; body oval to spherical . . . . (p. 201) *Helodidae* (in part)

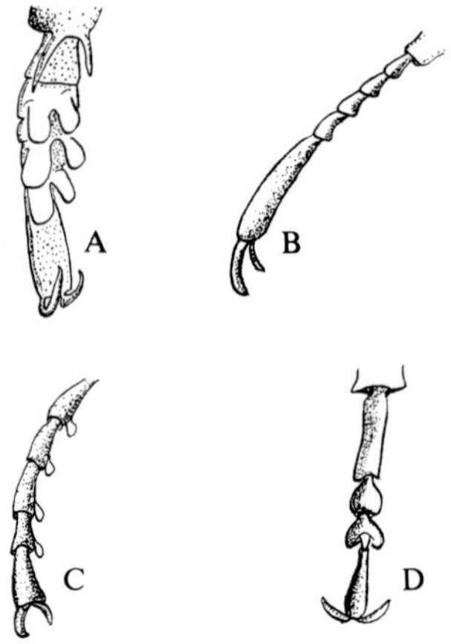

Figure 162 Tarsi modifications. A, paired ventral lobes; B, last tarsal segment nearly as long as others combined; C, ventral lobes; D, third segment bilobed and fourth hidden inside.

37a(29b) Tarsi with 1 or more ventral lobes (Fig. 162C). 38

37b Tarsi without ventral lobes, although segments may be bilobed or dilated (Fig. 162D) with hair pads beneath 41

38a(37a) Only fourth tarsal segment lobed ventrally; hind femora sometimes enlarged for jumping; 2-4 mm long . . . . (p. 201) *Helodidae* (in part)

38b Two or more tarsal segments or only third tarsal segment with prominent ventral lobes; hind femora not enlarged; 3-24 mm long 39

39a(38b) Tarsal segments 1-4 with prominent ventral lobes. . . . . (p. 210) *Cleridae* (in part)

39b Tarsal segment 3 or segments 1-3, 2-3, or 2-4 with prominent ventral lobes. 40

40a(39b)	Posterior corners of pronotum usually pointed (Fig. 182); prothorax loosely joined to mesothorax; posterior margin of prosternum elongated into spinelike projection between front coxae (Fig. 182A).... .... (p. 204) *Elateridae*		43a(41b)	Front coxae lie at right angles to main axis of body............ 44
40b	Pronotum, prothorax, and posternum not as above .... (p. 201) *Dascillidae*		43b	Front coxae do not lie at right angles to main axis of body.......... 45
			44a(43a)	Tarsi slender, first segment very short; pronotum often separated from base of elytra except at point of attachment (Fig. 188B); elytra never truncate.................... .... (p. 210) *Trogositidae*
41a(37b)	Front coxae conical, projecting prominently from coxal cavity (Fig. 163A)................ 42			
41b	Front coxae round (Fig. 163B), oval, or lying at right angles to main axis of body (Fig. 163C), and usually projecting only slightly........... 43		44b	First 3 tarsal segments about same size with brush of hair beneath, fourth segment smaller and without brush of hair; pronotum usually not separated from base of elytra; elytra often truncate and showing tip of abdomen...................... .... (p. 212) *Nitidulidae*

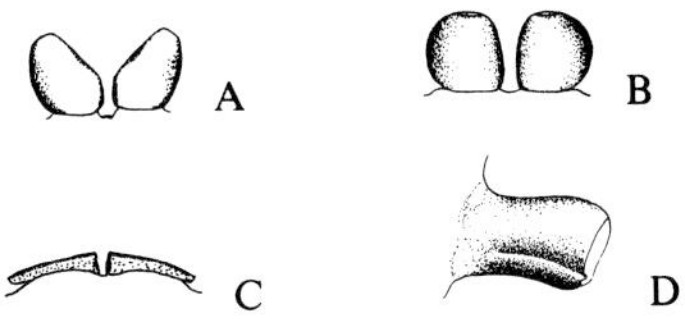

Figure 163 A-C, front coxae: A, conelike; B, rounded; C, transverse; D, grooved hind coxa.

			45a(43b)	Posterior margin of prosternum elongated into spinelike projection between front coxae (Fig. 182A).... .... (p. 203) *Buprestidae*
42a(41a)	Body oval, broadening posteriorly (Fig. 179D); antennae serrate; dorsum rarely hairy, not scaly; aquatic beetles .... (p. 202) *Psephenidae*		45b	Posterior margin of prosternum not elongated as above 46
			46a(45b)	Antennae with abrupt 2-segmented (rarely 3) club; head narrows behind eyes .... (p. 209) *Lyctidae*
42b	Body oval or broadly oval, not broadening posteriorly (Fig. 185); antennae clubbed; dorsum often noticeably hairy or scaly; terrestrial beetles .... (p. 207) *Dermestidae*		46b	Not with above combination of features 47
			47a(46b)	Body very flattened and sides parallel or body oval; middle coxal cavities open (Fig. 164D) .... (p. 213) *Cucujidae* (in part)

Order Coleoptera

47b   Body oval, elongate-oval, or cylindrical with parallel sides; middle coxal cavities closed (Fig. 164C) .. 48

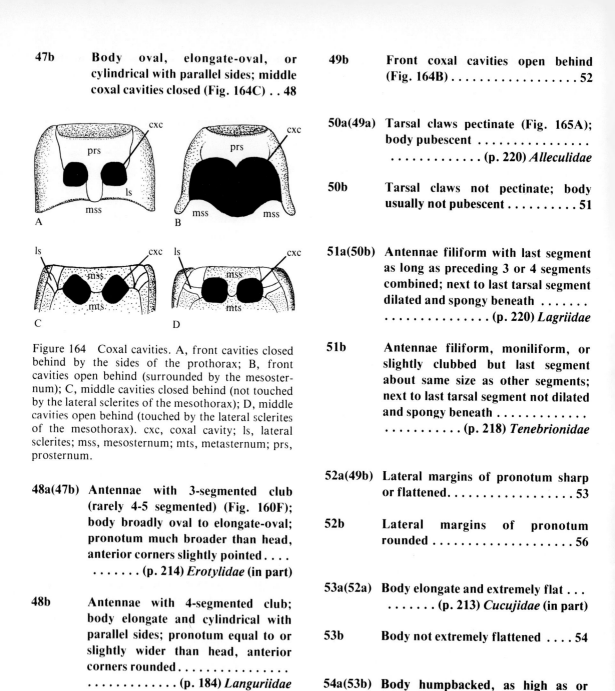

Figure 164 Coxal cavities. A, front cavities closed behind by the sides of the prothorax; B, front cavities open behind (surrounded by the mesosternum); C, middle cavities closed behind (not touched by the lateral sclerites of the mesothorax); D, middle cavities open behind (touched by the lateral sclerites of the mesothorax). cxc, coxal cavity; ls, lateral sclerites; mss, mesosternum; mts, metasternum; prs, prosternum.

48a(47b)   Antennae with 3-segmented club (rarely 4-5 segmented) (Fig. 160F); body broadly oval to elongate-oval; pronotum much broader than head, anterior corners slightly pointed.... .... (p. 214) *Erotylidae* (in part)

48b   Antennae with 4-segmented club; body elongate and cylindrical with parallel sides; pronotum equal to or slightly wider than head, anterior corners rounded................ .... (p. 184) *Languriidae*

49a(8b)   Front coxal cavities closed behind (Fig. 164A).................. 50

49b   Front coxal cavities open behind (Fig. 164B).................. 52

50a(49a)   Tarsal claws pectinate (Fig. 165A); body pubescent .... (p. 220) *Alleculidae*

50b   Tarsal claws not pectinate; body usually not pubescent 51

51a(50b)   Antennae filiform with last segment as long as preceding 3 or 4 segments combined; next to last tarsal segment dilated and spongy beneath .... (p. 220) *Lagriidae*

51b   Antennae filiform, moniliform, or slightly clubbed but last segment about same size as other segments; next to last tarsal segment not dilated and spongy beneath .... (p. 218) *Tenebrionidae*

52a(49b)   Lateral margins of pronotum sharp or flattened.................. 53

52b   Lateral margins of pronotum rounded 56

53a(52a)   Body elongate and extremely flat ... . (p. 213) *Cucujidae* (in part)

53b   Body not extremely flattened 54

54a(53b)   Body humpbacked, as high as or higher than wide (Fig. 204B); head bent downward (Fig. 204B); tip of abdomen pointed (Fig. 204) .... (p. 222) *Mordellidae*

184   Subclass Pterygota

54b	Without above combination of features 55	58b	At least outer portion or all of antennae filiform 59

55a(54b) Antennae filiform; 2 dents or depressions near posterior margin of pronotum; first segment of hind tarsi much longer than any other segment; 3-20 mm..................... ..... (p. 222) *Melandryidse* (in part)

55b Antennae with prominent 3-segmented club; no dents or depressions near posterior margin of pronotum; first segment of hind tarsi not greatly lengthened; 1-5 mm .... (p. 213) *Cryptophagidae*

56a(52b) Two dents or depressions near posterior margin of pronotum (Fig. 203C); first segment of hind tarsi much longer than any other segment .... (p. 222) *Melandryidae* (in part)

56b Not as above................. 57

57a(56b) Next to last segment of hind tarsi expanded and heart-shaped or bilobed, often with dense brush of hairs ventrally; pronotum widest in front and narrower than elytra at back; elongated body............ .... (p. 222) *Oedemeridae*

57b Not with above combination of features 58

58a(57b) Antennae serrate, flabellate, pectinate, or plumose (Figs. 160C, E; 202A)....... (p. 221) *Pyrochroidae*

59a(58b) Head not abruptly narrowed behind eyes; pronotum widest at middle and narrows toward front and back (Fig. 201A)..................... .... (p. 220) *Salpingidae*

59b Head abruptly narrowed behind eyes to form a slender neck (Figs. 205; 206) 60

60a(59b) Each tarsal claw split to base or toothed (Figs. 165B, C); abdomen with 6 ventral segments .... (p. 223) *Meloidae*

60b Tarsal claws not split or toothed; abdomen usually with 5 ventral segments................... 61

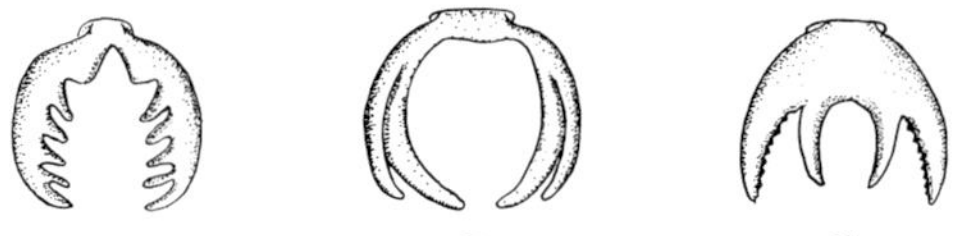

Figure 165 Tarsal claws. A, pectinate (Alleculidae); B, split to base (Meloidae); C, toothed (Meloidae).

61a(60b) Length 7-12 mm; eyes usually with slight notch on margin; hind coxae touching; pronotum without horn .. .. (p. 224) *Pedilidae*

61b Length under 6 mm; eyes without a notch; hind coxae separated; often with pronotal horn extending over head (Fig. 206A) .... (p. 224) *Anthicidae*

Order Coleoptera

62a(10a)	Tibia widened and armed with row of prominent flattened spines on outer margin (Fig. 180D); mandibles greatly flattened and project forward; first and fourth tarsal segments longer than second or third............. .... (p. 203) *Heteroceridae*		notum; pubescence consists of erect hairs; elongated and cylindrical .... (p. 221) *Ciidae*
62b	Not with above combination of features 63	66b	Head not concealed; tarsal segments variable in length; no horn; pubescence present or absent but hairs not erect; elongate-oval or oval 67
63a(62b)	Terminal 1-3 segments of antennae distinctly clubbed, or if antennae clavate the last 2-3 segments are clubbed 64	67a(66b)	Antennae clavate (segments 7-11 enlarged), last 2-3 segments with loose club; anterior corners of pronotum not prolonged; tarsal segmentation 4-4-4 or 3-4-4; usually brown to black .... (p. 220) *Mycetophagidae*
63b	Antennae variable but not clubbed, if clavate then the last 2-3 segments not distinctly clubbed 69	67b	Antennae not clavate but last 3 segments abruptly clubbed; anterior corners and margin of pronotum may extend forward to partly conceal head (Figs. 197A, B); tarsal segmentation actually or apparently 4-4-4; often red or yellow markings 68
64a(63a)	Antennae elbowed and with a distinct, large, solid club (Fig. 160G); elongate and cylindrical body...... .... (p. 233) *Scolytidae* (in part)		
64b	Antennae not elbowed although 1- to 3-segmented terminal club present; body shape variable 65	68a(67b)	Anterior corners of pronotum extended distinctly forward (Figs. 197A, B); pronotum usually with 2 longitudinal grooves (Fig. 197A) or 2 pits near posterior margin connecting to longitudinal grooves extending up to half the length of pronotum; tarsal segmentation actually 4-4-4 but sometimes appearing 3-3-3 due to small third segment .... (p. 217) *Endomychidae* (in part)
65a(64b)	Body broadly oval and highly convex (Fig. 195); shiny, dark .... (p. 215) *Phalacridae*		
65b	Body oval to elongate and cylindrical; color variable, often brightly patterned............. 66		
66a(65b)	Head concealed from above; first 3 tarsal segments short; horn sometimes present on head or pro-	68b	Anterior corners of pronotum very slightly or not extended forward; pronotum without longitudinal grooves or pits with connecting grooves although lateral margins

	may be very distinct and groovelike; tarsal segmentation actually 5-5-5 but appearing 4-4-4 due to small fourth segment .... (p. 214) *Erotylidae* (in part)
69a(63b)	Antennae usually clavate; body oval and greatly convex to give a pill shape (Fig. 179C); head nearly or completely concealed .... (p. 202) *Byrrhidae*
69b	Antennae sometimes clavate but usually filiform, serrate, pectinate, or flabellate; if body oval then not greatly convex; head sometimes concealed..................... 70
70a(69b)	Body elongate and parallel-sided or narrowing posteriorly; antennae usually much more than half body length (Fig. 207) (rarely shorter than half body length) and inserted into the notched area of eyes (Fig. 207A); commonly 3-50 mm long .... (p. 225) *Cerambycidae*
70b	Body usually oval or elongate oval, if elongate it does not narrow posteriorly; antennae shorter than half body length; eyes not notched; usually less than 12 mm long..... 71
71a(70b)	Body oval or egg-shaped, broadest posteriorly; head expanded into short, broad snout; elytra expose tip of abdomen (Fig. 211) .... (p. 232) *Bruchidae*
71b	Body not egg-shaped or broadest posteriorly; no short, broad snout; tip of abdomen not usually exposed .... (p. 227) *Chrysomelidae*
72a(10b)	Elytra short, covering about half of abdomen (Fig. 175A); tarsi rarely 2-segmented .... (p. 196) *Pselaphidae*
72b	Elytra cover abdomen; tarsi never 2-segmented, apparently 3-3-3 but actually 4-4-4 73
73a(72b)	Oval to elongate; 2 longitudinal grooves on pronotum (Fig. 197A) or 2 pits near posterior margin connected to longitudinal grooves extending up to half the length of the pronotum; anterior corners of pronotum extended and partly enclosing head (Fig. 197A) .... (p. 217) *Endomychidae* (in part)
73b	Broadly oval to spherical and greatly convex above (Fig. 196); pronotum without longitudinal grooves or pits with connecting grooves; anterior corners of pronotum not extended .. .. (p. 215) *Coccinellidae*
74a(1b)	No snout or broad muzzle present; antennae elbowed and clubbed (Fig. 160G); body elongated and cylindrical (Fig. 212); most are 1-3 mm long.................. .... (p. 233) *Scolytidae* (in part)
74b	Snout (Figs. 211C; 213; 214) or broad muzzle present (Fig. 211B); antennae variable; oval to elongate; 1-35 mm..................... 75

75a(74b) Broad, short muzzle present (Fig. 211B); antennae clubbed but not elbowed, rarely unclubbed; elytra often mottled with white, brown, or black pubescence .... (p. 232) *Anthribidae*

75b Short to long snout present (Figs. 211C; 213; 214); antennae clubbed and almost always elbowed, or sometimes filiform or moniliform; elytra with or without pubescence... 76

76a(75b) Snout generally curved (Figs. 213; 214); antennae clubbed and almost always elbowed (Fig. 213A) .... (p. 234) *Curculionidae*

76b Snout extends straight forward (Fig. 211C); antennae filiform or moniliform (Fig. 211C) .... (p. 232) *Brentidae*

SUBORDER ADEPHAGA

Family Rhysodidae

Rhysodid beetles are 5.5-8.0 mm long, dark reddish brown, and slender-bodied. The antennae are moniliform and the pronotum has three distinct longitudinal furrows. The four species in the U.S. occur under the bark of ash, beech, elm or pine.

Common Species

Omoglymmius americanus (Laporte) (Fig. 166)—6-8 mm; shiny dark reddish brown; elytral grooves with large punctures; eastern 1/2 U.S.

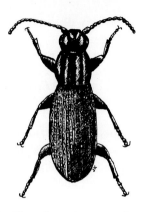

Figure 166 *Omoglymmius americanus* (Rhysodidae).

Family Carabidae—Ground Beetles and Tiger Beetles

This family of predaceous beetles is one of the largest in the order with nearly 40,000 species worldwide and about 3,100 in the U.S. and Canada. The tiger beetles are considered a subfamily by carabid taxonomists although many texts classify them as a separate family, the Cicindelidae, partly due to custom. Another group, the genus *Omophron,* has sometimes been placed in a separate family, Omophronidae.

Tiger beetles are long-legged and run and fly rapidly. They frequently fly low to the ground for varying distances. The adults are often colorful, sometimes iridescent, and always have distinct, sharp mandibles. The "S"-shaped larva has a hump with curved hooks on the fifth abdominal segment. Larvae are predaceous and occur in vertical tunnels beneath the ground. The larva's head protrudes from the circular tunnel opening to grasp a passing insect. Sandy beaches with scattered vegetation, stream banks, and sandy trails are common habitats for adults and larvae. Some also occur along forest edges, on clay soils, and in rotten stumps. The genus *Omus* contains black, wingless, nocturnal species that occur in the Far West.

Ground beetles are usually dark, shiny, and somewhat flattened although some are brilliant and/or iridescent. The legs are long and often the mandibles are well developed. The elytra of many are longitudinally grooved and the trochanters of some are large and kidney-shaped. Most species are nocturnal and occur under debris, wood, and rocks during the day. Adults are usually scavengers and larvae are predaceous. Some adults feed on insects, slugs, snails, or climb trees in search of caterpillars. One group eats seeds, particularly of grasses. The elongate larvae have ten distinct segments and taper toward each end. The head and mandibles are large, and the ninth abdominal segment has a pair of cerci and an anal tube. Adults of the bombardier beetles (genus *Brachinus*, having the tips of the elytra truncate), have an unusual and audible chemical defense mechanism. A chemical reaction occurs in glands near the tip of the abdomen; the temperature reaches 100 °C and the chemical is sprayed in a water vapor cloud at the aggressor.

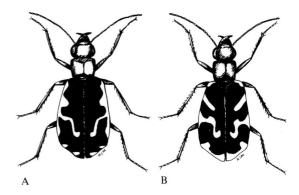

Figure 167 Tiger beetles (Carabidae). A, *Cicindela repanda;* B, *Cicindela tranquebarica.*

Common Species

Tiger Beetles

Cicindela oregona LeConte—12-14 mm; green or blue; white or dull yellow markings; long hairs near inner edge of eyes; labrum short; punctate elytra, edges sometimes faintly toothed; western North America (edges of lakes and streams).

Cicindela punctulata Olivier—10-13 mm; brownish black or slightly green with bronze sheen; row of small greenish or bluish depressions on elytra; small white markings present or absent along elytral margins; North America except Far West (attracted to lights; emits applelike odor).

Cicindela repanda Dejean (Fig. 167A)—12-14 mm; bronze or brownish with greenish cast; white markings; short labrum; North America east of Rocky Mts., British Columbia (riverbanks).

Cicindela sexguttata Fabricius—12-14 mm; brilliant green, blue-green, or violet; usually 9 spots on elytra, sometimes 6-10 or spots absent; elytra punctate; eastern North America (meadows, paths in open woods).

Cicindela tranquebarica Herbst (Fig. 167B)—12-14 mm; brownish, greenish, dark blue, or black; oblique white markings; throughout North America.

Ground Beetles

Bembidion patruele Dejean (Fig. 168A)—3-5 mm; black with greenish luster; head, pronotum, and legs reddish brown; elytra iridescent with orange or dull yellow patches; most of U.S. except southernmost states, Canada.

Bembidion quadrimaculatum (Linnaeus)—2-4 mm; black with bronze tinge; legs and basal antennal segments brownish yellow; each elytron with 2 yellow spots; east of Rocky Mts. south to TX.

Calathus ruficollis Dejean—8-10 mm; pale red; pronotum brighter red; elytra finely striated; Pacific coast states.

Calosoma calidum (Fabricius) (Fig. 168B). Fiery hunter—20-27 mm; black, sometimes green margins; elytra deeply grooved, 6 rows red or gold punctures; throughout North America.

Calosoma scrutator Fabricius. Fiery searcher—25-35 mm; black with violet tinge; head and prothorax with gold, green, or red margins; elytra iridescent green or blue-green, purple margins, grooved and punctate; throughout U.S., southern Canada (sometimes in trees).

Chlaenius sericeus Forster—12-17 mm; bright green; antennae and legs pale brownish yellow, antennae darker toward tip; fine yellowish pubescence on elytra; throughout North America (lake and stream margins).

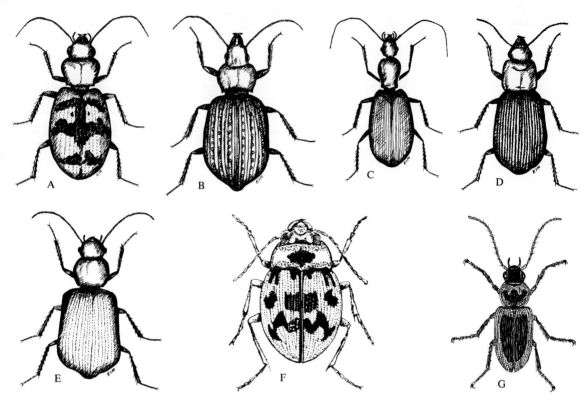

Figure 168 Ground beetles (Carabidae). A, *Bembidion patruele;* B, fiery hunter, *Calosoma calidum;* C, *Galerita janus;* D, *Harpalus pensylvanicus;* E, *Lebia atriventris;* F, *Omophron tessellatum;* G, *Stenolophus comma.*

Chlaenius tricolor Dejean—10-13 mm; head and pronotum green; antennae and legs brownish orange; elytra bluish black, deep and narrow grooves, punctures; covered with fine hairs; throughout U.S., southern Canada.

Galerita janus Fabricius (Fig. 168C)—16-23 mm; pronotum, legs, and base of antennae reddish brown; elytra bluish black; short pale yellow hairs on body; central and eastern U.S., Ontario (comes to lights).

Harpalus caliginosus (Fabricius) (Fig. 157A)—21-26 mm; black; antennae and tarsi reddish brown; elytra deeply grooved; throughout North America.

Harpalus pensylvanicus De Geer (Fig. 168D)—12-15 mm; shiny black; convex; reddish brown ventrally; elytra deeply grooved with few to many punctures; throughout North America (especially cotton fields, pastures).

Lebia atriventris Say (Fig. 168E)—6-10 mm; head and pronotum reddish yellow; elytra dark blue, dense punctures, shallow grooves; throughout North America (climbs trees). *L. grandis* Hentz similar but larger with deep striations, eastern U.S. and southern Canada.

Omophron tessellatum Say (Fig. 168F)—6-7 mm; pale brownish yellow; green metallic markings on head, pronotum, elytra; central U.S. and Canada (in moist sand near water, pour water over sand to force out; flies to lights).

Pasimachus depressus Fabricius—25-30 mm; flat; shiny or dull black; pronotum and elytra with blue margins; elytra smooth, ridge near base; east of Rocky Mts.

Scarites subterraneus Fabricius—14-21 mm; shiny black; mandibles large; head with 2 large depressions or parallel lines; elytra grooved and separated by convex ridges; U.S. except Northwest, south to TX, southern Ontario.

Stenolophus comma (Fabricius) (Fig. 168G)—6-7 mm; head black; pronotum and elytra reddish brown or yellowish brown; black patch on pronotum and black stripe on inner margin of each elytron, elytral stripe blunt anteriorly; central and southern U.S. (flies to lights). *S. lecontei* Chaudoir, seedcorn beetle, similar, 5-6 mm, dark color patch on elytra pointed anteriorly.

Family Haliplidae— Crawling Water Beetles

These small (2-6 mm long), broadly oval beetles are common in weedy areas of ponds and streams. They are brownish or reddish yellow with dark spots. The extremely long, flat and broad hind coxae cover much of the abdomen (Fig. 169B). Adults and larvae feed on algae. Some adults in the genus *Peltodytes* have two black spots on the posterior margin of the pronotum whereas those in the genus *Haliplus* lack these two spots.

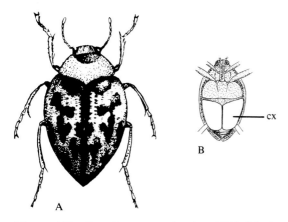

Figure 169 Crawling water beetles (Haliplidae). A, *Haliplus triopsis;* B, ventral view of a haliplid showing enlarged coxae. cx, coxa.

Common Species

Haliplus immaculicollis Harris—2.5-3.0 mm; reddish yellow with black spots; 2 impressions on posterior margin of pronotum; deep grooves on prosternum; most of North America.

Haliplus triopsis Say (Fig. 169A)—3-4 mm; dull yellow; anterior margin of pronotum with black patch; black spots on elytra; east of Rocky Mts., NM.

Peltodytes duodecimpunctatus (Say)—3-4 mm; dull yellow; 12 black spots on elytra; eastern 1/2 North America. *P. edentulus* LeConte similar but with black between eyes. *P. pedunculatus* Blatchley pale yellow and black, legs yellow except hind femora entirely black, NY to IA and south to GA and TX.

Family Dytiscidae— Predaceous Diving Beetles

These oval, predaceous beetles occur in most aquatic habitats. Many are yellowish brown to nearly black and they often have dark yellow patterns or the elytral and pronotal margins are yellowish brown. Dytiscids resemble the Hydrophilidae (water scavenger beetles) but are more streamlined and have short maxillary palpi, long filiform antennae, and move their hind legs simultaneously rather than alternately when swimming. Dytiscids migrate from pond to pond by flying and sometimes appear at lights. Air bubbles for breathing are stored beneath the elytra. Larvae, sometimes called water tigers, have a large head and long mandibles, two tarsal claws, eight abdominal segments that form a body tapering at both ends, and two tufted cerci on the tubular tip of the abdomen. Larvae are predaceous on aquatic invertebrates, fish, and tadpoles.

Common Species

Acilius semisulcatus Aubé (Fig. 170A)—12-14 mm; broadly oval; brownish yellow above, black below; dark "M"-shaped marking on top of head; 2 crosslines on pronotum; sides and tip of abdomen yellowish; most of U.S.

Agabus disintegratus (Crotch) (Fig. 170B)—7.5-8.5 mm; head and pronotum reddish yellow; pronotum with 2 black crossbands; elytra brownish yellow, each with 3-4 black longitudinal stripes; throughout North America.

Cybister fimbriolatus (Say)—30-33 mm; brown with greenish tinge; pronotum and elytra with yellow margins; pronotum sometimes with 5 grooves (♀); most of U.S.

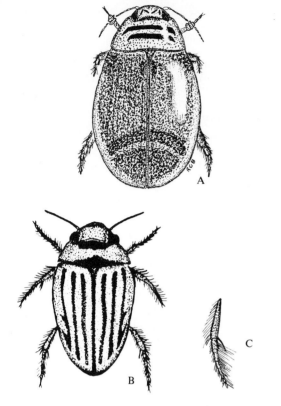

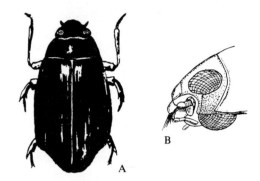

Figure 170 Predaceous diving beetles (Dytiscidae). A, *Acilius semisulcatus;* B, *Agabus disintegratus;* C, leg of a dytiscid.

Dytiscus fasciventris Say—25-28 mm; greenish black; pronotum and elytra with yellow margins; basal 2/3 each elytron with 10 deep grooves (♀); eastern 1/2 North America.

Hydroporus undulatus Say—4.0-4.5 mm; very convex; pale yellow to yellowish brown; pronotum with anterior and posterior margins black; elytra blackish and with irregular yellowish patches; hind coxal cavities touching; eastern 1/2 U.S., southeastern Canada.

Family Gyrinidae—Whirligig Beetles

Members of this aquatic family are oval, flattened, and black or dark metallic green. Adults are readily recognized by two characteristics: (1) each compound eye is widely separated into an upper half (above the water) and lower half (below the water) (Fig. 171B); and (2) they swim together in circular movements on the water's surface. Adults also swim underwater and fly at night. Adults and larvae are predaceous. The genus *Gyrinus* contains small species from 3.5—8.0 mm long.

Figure 171 Whirligig beetles (Gyrinidae). A, *Dineutus americanus;* B, divided compound eye of a gyrinid.

Common Species

Dineutus americanus (Fabricius) (Fig. 171A)—10-12 mm; shiny black with bronze sheen; outer margin of elytra curved inward in a weak "S" near apex; eastern U.S., southeastern Canada.

Dineutus assimilis (Kirby)—10-11 mm; shiny black with bronze sheen; legs brownish yellow; most of North America.

Dineutus ciliatus (Forsberg)—12-14 mm; black; pronotum and elytra with diffuse, bronzed, curved stripe on sides; legs dark brown; eastern U.S. south to FL, southwest to OK, southeastern Canada (small streams).

Dineutus discolor Aubé—11-13 mm; black on top; brown, straw-colored, or reddish below; distinctly narrowed toward head producing a triangular shape; eastern 1/2 U.S., southern Canada.

Gyrinus borealis Aubé—6.5-7.5 mm; shiny black; elytral margins bronzed; underside evenly dark brown to black; most of U.S., Canada. *G. maculiventris* LeConte similar, underside with light and dark blotches.

SUBORDER POLYPHAGA

Family Hydrophilidae— Water Scavenger Beetles

Most members of this family are aquatic although one subfamily is primarily terrestrial. They frequent quiet water with abundant vegetation and move their hind legs alternately when swimming. Many species fly to lights at night. Hydrophilids are typically black or brown, or occasionally dull green or yellowish. They have long, 4-segmented maxillary palpi and short, clubbed antennae. A few common and large species contain a long, backward-pointing spine on the metasternum. Adults feed chiefly on dead vegetation or are omnivorous, but the larvae are predators. Larvae are yellowish brown or gray, elongated, generally tapered at both ends, and have a single tarsal claw.

Common Species

Berosus striatus (Say)—4-5 mm; convex; greenish yellow; head black; pronotum with 2 black median stripes and 2 spots; elytra with scattered indistinct spots; elytra with longitudinal grooves containing fine punctures, ridges between grooves flat and coarsely punctured; most of U.S. except Southeast, southern Canada. *B. pantherinus* LeConte similar but pronotum with paired spots and elytra with 10 distinct spots.

Enochrus ochraceus (Melsheimer)—3.5-4.0 mm; shiny pale yellowish brown; pale spot in front of each eye; pronotum and elytra with pale margins, many punctures; eastern U.S., southeastern Canada.

Hydrophilus triangularis Say (Fig. 172A)—34-37 mm; shiny black with greenish tinge; abdomen with triangular yellowish or pale reddish lateral markings; prominent central ridge on ventral part of thorax; throughout North America (flies to lights).

Sphaeridium scarabaeoides (Linnaeus) (Fig. 172B)—5.5-7.0 mm; shiny black; elytra with faint reddish spot near base and apical 1/4 faint yellowish; throughout North America (terrestrial; in cow manure).

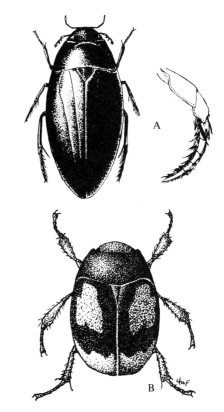

Figure 172 Water scavenger beetles (Hydrophilidae). A, *Hydrophilus triangularis* and a flattened hind leg; B, *Sphaeridium scarabaeoides*.

Tropisternus lateralis (Fabricius)—8.5-9.0 mm; convex; shiny greenish or bronze-black; margins of head, prothorax, and elytra yellow; legs yellowish, femora black at base; underside black; throughout North America.

Family Silphidae—Carrion Beetles

These beetles are moderately large, usually black with orange, red, or yellow markings, and have clubbed antennae. Adults and larvae feed on decomposing animal matter and the associated insect larvae. Species of the genus *Nicrophorus* excavate beneath dead rodents and snakes and the young develop on the partially buried carrion. *Nicrophorus* beetles are elongated, usually red and black, and have short, truncated elytra. *Silpha* beetles, the other common genus, are broadly oval, flattened, and the pronotum is very large.

Common Species

Nicrophorus americanus (Olivier)—27-35 mm; shiny black; elytra short, 2 broad orange-red bands on each; club of antennae, top of head, and pronotum orange-red; east of Rocky Mts. *N. marginatus* Fabricius similar, 20-27 mm, pronotum black with lighter margins, most of U.S., but primarily western 1/2, southern Canada.

Nicrophorus orbicollis Say (Fig. 173A)—20-25 mm; shiny blackish; orange-red or yellowish patch on basal 1/3 and apical 1/4 of elytra; last 3 antennal segments reddish brown; eastern 1/2 North America south to FL.

Nicrophorus tomentosus Weber—15-20 mm; shiny black; elytra short with 2 orange-red bands; pronotum with dense yellow pubescence; eastern 1/2 U.S., southeastern Canada.

Silpha americana Linnaeus (Fig. 173B)—15-25 mm; pronotum yellow with black or brown central patch; elytra brownish with 3 indistinct, irregularly branched, raised ribs on each elytron; east of Rocky Mts.

Silpha lapponica Herbst—9-13 mm; dull black; head and prothorax usually with dense yellowish pubescence; Rocky Mts. to Midwest.

Silpha ramosa Say—12-18 mm; velvety black, each elytron with 3 irregularly branched longitudinal ridges; pronotum finely punctate; western North America.

Family Scaphidiidae— Shining Fungus Beetles

These small (2-7 mm long), oval, convex beetles have a large pronotum and are usually shiny black; a few have red markings. The truncated elytra expose the pointed tip of the abdomen. The beetles feed on fungi and occur in decaying leaves and wood and under loose bark.

Common Species

Eubaeocera apicalis LeConte—1.4 mm; shiny black; antennae, legs, and tips of elytra and abdomen dark brown; each elytron with deep line along base and inner margin; eastern 1/2 U.S.

Scaphidium quadriguttatum Say (Fig. 173C)—3.5-4.5 mm; shiny black; each elytron with reddish or yellowish spot in basal 1/3 and apex, or only basal spot or without spots; eastern 1/2 U.S.

Family Staphylinidae— Rove Beetles

This family is very large with over 26,000 species worldwide and about 2,900 in North America. The very short elytra, slender body, and the habit of turning the tip of the abdomen up and over the body when running usually are reliable identification characteristics for rove beetles. The crossed mandibles are prominent and some species have hairy bodies. Rove bee-

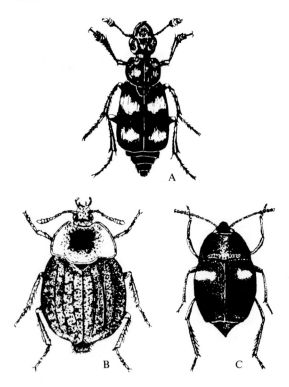

Figure 173 A and B, carrion beetles (Silphidae): A, *Nicrophorus orbicollis;* B, *Silpha americana.* C, a shining fungus beetle, *Scaphidium quadriguttatum.*

tles are predators and scavengers frequenting carrion, dung, fungi, and decomposing plant material; a few are parasites of other insects. A number of species live in bird and mammal nests and others occur as guests or predators in ant and termite nests. Species in the genus *Diaulota* occur in tide pools along the California coastline.

Common Species

Pelecomalium testaceum Mannerheim (Fig. 174A)—3.5-5.0 mm; head and abdomen black; pronotum, elytra, and sides brownish yellow; punctures sparse, contain short setae; Far West.

Philonthus cyanipennis (Fabricius)—12-15 mm; shiny black; elytra metallic blue, green, or purple; fine punctures dense on elytra, sparse on abdomen; throughout North America.

Philonthus lomatus Erichson (Fig. 174B)—6.5-8.0 mm; shiny black with reddish tinge, although punctures and elytra occasionally brownish yellow; legs yellowish; most of U.S., southern Canada.

Scopaeus exiguus Erichson (Fig. 174C)—2-3 mm; black with reddish tinge; antennae and legs pale yellow; pronotum dark yellowish; abdomen blackish, paler at tip; eastern U.S. *S. concavus* Hatch similar but has prominent "U"-shaped depression on 5th abdominal sternite (♂); western U.S.

Staphylinus cinnamopterus Gravenhorst—12-14 mm; shiny dark brown; elytra brownish red; pronotum and elytra with coarse, dense punctures and few hairs; abdomen black with reddish tip; most of U.S. except Pacific Northwest, southeastern Canada.

Staphylinus maculosus (Gravenhorst)—19-26 mm; brown with reddish brown spots; dorsum of abdo-

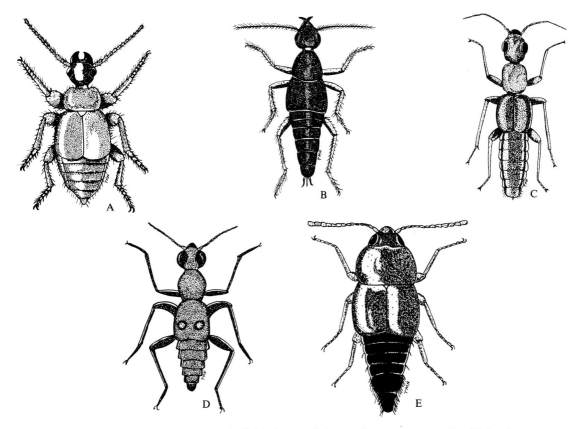

Figure 174 Rove beetles (Staphylinidae). A, *Pelecomalium testaceum;* B, *Philonthus lomatus;* C, *Scopaeus exiguus;* D, *Stenus comma;* E, *Tachyporus jocosus.*

men pale brown and spotted; elytra with dense punctures; eastern U.S., southeastern Canada.

Staphylinus maxillosus (Linnaeus). Hairy rove beetle—10-21 mm; shiny black; elytra with band of yellowish gray or yellowish brown hairs; abdominal segments 2-3 or 2-4 with dense cover of yellowish gray or yellowish brown hairs; throughout North America.

Stenus comma LeConte (Fig. 174D)—3.5-5.0 mm; shiny black; tarsi and most of antennae dark brown; eyes large; red or orange spot near middle of each elytron; throughout North America.

Tachinus fimbriatus Gravenhorst—7-9 mm; broad; head and pronotum shiny black; elytra reddish brown with dark tips, as broad as long; most of North America.

Tachyporus jocosus Say (Fig. 174E)—3-4 mm; shiny black with reddish tinge; pronotum, legs, and elytra reddish yellow or yellowish brown; antennae dull yellow with dark tips; throughout North America.

Family Pselaphidae— Shortwinged Mold Beetles

This family of short-winged beetles resembles the rove beetles (Staphylinidae) but the members have a wider abdomen and a narrower thorax and head. All are 5.5 mm or less in length. The species occur under stones, loose bark, moss, forest litter, and in ant nests. They feed on mold, mites, and sometimes ant larvae. Those in ant nests secrete substances attractive to ants.

Common Species

Cylindrarctus longipalpis (LeConte)—2 mm; reddish brown with paler abdominal margins; palpi long with enlarged flattened tips; stiff hairs; most of North America.

Tmesiphorus costalis LeConte (Fig. 175A)—3.0-3.5 mm; shiny dark brown to black; short yellowish hairs pressed against body; clubbed antennae over 1/2 body length (♂); most of U.S. (sometimes in ant nests).

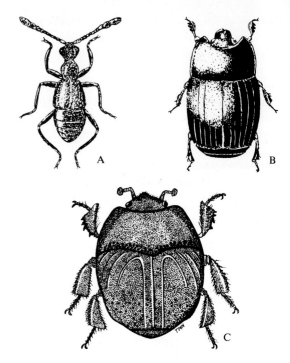

Figure 175 A, a shortwinged mold beetle, *Tmesiphorus costalis* (Pselaphidae); B and C, hister beetles (Histeridae): B, *Platysoma carolinum*; C, *Saprinus pennsylvanicus*.

Family Histeridae—Hister Beetles

This family contains small (0.5-10.0 mm long), typically shiny black or sometimes bronze or green beetles. The pronotum is very large, sometimes partially covering the small head, and the shortened, truncated elytra leave one or two abdominal segments exposed. Most species are predaceous on other insects but many are scavengers and occur in carrion and dung. Some are extremely flattened and occur under loose bark; others infest ant nests.

Common Species

Hister abbreviatus Fabricius—3.5-5.5 mm; shiny black; 2 grooves parallel pronotal margin, outer groove shorter; many deep grooves on elytra; most of U.S.

Hololepta quadridentata (Fabricius)—7-10 mm; shiny black; rectangular shaped; elytra with only 1 complete groove and 1 incomplete groove near outer margin; abdomen coarsely punctate; southern U.S.

Platysoma carolinum (Paykull) (Fig. 175B)— 3-4 mm; very flattened; shiny black; 5 shallow grooves on elytra, inner 2 only on apical 1/2; most of U.S. (under bark).

Saprinus pennsylvanicus (Paykull) (Fig. 175C)— 4-5 mm; metallic green or bronze-green; apical 1/2 elytra with coarse punctures, elytral groove closest to inner margin is arched; east of Rocky Mts. *Euspilotus assimilis* (Paykull) similar but is black, eastern 1/2 U.S., southeastern Canada.

Spilodiscus biplagiatus (LeConte)—5.0-6.5 mm; black with 2 red elytral patches; eastern U.S., southeastern Canada.

Family Lucanidae—Stag Beetles

Stag beetles have long mandibles and lamellate antennae whose terminal segments cannot be closed as can those of the similar scarab beetles (Scarabaeidae). Lucanids occur chiefly in wooded areas and especially in decaying logs. Some adults fly to lights. The larvae, which are similar to scarab white grubs, feed on decaying wood.

Common Species

Dorcus parallelus Say—15-26 mm; dark brown to blackish; head and pronotum shiny; mandibles with large central tooth; elytra deeply grooved and punctured; eastern 1/2 U.S., southeastern Canada.

Lucanus elaphus Fabricius (Fig. 176A). Giant stag beetle—28-40 mm, excluding mandibles; shiny reddish brown; antennae and legs blackish; head wider than pronotum; mandibles extremely large, pronged, about 12-15 mm long (♂); NC and VA west to IL, IN, and OK.

Platycerus depressus LeConte—13 mm; black; western U.S., southern Canada (in dead aspen).

Pseudolucanus capreolus (Linnaeus) (Fig. 176B)— 22-35 mm; shiny, dark reddish brown; mandibles with 1 tooth on inner edge (♂); head equal or broader than pronotum; elytra smooth; eastern 1/2 U.S., southeastern Canada.

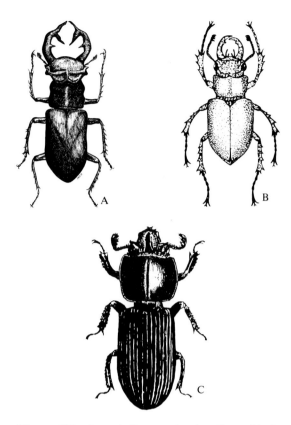

Figure 176 A and B, stag beetles (Lucanidae): A, giant stag beetle, *Lucanus elaphus;* B, *Pseudolucanus capreolus.* C, a bess beetle, *Popilius disjunctus* (Passalidae).

Family Passalidae—Bess Beetles

The passalids are shiny, blackish beetles with deeply grooved elytra and a horn on the head. These beetles are gregarious and both adults and larvae occur together in galleries inside decaying logs. One species, *Popilius disjunctus* (Illiger) (Fig. 176C), is 30-35 mm long and occurs in the eastern 1/2 of the U.S.; the two remaining North American species (genus *Passalus*) are limited to southern Texas.

Family Scarabaeidae—Scarab Beetles

Scarab beetles form a very large family (20,000 species worldwide and about 1,380 in North

America) of robust, convex beetles with lamellate antennae (Fig. 160D). The body sizes range from 2-70 mm in length in North America, some are metallic, and others are very hairy ventrally. Scarabs are generally nocturnal and frequently attracted to lights. This family may be divided into two feeding groups: (1) both larvae and adults feed on carrion, dung, skin, and feathers; (2) adults feed on leaves and flowers and larvae feed on roots, sap, and decaying wood. Many species in the latter group are important agricultural pests. The stout larvae have a curved body that is often wrinkled and the thoracic legs are well developed (Fig. 178D).

The genus *Phyllophaga* contains about 131 species of June (or May) beetles. The adults fly noisily at night and feed on leaves of many kinds of trees. The larvae, called white grubs (Fig. 178D), cause damage to plant roots and are very common in gardens and under sod. The skin beetles, subfamily Troginae, have been classified in some texts as a family, Trogidae.

Common Species

Subfamily Scarabaeinae—Dung Beetles and Tumblebugs; hind legs insert closer to tip of abdomen than to middle legs, hind tibia with 1 spur at tip; found under dung and carrion; tumblebugs form a piece of dung into a ball, roll it with hind legs to a suitable site, bury it, then lay eggs in it.

Canthon pilularius (Linnaeus) (Fig. 177A)—11-19 mm; black with bluish, greenish or coppery tinge; pronotum and elytra with dense granules; most of U.S.

Copris fricator Fabricius (Fig. 177B)—13-18 mm; black; top of head with horn (♂) or tubercle (♀); pronotum coarsely and densely punctate, median groove present; 8 grooves on each elytron; eastern 1/2 North America.

Onthophagus hecate Panzer—5-9 mm; black with purplish tinge and scattered, short gray hairs; pronotum granular, projects anteriorly over head in scooplike fashion (♂); elytra grooved; throughout North America.

Subfamily Aphodiinae—Aphodian Dung Beetles; antennae 9-segmented, 2 spurs at tip of tibia; under dung and debris.

Aphodius distinctus (Müller) (Fig. 177C)—5.0-6.5 mm; shiny; head and pronotum black; elytra yellowish with black spots and fine grooves; throughout North America.

Aphodius fimetarius (Linnaeus)—6-8 mm; shiny black; anterior corners pronotum reddish yellow; elytra reddish; head with 3 tubercles (♂); throughout North America.

Subfamily Geotrupinae—Earthboring Dung Beetles; 11-segmented antennae with 3-segmented clubs; front tibiae very broad and scalloped or toothed; under dung.

Eucanthus lazarus (Fabricius)—6-12 mm; light or dark brown; pronotum with large central swelling; elytra with large deep grooves; east of Rocky Mts., AZ.

Geotrupes splendidus (Fabricius) (Fig. 177D)—12-18 mm; broadly oval; metallic green, purple, or bronze; head surface rough, central tubercle; eastern 1/2 U.S., southeastern Canada.

Subfamily Troginae—Skin Beetles; dorsal surface very rough, convex, grayish brown; in and near old and dry carrion, skin, feathers, nests; most species in genus *Trox*.

Trox monachus Herbst (Fig. 177E)—12-16 mm; dark brown to blackish; head with 2 tubercles; pronotum ridged and with tubercles; elytra with large, round, widely separated tubercles in rows; most of U.S.

Subfamily Melolonthinae—June (or May) Beetles, Chafers; tarsal claws of hind legs equal in size and toothed, or tips of hind tibiae lack spurs. *Hoplia* spp. usually 6-9 mm long, somewhat flattened, silvery scales ventrally, hind legs with 1 claw; *Serica* spp. shaped like small June beetles *(Phyllophaga)* but elytral grooves evenly spaced in former.

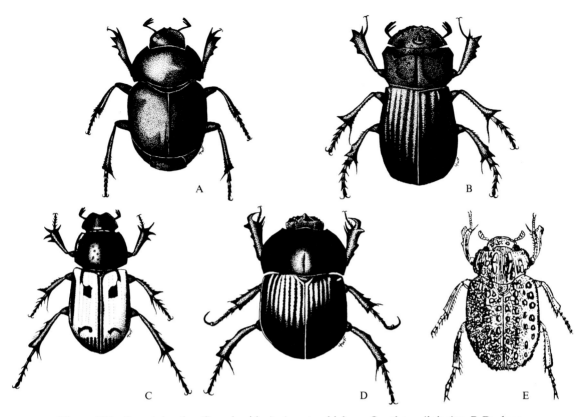

Figure 177 Scarab beetles (Scarabaeidae). A, a tumblebug, *Canthon pilularius*. B-D, dung beetles: B, *Copris fricator;* C, *Aphodius distinctus;* D, *Geotrupes splendidus*. E, a skin beetle, *Trox monachus*.

Diplotaxis tenebrosa Fall—7.5-10 mm; black; elytra with small markings connected in netlike pattern; western U.S., southwestern Canada.

Macrodactylus subspinosus (Fabricius) (Fig. 178A). Rose chafer—8-10 mm; slender; reddish brown with dull yellow hairs or scales; legs very long, tarsi as long as femora and tibiae combined; throughout U.S.

Phyllophaga ephilida (Say)—14-18 mm; shiny brownish yellow, head and pronotum darker; front of head not notched when viewed from above; chiefly southern U.S.

Phyllophaga fervida (Fabricius) (Fig. 178B)—19-20 mm; shiny, dark reddish brown to reddish black; front of head notched when viewed from above; pronotum widest at middle, punctured; elytra weakly punctate, surface roughens posteriorly; most of U.S. *P. errans* (LeConte) similar, 15-19 mm, Far West.

Phyllophaga fusca (Froelich) (Fig. 178C)—17-24 mm; shiny, dark reddish brown to reddish black; front of head broadly but shallowly notched when viewed from above; pronotum widest posteriorly; throughout U.S., southern Canada.

Phyllophaga rugosa (Melsheimer)—18-23 mm; shiny reddish brown or reddish black; body shape similar to *P. fusca;* front of head deeply notched when viewed from above; pronotum widest at middle; Rocky Mts. eastward.

Polyphylla decimlineata (Say). Tenlined June beetle—25-35 mm; brown with yellowish and white scales; 2 white stripes on head, 3 on pronotum, 5 on each elytron; Rocky Mts., Southwest. *P. crinita* LeConte similar, long erect hairs on head and pronotum, Pacific coast states.

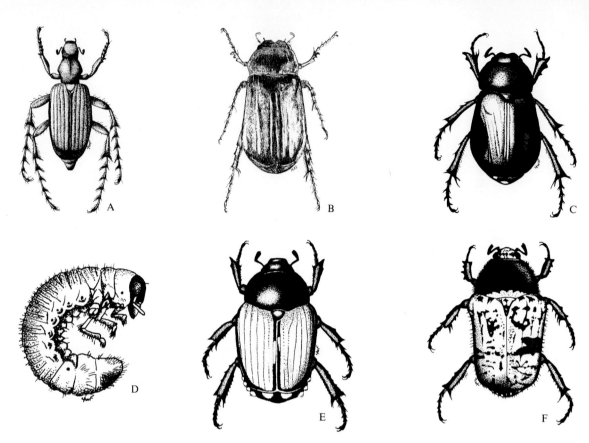

Figure 178 Scarab beetles (Scarabaeidae). A, rose chafer, *Macrodactylus subspinosus;* B-D, June beetles: B, *Phyllophaga fervida;* C, *Phyllophaga fusca;* D, a white grub (*Phyllophaga* sp.); E, Japanese beetle, *Popillia japonica;* F, bumble flower beetle, *Euphoria inda.*

Serica sericea Illiger—8-10 mm; dull black or purplish brown, very iridescent; most of U.S., southern Canada (common at lights). *S. anthracina* LeConte similar, 7-8 mm, not iridescent, Far West.

Serica vespertina (Gyllenhal)—8-11 mm; shiny yellowish brown to dark brown; each deep groove of elytra with 2 rows of punctures; eastern North America.

Subfamily Rutelinae—Shining Leaf Chafers; hind legs closer to middle legs than to tip of abdomen, outer claw of hind tarsi larger than inner claw, 2 spurs on tip of hind tibia; usually shining.

Anomala undulata Melsheimer—8.0-9.5 mm; convex and oval; shiny brownish yellow; center of pronotum blackish with greenish bronze luster; elytra with membranous margins; elytra color variable, usually with crossband of dark brown oval dots near middle and in apical 1/3, but could cover all elytra or be nearly absent; eastern U.S.

Cotalpa lanigera (Linnaeus)—20-26 mm; body shaped like *Phyllophaga* spp. (June beetles); metallic dark yellow, whitish yellow, or greenish yellow; ventral side dark with long dense hairs, eastern 1/2 U.S.

Pelidnota punctata Linnaeus—17-25 mm; shiny light brown to brownish yellow; pronotum with black spot on each side; elytra with 3 black spots on each side; ventral surface blackish or dark green; throughout U.S., southern Canada.

Popillia japonica Newman (Fig. 178E). Japanese beetle—8-12 mm; shiny dark green; elytra brownish orange with green margins; ventral surface blackish with white pubescent spots on sides of abdomen; tip of abdomen with 2 white pubescent spots dorsally; eastern U.S.

Subfamily Dynastinae—Elephant, Hercules, and Rhinoceros Beetles; primarily South and Southwest; horn on head and/or pronotum (♂) or absent (♀).

Strategus spp. (elephant beetles) are brown, 38-50 mm long, head with 3 horns, southern U.S. north to NY and west to TX; *Dynastes* spp. (Hercules beetles) are 50-70 mm, long horn over head, *D. tityus* (Linnaeus) in East, *D. granti* Horn in Southwest; rhinoceros beetles include *Phileurus* spp., about 25 mm long, 2 horns on head, southern 1/2 U.S.

Xyloryctes jamaicensis (Drury). Rhinoceros beetle—25-30 mm; shiny, dark reddish brown or brownish black; ventral surface with thick red-brown hairs; 1 curved horn on head (♂) or tubercle (♀); eastern U.S. west to AZ.

Subfamily Cetoniinae—Flower Beetles; tarsal claws equal size and untoothed, body flattened from above; *Euphoria* spp. hairy and resemble bumble bees.

Cotinus nitida (Linnaeus). Green June beetle—20-23 mm; narrowed posteriorly; velvety green; head, tibiae, and metasternum metallic green; pronotum and elytra brownish yellow laterally; eastern 1/2 U.S. *C. texana* Casey similar, 20-34 mm; coppery green, TX to so. CA.

Euphoria inda (Linnaeus) (Fig. 178F). Bumble flower beetle; 13-16 mm; brownish yellow; head and pronotum blackish with bronze luster, sometimes spotted; most of body covered with dense yellow pubescence; eastern U.S. west to ID, NM.

Trichiotinus piger (Fabricius)—9-11 mm; head and pronotum bronze-black, with dense yellow hairs; elytra dark orange-brown to blackish, sparse hairs, white markings; tip of abdomen with white stripe on each side; eastern 1/2 U.S., southeastern Canada. *T. affinis* (Gory and Percheron) similar, found with *T. piger,* shiny, 9-10 mm.

Family Helodidae—Marsh Beetles

These oval beetles are 2-4 mm long. Some have enlarged hind femora for jumping. Adults occur on vegetation and decaying material in wet areas.

Common Species

Cyphon collaris Guérin-Méneville (Fig. 179A)—3.0-5.5 mm; shiny black with fine pubescence; reddish yellow markings on sides of elytra; eastern 1/2 U.S.

Family Dascillidae—Softbodied Plant Beetles

Members of this family are pubescent, softbodied beetles 3-14 mm long. The mandibles

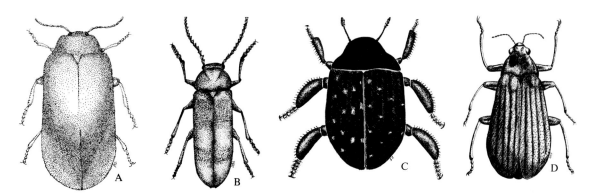

Figure 179 A, a marsh beetle, *Cyphon collaris* (Helodidae); B, a softbodied plant beetle, *Dascillus davidsoni* (Dascillidae); C, a pill beetle, *Byrrhus americanus* (Byrrhidae); D, a waterpenny beetle, *Psephenus herricki* (Psephenidae).

are prominent in many species. Adults are occasionally collected on vegetation near water.

Common Species
Dascillus davidsoni LeConte (Fig. 179B)—10-14 mm; shiny brown or blackish; fine gray pubescence on elytra, 2 irregular brown bands present or sometimes absent; CA.

Family Byrrhidae—Pill Beetles
These beetles are convex, black, and 5-10 mm long. The head is concealed from above. When disturbed these hard-bodied beetles retract their legs and antennae, forming a compact "pill." Habitats include along sandy shorelines, beneath logs and stones, and among grass roots. Adults and larvae are plant feeders.

Common Species
Byrrhus americanus LeConte (Fig. 179C)—8.5-9.5 mm; broadly oval, very convex; black; dense gray pubescence; elytra with 3-4 black lines; northern 1/2 U.S. except Far West, most of Canada.

Family Psephenidae—Waterpenny Beetles
Only six species of these plant-feeding water beetles occur in the U.S. The common family name refers to the flat, circular larvae. Larvae occur under stones in rapidly flowing water (e.g., shorelines, riffles) while adults crawl over rocks and vegetation along streams.

Common Species
Psephenus herricki (DeKay) (Fig. 179D)—4-6 mm; brownish black or dull black; head and thorax darkest; fine pubescence and punctures; eastern North America. Remaining *Psephenus* species occur in Pacific coast states.

Family Ptilodactylidae
These small (4-6 mm long), brownish beetles usually have the head hidden from above. The antennae bear a slender process on most segments. Ptilodactylids are found on vegetation near water.

Common Species
Ptilodactyla serricollis (Say) (Fig. 180A)—4-6 mm; brown to blackish; pubescent; eastern 1/2 U.S.

Family Dryopidae—Longtoed Water Beetles
Beetles in this primarily aquatic family are 1-8 mm long, gray or brown, and have long legs with large tarsal claws. The head, with its very

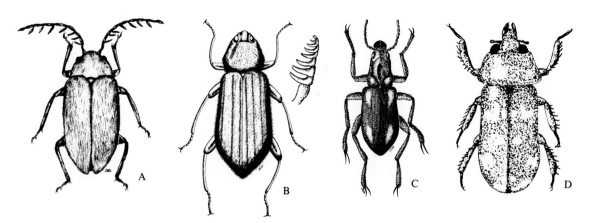

Figure 180 A, *Ptilodactyla serricollis* (Ptilodactylidae); B, a longtoed water beetle, *Helichus lithophilus,* and antennae (Dryopidae); C, a riffle beetle, *Stenelmis quadrimaculata* (Elmidae); D, a variegated mud-loving beetle, *Neoheterocerus pallidus* (Heteroceridae).

short antennae, is usually concealed in the prothorax. Adults occur on stones, debris, and stream bottoms in rapidly moving water such as riffles and shorelines. The aquatic larvae are flattened; a few are terrestrial.

Common Species
Helichus lithophilus (Germar) (Fig. 180B)—5-6 mm; dark reddish brown with bronze luster; dense covering of fine pubescence; tip of abdomen pale red beneath; east of Rocky Mts.

Family Elmidae—Riffle Beetles
These beetles are oval to cylindrical, 1-8 mm long, and the elytra may be smooth or highly ridged. The legs are long with large tarsal claws. Most species occur on rocks and debris in riffles of streams. A few are in ponds or are terrestrial.

Common Species
Stenelmis quadrimaculata Horn (Fig. 180C)—2.7-3.5 mm; dark reddish brown to black; pronotum often grayish; 2 oblong yellowish white patches near base and apical 1/3 of elytra; white, waxy pubescence on body; eastern U.S.

Family Heteroceridae—Variegated Mudloving Beetles
Members of this family are flattened above, 1-8 mm long, and often variegated with yellowish bands or spots. The flat mandibles extend forward and the outer margins of the front and middle tibiae have a row of distinct spines. These beetles inhabit burrows along shorelines and often fly to lights. Adults feed on zooplankton or are omnivorous.

Common Species
Neoheterocerus pallidus (Say) (Fig. 180D)—6-7 mm; black; covered with brownish and yellowish hairs, latter forming 3 indistinct crossbands on elytra; most of U.S.

Family Rhipiceridae—Cedar Beetles
Cedar beetles, brown insects with orange antennae, resemble June beetles (Scarabaeidae). The larvae parasitize cicadas. The five North American species are in the genus *Sandalus*.

Family Buprestidae—Flatheaded or Metallic Wood Borers
Buprestids are a large family (about 15,000 species worldwide and 720 species in North America) of hard-bodied, often metallic or brightly colored beetles. The body usually has a characteristic shape (Fig. 181). This family resembles the click beetles (Elateridae) but members lack the pointed posterior corners of the pronotum and the movable prothorax of the click beetle. Adults frequent flowers, leaves, and tree trunks and limbs that occur in full sunlight (e.g., the southern edge of a woods). Dead or dying trees and especially freshly cut wood attract these beetles. Larvae, known as flatheaded wood borers, typically have an expanded and flattened anterior area (Fig. 181C). The winding, grass-filled galleries under bark, and oval holes in wood are signs of these larvae, many of which are serious pests of orchards and forests. A few species produce twig galls and some are leaf miners (genus *Brachys*).

A few of the common genera are identified as follows: *Agrilus* species have long, narrow bodies and the first segments of the hind tarsi are as long as the three following segments combined; *Chrysobothris* species have the coxae of the hind legs flattened and distinctly widened at the bases, the prosternum is broadly diamond-shaped behind the procoxae, and the lateral lobes of the third segments of the hind tarsi are equal in length to the second segments; *Dicerca* species have the tips of the elytra elongated and diverging.

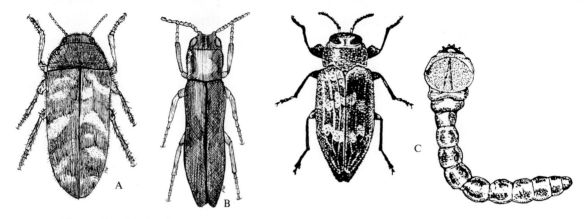

Figure 181 Flatheaded or metallic wood borers (Buprestidae). A, *Acmaeodera pulchella;* B, rednecked cane borer, *Agrilus ruficollis;* C, flatheaded apple tree borer, *Chrysobothris femorata* adult and larva.

Common Species

Acmaeodera pulchella (Herbst) (Fig. 181A)—5.5-12.0 mm; cylindrical; shiny, dark bronze-brown; pronotum with orange or yellow spot near posterior corners; elytra black with orange- yellow markings; many short brown hairs; most of U.S., southern Canada.

Agrilus anxius Gory. Bronze birch borer—6-13 mm; dark green-bronze; front of head greenish or copper-bronze; front of pronotum coppery; apex of elytra granular; most of U.S. and Canada (on birch).

Agrilus bilineatus (Weber). Twolined chestnut borer—6-12 mm; greenish black; dense yellow pubescence on sides of pronotum and elytra, forms 2 longitudinal stripes on elytra; east of Rocky Mts.

Agrilus ruficollis (Fabricius) (Fig. 181B). Rednecked cane borer—5-8 mm; bluish black; head and thorax coppery red; east of Rocky Mts.

Anthaxia aeneogaster Castelnau and Gory—4-5 mm; flat; dull bronze-black dorsally, brassy green ventrally; Rocky Mts. westward.

Buprestis apricans Herbst. Turpentine borer—24-26 mm; grayish bronze with greenish metallic luster; each elytron with 8 rows large punctures; NC to TX (on pines).

Buprestis aurulenta Linnaeus. Golden buprestid—14-19 mm; iridescent green or bluish green; elytral margins and central strip golden or coppery; Rocky Mts. westward.

Chalcophora virginiensis (Drury)—20-38 mm; shiny bronze-black; punctures and ventral side brassy; pronotum and elytra roughly sculptured; irregular lighter areas on elytra; eastern 1/2 U.S. and Canada. *C. liberta* (Germar) similar, 19-25 mm, bright coppery or brassy, blackish brown markings.

Chrysobothris femorata (Olivier) (Fig. 181C). Flatheaded appletree borer—7-16 mm; dark bronze, sometimes brassy, coppery, or greenish luster; usually 2 irregular crossbands on elytra; coarsely and densely punctate; throughout North America.

Dicerca divaricata (Say)—15-21 mm; brown or gray with brassy, coppery, or greenish bronze luster; apex of elytra prolonged and diverging; throughout North America.

Melanophila fulvoguttata (Harris). Hemlock borer—9-12 mm; shiny black with brassy luster; apical 1/2 elytra with 6-8 small orange-yellow spots in a circle; body rough and punctated; northern 1/2 U.S. south to AZ, southern 1/2 Canada (on cut pine logs, spruce, hemlock).

Family Elateridae—Click Beetles

Click beetles are elongated, flattened insects with a large, movable prothorax whose posterior corners are usually pointed. The body is often greatly tapered posteriorly. The family is large containing over 7,000 species worldwide and nearly 800 species in North America. The spinelike prosternal process (Fig. 182A) that fits into a socket in the mesosternum and the loosely joined prothorax and mesothorax enable the beetle to flip over when it is upside down. The beetle arches its body and then quickly straightens it, forcing the spine back into the socket and snapping the pronotum and elytral base against the supporting surface; the beetle flips over with an audible clicking sound. Adults feed on vegetation and are especially common under loose bark of decaying trees.

Larvae, called wireworms, are cylindrical and elongated, usually reddish brown, and have a hard and shiny cuticle. Wireworms occur chiefly in the soil and many are crop pests due to their feeding on roots, seeds, and stems. Some species inhabit rotting wood or occur beneath bark and feed on other insects.

Common Species

Agriotes mancus (Say). Wheat wireworm—7-9 mm; yellowish brown to dark brown; pronotum broader than long; dense coarse punctures in deep elytral grooves; posterior corners of pronotum and sides of elytra dull yellow; short dull yellow pubescence; east of Rocky Mts.

Alaus oculatus (Linnaeus). Eyed click beetle—25-45 mm; shiny black with sparse gray-white scales; pronotum with 2 large oval eyespots margined with gray-white scales; east of Rocky Mts. (in decaying logs). *A. melanops* LeConte similar, 25-36 mm, without scales on pronotum, Far West. *A. myops* (Fabricius) also similar, 24-38 mm, sparse pubescence, small pronotal eyespots with indistinct margins, chiefly southern U.S.

Ampedus collaris (Say)—8-9 mm; shiny black; prothorax bright red; antennae and legs dark brown; eastern North America.

Conoderus vespertinus (Fabricius) (Fig. 182B). Tobacco wireworm—7-10 mm; yellowish with dark reddish brown or blackish markings as illustrated; eastern 1/2 U.S.

Ctenicera inflata (Say) (Fig. 182C)—8-11 mm; bronze-black; relatively broad, convex; legs and sides sometimes reddish; dense yellowish pubescence pressed against body; 3rd antennal segment > 2x length of 2nd and longer than 4th; most of U.S., southern Canada.

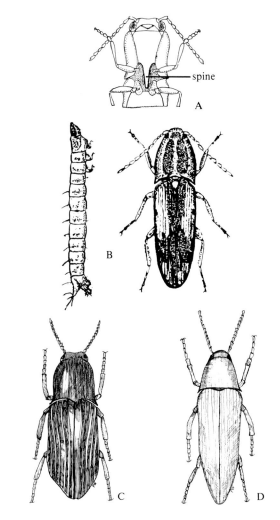

Figure 182 Click beetles (Elateridae). A, spine on prosternum of click beetles (also on buprestids); B, tobacco wireworm, *Conoderus vespertinus*, larva and adult; C, *Ctenicera inflata*; D, *Melanotus communis*.

Ctenicera lobata tarsalis (Melsheimer)—9-12 mm; shiny black; elytra dull yellow, lateral and inner margins with narrow black line; 2nd antennal segment very small, 3rd triangular and larger than 4th; eastern U.S. *C. pruinina* (Horn), Great Basin wireworm, similar but all black, elytra deeply grooved, Pacific Northwest, Great Basin.

Limonius californicus (Mannerheim). Sugarbeet wireworm—7-10 mm; antennae and femora dark brown; elytra reddish brown; dense white or yellowish pubescence; dorsum coarsely punctured; western North America.

Melanotus communis (Gyllenhal) (Fig. 182D)—11-15 mm; reddish brown; sparse pubescence; finely punctate; 3rd antennal segment twice length of 2nd and together as long as or longer than 4th; eastern 1/2 North America (very common; flies to lights). *M. similis* (Kirby) similar shape, 13-17 mm, dark brown to blackish, most of North America.

Family Lampyridae—Fireflies

The broad, flat pronotum generally covers the head from above in this family. The elytra are relatively soft. Light is produced in organs located in the yellow or greenish areas near the end of the abdomen. This biolumenescence is caused by an enzyme, luciferase, reacting with a substance, luciferin. The color and flashing frequency of the light vary by species and are associated with mating. Most of the individuals flying at night are males; the females of most species fly very little and remain in the vegetation while flashing. Larvae are also able to glow; they and the short-winged or wingless females are called glowworms. Not all species produce light. Adults and larvae are predaceous on insect larvae, slugs, and snails.

Common Species

Ellychnia corrusca (Linnaeus)—10-14 mm; black or brownish black; reddish yellow curved patch near but not reaching side margins of pronotum; elytra with brown or yellowish pubescence and 3-4 ridges; no light organs; most of U.S. (diurnal).

Lucidota atra (Fabricius) (Fig. 183A)—8-11 mm; black; sides of pronotum yellowish with red-orange marking above; elytra granular, 4 low ridges in basal 2/3; eastern North America (in trees; weak light).

Microphotus angustus LeConte—10-15 mm; pronotum and elytra grayish brown and abdomen pinkish (♂); ♀ larvalike, pinkish, flattened, tiny elytra; CA, OR, CO (dry grass of foothills).

Photinus scintillans (Say)—5-8 mm; dark brown; pronotum reddish with dull yellow margin and black central spot; elytra outer margin and central stripe yellow; ♀ very short-winged; eastern U.S. southwest to KS and TX. *P. pyralis* (Linnaeus) similar, 10-14 mm, elytra slightly wrinkled, ♀ long-winged.

Photuris pennsylvanicus (De Geer) (Fig. 183B)—11-15 mm; head yellowish; pronotum yellowish with reddish central area and dark middorsal stripe; elytra dark brown with yellowish stripes and margins; eastern U.S. southwest to KS and TX.

Family Cantharidae—Soldier Beetles

Beetles in this family generally have dark yellow, orange, or red elongated bodies. The elytra are pliable or leatherlike and the eyes bulge outward. The beetles are common on flowering plants. Adults and larvae are chiefly predaceous on other insects although some adults feed on nectar and pollen or are omnivorous.

Common Species

Cantharis bilineatus Say—6-8 mm; antennae black except base, 2 oblique black pronotal bands, elytra black except side margins, and legs black; remainder of body reddish yellow; eastern 1/2 U.S.

Cantharis divisa LeConte—6-8 mm; black; thorax yellow with 2 large black spots in central area; CA to British Columbia.

Chauliognathus pennsylvanicus (De Geer) (Fig. 183C)—9-12 mm; pronotum and elytra dull yellow-orange with a dark oval area on posterior 1/2 of each elytron; rest of body blackish; east of Rocky Mts., AZ (on goldenrod and Queen Anne's lace in late summer).

Podabrus tomentosus (Say) (Fig. 183D)—9-12 mm; elytra black with grayish pubescence; head, antennal base, prothorax, and legs reddish yellow; most of U.S. (on giant ragweed along streams).

Trypherus latipennis (Germar)—6-7 mm; elytra short, covering 1/2 or less of body; hind wings extend to tip of abdomen (hind wings short in similar rove beetles [Staphylinidae]); dorsal surface black; elytral tips, pronotal margins, and ventral side yellowish; eastern 1/2 U.S. (on catnip).

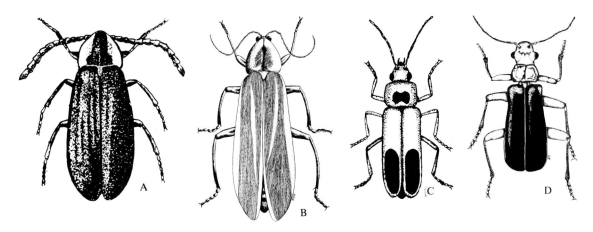

Figure 183 A and B, fireflies (Lampyridae): A, *Lucidota atra;* B, *Photuris pennsylvanicus.* C and D, soldier beetles (Cantharidae): C, *Chauliognathus pennsylvanicus;* D, *Podabrus tomentosus.*

Family Lycidae— Netwinged Beetles

Members of this family usually have soft, ridged, netlike wings and a large pronotum that often covers most of the head. A few western species (genus *Lycus*) have a short snout. Adults feed on decaying vegetation and larvae are predaceous. Many species are brightly colored and occur in wooded areas with dense undergrowth.

Common Species

Calopteron reticulatum (Fabricius) (Fig. 184)—10-19 mm; orange-yellow; base and posterior 1/3 of elytra with broad black crossband; pronotum with black middorsal stripe; eastern U.S., southeastern Canada.

Calopteron terminale (Say)—8-17 mm; orange-yellow; posterior 1/3 of elytra with purplish black band; most of U.S.

Dictyopterus aurora (Herbst)—6-11 mm; black; pronotum with orange-red sides; most of North America.

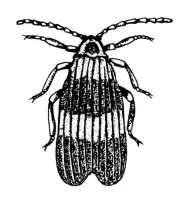

Figure 184 A netwinged beetle, *Calopteron reticulatum* (Lycidae).

Family Dermestidae— Dermestid Beetles

This group consists of convex, oval, or elongate-oval beetles 2-12 mm long. The antennae are clubbed and the body is hairy or scaly. Larvae are light brown and covered with long hairs (Fig. 185A). Most species are scavengers; the larvae feed on leather, furs, museum specimens (including insects), woolen and silk products, rugs, stored foods, and carrion.

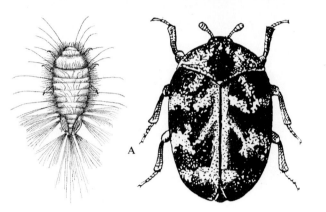
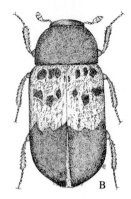

Figure 185 Dermestid beetles (Dermestidae). A, carpet beetle, *Anthrenus scrophulariae*, larva and adult; B, larder beetle, *Dermestes lardarius*.

Common Species

Anthrenus scrophulariae (Linnaeus) (Fig. 185A). Carpet beetle—2.0-3.5 mm; broadly oval, convex; black; white scales cover sides of pronotum and form zigzag irregular crossbands; dark red or dull yellow stripe along inner margins of elytra; throughout North America (adults on flowers).

Attagenus megatoma (Fabricius). Black carpet beetle—3.5-5.0 mm; oblong; convex; head and pronotum black; elytra reddish brown to black; pubescence sparse; throughout North America.

Dermestes caninus Germar—7.0-8.5 mm; black; pronotum with dense yellow and white pubescence; gray, reddish brown, and black pubescence on elytra; throughout North America (used to clean flesh from bones of decaying animals).

Dermestes lardarius Linnaeus (Fig. 185B). Larder beetle—6.0-7.5 mm; elongate-oblong; black; basal 1/2 elytra with dense yellowish, light brown, or gray hairs, 2 black patches at base, 6 black spots near middle of basal 1/2; throughout North America (on stored meat, cheese, insect collections).

Trogoderma ornata Say—2.0-2.5 mm; oblong-ovate; shiny black with scattered yellow pubescence; black elytra with red patches at base and smaller patches apically; throughout North America (also in stored cereals, seeds, nuts).

Family Ptinidae—Spider Beetles

Spider beetles are 1-5 mm long, the pronotum is very narrow, and the legs are quite long. The head is partially or completely hidden from above. These beetles often resemble spiders. They inhabit dried plant and animal materials such as stored grain, carcasses, and museum specimens.

Common Species

Ptinus fur (Linnaeus) (Fig. 186A). Whitemarked spider beetle—2.5-4.5 mm; dull reddish yellow or pale brown (♂), or dark brown (♀); body oval (♀) or elongate-oval (♂); pronotum with pale yellowish markings; elytra with irregular patches of white scales sometimes forming 4 spots; throughout North America.

Family Anobiidae—Deathwatch and Drugstore Beetles

Beetles in this family are 1-9 mm long, the pronotum extends hoodlike over the head to conceal it, and the body is usually cylindrical but can range to nearly spherical. The last three segments of the antennae are usually lengthened and enlarged although a few species have serrate or pectinate antennae. Adults and larvae live and feed on dry plant materials such as dead twigs and logs, furniture, seeds, cereals, tobacco, and museum specimens. The name "deathwatch" is derived from the ticking noise a few woodboring species make, once said by some to foretell a death.

Common Species

Lasioderma serricorne (Fabricius). Cigarette beetle—2-3 mm; elongate-oval, convex; dull reddish yellow or brownish red; antennae serrate, 2nd and 3rd segments shorter than 1st, 11th segment oval; throughout North America (indoors).

Ptilinus ruficornis Say—3.0-4.5 mm; elongate, cylindrical; black or dark reddish brown; antennae and legs reddish yellow; antennae fan-shaped (♂), branch of 4th segment 6-7 times length of segment; eastern U.S. (on dead branches of oak and maple).

Stegobium paniceum (Linnaeus) (Fig. 186B). Drugstore beetle—2.5-3.5 mm; oblong, convex; reddish brown with yellowish pubescence; sides of pronotum serrated anteriorly; elytra finely grooved and punctured; throughout North America (indoors).

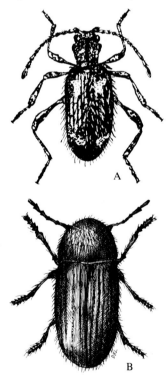

Figure 186 A, whitemarked spider beetle, *Ptinus fur* (Ptinidae); B, drugstore beetle, *Stegobium paniceum* (Anobiidae).

Family Bostrichidae— False Powderpost Beetles

These cylindrical and elongated beetles are usually 3-20 mm long and the head of most species is bent downward so it is almost hidden from above. A few species primarily in the western states have a readily visible head and large mandibles. The pronotum in this family has pointed tubercles anteriorly and the tips of the elytra are bent downward in many species. Bostrichids bore into living trees, dead branches, and seasoned lumber. One species in the western states, *Scobicia declivis* (LeConte) sometimes bores into lead telephone cables (it does not eat the lead) causing a short circuit from the moisture that enters the holes.

Common Species

Amphicerus bicaudatus (Say) (Fig. 187A). Apple twig borer—6.0-11.5 mm; reddish brown to brownish black; pronotum round, many tubercles on anterior 1/2; elytra slope downward at apex, 2 spines curve inward near apex of elytra; eastern U.S. west to CO, NM (burrow into fruit tree twigs). *A. cornutus* (Pallas) similar, 11-13 mm, Southwest.

Melalgus confertus (LeConte) (Fig. 187B)—14-15 mm; black; elytra brown; head not concealed by pronotum, mandibles large; pronotum narrowest at base; Far West (in orchards).

Family Lyctidae— Powderpost Beetles

Lyctid beetles are 2-7 mm long, narrow and somewhat flattened, and brown or black. The head is visible from above and the antennal club is 2-segmented. These beetles bore into dried wood and wood products such as furniture, beams, and hardwood floors, reducing them to powder.

Common Species

Lyctus linearis (Goeze)—25-30 mm; reddish brown; elytra with 1 row of punctures; throughout North America; *L. cavicollis* LeConte similar, > 1 row of punctures on elytra, Pacific coast states.

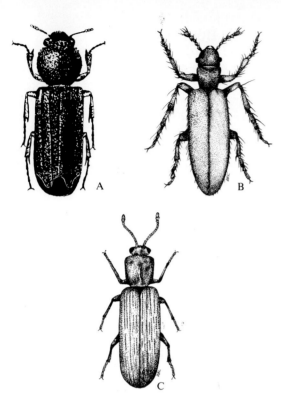

Figure 187 A and B, false powderpost beetles (Bostrichidae): A, apple twig borer, *Amphicerus bicaudatus;* B, *Melalgus confertus.* C, southern lyctus beetle, *Lyctus planicollis* (Lyctidae).

Lyctus planicollis LeConte (Fig. 187C). Southern lyctus beetle—4-6 mm; blackish with short yellow-white hairs; pronotum as long as wide; each elytron with 11 ridges that contain hairs, double row of punctures in grooves; most of U.S., chiefly South.

Family Trogositidae (Ostomidae)— Trogositid Beetles

This group has two different forms: one form (subfamily Ostominae) is oval, the head is much narrower than the pronotum, and the elytra have long, erect hairs; the other form (subfamily Tenebroidinae) is elongate, the head is about equal to the width of the pronotum, and the pronotum is widely separated from the elytral base. Adults are generally metallic green, blue, or dark brown to black. Adults and larvae commonly occur under bark and feed on other insects and possibly fungi.

Common Species

Calitys scabra (Thunberg) (Fig. 188A)—9-10 mm; reddish brown; head about 1/4 greatest width of pronotum; front coxal cavities closed; Far West.

Temnochila virescens (Fabricius)—10-13 mm; elongated and flattened; metallic green or blue; Far West.

Tenebroides corticalis (Melsheimer)—7-8 mm; dull black; shape similar to *T. mauritanicus;* 8th antennal segment smaller than 9th; eastern 1/2 U.S.

Tenebroides mauritanicus (Linnaeus) (Fig. 188B). Cadelle—6-11 mm; flattened; shiny dark brown to black; pronotum not grooved; throughout North America (in grain, flour).

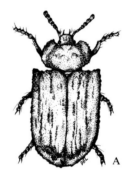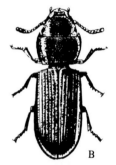

Figure 188 Trogositid beetles (Trogositidae). A, *Calitys scabra;* B, *Tenebroides mauritanicus.*

Family Cleridae— Checkered Beetles

The beetles in this family are often brightly colored with orange, red, or blue, and the body is very hairy. The last segments of the palpi are dilated and there are membranous lobes beneath the tarsi. Most adults and larvae are predaceous on other insects, especially wood-inhabiting beetles such as bark beetles (Scolytidae). Many clerids occur on flowers and foliage. Species in the genus *Necrobia,* known as ham beetles, occur on spoiled meat, fish, and cheese, feeding on insect scavengers that are present.

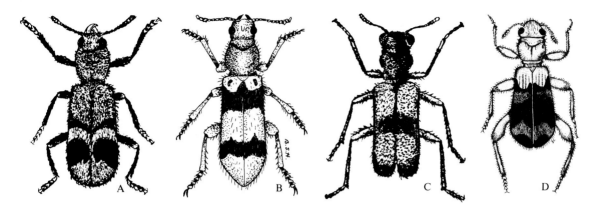

Figure 189 Checkered beetles (Cleridae). A, *Enoclerus nigripes;* B, *E. sphegeus;* C, *Phyllobaenus pallipennis;* D, *Thanasimus dubius.*

Common Species

Enoclerus nigripes (Say) (Fig. 189A)—5-7 mm; dull red or reddish brown; apical 2/3 elytra black with 2 whitish crossbands; eastern U.S. (in forests).

Enoclerus sphegeus (Fabricius) (Fig. 189B)—9-10 mm; face yellow; pronotum and elytra brownish black, elytra with 2 apical yellow or whitish crossbands; abdomen red; western 1/2 U.S. (in forests, May-June).

Necrobia rufipes (De Geer). Redlegged ham beetle—3.5-6.0; shiny metallic blue or green; basal segments of antennae and legs reddish brown; most of North America (on drying carrion, bones, fish, ham, cheese). *N. ruficollis* (Fabricius), redshouldered ham beetle, similar, 4-5 mm, head and posterior 3/4 elytra metallic blue or green, pronotum and base of elytra reddish brown.

Phyllobaenus pallipennis (Say) (Fig. 189C)—3.5-5.0 mm; bronze-black; antennae, elytra and legs brownish yellow; elytra with apex, side margins and central crossband faintly dark brown; pronotum widest at middle; most of U.S.

Thanasimus dubius (Fabricius) (Fig. 189D)—7.5-9.0 mm; head, pronotum, and basal area of elytra reddish brown; remainder of elytra black with 2 irregular bands of light pubescence; eastern 1/2 U.S.

Trichodes nutalli (Kirby)—8-11 mm; dark blue, greenish blue, or purple; elytra with 3 reddish yellow crossbands; east of Rocky Mts., ID, British Columbia. *T. ornatus* Say similar, 6-7 mm, 3 yellow irregular crossbands on elytra.

Family Melyridae— Softwinged Flower Beetles

Members of this family have prominent front coxae, the antennae insert in front of the head above the mandibles, and the body may have a soft appearance from the numerous erect hairs. Many species have a fleshy lobe between the tarsal claws and some have orange, eversible sacs along the sides of the abdomen that are sometimes visible. Melyrids chiefly are insect predators (e.g., *Collops* species) but the food preferences vary within the family.

Common Species

Attalus scincetus (Say)—2.5-3.0 mm; dull yellow; back of head, pronotum (except posterior corners) and scutellum blackish; eastern 1/2 U.S.

Collops bipunctatus Say—5-6 mm; pronotum pale red with 2 oblique markings; elytra blue-black; western 1/2 U.S.

Collops quadrimaculatus (Fabricius) (Fig. 190A)—4-6 mm; pronotum and elytra reddish yellow, each elytron with 2 large blue-black patches; rest of body brownish black; east of Rocky Mts.

Collops vittatus (Say)—4-5 mm; black; pronotum and sides of each elytron reddish yellow or dull orange; remainder of elytra dark blue; pronotum may have central dark area; hairy and punctate; most of U.S. and Canada.

Malachius aeneus (Linnaeus) (Fig. 190B)—5-6 mm; front of head, anterior corners pronotum, and all but base and central strip of elytra reddish brown to orange; remainder dark metallic green; east of Rocky Mts., OR, WA, British Columbia.

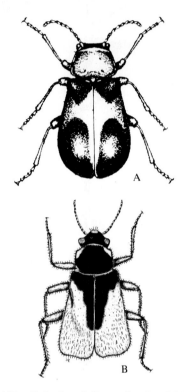

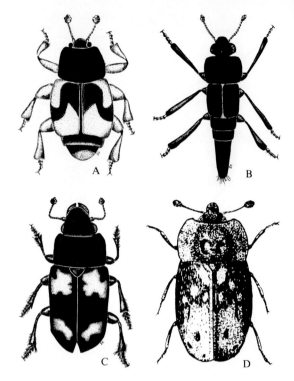

Figure 190 Softwinged flower beetles (Melyridae). A, *Collops quadrimaculatus;* B, *Malachius aeneus.*

Figure 191 Sap beetles (Nitidulidae). A, driedfruit beetle, *Carpophilus hemipterus;* B, *Conotelus obscurus;* C, *Glischrochilus fasciatus;* D, *Omosita colon.*

Family Nitidulidae—Sap Beetles

Sap beetles are 1.5-12 mm long, somewhat flattened, the last three abdominal segments are clubbed, and the elytra are sometimes shortened leaving the tip of the abdomen exposed. Adults and larvae feed primarily on decaying plant materials, including damaged and fermenting fruit and vegetables, sap, and fungi beneath bark. A few are predaceous on other insects. *Glischrochilus* species (Fig. 191C) are referred to as picnic beetles because they are attracted to food and beverages at picnics.

Common Species

Carpophilus dimidiatus (Fabricius). Corn sap beetle—2.5-5.0 mm; brownish yellow to black; squarish pronotum; orange spot on each elytron; most of U.S. except Rocky Mts. and northcentral states (in corn fields).

Carpophilus hemipterus (Linnaeus) (Fig. 191A). Driedfruit beetle—2.5-4.5 mm; black; elytra with dull brownish yellow bands; most of U.S., especially South.

Conotelus obscurus Erichson (Fig. 191B)—2.5-5.0 mm; very slender; short wings; antennae (except club) and legs brownish yellow; east of Rocky Mts. (in flowers). *C. stenoides* Murray similar, elytra wide with fine grooves and coarse punctures, southern U.S. (on corn).

Glischrochilus fasciatus (Olivier) (Fig. 191C)—5-7 mm; shiny black; elytra with four yellow or reddish patches; throughout North America.

Glischrochilus sanguinolentus (Olivier)—3.5-6.5 mm; shiny black; elytra red with 2 black spots, apex black; eastern 1/2 North America. *G. quadrisignatus* (Say) similar, reddish brown, elytra with 4 pale yellow spots.

Nitidula bipunctata (Linnaeus)—5-6 mm; black; reddish spot on apical 1/2 of each elytron; fine pubescence; throughout North America (on bones and skins of dead animals).

Omosita colon (Linnaeus) (Fig. 191D)—2-3 mm; brownish black; margin of thorax yellow; each elytron with 3-4 yellow spots near base and large patch on apical 1/2 that contains a dark spot; east of Rocky Mts.

Family Cucujidae—
Flat Bark Beetles

The unusual flatness of most of these beetles is a good field identification characteristic. Cucujids are 2-12 mm long. Most adults and larvae live under bark where they prey on bark beetles, other small insects, and mites; others are parasitic or infest stored grain and similar foods.

Common Species

Cathartus quadricollis (Guérin-Méneville). Square-necked grain beetle—2.5-3.5 mm; reddish brown; pronotum longer than wide, rectangular to square, margins not toothed; throughout North America (in stored foods).

Catogenus rufus Fabricius—5-11 mm; dark reddish brown; crossgroove on head behind eyes; pronotum with depressed central area; eastern U.S.

Cucujus clavipes Fabricius—10-14 mm; entirely red or dull red; most of North America (under bark).

Oryzaephilus surinamensis (Linnaeus) (Fig. 192A)—Sawtoothed grain beetle—2.5 mm; dark brown to reddish brown; pronotum with 6 teeth on each side and 3 ridges on dorsum; throughout North America (in stored foods). *O. mercator* (Fauvel), merchant grain beetle, similar but area directly behind eye is < 1/2 vertical diameter of eye whereas it is > 1/2 in *O. surinamensis,* on cereals but not grain.

Telephanus velox Haldeman (Fig. 192B)—3.5-5.0 mm; yellowish brown; head and apical 1/3 of elytra dark brown; very pubescent and coarsely punctate; east of Rocky Mts. (under stones, debris).

Uleiota dubius (Fabricius) (Fig. 192C)—4-6 mm; very flattened; brownish black or pale brown, elytra margins paler; pronotal margin serrated; most of U.S., southern Canada (under bark).

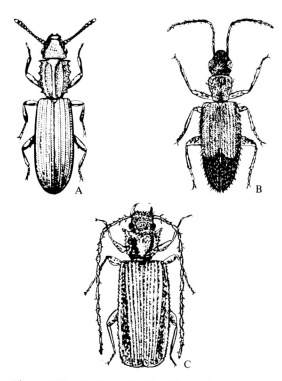

Figure 192 Flat bark beetles (Cucujidae). A, sawtoothed grain beetle, *Oryzaephilus surinamensis;* B, *Telephanus velox;* C, *Uleiota dubius.*

Family Cryptophagidae—
Cryptophagid Beetles

These beetles are 1-5 mm long, yellowish brown or black, and usually covered with silky pubescence. The body is oval and convex and the antennal club is 3-segmented. Cryptophagids live beneath leaves and wood chips, in rotting logs, and on fungi and flowers.

Common Species

Anchicera ephippiata Zimmermann (Fig. 193A)—1-2 mm; shiny black head and pronotum; elytra reddish yellow with 2 dark patches sometimes extending into wide band; most of U.S.

Cryptophagus acutangulus Gyllenhal—2.5 mm; flattened; pale brownish yellow, head and thorax darker; lateral margin of pronotum bowed inward on apical 1/2 leaving a projection at middle and on anterior margin; throughout North America (sometimes in stored food).

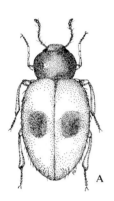
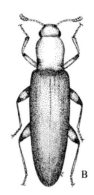

Figure 193 A, a cryptophagid beetle, *Anchicera ephippiata* (Cryptophagidae); B, clover stem borer, *Languria mozardi* (Languriidae).

Family Languriidae—Languriid Beetles

These elongated beetles are 5-10 mm long and black or blue-black with red or orange markings. Adults are pollen and leaf feeders and the larvae are stem borers.

Common Species

Languria mozardi Latreille (Fig. 193B). Clover stem borer—5.0-7.5 mm; shiny blue-black; head, pronotum, and ventral side red; east of Rocky Mts., AZ (on clover, alfalfa).

Family Erotylidae—Pleasing Fungus Beetles

Erotylids are 3-20 mm long and generally oval with black, reddish, and yellowish markings. The antennae have a large 3- or 4-segmented club. The tarsi appear 4- or 5-segmented. Adults and larvae occur in decaying wood, under bark, on fungi, or in debris, and they feed on fungi.

Common Species

Ischyrus quadripunctatus (Olivier) (Fig. 194A)—7-8 mm; head black; pronotum yellowish red or yellow with 4 black spots; elytra reddish yellow or yellow with black patches as illustrated; eastern U.S.

Megalodacne heros (Say) (Fig. 194B)—18-21 mm; shiny black; elytra with 2 red patches at base and 2 on apical 1/3; eastern 1/2 U.S. *M. fasciata* (Fabricius) similar, 9-15 mm.

Triplax festiva Lacordaire—5.0-6.5 mm; shiny black; pronotum, scutellum, and central crossband of elytra reddish yellow; eastern 1/2 U.S., chiefly South.

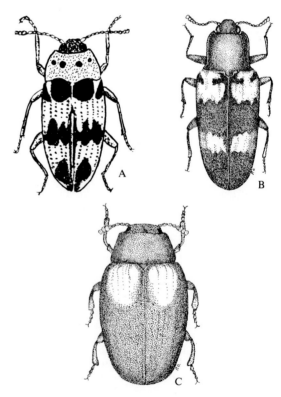

Figure 194 Pleasing fungus beetles (Erotylidae). A, *Ischyrus quadripunctatus;* B, *Megalodacne heros;* C, *Tritoma humeralis.*

Tritoma humeralis Fabricius (Fig. 194C)—3-4 mm; shiny black; elytra with reddish yellow patches near base; eastern U.S. *T. sanguinipennis* (Say) similar, 4-5 mm, elytra red.

Family Phalacridae— Shining Fungus Beetles

These brownish beetles are very convex, shiny, and 1-3 mm in length. The body is broadly oval to nearly spherical. Phalacrids are common on goldenrod, daisy, Queen Anne's lace, and many other flowers.

Common Species

Phalacrus simplex LeConte (Fig. 195A)—2.0-2.3 mm; red-brown or blackish; antennae and legs pale; antennae insert under shelf between eyes so bases not visible; central and western U.S.

Stilbus apicalis (Melsheimer)—1.9-2.4 mm; oval; shiny black except tips of elytra abruptly reddish; eastern 1/2 U.S.

Stilbus nitidus (Melsheimer) (Fig. 195B)—1.2-1.5 mm; very shiny orange-brown; eastern 1/2 U.S.

Family Coccinellidae— Lady Beetles

Lady beetles are broadly oval, convex insects that often are brightly colored. There are about 5,000 species worldwide and over 400 species in North America. Some lady beetles are similar to the leaf beetles (Chrysomelidae) but differ by having three distinct tarsal segments instead of appearing to have four as do leaf beetles. Adults and larvae are predaceous, commonly found feeding on aphids and sometimes scale insects, mealybugs, and mites. Two common species in the genus *Epilachna* are plant feeders and garden pests. Lady beetle adults hibernate in the winter and some are found in large aggregations under debris or on boulders at high elevations. Larvae (Fig. 196F) are covered with spines or tubercles and often are dark with bright bands of color.

Only a few of the many common lady beetles are described below. The genus *Scymnus* contains tiny pubescent beetles, usually 1-3 mm long, and marked with red or yellow.

Common Species

Adalia bipunctata (Linnaeus) (Fig. 196A). Twospotted lady beetle—4.0-5.5 mm; head black with frontal area yellowish; pronotum with broad black "M"-shaped patch in center and pale yellowish white margins; each elytron reddish with round black spot in center; throughout North America.

Chilocorus stigma (Say) (Fig. 196B). Twicestabbed lady beetle—4-5 mm; very convex, spherical; shiny black; 1 round red spot in center of each elytron; abdominal segments red; most of U.S. and Canada.

Coccinella transversoguttata richardsoni Brown (Fig. 196C). Transverse lady beetle—6.0-7.5 mm; head black, white spot on each side near eyes; pronotum black with spot on anterior corners; elytra red or yellow with black band and spots as shown; throughout North America.

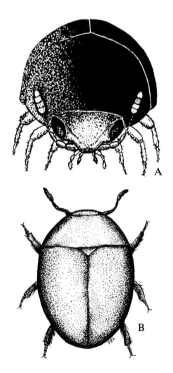

Figure 195 Shining fungus beetles (Phalacridae). A, *Phalacrus simplex;* B, *Stilbus nitidus.*

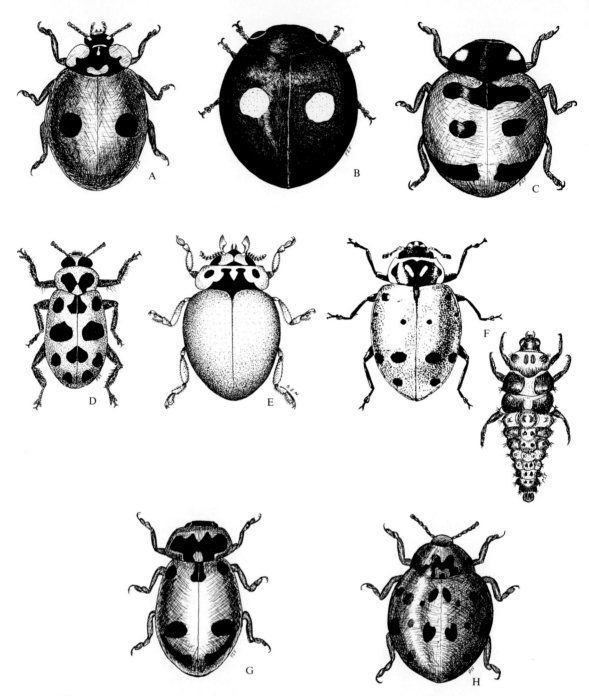

Figure 196 Lady beetles (Coccinellidae). A, twospotted lady beetle, *Adalia bipunctata;* B, twice-stabbed lady beetle, *Chilocorus stigma;* C, transverse lady beetle, *Coccinella transversoguttata richardsoni;* D, *Coleomegilla maculata fuscilabris;* E, *Cycloneda munda;* F, convergent lady beetle, *Hippodamia convergens,* adult and larva; G, *H. parenthesis;* H, *Olla abdominalis.*

Coleomegilla maculata fuscilabris Mulsant (Fig. 196D)—5-7 mm; elongate oval; head black; pronotum and elytra pinkish red or sometimes yellowish with irregular black spots; throughout North America.

Cycloneda munda (Say) (Fig. 196E)—4-5 mm; very convex; head black with white on front (♂) or 2 white spots (♀); pronotum black bordered with white, anterior margin with 3 white dashes toward posterior margin; elytra reddish yellow or red, without spots; throughout U.S., chiefly western states. *C. sanguinea* (Linnaeus) similar, east of Rocky Mts., AZ.

Epilachna varivestis Mulsant. Mexican bean beetle—6-7 mm; brownish yellow; each elytron with 3 basal, 3 central and 2 apical spots; most of U.S., southeastern Canada (on beans; larva spiny and plump). *E. borealis* (Fabricius), squash beetle, similar, 7-10 mm, spots larger, only 1 apical elytral spot, eastern 1/3 U.S.

Hippodamia convergens Guérin-Méneville (Fig. 196F). Convergent lady beetle—4-6 mm; head and thorax black with pale whitish margins; 2 whitish bars converge on prothorax; elytra reddish orange usually with 6 spots on each elytron, spots sometimes fewer in number or absent; throughout North America.

Hippodamia parenthesis (Say) (Fig. 196G)—4-5 mm; head mostly black; pronotum black with white irregular border; elytra yellowish red with black apical markings forming 2 parentheses; throughout North America.

Hippodamia tredecimpunctata tibialis (Say). Thirteenspotted lady beetle—4.5-5.2 mm; elongate oval; head black; pronotum black with pale yellowish border; elytra orange or reddish with 13 black spots; throughout North America.

Hyperaspis undulata (Say)—2.0-2.5 mm; shiny black; head yellow (♂); pronotum yellow laterally; each elytron with irregular, yellowish oval spot near center, and narrow marginal band; most of U.S., southern Canada. *H. signata* (Olivier) similar, 2.5-3.0 mm, elytra lack marginal bands, head and pronotum all black (♀); eastern 1/2 U.S.

Olla abdominalis (Say) (Fig. 196H)—4-5 mm; gray to pale yellow with black spots as shown; throughout North America (especially western states).

Paranaemia vittigera (Mannerheim)—4-5 mm; black; pronotum reddish with 2 large black patches; elytra reddish with 3 broad longitudinal stripes; Far West.

Family Endomychidae— Handsome Fungus Beetles

These small (chiefly 3-8 mm long), oval beetles are shiny and often bright orange or reddish with black markings. Many resemble lady beetles (Coccinellidae) but differ by having the head readily visible from above and the anterior corners of the pronotum generally extended forward. The tarsi may appear 3- or 4-segmented. Most species occur on fungi in decaying wood or fruit and under bark; they feed on fungi.

Common Species

Aphorista vittata (Fabricius) (Fig. 197A)—5-6 mm; brownish red or orange; flat pronotum with black on edges, sometimes 2 indistinct brownish spots; elytra with long central and 2 shorter lateral black bands; most of U.S., southern Canada.

Endomychus biguttatus Say (Fig. 197B)—3.5-5.0 mm; head, pronotum and legs shiny black; elytra reddish or orange with 2 smaller black spots before middle and 2 much larger patches toward apex; most of U.S. and Canada.

Family Byturidae— Fruitworm Beetles

The small (3.5-4.5 mm), oval, hairy beetles of this family are orange or pale brown in color. The second and third tarsal segments are lobed ventrally. There are only five species in the U.S. *Byturellus grisescens* (Jayne) may be swept from oaks in California and Oregon. *Byturus rubi* Barber (Fig. 197C), the eastern raspberry fruitworm, is brownish yellow with dense, pale yellow pubescence. The adults of this species and *B. bakeri* Barber, the western raspberry fruitworm, occur on flowers of raspberries and blackberries and the larvae feed on the fruit.

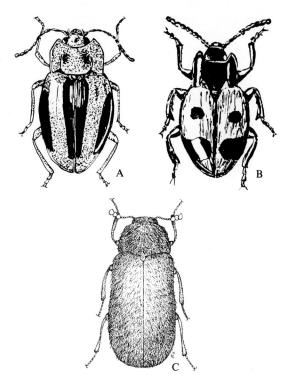

are shiny and cylindrical; they are sometimes called false wireworms and resemble click beetle larvae (Elateridae).

Members of the genus *Eleodes* (Fig. 198) are large black beetles that angle their abdomens upward when disturbed and may emit a foul-smelling black fluid. There are about 100 species of this western genus, many of which are quite common in semiarid and desert areas. The genus *Diaperis* contains colorful, spotted species that resemble lady beetles (Coccinellidae).

Figure 197 A and B, handsome fungus beetles (Endomychidae): A, *Aphorista vittata;* B, *Endomychus biguttatus.* C, eastern raspberry fruitworm, *Byturus rubi* (Byturidae).

Family Tenebrionidae— Darkling Beetles

The tenebrionids are a large family of beetles with about 15,000 species worldwide and 1,400 species in North America. The many common species tend to be dull black or brown and often resemble ground beetles (Carabidae). Darkling beetles have a 5-5-4 segmentation, the eyes are usually notched rather than entirely round, and the 11-segmented antennae are generally moniliform or filiform. The body varies from broadly oval to elongate, and smooth to very rough. These beetles are more common in the arid western states, occurring under rocks, debris, and loose bark. Larvae of many species

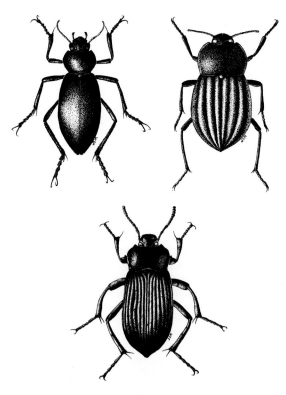

Figure 198 *Eleodes* spp. of darkling beetles (Tenebrionidae).

Common Species

Alobates pennsylvanica (De Geer) (Fig. 199A)— 20-23 mm; black; posterior corners of pronotum pointed; most of U.S. (under bark).

Bolitotherus cornutus (Panzer) (Fig. 199B)—10-12 mm; dark brown to dull black; body surface extremely rough; pronotum with 2 horns (♂) or tubercles (♀); elytra with 4 rows large tubercles; eastern 1/2 U.S. (in old fungi on logs).

Neomida bicornis (Fabricius)—3-4 mm; elongate-oval; dorsal side metallic bluish green; pronotum occasionally brownish; front of head with 2 short horns and top with 2 longer horns (♂); throughout North America (on old fungi).

Phloeodes pustulosus (LeConte) (Fig. 199C)—15-23 mm; flattened; grayish black with highly roughened and hardened dorsum; elytra sometimes whitish at base and apex; CA.

Platydema excavatum (Say) (Fig. 199D)—4.5-5.5 mm; broadly oval; shiny black; antennae and legs dark reddish brown; 2 horns between eyes (♂) with concave area between horns, or head with tubercles (♀); elytra deeply grooved; most of U.S. (under bark). *P. ruficorne* (Sturm) similar, without horn, dull black, antennae pale reddish yellow, eastern 1/3 U.S. *P. ellipticum* (Fabricius) without horn, 2 dull red spots at base of elytra, eastern 1/2 U.S.

Tenebrio molitor Linnaeus (Fig. 199E). Yellow mealworm—13-16 mm; shiny blackish or dark reddish brown; deeply grooved, intervals between grooves densely punctured; throughout North America (in stored grains, meal, flour). *T. obscurus* Fabricius, dark mealworm, similar, dull color, intervals between elytral grooves granular. *Neatus tenebroides* (Palisot de Beauvois) similar, eastern 1/2 U.S. (under bark).

Tribolium confusum Jacquelin duVal (Fig. 199F). Confused flour beetle—4.5-5.0 mm; flattened; reddish brown; antennae enlarged gradually into a club; throughout North America (in stored dried food, museum specimens). *T. castaneum* (Herbst), red flour beetle, similar, antennae enlarged abruptly at apex.

Uloma punctulata LeConte—7.0-8.5 mm; pale reddish brown; last segment antennae rounded at tip; eastern U.S., chiefly South (under pine bark). *U. impressa* Melsheimer similar, 11-12 mm, head with deep curved impression, eastern U.S. (under oak and beech bark).

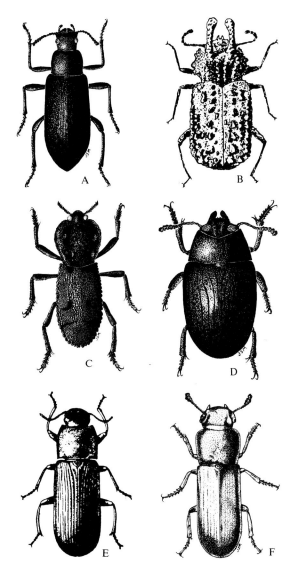

Figure 199 Darkling beetles (Tenebrionidae). A, *Alobates pennsylvanica;* B, *Bolitotherus cornutus;* C, *Phloeodes pustulosus;* D, *Platydema excavatum;* E, yellow mealworm, *Tenebrio molitor;* F, confused flour beetle, *Tribolium confusum.*

Family Alleculidae—Combclawed Beetles

These brownish or black, elongated and convex beetles resemble darkling beetles (Tenebrionidae) but have a row of small, blunt teeth on each tarsal claw which darkling beetles lack (Fig. 165A). Alleculids are 4-12 mm long and the body often is glossy due to its pubescence. Adults occur on flowering plants, fungi, and under bark, and many probably are pollen feeders. Larvae resemble the false wireworms (Tenebrionidae).

Common Species

Hymenorus niger (Melsheimer)—5-7 mm; black; anterior part of elytra brown to brownish black; legs brown to yellow brown; elytral grooves disappear posteriorly; eastern U.S. south to FL, west to MN and TX, southern Canada.

Isomira sericea (Say) (Fig. 200A)—5-6 mm; pale brownish yellow; dense fine pubescence; 4th segment of maxillary palpi long and slender; eastern 1/2 U.S.

Family Lagriidae—Longjointed Beetles

These slender beetles are 10-15 mm long and a dark metallic color. The 5-5-4 tarsal segmentation and enlarged apical segment of the antennae help characterize this family. Adults occur and feed on leaves or sometimes are found under bark.

Common Species

Arthromacra aenea (Say) (Fig. 200B)—9-14 mm; metallic green, coppery blue, or dark brown; antennae reddish brown; ventral side dark bronze; eastern U.S., southeastern Canada.

Family Salpingidae—Narrowwaisted Bark Beetles

The pronotum abruptly narrows posteriorly in many species of this family and the general body shape and black color resemble the ground beetles (Carabidae). Body length varies from 3-30 mm. Adults occur under bark, rocks, litter, and on vegetation; adults and larvae are predaceous.

Common Species

Pytho niger Kirby (Fig. 201A)—5.5-6.0 mm; black; pronotum flattened; head and pronotum with longitudinal depression; elytra grooved, many punctures, base with 2 depressions; northern 1/2 U.S., Canada.

Family Mycetophagidae—Hairy Fungus Beetles

These small (1.5-6.0 mm long), hairy beetles are flattened and broadly oval. They are brown to black and may be marked with red or orange. Members appear on shelf fungi and under moldy bark and vegetation, apparently feeding on fungi.

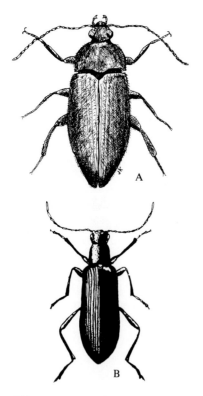

Figure 200 A, a combclawed beetle, *Isomira sericea* (Alleculidae); B, a longjointed beetle, *Arthromacra aenea* (Lagriidae).

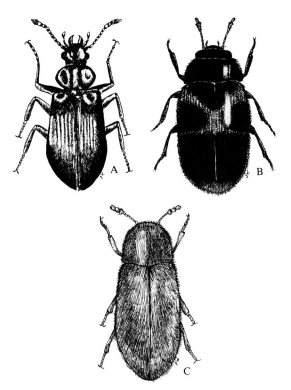

Common Species

Cis fuscipes Mellie (Fig. 201C)—2.0-5.5 mm; black to dark reddish brown; eastern 1/2 U.S.

Family Pyrochroidae— Firecolored Beetles

These flattened beetles have soft elytra, the head is narrowed between the large eyes, and many have prominent serrate or pectinate antennae. The family name is derived from the black color with contrasting red or yellow pronota. Adults occur on vegetation and the predaceous larvae live under bark.

Common Species

Dendroides cyanipennis Latreille (Fig. 202A)—9-13 mm; head, antennae, and elytra blackish; rest of body reddish yellow; most of U.S., southern Canada.

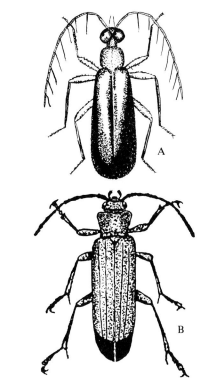

Figure 201 A, a narrowwaisted bark beetle, *Pytho niger* (Salpingidae); B, a hairy fungus beetle, *Mycetophagus punctatus* (Mycetophagidae); C, a minute treefungus beetle, *Cis fuscipes* (Ciidae).

Common Species

Mycetophagus punctatus Say (Fig. 201B)—4.0-5.5 mm; blackish with reddish yellow markings; antennae red-brown becoming blackish at apex; eastern 1/2 U.S.

Family Ciidae— Minute Treefungus Beetles

The cylindrical body of these brownish or black beetles is 0.5-6.0 mm long. The head is not visible from above and the antennae have a 3-segmented club. These beetles resemble the Scolytidae and Bostrichidae and occur under bark, in decaying wood, or in dry, hard fungi. Ciids feed on fungi.

Figure 202 A, a firecolored beetle, *Dendroides cyanipennis* (Pyrochroidae); B, wharf borer, *Nacerdes melanura* (Oedemeridae).

Neopyrochroa flabellata (Fabricius)—15-17 mm; head, basal portion of antennae, thorax, and legs reddish yellow; elytra black; most of U.S., southern Canada (in woods).

Family Oedemeridae— Oedemerid Beetles

Oedemerids are 5-20 mm long, have soft elytra, and the tarsal segmentation is 5-5-4. These slender beetles resemble the more common blister beetles (Meloidae) but differ partly because the next to last tarsal segment is expanded and very hairy beneath. Adults occur on flowers or foliage and are attracted to lights. Adults are pollen and nectar feeders; larvae live and feed in damp decaying wood (especially conifers).

Common Species

Nacerdes melanura (Linnaeus) (Fig. 202B). Wharf borer—8-12 mm; dull yellow; elytra deep purple at apex, with 8 elevated longitudinal lines; legs and ventral surface blackish; throughout U.S., southern Canada (in lumberyards, woodsheds).

Family Melandryidae—False Darkling Beetles or Melandryid Bark Beetles

These dark-colored beetles are 3-20 mm long and often slightly flattened and pubescent. They have a 5-5-4 tarsal segmentation and usually two dents near the posterior margin of the pronotum. Adults and larvae occur under dry bark and in dry wood and fungi; a few adults frequent flowers and leaves.

Common Species

Eustrophinus bicolor (Fabricius) (Fig. 203A)— 4.5-6.0 mm; convex; shiny black; antennal base and apex, legs, and ventral side reddish brown; each 1/2 posterior margin of pronotum distinctly "S"-shaped; most of U.S.

Osphya varians (LeConte) (Fig. 203B)—5-8 mm; black; margin and center of thorax reddish yellow; sparse, fine, gray prostrate hairs; most of U.S.

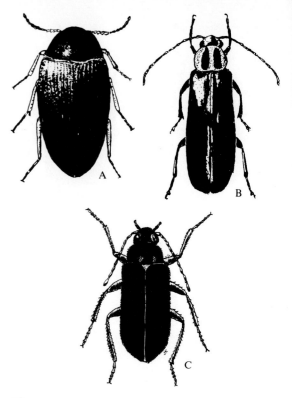

Figure 203 False darkling beetles (Melandryidae). A, *Eustrophinus bicolor;* B, *Osphya varians;* C, *Penthe obliquata.*

Penthe obliquata (Fabricius) (Fig. 203C)—11.5-14.0 mm; velvety black; scutellum with long yellow-orange hairs; eastern 1/2 U.S. *P. pimelia* (Fabricius) similar, scutellum black.

Pisenus humeralis (Kirby)—3-4 mm; convex; shiny; dark reddish brown or blackish; scattered yellowish pubescence; often 2 indistinct reddish spots at base of elytra; east of Rocky Mts.

Family Mordellidae— Tumbling Flower Beetles

The arched body, slightly shortened elytra, and sharply tapered, pointed abdomen readily identify members of this family. They are common on flowers and jump or tumble when disturbed. The adults and most larvae are plant feeders.

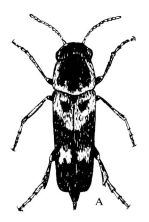 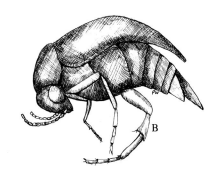 

Figure 204 Tumbling flower beetles (Mordellidae). A, *Glipa oculata;* B, *Mordella marginata;* C, *Mordellistena pustulata.*

Common Species

Glipa octopunctata (Fabricius)—6-7 mm; black or grayish black; pronotum with irregular yellowish pubescence on margins; elytra with 8 irregular spots of yellowish pubescence; eastern U.S. west to KS and TX.

Glipa oculata (Say) (Fig. 204A)—5-7 mm; head, thorax, and abdomen blackish with yellow and gray markings; antennae, tibiae, and tarsi dull red; eastern 1/2 U.S.

Mordella atrata Melsheimer—3-6 mm; dull black; brownish pubescence; margins of sternites gray; eastern 1/2 U.S. *M. albosuturalis* Liljeblad similar, 3-5 mm, base and inner margin of each elytron with silvery pubescence; western 1/2 U.S.

Mordella marginata Melsheimer (Fig. 204B)— 3.0-4.5 mm; black or dark gray; silvery pubescence scattered as short lines in various arrangements on pronotum and elytra; Rocky Mts. westward.

Mordellistena pustulata (Melsheimer) (Fig. 204C)— 2-3 mm; black; about 7 rows small silvery pubescent spots across elytra, base with lighter band; throughout North America.

Family Meloidae—Blister Beetles

The blister beetles are elongated beetles with a prothorax that is often narrower than the head and sometimes necklike. The wings and body are soft and frequently the tip of the abdomen is exposed. Adults are plant feeders and many larvae are predators of grasshopper eggs or parasites of bee eggs and larvae. The active, parasitic larvae gain access to bee nests by attaching themselves to foraging bees; they then develop into several distinct and more sedentary forms (hypermetamorphosis). All species contain a blistering substance, cantharadin. This material is extracted from a species in southern Europe, the Spanishfly, and used as a drug. Some species will secrete blistering materials or oily substances (latter in genus *Meloe*) as a defensive action.

Common Species

Epicauta albida (Say)—14-20 mm; yellowish or grayish; 2 dark longitudinal lines on pronotum; 1st segment of antennae as long as following 4 segments; western 1/2 U.S.

Epicauta maculata (Say). Spotted blister beetle—10-14 mm; black; gray pubescence; many small spots on elytra; Far West.

Epicauta pennsylvanica (De Geer) (Fig. 205A). Black blister beetle—7-15 mm; entirely dull black; sparse pubescence; most of U.S., southern Canada (on goldenrod, garden flowers).

Epicauta vittata (Fabricius) (Fig. 205B). Striped blister beetle—12-18 mm; dull yellow above with black or dark brown stripes; black beneath; U.S. west to MO, southeastern and southcentral Canada.

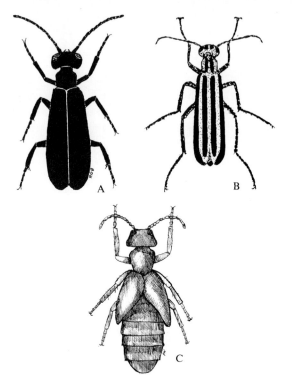

Figure 205 Blister beetles (Meloidae). A, *Epicauta pennsylvanica;* B, *E. vittata;* C, *Meloe angusticollis.*

Lytta cyanipennis (LeConte)—13-18 mm; metallic green or bluish; Rocky Mts. westward. *L. nuttallii* (Say), Nuttall blister beetle, similar, 16-28 mm.

Lytta immaculata (Say)—25-27 mm; black; gray or yellowish pubescence; Midwest to Rocky Mts.

Meloe angusticollis Say (Fig. 205C)—12-15 mm; head and pronotum blue-black; elytra and abdomen blackish with violet tones; elytra short, wrinkled, overlap, and point outward; northern U.S., Canada.

Family Anthicidae— Antlike Flower Beetles

Anthicids are 2-6 mm long and somewhat antlike in appearance. The greatly lowered head with its necklike posterior region and the 5-5-4 tarsal segmentation help distinguish the family. The pronotum of some extends hornlike over the head (i.e., genus *Notoxus* with the tarsi shorter than the tibiae, and genus *Mecynotar-sus* with the tarsi as long as or longer than the tibiae). Anthicids occur on flowers and foliage, on the ground, and a few frequent sand dunes.

Common Species

Anthicus cervinus LaFerté-Sénectere—2.4-2.7 mm; reddish brown; antennae and legs dull yellow; elytra with blackish markings on apical 1/2 which enclose a broad pale yellow cross; throughout North America.

Notoxus monodon Fabricius (Fig. 206A)—2.5-4.0 mm; dull brownish yellow; pronotal horn present; dense grayish hairs; elytra with black crossband on apical 1/2, 2 black spots near base; head and sides of pronotum sometimes blackish; throughout North America. *N. constricta* Casy similar, pronotal horn forked, CA (in orchards).

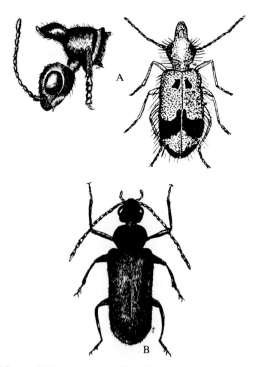

Figure 206 A, an antlike flower beetle, *Notoxus monodon* (Anthicidae); B, a pedilid beetle, *Pedilus lugubris* (Pedilidae).

Family Pedilidae—Pedilid Beetles

Members of this family resemble the Anthicidae but are more elongate (4-15 mm long) and cylindrical. The head narrows posteriorly into a neck and the pronotum is oval or rounded. Adults frequent flowers or foliage and are attracted to lights. Most of the species occur in the western states.

Common Species

Pedilus lugubris Say (Fig. 206B)—6-8 mm; black except 1st 2 antennal segments and front of head reddish brown; most of U.S., central Canada.

Family Cerambycidae— Longhorned Beetles or Roundheaded Wood Borers

Cerambycids are usually medium to large, elongated and cylindrical beetles with antennae over half the body length (the antennae of some species are much longer than the body). The eyes are notched (Fig. 207A) and many species are brightly colored. Some smaller cerambycids resemble the leaf beetles (Chrysomelidae) but the leaf beetles are smaller (usually less than 12 mm) and the antennae are less than half the body length. The longhorned beetles are a large family with about 20,000 species worldwide and over 1,200 in the U.S.

Most adults occur and feed on flowers and foliage during the day; some eat bark or stems. Numerous species are nocturnal, resting under bark or on trees during the day, and are often attracted to lights at night. Adults prefer to oviposit on solid wood of weakened, dying, dead, or freshly cut or fallen trees. Larvae are cylindrical, whitish, and have a rounded head-thorax area as contrasted with the flatheaded wood borers (Buprestidae). Cerambycid larvae bore into and feed on wood. Only a few of the common species are described below.

Common Species

Subfamily Prioninae—pronotum with distinct margin on sides and often margin is toothed; up to 76+ mm long.

Orthosoma brunneum (Forster)—22-48 mm; shiny; light reddish brown; deep impressions between antennae; 2-3 teeth on lateral margins of pronotum; eastern U.S., southeastern Canada.

Prionus californicus Motschulsky. California prionus—40-60 mm; shiny; dark reddish brown; 3 sharp teeth on lateral margins of pronotum; Far West.

Prionus laticollis (Drury) (Fig. 207A). Broadnecked root borer—22-47 mm; shiny black; antennae reaches apical 1/2 of elytra (♂) or 1/2 or less body length (♀); pronotum as wide as base of elytra, pronotum with 3 lateral teeth; eastern 1/2 North America. *P. imbricornis,* tilehorned prionus, similar, dark reddish brown.

Subfamily Lamiinae—last segment of maxillary palpi pointed at tip.

Monochamus maculosus Haldeman (Fig. 207B). Spotted pine sawyer—14-27 mm; dark reddish brown to blackish; bluish gray markings on elytra; antennae up to 2 1/2x body length; pronotum with large tooth on each side; throughout U.S. (chiefly Far West), southwest Canada.

Monochamus titillator Fabricius. Southern pine sawyer—20-30 mm; brown; elytra mottled with gray, brown and black pubescence; pronotum with large tooth on each side; apex of each elytron with a spine; most of U.S., southern Canada.

Saperda calcarata Say (Fig. 207C). Poplar borer—25-31 mm; dense covering of gray and yellow pubescence; front of head and scutellum yellow; pronotum with 3 yellowish longitudinal stripes; coarse punctures; tip of each elytron spined; throughout North America.

Saperda candida Fabricius. Roundheaded appletree borer—15-20 mm; light brown with 2 longitudinal stripes from head to elytral apex; ventral side white; eastern U.S., Canada.

Tetraopes tetrophthalmus (Forster). Red milkweed beetle—9-14 mm; red-orange; pronotum with 4 black spots; elytra with 6 black spots; eastern 1/2 North America (on milkweed). *T. melanurus* Schönherr similar, 2 wide crossbands on elytra, southern U.S. *T. femoratus* LeConte similar, 12-17 mm, antennae ringed with gray, 4 elytral spots, throughout U.S.

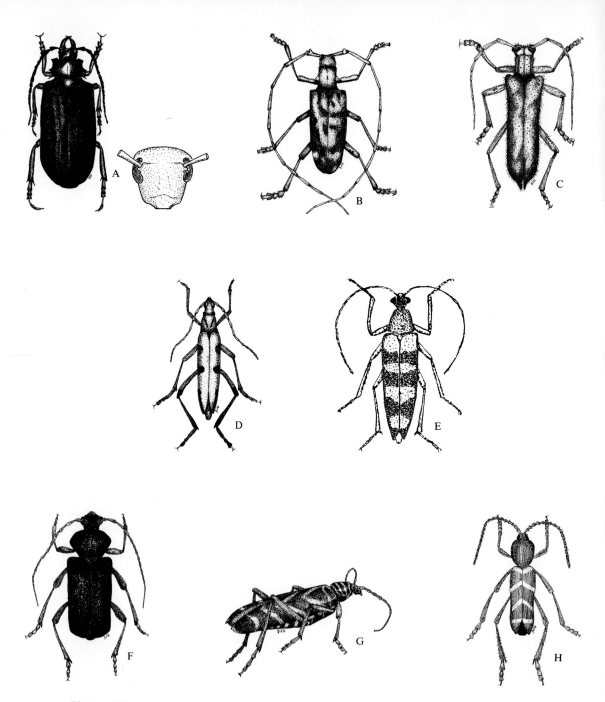

Figure 207 Longhorned beetles or roundheaded wood borers (Cerambycidae). A, broadnecked root borer, *Prionus laticollis,* and a notched eye; B, spotted pine sawyer, *Monochamus maculosus;* C, poplar borer, *Saperda calcarata;* D, *Strangalia famelica;* E, *Typocerus velutinus;* F, *Callidium antennatum hesperum;* G, locust borer, *Megacyllene robiniae;* H, redheaded ash borer, *Neoclytus acuminatus.*

Subfamily Lepturinae—apex of last segment of maxillary palpi blunt, elytra often greatly narrowed posteriorly, eyes oval or only slightly notched.

Enaphalodes rufulus (Haldeman). Red oak borer—22-28 mm; reddish brown, covered with brownish orange pubescence; pronotum usually with 2 small tubercles near center; apex of each elytron with 2 short spines; eastern U.S. west to MN and TX.

Strangalia famelica (Newman) (Fig. 207D)—12-14 mm; very slender; brownish yellow with yellowish pubescence; pronotum with 2 black stripes; black spots on elytra; eastern U.S.

Strophiona laeta (LeConte)—10-13 mm; yellow or golden; 3 wide crossbands on elytra; Far West (near oaks).

Typocerus velutinus (Olivier) (Fig. 207E)—8-14 mm; head, antennae, and pronotum black; pronotum with yellow pubescence; elytra reddish brown with yellow crossbands; eastern U.S.

Subfamily Cerambycinae—apex of last segment of maxillary palpi blunt, elytra not greatly narrowed posteriorly, eyes distinctly notched and partly surround base of antennae.

Callidium antennatum hesperum Casey (Fig. 207F)—9-14 mm; metallic blue or blue-black; rounded thorax with 2 depressions; legs black, femora large; throughout U.S. (early spring).

Megacyllene robiniae (Forster) (Fig. 207G). Locust borer—12-18 mm; black; head, pronotum, and elytra with yellow crossbands as shown; Rocky Mts. eastward (on goldenrod in fall).

Neoclytus acuminatus (Fabricius) (Fig. 207H). Red-headed ash borer—6-18 mm; head, thorax, and middle and hind legs reddish; elytra light brown with darker apex, 4 crossbands of fine yellow hairs; eastern U.S., southeastern Canada.

Family Chrysomelidae—Leaf Beetles

Leaf beetles are a large family of very common and often brightly colored insects. There are over 25,000 species worldwide and about 1,460 species in North America. The body shape of many resembles lady beetles (Coccinellidae) but the tarsal segmentation of lady beetles is 3-3-3 whereas that of leaf beetles appears as 4-4-4. Leaf beetles also resemble some longhorned beetles (Cerambycidae) but the former are usually smaller (1-12 mm long in the U.S.) and the antennae are nearly always less than half the body length.

Adults feed on flowers and foliage; most larvae eat foliage (includes leaf mining) and others bore into stems and roots. Larvae tend to be soft-bodied and brightly colored. Some bear spines or are partially covered with their own excrement. Many chrysomelids are important agricultural pests. Only a few of the common species are described below.

Common Species

Subfamily Cassidinae—Tortoise Beetles; broadly oval to circular, head usually concealed from above, some resemble small turtles or lady beetles, many brilliantly colored, some flattened, often on morning glory.

Agriconota bivittata (Say) (Fig. 208A)—4.5-6.0 mm; pronotum pale yellow with large brown triangular area; each elytron pale yellow with 5 blackish longitudinal stripes; most of U.S.

Chelymorpha cassidea (Fabricius) (Fig. 208B). Argus tortoise beetle—8.0-11.5 mm; reddish or yellow; pronotum with crossrow of 4 black dots and 2 larger ones posteriorly; elytra with 13 black dots; most of U.S.

Deloyala guttata (Olivier) (Fig. 208C). Mottled tortoise beetle—5-6 mm; margins of pronotum and elytra translucent pale yellow except at anterior corners and apex of elytra; elytra blackish with irregular yellow patches; most of U.S.

Metriona bicolor (Fabricius) (Fig. 208D). Golden tortoise beetle—5-6 mm; brilliant brassy gold when alive but dull reddish yellow when dead; margins of pronotum spread out and flattened; elytra with 2 round, depressed spots; eastern U.S.

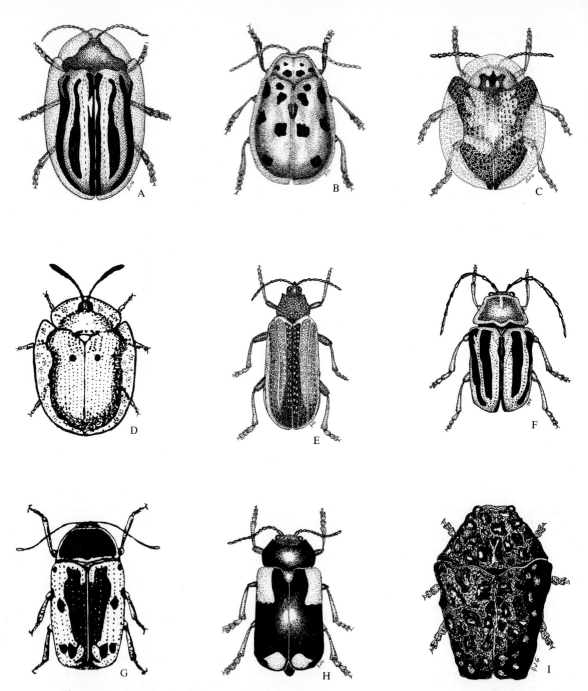

Figure 208 Leaf beetles (Chrysomelidae). A, *Agriconota bivittata;* B, argus tortoise beetle, *Chelymorpha cassidea;* C, mottled tortoise beetle, *Deloyala guttata;* D, golden tortoise beetle, *Metriona bicolor;* E, *Odontota dorsalis;* F, *Pachybrachis* sp.; G, *Cryptocephalus* sp.; H, *Babia quadriguttata;* I, *Neochlamisus gibbosa.*

Subfamily Hispinae—Leafmining Leaf Beetles; 4-7 mm, mouth directed downward or posteriorly rather than forward, elytra often distinctly ridged and with punctures between ridges; larvae are leaf miners.

Microrhopala vittata (Fabricius)—5-7 mm; head and pronotum reddish; elytra blackish or brown, with 2 reddish longitudinal stripes, 8 rows of punctures; eastern U.S. (on goldenrod).

Odontota dorsalis (Thunberg) (Fig. 208E)—6.0-6.5 mm; orange-red with head and middorsal elytral stripe black; elytra ridged; eastern 1/2 U.S. (on locust trees). *O. scapularis* (Olivier) similar shape, pronotum with 2 lateral stripes, anterior corners elytra reddish, AZ eastward.

Subfamilies Chlamisinae, Clytrinae, and Cryptocephalinae—Casebearing Leaf Beetles; pronotum covers head up to eyes, elytra expose last abdominal segment, larvae carry protective case often of own excrement. Members of genus *Pachybrachis* (Fig. 208F) have pronotum with distinct posterior margin and non-scalloped lateral margins; those in genus *Cryptocephalus* (Fig. 208G) have pronotum without distinct posterior margin and lateral margins are weakly scalloped.

Babia quadriguttata (Olivier) (Fig. 208H)—3.0-5.5 mm; shiny black; 2 reddish spots at base and apex of elytra; antennae short and weakly serrated; throughout U.S.

Neochlamisus gibbosa (Fabricius) (Fig. 208I)—2.5-4.5 mm; bronze-brown, elytra sometimes blackish; pronotal surface rough, large 2-pronged tubercle in center; elytra with many tubercles; pulls in legs and rolls off plants if disturbed; eastern U.S.

Subfamily Donaciinae—metallic green, blue, coppery, or black, long antennae, adults and larvae on aquatic or shoreline plants like water lilies or sedges; common genus is *Donacia* (Fig. 210A).

Subfamily Alticinae—Flea Beetles; dark blue, green, or black, jumping behavior, hind femora enlarged, adults eat many tiny holes in leaves.

Chaetocnema pulicaria Melsheimer. Corn flea beetle—1.5-2.5 mm; shiny black with bronze or blue-green luster; basal 3-4 antennal segments orange; tibiae and tarsi brownish yellow; most of U.S. (on corn, grasses).

Disonycha triangularis (Say) (Fig. 209A). Threespotted flea beetle—4.5-7.0 mm; shiny blue or blue-black; pronotum yellow with 3 black dots; most of U.S. *D. xanthomelas* (Dalman), spinach flea beetle, similar but pronotum without dots.

Epitrix cucumeris (Harris) (Fig. 209B). Potato flea beetle—1.5-2.5 mm; shiny black; hairy; antennae and legs brownish orange; throughout U.S. *E. hirtipennis* (Melsheimer), tobacco flea beetle, similar shape, dull reddish yellow, sometimes darker band across middle of elytra.

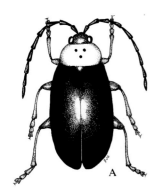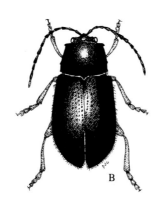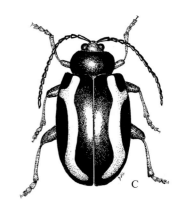

Figure 209 Leaf beetles (Chrysomelidae). A, threespotted flea beetle, *Disonycha triangularis*; B, potato flea beetle, *Epitrix cucumeris*; C, *Systena blanda*.

Systena blanda Melsheimer (Fig. 209C)—3.0-4.5 mm; head and prothorax reddish brown; elytra pale yellow, brown, or blackish, and with 2 lighter stripes; throughout U.S.

Subfamily Criocerinae—head narrowed posteriorly to form neck, elytral punctures in rows.

Crioceris asparagi (Linnaeus) (Fig. 210B). Asparagus beetle—6-7 mm; flattened; shiny; head dark metallic blue; pronotum red, 2 bluish spots; elytra bluish green with yellow patches; most of U.S. (on asparagus). *C. duodecimpunctata* (Linnaeus), spotted asparagus beetle, similar shape, head and pronotum reddish brown, each orange-red elytron with 6 black spots, eastern U.S.

Lema trilineata (Olivier). Threelined potato beetle—6.0-7.5 mm; reddish yellow; 2 small black spots on pronotum; elytra with 3 longitudinal black stripes; pronotum distinctly constricted in middle; eastern U.S. (on potatoes).

Subfamily Chrysomelinae—broadly oval, convex, head inserted into pronotum up to eyes, brightly colored. *Phyllodecta* species are small, metallic blue or purple, feed on willow and poplar.

Calligrapha philadelphica (Linnaeus) (Fig. 210C)—8-9 mm; very convex; head and pronotum dark olive green, metallic; antennae and legs dark reddish brown; elytra yellowish white with dark markings as shown; eastern 1/2 U.S.

Chrysomela scripta (Fabricius) (Fig. 210D)—7.0-9.5 mm; prothorax blackish with orange lateral margins; each yellowish elytron with 7 purplish spots, pattern variable; most of North America (on willow, poplar).

Leptinotarsa decemlineata (Say) (Fig. 210E). Colorado potato beetle—5.5-11 mm; dull yellow; pronotum spotted; elytra with 10 longitudinal black lines as shown; throughout U.S., southern Canada (on potatoes; larva humped, brick red).

Subfamily Galerucinae—soft elytra, pronotum usually with distinct lateral margins.

Acalymma trivittatum (Mannerheim) (Fig. 210F). Western striped cucumber beetle—4-6 mm; pronotum orange-yellow; elytra pale yellow with 3 black longitudinal stripes; base of antennae and femora pale yellow; Far West. *A. vittatum* (Fabricius), striped cucumber beetle, similar, antennae all black, eastern U.S. west to CO and NM, eastern Canada.

Ceratoma trifurcata (Forster)—4.0-8.5 mm; yellow or dull red; head black or blue-black; elytra with blackish border and 3 pairs large blackish spots, spots sometimes connected or nearly absent; east of Rocky Mts.

Diabrotica longicornis (Say). Northern corn rootworm—5-6 mm; pale green or greenish yellow; sometimes reddish brown tinge on head and thorax; east of Rocky Mts.

Diabrotica undecimpunctata howardi Barber (Fig. 210G). Spotted cucumber beetle, southern corn rootworm—6-7 mm; greenish yellow; head black; elytra with 12 black spots; bases of antennae and legs pale yellow; most of U.S. (primarily eastern). *D. undecimpunctata undecimpunctata* Mannerheim, western spotted cucumber beetle, similar, antennae, legs, and body all black, Far West.

Subfamily Eumolpinae—convex, oblong, 3rd tarsal segment bilobed, metallic or yellowish and spotted.

Chrysochus auratus (Fabricius)—8-11 mm; iridescent blue-green with coppery tinge; eastern 1/2 U.S. (on milkweed, dogbane). *C. cobaltinus* LeConte similar, 9-10 mm, darker metallic blue, Far West.

Colaspis brunnea (Fabricius) (Fig. 210H). Grape colaspis—4-6 mm; dull brownish yellow; outer segments antennae often black; legs pale yellowish; most of U.S. (on grapevines, cotton).

Paria fragariae Wilcox (Fig. 210I). Strawberry rootworm—3-4 mm; color and markings variable; head reddish yellow; pronotum black or yellow; elytra black, yellow, or yellow with 4 (sometimes 6) black spots; throughout North America.

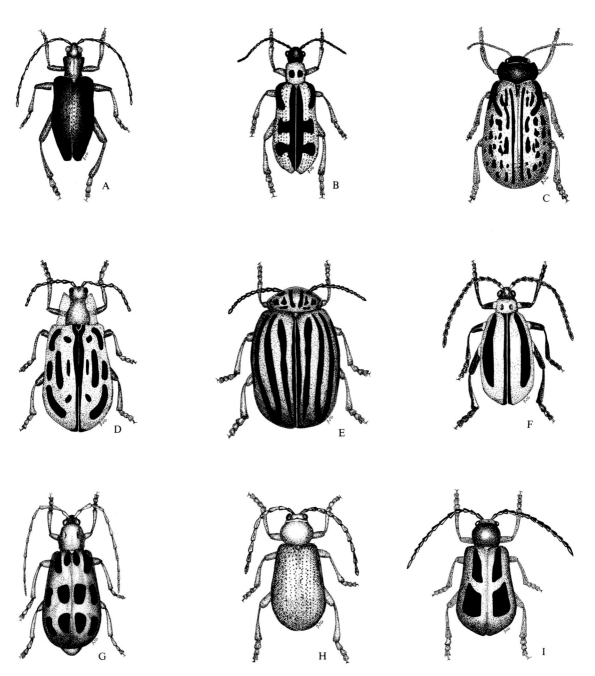

Figure 210 Leaf beetles (Chrysomelidae). A, *Donacia* sp.; B, asparagus beetle, *Crioceris asparagi;* C, *Calligrapha philadelphica;* D, *Chrysomela scripta;* E, Colorado potato beetle, *Leptinotarsa decemlineata;* F, western striped cucumber beetle, *Acalymma trivittatum;* G, spotted cucumber beetle or southern corn rootworm, *Diabrotica undecimpunctata howardi;* H, grape colaspis, *Colaspis brunnea;* I, strawberry rootworm, *Paria fragariae.*

Order Coleoptera 231

Family Bruchidae—Seed Beetles

These oval or egg-shaped beetles are generally less than 5 mm long and their elytra do not cover the tip of the abdomen. The head is extended into a short, broad snout and the antennae may be clubbed, serrate, or pectinate. Adults occur on flowers, leaves, and seed pods, and both adults and larvae feed inside stored seeds such as peas and beans.

Common Species

Acanthoscelides obtectus (Say) (Fig. 211A)—2.5-4.0 mm; blackish with brownish gray pubescence; antennae black with red-brown base and tip; elytra with short crossbands of brown pubescence; throughout North America.

Bruchus pisorum (Linnaeus). Pea weevil—4-5 mm; black with dense reddish brown and whitish pubescence; pronotum with small whitish triangle at middle of posterior margin; elytra with grayish, yellowish, and whitish patches, irregular white crossband on posterior 1/2 of elytra; whitish patch at tip of abdomen; most of North America (on peas in spring).

Family Anthribidae— Fungus Weevils

Fungus weevils are generally less than 10 mm long and have a short, broad, and flat beak. The antennae are clubbed (but not elbowed) or sometimes filiform. The tip of the abdomen is not exposed as with the similar seed beetles (Bruchidae). Anthribids occur beneath old bark, on dead branches, and sometimes in weedy fields. Wood, fungi, seeds, and pollen serve as food for larvae and probably adults to some extent.

Common Species

Araecerus fasciculatus (De Geer). Coffee bean weevil—2.5-4.5 mm; dark brown with yellow or brown pubescence; antennae and legs reddish brown; throughout most of North America (in seeds, berries, dried fruit).

Euparius marmoreus (Olivier) (Fig. 211B)—3.5-8.5 mm; yellowish brown and pubescent; dense pale brown and grayish yellow pubescence on snout, anterior 1/2 pronotum, and a patch in center of elytra; posterior part of elytral patch with black spots; legs ringed with gray and dark brown; eastern 1/2 North America.

Family Brentidae—Brentid Weevils

These elongated and cylindrical weevils are 10-30 mm long and have a snout that extends forward and does not curve downward. Adults occur under bark or in damaged wood and feed on fungi, sap, or insects. Five of the six North American species are confined to Florida, Texas, and California, and the sixth species inhabits the eastern U.S.

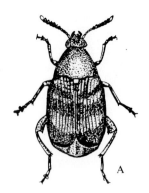
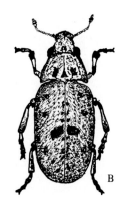
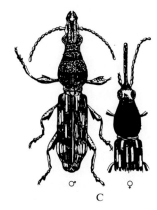

Figure 211 A, a seed beetle, *Acanthoscelides obtectus* (Bruchidae); B, a fungus weevil, *Euparius marmoreus* (Anthribidae); C, oak timberworm, *Arrhenodes minutus* (Brentidae).

Common Species

Arrhenodes minutus (Drury) (Fig. 211C). Oak timberworm—7-17 mm; dark reddish brown; elytra with narrow and elongate yellowish spots; beak longer than head and very slender (♀) or as long as head and broad (♂); eastern U.S.

Family Scolytidae—Bark Beetles

Bark beetles are generally 1-3 mm long (some reach 9 mm in length), very cylindrical, and brown or black in color. The antennae are clubbed and elbowed (Fig. 160G) and the head may be lowered and not visible from above. The prothorax is very large and often the posterior abdominal segments are curved downward to form a broad concavity (Fig. 212A). Adults and larvae live under tree bark and either (1) mine the wood's surface and the underside of the bark (bark or engraver beetles), (2) bore into and feed on the wood, or (3) bore deeply into the wood to cultivate and feed on fungi which will grow on the tunnel walls (ambrosia or timber beetles). The larvae of surface feeders chew a maze of characteristic tunnels (galleries) on the wood and bark, thus creating an engraved appearance. The new adults emerge through tiny holes they have chewed in the bark. Bark beetles as a group are often considered to be the most destructive forest insects. The Dutch elm disease of American elms is transmitted by *Scolytus multistriatus* (Marsham).

Common Species

Dendroctonus frontalis Zimmerman. Southern pine beetle—2.2-4.2 mm; dark brown to black; front of head with 2 tubercles separated by a groove; elytra over 2x length of pronotum, not concave at tips, long hairs on sides and tips; southern U.S. west to AZ (on pine). *D. brevicomis* LeConte similar, head tubercles smaller, area near front margin of pronotum with narrow transverse elevation (♀) or depression (♂), Far West.

Dendroctonus valens LeConte. Red turpentine beetle—5.5-9.0 mm; light reddish brown to dark brown; most of U.S. except southeastern and Gulf coast states (on pines primarily). *D. terebrans* (Olivier), black turpentine beetle, similar, 5-10 mm, dark reddish brown to black, eastern 1/3 and southern U.S.

Ips grandicollis (Eichhoff) (Fig. 212A). Southern pine engraver—2.8-4.7 mm; dark reddish brown to black; tips of elytra deeply concave, punctured, with 5 teeth on each side; eastern 1/2 North America (on pines). *I. calligraphus* (Germar) similar, 3.5-6.5 mm, elytral concavity with 6 teeth on each side.

Scolytus rugulosus (Ratzeburg) (Fig. 212B). Shothole borer—1.7-2.9 mm; grayish black; elytra reddish brown at apex; most of U.S., primarily eastern (chiefly on fruit trees). *S. multistriatus* (Marsham), smaller European elm bark beetle, similar, 3 mm, ventral side abdomen concave and with a spine, on American elm.

Xyleborus saxesensi (Ratzeburg)—1.5-2.5 mm; pale brown to black; scutellum conical (♀); tip of elytra steeply sloped, with rows of pointed granules and 2 tubercles in center; humpbacked appearance (♂); most of U.S., southern Canada.

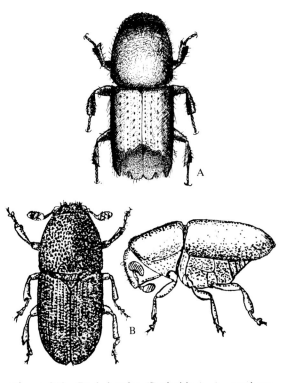

Figure 212 Bark beetles (Scolytidae). A, southern pine engraver, *Ips grandicollis;* B, shothole borer, *Scolytus rugulosus.*

Family Curculionidae—Weevils or Snout Beetles

Most weevils have a well-developed, downward-curved snout. The antennae of most species are elbowed and clubbed and the first segment often fits into a groove in the side of the snout (Fig. 213A). This family is the largest in the order with over 40,000 species worldwide and 2,500 species in North America. The great majority occur and feed on all parts of plants and many species are important pests. There are about 42 subfamilies of weevils but only a few will be described below.

Common Species

Subfamily Cyladinae

Cylas formicarius elegantulus (Summers) (Fig. 213B). Sweetpotato weevil—5-6 mm; reddish brown; elytra bluish black; antlike in appearance; antennae straight; pronotum greatly constricted in posterior 1/3; southern U.S. (larvae in sweet potatoes).

Subfamily Apioninae—1.0-4.5 mm; antennae not elbowed; abdomen deeply grooved and broadly oval giving a pear-shaped body; gray or black; commonly on legumes.

Subfamily Rhynchitinae—Toothnosed Snout Beetles

Rhynchites bicolor (Fabricius). Rose curculio—6 mm; red; snout and ventral sides of body black; elytra very broad; most of U.S. (in rose flowers).

Subfamily Ithycerinae

Ithycerus noveboracensis (Forster). New York Weevil—12-18 mm; shiny black; gray and brown pubescence; scutellum yellowish; eastern 1/2 North America (on beech, hickory, oak).

Subfamily Thylacitinae—beak stout, squarish and widened toward tip; 1st antennal segment passes below eye when pressed against head.

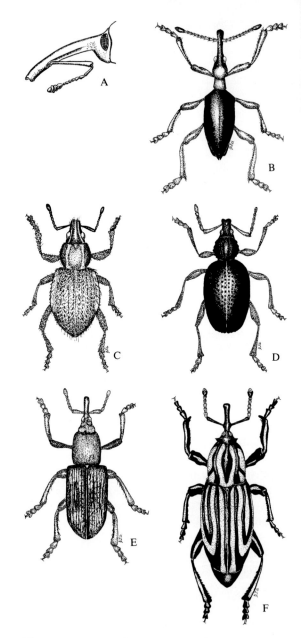

Figure 213 Weevils or snout beetles (Curculionidae). A, lateral view of the head; B, sweetpotato weevil, *Cylas formicarius elegantulus;* C, a whitefringed beetle, *Graphognathus leucoloma;* D, strawberry root weevil, *Otiorhynchus ovatus;* E, rice weevil, *Sitophilus oryzae;* F, maize billbug, *Sphenophorus maidis.*

Graphognathus leucoloma (Boheman) (Fig. 213C). A whitefringed beetle—9-12 mm; blackish; dense covering of gray or gray-brown scales; pronotum with 2 whitish longitudinal stripes; elytra with long hairs and whitish stripe on each lateral margin; southern U.S.

Pantomorus cervinus (Boheman). Fuller rose weevil—7-9 mm; pale brown; elytra with whitish and oblique stripe on sides; without hind wings; Far West.

Subfamily Otiorhynchinae—beak as in subfamily Thylacitinae; 1st antennal segment passes above middle of eye when pressed against head.

Otiorhynchus ovatus (Linnaeus) (Fig. 213D). Strawberry root weevil—5-6 mm; shiny black; antennae and legs reddish brown; sparse yellowish pubescence; pronotum nearly round and ridged; elytra grooved, with punctures and rough surface; most of U.S. (on strawberries).

Subfamily Rhynchophorinae—Billbugs and Grain Weevils; antennae insert near eyes, 1st segment extends behind eye, 1/2 or more of antennal club enlarged and shiny.

Rhodobaenus tredecimpunctatus (Illiger). Cocklebur weevil—7-11 mm; red; pronotum with 5 black spots, elytra with 8; primarily eastern 1/2 U.S.

Sitophilus granarius (Linnaeus). Granary weevil—3-4 mm; shiny reddish brown to blackish; pronotum with sparse elongate punctures; elytra deeply grooved, single rows of punctures; throughout North America (in stored grain, cereals).

Sitophilus oryzae (Linnaeus) (Fig. 213E). Rice weevil—2.1-2.8 mm; dull brown; pronotum with dense rounded punctures; elytra with 4 reddish spots, deeply grooved, double rows deep punctures; throughout North America (in stored grain, under bark, on leaves).

Sphenophorus maidis Chittenden (Fig. 213F). Maize billbug—11-15 mm; shiny reddish brown to blackish; grayish yellow crusty stripes between black stripes; southern U.S. (on corn).

Sphenophorus zeae Walsh—6.5-9.0 mm; shiny reddish brown to blackish; body shape like *S. maidis*; pronotum with 3 raised longitudinal bands, central band widest in middle and lateral bands wavy, area between bands coarsely punctured; elytra with many rows alternating coarse and fine punctures; eastern 1/2 U.S. (on grasses, corn). *S. parvulus* Gyllenhal, bluegrass billbug, similar shape, 5-8 mm, black with gray crust, pronotum without raised bands.

Subfamily Cryptorhynchinae—snout held at rest in groove under prosternum; genus *Conotrachelus* contains spp. with rough and bumpy elytra and elytra much broader than pronotum.

Conotrachelus nenuphar (Herbst) (Fig. 214A). Plum curculio—4.5-6.5 mm; dark brown; brownish yellow pubescence; whitish yellow hairs form 2 short irregular bands on pronotum and crossband on posterior 1/3 of elytra; elytra with broken ridges and tubercles; east of Rocky Mts. (on fruit trees, hawthorn).

Subfamily Baridinae—each side of mesothorax expanded and visible from above as a bump between corners of elytra and pronotum.

Trichobaris trinotata (Say) (Fig. 214B). Potato stalk borer—3-4 mm; black; dense white scales or prostrate pubescence on body except for head, posterior corners pronotum, and scutellum; fine elytral grooves; east of Rocky Mts., AZ, CA.

Subfamily Cleoninae—spp. in common genus *Lixus* are elongated and cylindrical, 10-15 mm, curved beak nearly length of prothorax, often on plants near water.

Lixus concavus Say (Fig. 214C). Rhubarb curculio—10.0-13.5 mm; black; antennae and tarsi reddish brown; sparse, short grayish pubescence; often orange-brown pollen on fresh specimens; pronotum with large depression which extends onto elytra; most of U.S. (on rhubarb, curled dock).

Subfamily Hyperinae—on clover and alfalfa; all 7 spp. in genus *Hypera*.

Hypera postica (Gyllenhal) (Fig. 214D). Alfalfa weevil—3-5 mm; brown, gray-brown, or grayish black; dark brown middorsal stripe on elytra, other irregular spots; throughout U.S.

Hypera punctata (Fabricius) (Fig. 214E). Clover leaf weevil—5.0-8.5 mm; black; dense brown, yellow-brown, and gray scales give mottled and stippled appearance; snout short; throughout U.S.

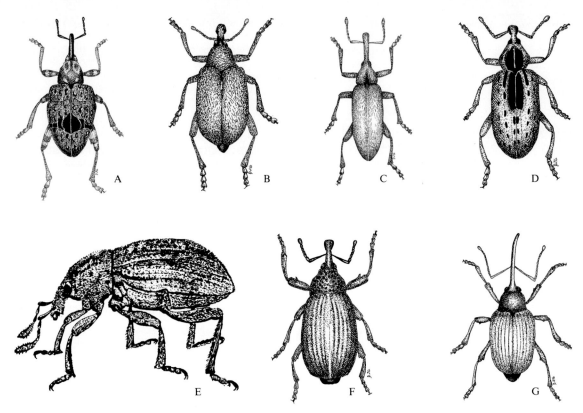

Figure 214 Weevils or snout beetles (Curculionidae). A, plum curculio, *Conotrachelus nenuphar;* B, potato stalk borer, *Trichobaris trinotata;* C, rhubarb curculio; *Lixus concavus;* D, alfalfa weevil, *Hypera postica;* E, clover leaf weevil, *H. punctata;* F, boll weevil, *Anthonomus grandis;* G, an acorn weevil, *Curculio sulcatulus.*

Subfamily Anthonominae— >100 spp. in genus *Anthonomus,* snout long.

Anthonomus grandis Boheman (Fig. 214F). Boll weevil—4.0-7.5 mm; reddish brown to blackish; pale yellow scalelike hairs irregularly cover body; southern U.S. west to TX (on cotton).

Anthonomus signatus Say. Strawberry weevil—2-3 mm; black; antennae and legs reddish brown; scutellum whitish; elytra reddish brown, 2 black patches on posterior 1/2; beak longer than head and pronotum; pronotum coarsely punctured; east of Rocky Mts.

Subfamily Curculioninae—Acorn and Nut Weevils; very long and slender snout, all North American spp. in genus *Curculio,* larvae in acorns, pecans, hickory nuts, other nuts.

Curculio sulcatulus (Casey) (Fig. 214G)—6-8 mm; dark or reddish brown; thin covering of short gray-brown scales and hairs; eastern 1/2 U.S. (on oaks). *C. caryae* (Horn), pecan weevil, similar, 7-9 mm, dull brown with yellowish hairs, outer 1/4 of femora with distinct rectangular tooth projecting at right angle (on hickory, pecan).

GENERAL REFERENCES

Arnett, R.H., Jr. 1967. Present and Future Systematics of the Coleoptera in North America. Ann. Ent. Soc. Amer. 60:162-170.

———. 1968. *The Beetles of the United States (A Manual for Identification).* Amer. Ent. Institute, Ann Arbor, Mich. 1,112 pp.

Baker, W.L. 1972. Eastern Forest Insects. U.S. Dept. Agric. For. Serv. Misc. Pulb. 1175. 642 pp.

Blatchley, W.S., and Leng, C.W. 1916. *Rynchophora or Weevils of North Eastern America*. Nature Publ. Co., Indianapolis, Ind. 682 pp.

Böving, A.G., and Craighead, F.C. 1930. An Illustrated Synopsis of the Principle Larval Forms of the Order Coleoptera. J. Ent. Soc. Amer. 11:1-351.

Dillon, E.S., and Dillon, L.S. 1972. *A Manual of Common Beetles of Eastern North America* (in 2 vol.). Dover, N.Y. 984 pp. (1961 reprint).

Evans, G. 1975. *The Life of Beetles*. Allen and Unwin, London. 232 pp.

Hatch, M.H. 1953-1973. The Beetles of the Pacific Northwest. Univ. Wash. Publ. Biol. 16: Part I (1953), 1-340; Part II (1957), 1-384; Part III (1961), 1-503; Part IV (1965), 1-268; Part V (1973), 1-662.

Hodek, I. 1973. *Biology of Coccinellidae*. W. Junk N.V., The Hague, Netherlands. 260 pp.

Jaques, H.E. 1951. *How to Know the Beetles*. Wm. C. Brown Company Publishers, Dubuque, Iowa. 372 pp.

Kissinger, D.G. 1964. *Curculionidae of America North of Mexico. A Key to the Genera*. Taxonomic Publ., South Lancaster, Mass. 143 pp.

Leech, H.B., and Chandler, H.P. 1956. Aquatic Coleoptera. In *Aquatic Insects of California*, R.L. Usinger (ed.). Univ. California Press, Berkeley. Pp. 293-371.

Leech, H.B., and Sanderson, M.W. 1959. Coleoptera. In *Fresh-Water Biology*, W.T. Edmondson (ed.). Wiley, N.Y. Pp. 981-1023.

Leng, C.W. et al. 1920-1948. *Catalogue of the Coleoptera of America North of Mexico*. J.D. Sherman, Mt. Vernon, N.Y. Original Catalogue, 1920. 470 pp. First Supplement, 1927, by C.W. Leng and A.J. Mutchler, 78 pp. Second and Third Supplements, 1933, by C.W. Leng and A.J. Mutchler, 112 pp. Fourth Supplement, 1939, by R.E. Blackwelder, 146 pp. Fifth Supplement, 1948, by R.E. Blackwelder and R.M. Blackwelder, 87 pp.

Linsley, E.G. 1961-1964. The Cerambycidae of North America. Univ. Calif. Publ. Ent. Part I, 1961, 18:1-135. Part II, 1962, 19:1-103. Part III, 1962, 20:1-188. Part IV, 1963, 21:1-165. Part V, 1964, 22:1-197.

Linsley, E.G., and Chemsak, J.A. 1972. Cerambycidae of North America. Part VI, no. 1. Taxonomy and Classification of the Subfamily Lepturinae. Univ. Calif. Publ. Ent. 69:1-138.

Peterson, A. 1951. *Larvae of Insects. Part II. Coleoptera, Diptera, Neuroptera, Siphonaptera, Mecoptera, Trichoptera*. Edwards Bros., Ann Arbor, Mich. 416 pp.

Thiele, H.-U. 1977. *Carabid Beetles in Their Environments*. Zoophysiology and Ecology Series, vol. 10. Springer-Verlag, N.Y. 380 pp. (approx.)

A Catalog of the Coleoptera of America North of Mexico. In preparation. U.S. Dept. of Agric. and Smithsonian Institution, Washington, D.C.

North American Beetle Fauna Project. In preparation: *Checklist of Beetles of Canada, United States, Mexico, Central America and the West Indies.* Also: identification manuals (by regions) for beetles of most of North America. Biological Research Institute of America, Box 108, Rensselaerville, N.Y.

ORDER STREPSIPTERA[25]
Twistedwinged Parasites

Strepsipterans are small (0.5-4.0 mm long) insects that are usually internal parasites of other insects. The two sexes differ markedly in appearance. The free-living males have very short, clublike, front wings and the membranous hind wings are large, fanlike, and contain only a few veins (Fig. 215A). The antennae often have distinct, elongated projections on some segments. The eyes bulge out to the sides. Females are wingless, usually lack antennae, eyes, and legs, and the body segmentation is vague (Fig. 215B).

A female remains in her insect host with only her head protruding. Males leave the host and must locate the females for mating. One to two thousand tiny, newly hatched larvae leave the female's body and escape from the host by falling onto plants or the ground. These active larvae have well-developed legs and after locating hosts they bore into host larvae and molt to legless, inactive forms (hypermetamorphosis).

25. Strepsiptera: *strepsi*, twisted; *ptera*, wings.

Larvae have a holometabolous type of development. A parasitized host is not normally killed.

Many beetle taxonomists classify all the North American strepsipterans as a family of beetles, the Stylopidae, due partly to life cycle similarities. Some beetles in the family Rhipiphoridae are quite similar to and easily mistaken for male strepsipterans. However, the latter lack trochanters, the wing venation is different, and the front wings are narrow and clubbed rather than like the small but broad front wings (elytra) of rhipiphorid beetles.

Strepsipterans are taken primarily by collecting them in the host. Males can then be reared. Parasitized hosts may have enlarged or otherwise distorted abdomens, or part of the parasites may protrude between the abdominal segments. Males occasionally fly to lights. All specimens can be preserved in 70% alcohol.

Species: North America, 60; world, > 300.
Families: North America, 7.

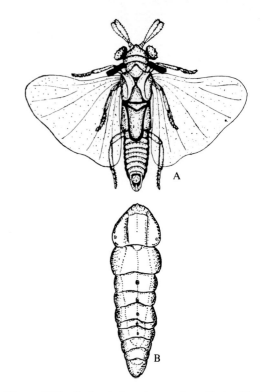

Figure 215 Twistedwinged parasites (Stylopidae). A, male *Stylops* sp.; B, female stylopid.

KEY TO COMMON FAMILIES OF STREPSIPTERA

1a   Tarsi 4-segmented; antennae 4- or 6-segmented, lateral projections on the third segment . . . . (p. 238) *Stylopidae*

1b   Tarsi 3-segmented; antennae 7-segmented, lateral projections on the third, fourth, and fifth segments . . . . . (p. 238) *Halictophagidae*

Family Stylopidae
This is the largest family of strepsipterans. Species in the genus *Stylops* (Fig. 215A) occur on *Andrena* bees (Andrenidae), *Pseudoxenos* species parasitize wasps in the families Eumenidae and Sphecidae, and *Xenos* species are found in *Polistes* wasps (Vespidae).

Family Halictophagidae
Members of this family parasitize leafhoppers, planthoppers, treehoppers, and pygmy grasshoppers. *Halictophagus* species in leafhoppers are encountered most commonly.

GENERAL REFERENCES
Bohart, R.M. 1941. A Revision of the Strepsiptera with Special Reference to the Species of North America. Univ. Calif. Pub. Ent. 7:91-160.

ORDER NEUROPTERA[26]
Dobsonflies, Alderflies, Snakeflies, Lacewings, Antlions, Mantispids, and Others

Neuropterans are weak-flying insects with four similar, membranous, large wings containing many longitudinal veins and crossveins. Adults and larvae have chewing mouthparts and feed on other insects, insect and spider eggs, and mites. Adults are terrestrial and the larvae occupy either terrestrial or aquatic habitats while undergoing a holometabolous type of development.

The order Neuroptera is commonly divided into the suborders Megaloptera (or Sialodea), Raphidiodea, and Planipennia, and this arrangement will be followed in this book. Some taxonomists believe these suborders should be elevated to an ordinal category and in certain references the following classification is used: Order Megaloptera (= Suborder Megaloptera), Order Raphidiodea or Raphidioptera (= Suborder Raphidiodea) and Order Neuroptera (= Suborder Planipennia).

SUBORDER MEGALOPTERA
Alderflies, Dobsonflies, and Fishflies

The adults of this group are usually nocturnal, although *Nigronia* species are diurnal. The chewing mouthparts project in front of the head but adults are not known to feed. The long, moderately heavy-looking wings are held almost flat over the body. Females deposit masses of several thousand eggs on rocks and branches above the water and the active larvae crawl into the water to live.

The aquatic larvae have chewing mouthparts with large mandibles. They employ large, paired gills on each abdominal segment for breathing (Figs. 217A; 218C). The predaceous larvae feed on aquatic insects and may require up to two or three years to develop. Certain species may temporarily leave the water to molt or capture prey. Pupation occurs on land inside earthen cells constructed in soil or under plant debris or stones in damp areas such as riverbanks.

Adult dobsonflies and fishflies may be pinned but alderflies are more soft-bodied and are generally placed in 70-80% alcohol, as are all larvae.

SUBORDER RAPHIDIODEA
Snakeflies

Snakeflies inhabit the western part of the U.S. The head of the adult is elongated and tapered and the prothorax is elongated and cylindrical, giving the head-prothorax combination a snakelike appearance (Fig. 219A). The front legs attach at the posterior part of the prothorax and the wings are held rooflike over the body. Eggs are deposited singly under loose tree bark or in cracks of the bark. The flat larvae remain on the bark, feeding on woodboring insects, aphids, and caterpillars, and if disturbed they wriggle backwards to escape. Adults may be pinned and larvae are preserved in 70-80% alcohol.

26. Neuroptera: *neuro*, nerve; *ptera*, wings.

SUBORDER PLANIPENNIA
Dustywings, Spongillaflies, Lacewings, Mantispids, Antlions, and Owlflies

Adults are characterized by their many-veined, lacy wings (Fig. 216). The wings are large for the body size and are held rooflike over the body (Figs. 221A, C). The chewing mouthparts project downward. Eggs may be deposited in sand (e.g., antlions), cemented to a surface, or adhered to the tip of a filamentous stalk (e.g., lacewings) produced by secretions from the female and attached to a leaf.

Larvae have two distinct, elongated, piercing-sucking jaws (Figs. 221B, C; 222B). A mandible and maxilla form the upper and lower part of each jaw respectively, and blood of the prey is sucked into the resulting hollow tube. Most larvae have an anal proleg on the terminal (10th) abdominal segment which is used for locomotion and spinning silk when making a cocoon. The life cycle is one year or less. All larvae are terrestrial except those in the family Sisyridae which are parasitic on freshwater sponges.

brown lacewings are often kept in 70-80% alcohol since they are not highly sclerotized and tend to shrivel. The green color of green lacewings will be lost in alcohol. Larvae are preserved in 70-80% alcohol.

Species: North America, 338; world, 4,600.
Families: North America, 15.

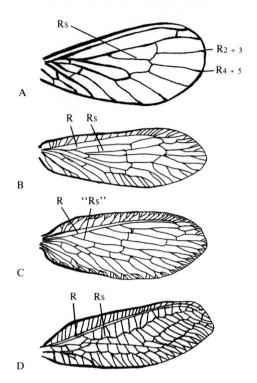

Figure 216 Front wings of neuropterans. A, dustywings (Coniopterygidae); B, spongillaflies (Sisyridae); C, brown lacewings (Hemerobiidae); D, green lacewings (Chrysopidae).

KEY TO COMMON FAMILIES OF NEUROPTERA

1a	Prothorax long and slender (Fig. 219)	2
1b	Prothorax normal size, not elongated	3
2a(1a)	Front legs raptorial (broad coxa and femur, femur with spines) and attached at anterior end of prothorax (Fig. 219B); mantidlike insects .... (p. 245) *Mantispidae*	
2b	Front legs not raptorial, and attached at posterior end of prothorax (Fig. 219A); western U.S. .... (p. 243) *Raphidiidae*	
3a(1b)	Base of hind wings broader than base of front wings, enlarged anal area of hind wings folded fanlike at rest ...	4

240 Subclass Pterygota

3b	Base of front and hind wings similar in size, hind wings without enlarged anal area that is folded fanlike at rest......................5	7b	Antennae short, length equal to combined head and thorax; resemble damselflies (Fig. 222A)............(p. 245) *Myrmeleontidae*
4a(3a)	Ocelli present; 4th tarsal segment cylindrical; wingspan greater than 40 mm, wings smoky or clear ....(p. 241) *Corydalidae*	8a(6b)	Radius of front wing with 3 or more branches (Fig. 216C); small, brown, hairy insects....................(p. 244) *Hemerobiidae*
4b	Ocelli absent; 4th tarsal segment dilated and bilobed (Fig. 218B); wingspan 20-40 mm, wings usually smoky..........(p. 242) *Sialidae*	8b	Radius of front wing with 1 branch (Rs) (Figs. 216B, D); size and color variable 9
5a(3b)	Wings covered with white, waxy powder; relatively few veins on wings, radial sector (Rs) with 2 branches (Fig. 216A)............(p. 243) *Coniopterygidae*	9a(8b)	Radial sector (Rs) with a zigzag appearance (Fig. 216D); wings often pale green or yellow; body length more than 8 mm; common insects ....(p. 244) *Chrysopidae*
5b	Wings not covered with white, waxy powder; wings with many veins, radial sector (Rs) with more than 2 branches (Figs. 216B, C, D) 6	9b	Radial sector (Rs) without a zigzag appearance (Fig. 216B); body length 6-8 mm; brown, uncommon insects ....(p. 244) *Sisyridae*
6a(5b)	Antennae clubbed or knobbed at tip (Figs. 222A, C); abdomen long and slender and insects resemble damselflies or dragonflies 7		
6b	Antennae not clubbed or knobbed at tip; not resembling damselflies or dragonflies..................8		
7a(6a)	Antennae long, nearly length of the body; resemble dragonflies (Fig. 222C)...(p. 246) *Ascalaphidae*		

SUBORDER MEGALOPTERA

Family Corydalidae— Dobsonflies and Fishflies

Adults are large insects, some with a wingspan up to 130 mm. They have three ocelli and the fourth tarsal segment is cylindrical rather than bilobed. These characteristics separate the family from the Sialidae, or alderflies. *Corydalus* and another genus contain the largest species, the dobsonflies. *Chauliodes, Nigronia, Protochauliodes* and other genera containing species with a wingspan generally less than 54 mm are referred to as fishflies. Their antennae are often serrate or pectinate. Adults in this family are most commonly collected at night near lights of various types. Sweeping vegetation near water during the day may produce specimens.

The larvae are aquatic, have eight pairs of filaments on the sides of the abdomen and possess two anal prolegs (Fig. 217A). Individuals (sometimes called hellgrammites) are usually found under rocks and on the bottom rubble of swift-flowing streams.

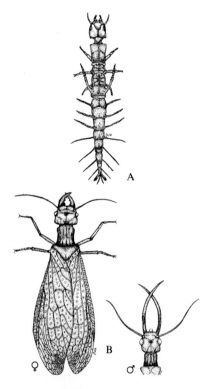

Figure 217 Dobsonflies and fishflies (Corydalidae). A, a fishfly larva, *Chauliodes rastricornus;* B, dobsonfly, *Corydalus cornutus,* female and head of male.

Common Species
Chauliodes rastricornus Rambur (Fig. 217A)—wingspan 50-80 mm; serrate (♀) or pectinate (♂) antennae; raised black markings on head and thorax; prothorax longer than wide; larva with spiracles of 8th abdominal segment on the ends of a pair of long unequal tubules, abdomen with black middorsal stripe; eastern and central U.S. and Canada (prefers lakes, ponds, swamps).

Corydalus cornutus (Linnaeus) (Fig. 217B). Dobsonfly—wingspan 90-130 mm; dull gray; crossed mandibles are half the body length (♂); larva with short gill tufts at base of 1st 7 pairs of abdominal filaments; eastern North America.

Nigronia fasciata (Walker)—wingspan 40-75 mm; blackish body; broad white band across middle of each wing nearly reaching hind margin; eastern and central U.S. (active during daytime).

Protochauliodes minimus Davis—wingspan 42-50 mm; dark smoky color with wing mottling present but indistinct; western North America.

Family Sialidae—Alderflies
Adults are medium-sized insects with a wingspan of 20-40 mm. They lack ocelli and the fourth tarsal segment is dilated and bilobed (Fig. 218B). Adults are most active during the middle of the day and may be caught flying near aquatic habitats. Lights sometimes attract adults at twilight.

Larvae have seven pairs of abdominal filaments, lack anal prolegs and have a central caudal filament (Fig. 218C). Individuals occur in all varieties of freshwater situations although they are more common in the slow-moving portions of streams and muddy bottoms of ponds.

Common Species
Sialis californica Banks—11-16 mm; black body; wings often iridescent black; eye ring and raised lines and dots on head all yellowish; genital plate under tip of abdomen produced into 2 bulbous hooks posteriorly (Fig. 218D); WA to CA, southwestern Canada.

Sialis infumata Newman (Fig. 218A). Smoky alderfly—11-13 mm; blackish body; narrow orange margin around eyes; apex of lateral plates at tip of abdomen produced into large knobs (♂) (Fig. 218E); eastern North America.

Sialis mohri Ross—11-13 mm; many fine hairs giving a velvety appearance; area between eyes concave; 6-10 crossveins in costal area of front wing; terminal plate at tip of abdomen produced into a pair of long ventral arms that converge at tips (♂) (Fig. 218F); 9th sternite large, triangular, and distinctly sclerotized (♀); eastern North America.

Sialis velata Ross—11-13 mm; similar to *S. infumata* but lateral plates not knobbed; eastern, central and southwestern U.S., southern Canada.

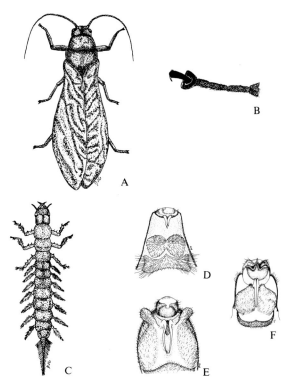

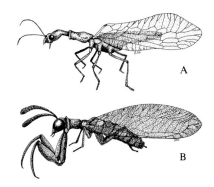

Figure 218 Alderflies (Sialidae). A, smoky alderfly, *Sialis infumata;* B, tarsus of an alderfly; C, larva of an alderfly; D-F, ventral view of tip of abdomen of males: D, *Sialis californica;* E, *S. infumata;* F, *S. mohri*.

SUBORDER RAPHIDIODEA

Family Raphidiidae—Snakeflies

The family description is the same as that of the suborder Raphidiodea. All U.S. species occur in the genus *Agulla* (Fig. 219A) and the identification of common species is based on genitalia morphology which is beyond the scope of this book. Adults may be collected in forests and shrubby areas from the Rocky Mts. to the west coast and occasionally Texas. Larvae are found under bark of trees, and especially under the loose bark of eucalyptus trees.

Figure 219 A, a snakefly, *Agulla* sp. (Raphidiidae); B, a mantispid, *Mantispa brunnea* (Mantispidae).

SUBORDER PLANIPENNIA

Family Coniopterygidae—Dustywings

Adults are 2.5-5.0 mm long and covered with a white, waxy powder (Fig. 220A). The antennae are long and slender, and the wing venation is reduced (Fig. 216A). The small larvae have five simple eyes on each side of the head and needle-like jaws. Homopterans and mites comprise their diet.

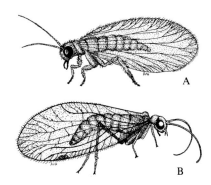

Figure 220 A, a dustywing, *Semidalis angusta* (Coniopterygidae); B, a spongillafly, *Climacia areolaris* (Sisyridae).

Order Neuroptera

Common Species

Semidalis angusta (Banks) (Fig. 220A)—3 mm; dark brown or yellow; wings mealy; AR to TX, CA (especially on citrus trees).

Semidalis vicina (Hagen)—2.5-3.0 mm; head and antennae pale yellowish brown; thorax and abdomen dark brown; tibiae dark and 1/3 longer than femora; eastern U.S.

Family Sisyridae—Spongillaflies

Adults are brown insects, 6-8 mm long, and have large eyes and long antennae (Fig. 220B). Larvae are hairy with long, bristlelike antennae and a piercing bristle on the end of each long jaw. The larvae are parasitic on freshwater sponges, especially sponges in the genus *Spongilla*, and adults are found on bushes near water or at lights.

Common Species

Climacia areolaris (Hagen) (Fig. 220B)—Front wing 5 mm; antennae dark brown; head and legs light yellow; front wing with a spot at apex of stigma; eastern and central U.S.

Family Hemerobiidae—Brown Lacewings

Adults are small, hairy, usually brown individuals with long antennae (Fig. 221A). The wingspan of most species ranges from 12 to 22 mm, and the hairy wings have only a few crossveins (Fig. 216C). Larvae are smooth and their jaws are short and incurved (Fig. 221B). They commonly feed on aphids (and are sometimes called aphidwolves), mealybugs and other homopterans, and mites. Larvae may be collected where their prey occur and adults are found most commonly in wooded areas.

Common Species

Hemerobius pacificus Banks (Fig. 221A)—10 mm; light to dark brown; black eyes; wings cloudy brown; west coast to NM.

Hemerobius stigmateris Fitch—front wing 7-8 mm; median longitudinal pale stripe on pronotum very narrow or absent; throughout U.S., southern Canada.

Megalomus moestus (Banks)—15-17 mm; reddish or rusty brown; wings clouded toward tips; Far West.

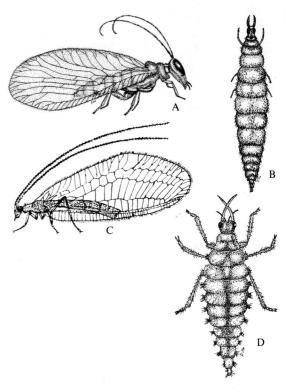

Figure 221 A and B, brown lacewings (Hemerobiidae): A, *Hemerobius pacificus;* B, larva of a brown lacewing. C and D, green lacewings (Chrysopidae): C, goldeneye lacewing, *Chrysopa oculata;* D, larva of a green lacewing.

Family Chrysopidae—Green Lacewings

Adults are small- to medium-sized insects with slightly angular, lacy wings (Figs. 216D; 221C). The wingspan of most is less than 35 mm. The head is small, the antennae long, and the large compound eyes are often golden. Thoracic glands are present on some adults and they emit a disagreeable odor when the insect is handled. Adults may be collected in weedy fields, on foliage of trees and shrubs, and at lights at night. They feed on pollen, nectar, and aphid honeydew.

Larvae have long, sharp, sicklelike jaws and body tubercles with hair tufts (Fig. 221D). Some species carry debris (e.g., lichens) which almost covers them. Larvae are predaceous on aphids (larvae are sometimes called aphidlions) and other insects and mites, and are occasionally sold as a biological control agent of aphids.

Common Species
Chrysopa carnea Stephens—9-15 mm; pale green or yellowish; side of head tinged with red; black band on margin of clypeus; yellow middorsal stripe on thorax; throughout North America.

Chrysopa coloradensis Banks—14-17 mm; pale green; pale reddish stripes on sides of thorax; Rocky Mts. westward.

Chrysopa oculata Say (Fig. 221C). Goldeneye lacewing—14-21 mm; pale green; black crescent under each eye; dark band around antennal sockets; green veins and black crossveins on wing; east of Rocky Mts.

Chrysopa rufilabris Burmeister—12-15 mm; yellowish green; red stripe under each eye down to mouth; pale yellow median dorsal stripe; throughout U.S.

Meleoma emuncta (Fitch)—13-17 mm; green; thorax and abdomen with yellow middorsal stripe; 3-pronged horn on front of face; elongated 3rd antennal segment; Rocky Mts. westward, northeastern and northcentral U.S., southern Canada.

Family Mantispidae—Mantispids

Adults have a wingspan of 10-55 mm. The prothorax is greatly elongated and the front legs are raptorial (broad coxae and femora, femora with spines) and attached to the front of the prothorax which gives an appearance similar to the mantids (Fig. 219B). Unlike most neuropterans the wings are folded flat at rest and not held rooflike over the body. Adults are collected in brushy and weedy fields and occasionally on tree leaves. Young larvae are active predators until sufficient food is found (spider eggs, wasp larvae) whereupon they assume an immobile and parasitic, grublike mode of life.

Common Species
Mantispa brunnea Say (Fig. 219B)—22 mm to wing tips; black or brown mixed with yellow; dark wings; throughout U.S.

Mantispa interrupta Say—19-20 mm to wing tips; greenish brown or brown; front wing with 1 dark brown spot near apex and 2 spots posterior to stigma; eastern U.S. southwest to NM.

Family Myrmeleontidae—Antlions

The nocturnal adults are long, slender insects with very long and delicate wings (Fig. 222A). The adults resemble damselflies but have short, knobbed antennae (antennae are bristlelike in damselflies). The wings may be spotted or clear and the adult's flight is slow and fluttery. Antlions are most common in the southern and southwestern U.S. and are commonly collected at lights near the larval habitat.

The larvae (Fig. 222B) are called antlions (or sometimes doodlebugs) and the 40-60 mm diameter cone-shaped pits of many species are common sights in sandy areas along rivers and lakes, trails, and in grasslands and dunes. The backward-moving larvae have a large abdomen but a narrow head and thorax. Large piercing-sucking, sicklelike jaws protrude from the head. The larva lies buried at the bottom of its pit with only the tip of its jaws showing. The loose sand causes an ant or other insect to fall into the pit where it is captured and pulled under the sand. The head-jaw combination of the larva is used to make the pit, throw sand on escaping prey, suck insect blood, and eject the shriveled prey from the pit. Some species do not build pits but remain buried in sand or debris waiting for prey. To capture larvae that dig pits, the sand under the pit should be scooped up and placed in a strainer or quickly spread out over a hard surface.

Common Species

Brachynemurus abdominalis (Say) (Figs. 222A, B)—20-35 mm; pronotum yellow with pair of brown dorsolateral bands; paired claws at tips of tibiae as long as tarsal segments 1 and 2 combined; 1st tarsal segment of hind leg short; larva with dark narrow stripes on head and pronotum and abdomen without median dark spots on tergum; throughout U.S. in grassland areas.

Brachynemurus sackeni Hagen—18-30 mm; pale reddish to dark brown; abdomen banded with large pale spots on tergites 2 and 3; western U.S.

Myrmeleon crudelis Walker—Front wing 22-29 mm; brownish; head with 2 dark spots above mouthparts, top of head reddish brown; thorax dark with pale sides; tarsal claws nearly straight; short white hair on legs and thorax; Atlantic and Gulf coast states.

Myrmeleon immaculatus De Geer—Front wing 22-36 mm; pronotum gray with 2 faint yellow spots in front; front wings with row of irregular light and dark areas along the radius (major longitudinal vein); larva pinkish gray, 2 pairs brown spots on head and 3 pairs long, black spots on abdomen, mandibles as long as or longer than length of head capsule; throughout North America.

Family Ascalaphidae—Owlflies

Adults are large insects, 40-50 mm long, and resemble slender dragonflies (Fig. 222C). However, the antennae are long and slender and not bristlelike as with dragonflies. The eyes are large and divided, and there are many long, fine hairs on the head. This group is more common in the southern and southwestern U.S. Larvae resemble antlions but they wait for prey without digging a pit.

Common Species

Ulolodes macleayana hageni Van der Weele (Fig. 222C)—Front wing 25-29 mm; wing stigma black; posterior margin of hind wings even and not incut; antenna reaches to or surpasses wing apex; Atlantic and Gulf coast states.

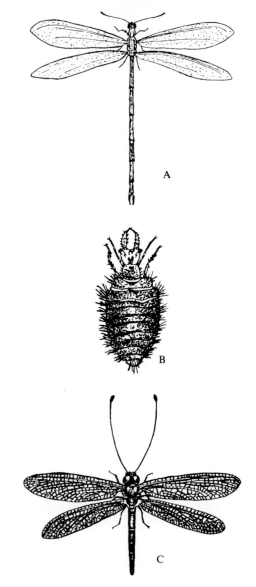

Figure 222 A and B, an antlion, *Brachynemurus abdominalis* (Myrmeleontidae): A, adult; B, larva. C, an owlfly, *Ulolodes macleayana hageni* (Ascalaphidae).

GENERAL REFERENCES

Banks, N. 1906. A Revision of the Nearctic Coniopterygidae. Proc. Ent. Soc. Wash. 8:77-86.

———. 1927. Revision of Nearctic Myrmeleontidae. Bull. Mus. Comp. Zool., Harvard Univ., 68:1-84.
Carpenter, F.M. 1936. Revision of the Nearctic Raphidiodea (Recent and Fossil). Proc. Amer. Acad. Arts Sci. 71:89-157.
———. 1940. A Revision of the Nearctic Hemerobiidae, Berothidae, Sisyridae, Polystoechotidae, and Dilaridae (Neuroptera). Proc. Amer. Acad. Arts Sci. 74:193-280.
Chandler, H.P. 1956. Megaloptera. In *Aquatic Insects of California,* R.L. Usinger (ed.). Univ. Calif. Press, Berkeley. Pp. 229-233.
Gurney, A.B., and Parfin, S. 1959. Neuroptera. In *Fresh-Water Biology,* W.T. Edmondson (ed.). Wiley, N.Y. Pp. 973-980.
Rehn, J.W. 1939. Studies in North American Mantispidae (Neuroptera). Trans. Amer. Ent. Soc. 65:237-263.
Ross, H.H. 1937. Studies of Nearctic Aquatic Insects. I. Nearctic Alder Flies of the Genus *Sialis.* Ill. Nat. Hist. Surv. Bull. 21:57-78.
Smith, R.C. 1922. The Biology of the Chrysopidae. N.Y. (Cornell) Agric. Expt. Sta. Mem. 58:1285-1377.
Stange, L.A. 1970. Revision of the Ant-Lion Tribe Brachynemurini of North America (Neuroptera: Myrmeleontidae). Univ. Calif. Pub. Ent. 55:1-192.

ORDER MECOPTERA[27]
Scorpionflies

Scorpionflies are small- to medium-sized, slender insects that typically have a distinctly elongated face (Fig. 223). The face projects downward and terminates with chewing mouthparts. Most scorpionflies have four long and narrow, membranous wings (often patterned) of similar size and venation; a few have very small wings and one species lacks them. The name "scorpionfly" is derived from males in the family Panorpidae; their bulblike genital segment at the tip of the abdomen is carried up over the back, resembling the sting of a scorpion (Fig. 223B). However, scorpionflies do not sting or bite.

Adults feed primarily on dead insects and less often on other animal debris. Deciduous woods with dense vegetation are common habitats for adults of many species, although they may also be collected from weedy fields, moss, and on the surface of snow. Eggs are laid in or on the ground. Development is holometabolous and most larvae resemble caterpillars in having thoracic legs, up to eight fleshy prolegs on the abdomen, and spines on the body. Most are scavengers and feed on dead insects and other animal debris. Adults are pinned and larvae are placed in 70% alcohol.

Species: North America, 67; world, 451. Families: North America, 5.

KEY TO COMMON FAMILIES OF MECOPTERA

1a   Each tarsus with one large claw .... (p. 248) *Bittacidae*

1b   Two claws on each tarsus 2

2a(1b)   Wings clear to yellowish, banded or spotted with brown; head elongated and beaklike ... (p. 248) *Panorpidae*

2b   Wings rudimentary or, if fully formed, uniformly yellowish brown; head elongation variable......... 3

3a(2b)   Wings fully formed (males) or small to rudimentary (females); body and wings yellowish brown, head not elongated.... (p. 248) *Panorpodidae*

3b   Wings slender, hooklike (males) or rudimentary and padlike (females); small, blackish or brown insects with elongated head.... (p. 248) *Boreidae*

27. Mecoptera: *meco,* long; *ptera,* wings.

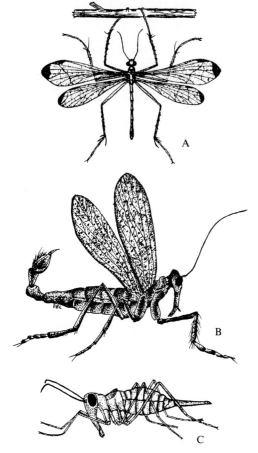

Figure 223 A, a hangingfly, *Bittacus apicalis* (Bittacidae); B, a scorpionfly, *Panorpa nebulosa* (Panorpidae); C, a snow scorpionfly, *Boreus californicus* (Boreidae).

Family Bittacidae—Hangingflies

The hangingflies are long-legged insects that resemble large crane flies (Tipulidae). The wings narrow greatly toward their bases. Adults typically hang by their front legs from vegetation and capture passing insects with their hind feet. One species on the west coast is wingless (*Apterobittacus apterus* [MacLachlan]).

Common Species

Bittacus apicalis Hagen (Fig. 223A)—15 mm; wingspread 30 mm; pale yellowish brown; wing tips dark; hangs with wings outstretched; eastern U.S. southwest to OK (associated with jewel weed and nettle undergrowth in deciduous woods). *B. stigmaterus* Say similar, light brown, wings not outstretched when hanging. *B. strigosus* Hagen similar, pale yellowish brown, wings yellowish with most crossveins margined with gray.

Family Panorpidae—Scorpionflies

This family is the largest (39 species in North America) and most commonly encountered group. The body is typically yellowish brown and the wings are spotted or banded. The bulbous genitalia of males resemble the sting of a scorpion. All species are in the genus *Panorpa*.

Common Species

Panorpa nebulosa Westwood (Fig. 223B)—15 mm; yellowish brown; wing markings reduced to numerous small spots; eastern U.S., southeastern Canada. *P. helena* Byers similar, dark yellowish brown, wings yellowish with 3 transverse blackish brown bands and 3 smaller spots, wooded areas of eastern U.S. *P. debilis* Westwood similar, wings pale yellow with 2 brown bands and several spots (in northeast, wings with scattered spots only).

Panorpa nuptialis Gerstaecker—20-25 mm; dark reddish brown; wings yellow-brown with 3 broad, blackish brown transverse bands; southcentral U.S. (in grasslands, cultivated fields, forest borders).

Family Panorpodidae

This small family contains one genus, *Brachypanorpa*. The head of adults is not elongated.

Family Boreidae—Snow Scorpionflies

Members of this family have very small wings and are unable to fly although they will jump. Moss is a preferred habitat and food of larvae and adults. Boreids often appear on snow on warm winter days.

Common Species

Boreus californicus Packard (Fig. 223C)—3 mm; black; western North America (December to February). *B. brumalis* Fitch similar, eastern U.S. (most common eastern sp.; November to March). *B. nivoriundus* Fitch dull brown, eastern U.S.

GENERAL REFERENCES

Byers, G.W. 1954. Notes on North American Mecoptera. Ann. Ent. Soc. Amer. 47:484-510.

Carpenter, F.M. 1931. Revision of Nearctic Mecoptera. Bull. Mus. Comp. Zool., Harvard Univ., 72:205-277.

———. 1931. The Biology of the Mecoptera. Psyche 38:41-55.

Setty, L.R. 1940. Biology and Morphology of Some North American Bittacidae. Amer. Midl. Nat. 23:257-353.

ORDER TRICHOPTERA[28]
Caddisflies[29]

Caddisfly adults are generally brown or gray holometabolus insects that frequent vegetated habitats associated with water. Caddisfly larvae are small- to medium-sized individuals that are common inhabitants of lakes and streams.

Adult caddisflies have four wings and resemble moths in size and shape. However, they differ from moths in the following ways: (1) most have a fairly dense mat of hairs rather than scales on the body and wing surfaces (Fig. 229A); (2) the antennae are very slender, many jointed, and may be much longer than the body (Fig. 228A); (3) the wings are held together at an angle (rooflike) over the body (Fig. 231A); (4) the maxillary and labial palpi are well developed (Fig. 224A); and (5) there is no coiled sucking tube for liquid uptake but instead a short, sucking tongue that is part of a rudimentary chewing mouthpart. The wings contain a reduced number of crossveins (Fig. 224B) which helps distinguish Trichoptera from Neuroptera (Fig. 216).

Adults ingest liquid food such as nectar and live one to two months. The complete life cycle requires about one year. Adults of most species are inactive during the day, remaining concealed in damp woods and dense foliage bordering aquatic habitats, and flight activity begins at dusk. Females deposit many hundreds of eggs in gelatinous strings or masses. Some species dive below the water surface to oviposit on stones, vegetation or submerged debris. Others oviposit on plants and objects protruding above the water or along the shoreline.

A larval caddisfly has a distinct sclerotized head capsule and pronotum, one pair of single-facet eyes, and minute, single-segmented antennae. The thoracic legs are well developed and a pair of hooklike anal legs occurs on the tip of the abdomen. The anal hooks are used to maintain a hold within a case. Filamentous gills may be visible on the abdomen. Larvae have chewing mouthparts and feed on aquatic vascular plants, algae, diatoms, crustaceans, and immature aquatic insects such as midge and blackfly larvae.

The larvae exhibit distinct modes of living in the water. The majority construct diverse types of cases made of leaf sections (Fig. 232C), twigs (Fig. 232B), sand grains, tiny pebbles, and other debris. These materials are held together by silk strands produced by modified salivary glands. The larva, with only its head and thorax protruding from the case, will crawl, feed on plants or capture food, and pupate

---

28. Trichoptera: *tricho,* hair; *ptera,* wings.

29. Caddis: cotton or silk; tape or ribbon of worsted yarn. Hickin (1968) suggests that the name was derived from caddice men who were English vendors of ribbons, braids and other miscellanea. They pinned their wares on their coats not unlike the caddisworm that patches together its case.

within this case. A second group of larvae does not construct cases but rather spins fragile, silken nets varying from tubular (Fig. 226B) to nearly flat configurations. The nets are attached to plants, rocks, or other supports. Edible food washed into the tubes or nets is captured by the agile larva that lies in wait inside the tube or in a retreat attached to the net. There are a few larval forms that are free-living and do not use a case or shelter.

The mature larva spins a cocoon around itself or seals the opening of its case and molts into the pupal stage. When the pupa is mature, it cuts its way out of the cocoon or case, swims to the surface and crawls out of the water. The pupa then attaches itself to a support and molts into the adult stage. Caddisfly larvae are an important source of food for fish.

Caddisfly larvae may be collected from virtually all marshes, lakes, and streams, although they are more common in cold, unpolluted water. Collect shoreline and drifting debris, and emergent weeds from the stream or lake bottom. Turn over stones and logs in rapids and riffles of streams. In small, cold, clear streams caddisfly cases sometimes cover the stream bed. Debris may be placed in a white enamel pan or on a cleared area of the ground in order to separate the moving larvae. At night, adults may be collected at lights, especially lamps emitting large amounts of ultraviolet or blue light. During the day collect in cool, dark, sheltered areas such as bark crevices and the undersides of bridges and ledges, and vegetation extended over water and along the shoreline. Adults and larvae should be placed in 70-80% alcohol for preservation. Although adults are sometimes pinned through the thorax, the resulting shrinkage and pin damage makes it difficult to determine the presence and type of thoracic warts or tubercles which form a part of keys used for identification.

Species: North America, 1,200; world, 10,000.
Families: North America, 18.

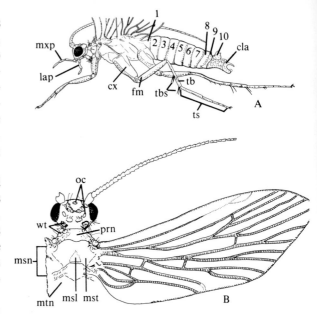

Figure 224 General structures of a caddisfly, *Rhyacophila lobifera* (Rhyacophilidae). A, lateral view; B, dorsal view. cla, clasper; cx, coxa; fm, femur; lap, labial palpus; msl, mesoscutellum; msn, mesonotum; mst, mesoscutum; mtn, metanotum; mxp, maxillary palpus; oc, ocelli; prn, pronotum; tb, tibia; tbs, tibial spurs; ts, tarsus; wt, warts.

KEY TO COMMON FAMILIES OF TRICHOPTERA

This key to adults is modified from that of Ross (1944). Most adult males may be recognized by the presence of a pair of claspers of varying size and shape below the tenth abdominal tergite (Fig. 224A). The female genital aperture is usually on the ninth sternite. Keys to larvae (families and genera) are found in Pennak (1953) and Ross (1959); Denning (1956) and Ross (1944) provide keys to species as well.

1a   Mesoscutellum with posterior portion forming a flat triangular area

	with steep sides, and mesoscutum without warts (Fig. 229C); front tibiae with 1 spur (tbs, Fig. 224A) or none; very hairy individuals with some wing hairs clubbed; antennae shorter than front wings (Fig. 229A); length 6 mm or less............. .... (p. 254) *Hydroptilidae*
1b	Mesoscutellum either evenly convex, without a triangular portion set off by steep sides, or mesoscutum with warts (Fig. 232E); front tibial spurs variable in number; no wing hairs clubbed; antennae usually as long as or longer than wings; length 5-40 mm 2
2a(1b)	Ocelli present (Fig. 224B)........ 3
2b	Ocelli absent................. 8
3a(2a)	Maxillary palpi 3-segmented (Fig. 231E); males .... (p. 255) *Limnephilidae* (in part)
3b	Maxillary palpi 4- or 5-segmented.. 4
4a(3b)	Maxillary palpi 4-segmented; males ..... (p. 255) *Phryganeidae* (in part)
4b	Maxillary palpi 5-segmented...... 5
5a(4b)	Maxillary palpi with fifth segment 2 or 3 times as long as fourth (Fig. 226D).................... .... (p. 252) *Philopotamidae*
5b	Maxillary palpi with fifth segment not more than 1 1/3 times as long as fourth 6
6a(5b)	Maxillary palpi with second segment short and similar to first; labrum evenly rounded and fairly wide (Fig. 225C).................... .... (p. 252) *Rhyacophilidae*
6b	Maxillary palpi with second segment much longer than first; labrum long and tonguelike (Fig. 230B) 7
7a(6b)	Front tibiae with 2 or more spurs; middle tibiae with 4 spurs (Fig. 224A); females............. ..... (p. 255) *Phryganeidae* (in part)
7b	Front tibiae with 1 spur or none; middle tibiae with 2 or 3 spurs; females...................... .... (p. 255) *Limnephilidae* (in part)
8a(2b)	Terminal segment of maxillary palpi without cross striations and usually of same length and general structure as the preceding segment; pronotum with a lateral pair of erect platelike warts separated by a deep notch (Fig. 232E); antennae very long and slender .... (p. 256) *Leptoceridae*
8b	Terminal segment of maxillary palpi with close cross striations and much longer than the preceding segment; pronotum without combination of platelike warts and deep notch; antennae variable in length......... 9
9a(8b)	Mesoscutum without warts (Fig. 228C); front tibiae lack preapical spurs... (p. 253) *Hydropsychidae*
9b	Mesoscutum with a pair of small warts (Fig. 227D); front tibiae with or without preapical spurs 10

10a(9b) Front tibia usually with a preapical spur . . . . (p. 253) *Polycentropodidae*

10b Front tibia without a preapical spur . . . . (p. 253) *Psychomyiidae*

Family Rhyacophilidae— Primitive Caddisflies

Adults of this family are 3-14 mm long and have 5-segmented maxillary palpi and short antennae. Larvae are found in cold, fast-flowing water such as mountain streams of the western states. In the eastern U.S. some species are found in cold springs. Larvae are free-living and recognized by a membranous mesonotum and metanotum, a sclerotized dorsal plate on the ninth abdominal segment, and a long anal leg and claw (Fig. 225B). A stone case is built only when the mature larva is ready to pupate.

Family Philopotamidae— Fingernet Caddisflies

The uncommon adults are small (6-9 mm long) and the fifth segment of the maxillary palpus is 2-3 times longer than the fourth (Fig. 226D). Females of the eastern species, *Dolophilodes distinctus* (Walker), are wingless (Fig. 226A) and both sexes of this species are active throughout the year.

Larvae frequent rapid streams of hilly or mountainous areas and may be recognized by the membranous metanotum and the T-shaped membranous labrum (Fig. 226C). These insects construct groups of silken nets shaped like fingers or elongated funnels (Fig. 226B).

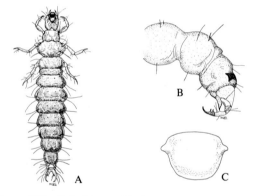

Figure 225 Primitive caddisflies (Rhyacophilidae). A, larva; B, tip of abdomen of *Rhyacophila lobifera* larva; C, labrum of adult *R. lobifera*.

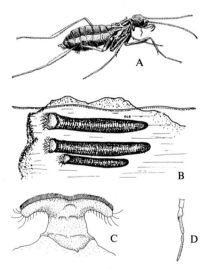

Figure 226 Fingernet caddisflies (Philopotamidae). A, *Dolophilodes distinctus;* B, silken nets of philopotamid larvae; C, "T"-shaped labrum of a philopotamid larva; D, maxillary palpus of an adult philopotamid.

Common Species

Rhyacophila fuscula (Walker)—10-12 mm; long claspers with ventral margin almost straight from base to apex (♂); apex of 8th abdominal sternite deeply notched (♀); eastern North America.

Rhyacophila lobifera Betten (Figs. 224; 225B, C)— 11-13 mm; dark bluish gray; scattered yellowish patches of hair; legs yellow to greenish; claspers long with wide notch (♂); northcentral U.S.

Common Species

Chimarra aterrima Hagen—6-8 mm; brownish black; sides of abdomen creamy white; upper corner of claspers bluntly pointed and posterior clasper margin flat and not concave (♂); eastern North America.

Chimarra obscura (Walker)—similar to *C. aterrima* except upper portion of clasper elongated into a narrow, curved, fingerlike lobe and the ventromesal process of the 9th sternite is long and narrow (♂); eastern U.S.

Family Polycentropodidae— Trumpetnet Caddisflies

Adults range in length from 4-11 mm. The antennae are stout and the terminal segment of the maxillary palpus is much longer than the preceding one. Preapical spurs occur on all tibiae of most species and the wings are covered with a dense pubescence.

Most larvae (Fig. 227A) will be encountered in fast-flowing streams although others occur in lakes. Larvae may be recognized by a membranous mesonotum and metanotum and a sclerotized labrum that is widest near its base (Fig. 227B). The anal appendages are long and often widely diverging. Larvae occupy retreats such as long silken tunnels attached to aquatic plants and stones or constructed in the sandy bottom of a stream. The nets associated with plants collapse when removed from the water.

Common Species

Polycentropus cinereus Hagen—7-9 mm; brown; wings with checkerboard brown and light mottling; both pairs of wings with R_{1-5} veins present and R_2 branching near margin of wing; eastern, midwestern, and northwestern U.S., southern Canada.

Family Psychomyiidae— Tubemaking Caddisflies

Larvae and adults of this small family occur in the same type of habitats as species of Polycentropidae and in general are quite similar in appearance. Adults are generally less than 7 mm long (Fig. 227C) and the front tibiae lack preapical spurs.

The labium of the larva projects beyond the maxillae and is long and pointed. Larvae inhabit permanent silken tunnels attached along their length to the surface or in cracks of submerged stones and wood.

Common Species

Psychomyia flavida Hagen (Fig. 227D)—4-6 mm; straw color with purplish tinge on some areas; 10th abdominal tergite divided into two large, flaplike lateral lobes (♂); northern CA, Rocky Mts. to east coast (in swift, cold streams).

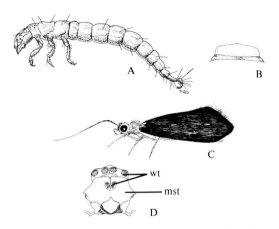

Figure 227 A and B, trumpetnet caddisflies (Polycentropodidae): A, larva; B, labrum of a larva. C and D, tubemaking caddisflies (Psychomyiidae): C, adult; D, thorax of *Psychomyia flavida* adult. mst, mesoscutum; wt, warts.

Family Hydropsychidae— Netspinning Caddisflies

Adults of this common family are 6-19 mm long and have 5-segmented maxillary palpi with the terminal segments longer than the preceding one. Adults lack ocelli, warts on the mesonotum (Fig. 228C), and preapical spurs on the front tibiae.

The wormlike, active larvae are most abundant in streams. Larvae exhibit rows of bushy abdominal gills and each thoracic segment has a dorsal plate (Fig. 228B). Shelter is sought in debris and under stones where a concave net is spun, facing upstream, in front of this retreat in order to capture food. Nets of some species are built under the brink of a waterfall, causing a brown coating of sediment to form on the rocks.

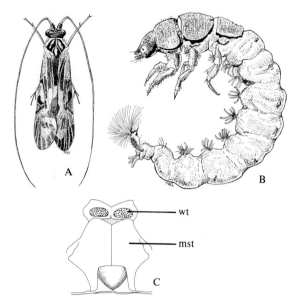

cream; hind wing gray; claspers with apical segment long, sinuate and pointed (♂); northcentral states (especially in large rivers), south to TX, west to CO.

Macronema zebratum Hagen (Fig. 228A)—15-18 mm; head and thorax metallic bluish brown; mouthparts and legs yellow; front wings brown with yellow markings; primarily eastern U.S.

Potamyia flava (Hagen)—10-11 mm; yellow-brown with pinkish tinge; very long, slender antennae and lacking spurs on front tibiae (♂) or shorter antennae and with spurs (♀); midwestern and southern U.S. (prefers large, slow rivers).

Family Hydroptilidae—Microcaddisflies

Adults of this family are small (1-6 mm long), very hairy, and have a mottled color pattern. The posterior portion of the mesoscutellum forms a flat, triangular area with steep sides (Fig. 229C) and is not evenly convex. There are no warts on the mesoscutum.

Larvae prefer streams although they occupy most types of aquatic habitats and occur throughout much of North America. Each thoracic segment of a larva has a dorsal plate (Fig. 229B) but, unlike species of Hydropsychidae, the abdomen lacks gills and generally is quite broad. Only the last instar builds cases.

Figure 228 Netspinning caddisflies (Hydropsychidae). A, *Macronema zebratum;* B, *Hydropsyche simulans* larva; C, thorax of *H. simulans* adult. mst, mesoscutum; wt, warts.

Common Species

Cheumatopsyche analis (Banks)—9-12 mm; color range from light brown with many small light spots on wings and 2 moderately large, white areas along anal margin (spring phase), to dark brown with very few spots on wings (summer phase); northern U.S. and southern Canada from Pacific to Atlantic coasts, southern U.S.

Cheumatopsyche campyla Ross—10-12 mm; brown; freckled wings with large light spot on anal margin near tip; throughout U.S., southern Canada.

Hydropsyche occidentalis Banks—10-11 mm; head and thorax with dense, short white hair; front wings brown with dense patches of white freckles; apical segment of clasper short and nearly square (♂); western U.S.

Hydropsyche orris Ross—12-13 mm; similar to *H. simulans* but eyes smaller and each equal to less than half the distance between them (♂); northcentral (especially in large rivers) and southern U.S.

Hydropsyche simulans Ross (Figs. 228B, C)—13-15 mm; head and body with hair patches of variable colors; front wings mottled with gray, brown and

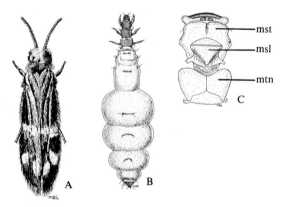

Figure 229 Microcaddisflies (Hydroptilidae). A, *Hydroptila hamata;* B, *Leucotrichia pictipes* larva; C, thorax of *L. pictipes* adult. msl, mesoscutellum; mst, mesoscutum; mtn, metanotum.

Common Species

Hydroptila angusta Ross—5 mm; like *H. hamata* but front femora light brown; 10th abdominal tergite wide and divided by deep cleft and upper part of clasper bladelike (♂); southwest to northcentral U.S.

Hydroptila hamata Morton (Fig. 229A)—2-4 mm; mottled gray and brown; head with patches of white hair; ocelli absent; wings hairy with white transverse band in middle and white spots on margin near apex; femora dark brown or black; hilly and mountainous areas of U.S.

Leucotrichia pictipes (Banks) (Figs. 229B, C)— 4.0-4.5 mm; dark brown to black; white bands on antennae and tarsi; white basal spot on front wing; throughout U.S. except Southeast.

Family Phryganeidae— Large Caddisflies

Adults range from 14-25 mm in length and the maxillary palpi are 4-segmented (males) or 5-segmented (females). Larvae have a membranous mesonotum and metanotum with a lateral tuft of long setae on each (Fig. 230C). Larvae are more likely to be collected in cold lakes and marshes. The cases typically are long, spiral- or ring-shaped, and manufactured from grass stems.

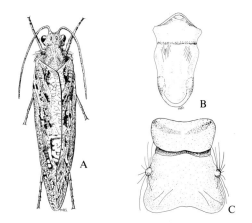

Figure 230 Large caddisflies (Phryganeidae). A, *Phryganea cinerea;* B, labrum of a phryganeid adult; C, prothorax and mesothorax of *Ptilostomis ocellifera* larva.

Common Species

Phryganea cinerea Walker (Figs. 230A, B)—21-25 mm; gray or brown; folded front wings with brown or gray patches along posterior margins forming triangular markings; hind wings uniformly gray or brown; claspers rounded and slightly upturned (♂); northern states and Canada from eastern seaboard to Rocky Mts.

Ptilostomis ocellifera (Walker) (Fig. 230C)—21-24 mm; yellowish brown; dark brown specks on front wing; 9th abdominal sternite either extended shelflike (♂) or broad (♀); northcentral to northeastern U.S., eastern Canada.

Family Limnephilidae— Northern Caddisflies

Adults in this large family of case-makers range from 7-23 mm in length, have ocelli (Fig. 224B) and the maxillary palpi are 3-segmented (males) (Fig. 231E) or 5-segmented (females). The larvae (Fig. 231C) occupy diverse habitats but are somewhat more common in marshes, ponds and slow-moving streams of the northern and western areas of the U.S. The antennae of larvae are located midway between the bases of the mandibles and the eyes (Fig. 231D) and the first abdominal tergite is humped. Larval cases vary widely in appearance (Fig. 231B) and may be constructed primarily of leaves, stems, moss, bark, sand or pieces of snail shells.

Common Species

Hesperophylax designatus (Walker) (Fig. 231A)— 20 mm; brown; wings with strong light and dark brown pattern and longitudinal silver stripe; northeastern and northcentral U.S., southeastern Canada.

Limnephilus rhombicus (Linnaeus) (Fig. 231B)— 19-20 mm; brownish yellow; wings with dark brown and cream colored oblique stripes; head and thorax with long silvery or brownish yellow bristles; northeastern North America, northcentral U.S.

Limnephilus submonilifer Walker (Figs. 231C-E)— 13-16 mm; slender; brown with variegated light and dark spots on wings; brushlike black spines on underside of front femur and tibia; head and thorax with stout bristles; northeastern and northcentral U.S., southern Canada (temporary ponds and marshes).

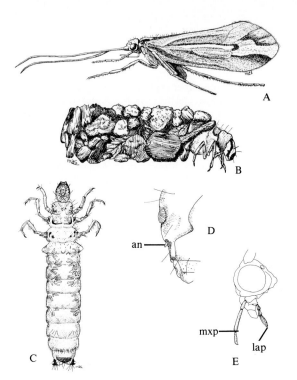

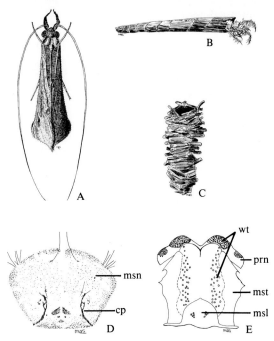

Figure 231 Northern caddisflies (Limnephilidae). A, *Hesperophylax designatus;* B, case of *Limnephilus rhombicus* larva; C, *L. submonilifer* larva; D, location of antenna on larval limnephilids; E, maxillary palpus of adult limnephilid. an, antenna; lap, labial palpus; mxp, maxillary palpus.

Figure 232 Longhorned caddisflies (Leptoceridae). A. *Mystacides sepulchralis;* B, case of *Triaenodes tarda* larva; C, case of *Oecetis cinerascens* larva; D, thorax of a larva; E, thorax of a leptocerid adult. cp, curved plate; msl, mesoscutellum; msn, mesonotum; mst, mesoscutum; prn, pronotum; wt, warts.

Family Leptoceridae— Longhorned Caddisflies

This family ranges across North America and its members are found in most habitats of caddisflies. Adults are slender insects, 5-17 mm long, with 5-segmented maxillary palpi in both sexes and quite long, slender antennae.

The larva's mesonotum is either membranous or has a pair of small, curved plates (Fig. 232D). The claws of the hind and middle legs are equal in length, and the antennae are at least eight times longer than wide. A "log cabin" type of case made from twigs (Fig. 232B) is constructed by some species in this family and the limnephilid family. Some cases are made from leaf and stem sections (Fig. 232C).

Common Species

Ceraclea maculata (Banks)—10 mm; dark brown to bright reddish brown; cerci long and pointed at apex and 10th abdominal tergite with bulbous base and fingerlike apex (♂); northeastern and northcentral U.S., southwest to TX.

Mystacides sepulchralis (Walker) (Fig. 232A)— 9 mm; blue-black; wings and thorax with metallic sheen; eastern and southern U.S., southeastern Canada.

Nectopsyche exquisita (Walker)—11-17 mm; head and thorax yellow brown and covered with white hair; front wings with brownish yellow crossbands and four square, black spots on posterior margin near apex; eastern and southern U.S., southeastern Canada.

Oecetis cinerascens (Hagen) (Fig. 232B)—11-13 mm; brown; pale hairs give grayish white effect; several dark spots on wing at vein forks; larva builds "log cabin" case; eastern and southern U.S., southern Canada.

Oecetis inconspicua (Walker)—10-12 mm; brown with reddish cast and no conspicuous markings; 10th abdominal tergite forms single, fairly long, straight rod (♂); throughout North America.

Triaenodes tarda Milne (Fig. 232C)—12-13 mm; yellow brown with cream and brown pattern; dorsal stripe elongated anteriorly and diamond-shaped posteriorly; northeastern and northcentral states, southwest to OK, southern Canada.

GENERAL REFERENCES

Betten, C. 1934. The Caddis Flies or Trichoptera of New York State. N.Y. St. Museum Bull. No. 292. 576 pp.

Denning, D.G. 1956. Trichoptera. In *Aquatic Insects of California,* R.L. Usinger (ed.). Univ. California Press, Berkeley. Pp. 237-270.

Hickin, N.E. 1968. *Caddis Larvae. Larvae of the British Trichoptera.* Associated Univ. Presses, Cranbury, N.J. 480 pp.

Pennak, R.W. 1953. *Fresh-Water Invertebrates of the United States.* Ronald Press, N.Y. 769 pp.

Ross, H.H. 1944. The Caddis Flies, or Trichoptera, of Illinois. Ill. Nat. Hist. Surv. Bull. 23:1-326.

———. 1959. Trichoptera. In *Fresh-Water Biology,* W.T. Edmondson (ed.). Wiley, N.Y. Pp. 1024-1049.

Wiggins, G.B. *Larvae of the North American Caddisfly Genera (Trichoptera).* Univ. Toronto Press, Toronto. 410 pp.

ORDER LEPIDOPTERA[30]
Butterflies and Moths

Adult Lepidoptera are small to large (2-300 mm wingspread) insects with minute, often powder-like scales that cover the wings and body of most species. Lepidopterans usually have four large wings although some female moths are short-winged or wingless. The compound eyes are large and the antennae are long. The mouthparts are generally reduced except for the maxillae, which are greatly elongated and fused to form a sucking tube (proboscis), and the extended labial palpi (Fig. 233). The proboscis is extended to suck up liquid food and coiled when not in use. Some moths have no proboscis. Certain moth species possess membranous auditory organs (tympana) on the metathorax or abdomen. Vision is relatively good and many individuals respond to motion, flower shapes, wing patterns, and blue-violet and ultraviolet light. Immatures or caterpillars have a holometabolous type of metamorphosis. Adult Trichoptera (caddisflies) are similar to primitive moths but usually lack scales (although they have many hairs), a coiled proboscis, and usually have very long, slender antennae.

The scales of Lepidoptera are flat, multi-shaped pieces of cuticle. Scale colors are derived from pigments and scale physical structure. The latter affects light transmission and diffraction due to the shinglelike arrangement of scales, and the ridges and thin, layered walls of scales. Scales are loose and slippery, a characteristic which sometimes enables the lepidopteran to escape from a predator. Males often have patches of specialized scent scales which emit pheromones involved in mating.

The differences between butterflies and moths are subject to numerous exceptions but generally are as follows. (1) The tip of a butterfly's antenna is knobbed (Fig. 236C) except for skippers which have hooked antennae (Fig. 236B); antennae of most moths are tapered to the tip and are feathery, thickened, or hairlike (Figs. 236A, D, E). (2) The front and hind wings of butterflies are held together in flight by a lobed process at the base of the hind wing which grips the underside of the front

---

30. Lepidoptera: *lepido,* scale; *ptera,* wings.

wing (Fig. 239A); most moths have at least one stiff bristle (frenulum) at the base of the hind wing that hooks into the underside of the front wing (Fig. 237). (3) Butterflies are diurnal (a few rare species are nocturnal) but most moths are nocturnal although a small number fly during the day.

lepidopteran but is a sawfly larva (Hymenoptera). A caterpillar's head is usually distinct from the rest of the body and has an inverted "Y"-shaped suture on the face, tiny 3-segmented antennae, and zero to six pairs of ocelli around each antenna. The larvae of some beetles and wasps resemble those of certain moths but the former lack adfrontal sclerites that surround the frons on the head (Fig. 234). The body of caterpillars may be naked or have a covering of hairs (setae) of varying density.

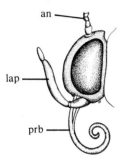

Figure 233 Lateral view of a lepidopteran head and mouthparts. an, antenna; lap, labial palpus; prb, proboscis.

Adult Lepidoptera feed primarily on flower nectar and are significant pollenizers. Secondary foods include sugary secretions from insects (honeydew), juices of decaying fruit, tree sap, manure liquids, etc. On warm days adults commonly cluster on the edges of mud puddles to ingest water and dissolved sodium. The mouthparts of some species are so reduced that feeding does not occur and the adult's stored energy is carried over from the larval stage.

Females usually deposit eggs (about 100-1,000) in masses on food plants or disperse them on or near the plant. Eggs hatch in a few days to weeks or may remain physiologically inactive through the summer, fall, or winter.

Larvae (caterpillars) have chewing mouthparts and most have prolegs which are short, fleshy projections on the ventral side of the abdomen (Fig. 234). Prolegs, used for clinging or walking, vary from two to five pairs or none. Hooks (crochets) are on the underside of the prolegs. A larva with more than five pairs of prolegs (which also lack crochets) is not a

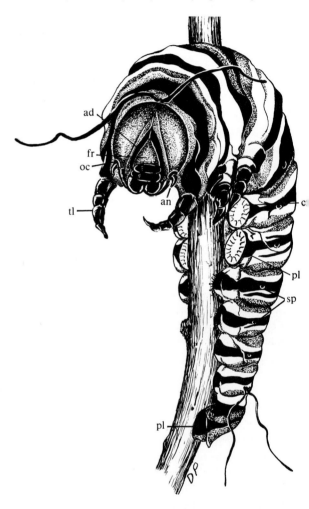

Figure 234 Monarch butterfly larva, *Danaus plexippus* (Danaidae). ad, adfrontal sclerite; an, antenna; cr, crochets; fr, frons; oc, ocelli; pl, proleg; sp, spiracles; tl, thoracic leg.

Nearly all caterpillars are plant feeders and most are restricted to one or a small group of food plants. Many larvae are serious pests, feeding on or inside foliage, fruit, stems, roots, wood, stored grain and flour, beeswax, and fabrics. A few caterpillars are predators on aphids, leafhoppers, mealybugs, scales, and immature ants. Larvae typically require several weeks to two months to develop but some take up to two years. Caterpillars usually remain on the food plant or in the general vicinity and may disperse individually or in masses when ready to pupate or due to crowding or starvation. Silk may be produced by glands in the head of larvae and spun out of the spinnerets in the mouth. The silken threads are used by the larva to adhere to surfaces, to construct cases, nests, and shelters, and to form a cocoon just before the larva molts into the pupal stage. The silkworm, *Bombyx mori* (Linnaeus), has been reared for centuries and the threads of its cocoon used for silk clothing.

Prior to pupating the mature larva may seek shelter in soil, debris, under bark and inside foliage, or it may fasten itself with silk to vegetation. While many moth larvae often spin silken cocoons around themselves, sometimes incorporating larval hair, secretions, leaves, or debris, many moth larvae do not spin cocoons (Fig. 235A). Most butterfly larvae do not form cocoons and the pupa (chrysalis) hangs from a silk pad spun by the larva (Fig. 235B). Caterpillars in the Parnassiinae (a subfamily of the Papilionidae) and Satyridae and Hesperiidae families of butterflies do spin cocoonlike shelters. The pupal stage lasts from a week to many months, depending on the season and species. Pupae are inactive (or able to move only a few abdominal segments) and do not feed. Adults emerge from the pupal case and cocoon in various ways: breaking the case by swelling from air intake; cutting with structures on the head, mouthparts, or wing bases; or secreting materials that partially dissolve the cocoon. A newly emerged adult pumps blood into the crumpled wings to force them open and they soon stiffen.

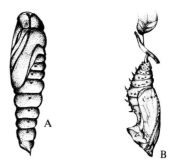

Figure 235 A, moth pupa; B, butterfly chrysalis.

Lepidoptera are well known for color patterns which give them protection against enemies (adaptive coloration). Mimicry between species, genera, and families of Lepidoptera is well documented. Some patterns help to conceal (cryptic colors) and others are bright, conspicuous colors that "warn" potential predators that the lepidopteran (adult or larva) may be distasteful. Some species have eyespots, specific areas of conspicuous circles on the wings of adults or bodies of larvae, that can be exposed suddenly and may startle and momentarily disuade predators. Other adults and larvae produce repellent odors and some larvae bristle with hollow, barbed hairs that contain toxic chemicals.

Certain male butterflies, singly or in groups, defend small territories such as perches or hilltops. Courtship may involve color and pattern recognition and fairly elaborate flying displays by both sexes. Migration of lepidopterans is well documented (monarch, painted lady, and sulphur butterflies, noctuid moths, others). Lepidopterans are most numerous, colorful, and diversified in the tropics but they range from deserts to high mountains and north into arctic habitats.

The suborders of Lepidoptera have been classified over the years by various names and arrangements. The generally preferred arrangement places most families (and the vast majority of species) in the suborder Ditrysia and the seven remaining primitive families in three suborders, Dachnonypha, Monotrysia, and Zeugloptera. Some textbooks use the subordinal names Heteroneura (or Frenatae) for all but three primitive moth families, the latter referred to as the Homoneura (or Jugatae). Another arrangement places butterflies and skippers in the suborder Rhopalocera (knobbed or hooked antennae) and moths in the suborder Heterocera (variable antennae). Sometimes lepidopterans are placed in two ill-defined groups for convenience: the microlepidoptera (generally small to tiny moths and often with wide wing fringes) and macrolepidoptera (medium to large moths and all butterflies).

Butterflies may be collected from flowers, mud puddles, rotting fruit, tree sap, aphid honeydew, carrion, and animal droppings. Amyl acetate is a chemical that attracts some butterflies. Varied habitats such as marshes, dry fields, damp meadows, sunlit forest openings, and edges of woods should be investigated. Waiting for a butterfly to land in order to net it is usually more successful than chasing it.

Some moths are commonly seen flying during the day. However, most are attracted to lights at night and the use of incandescent lights, white or colored fluorescent lamps, and blacklights work well. All-night light traps are standard devices but the struggling specimens tend to damage themselves unless a quick-killing agent is used such as concentrated ethyl acetate. Dusk to midnight and again at dawn seem to be the major flying times although specimens may be caught throughout the night. "Sugaring" for moths at night involves making a partially fermented liquid bait and brushing it on a tree trunk. One type of bait consists of sugar, overripe pieces of fruit (e.g., bananas), and molasses mixed together. Some collectors add beer or rum. The bait is applied at chest level to about a 10 sq. cm area on trees along open woods or forest edges just before dark, and checked at half-hour intervals for feeding moths. The "sugaring" method is not as useful in arid climates. Excellent adult specimens of Lepidoptera are commonly obtained by rearing the larvae on their specific food plant. Small day-flying moths often fly near their food plant. Pupae are also collected and placed in a moderately humid environment to obtain adults.

Lepidoptera are captured with a net or by quickly placing an opened killing jar over them. Picking up butterflies and moths with the fingers should be avoided to reduce damage to the powdery scales. Many collectors prefer to stun the netted insect by holding it through the net between the thumbnail and forefinger and carefully pinching its thorax (recommended for butterflies only). The specimen is then manipulated into a killing jar (preferably without picking it up) and when it is dead it is inserted into an envelope for storage. If the netted specimen is not pinched, the killing jar in which it is placed must be at full strength to kill quickly the insect and reduce wing damage from fluttering. Some collectors use a hypodermic syringe to inject ethyl acetate or ethyl alcohol into large-bodied moths (inject into the rear of the thorax under the abdomen). On hot days it is important to keep the killing jar relatively dry to prevent the wings from becoming too damp or sticking to the jar.

The wings should be spread the same day the insect is captured (see section on spreading insect wings). If the specimens are stored before spreading the wings, the envelope containing the specimens should be placed in a relaxing chamber to make the insects pliable. Specimens may be kept in a pliable or "fresh" condition for spreading by refrigerating them in a plastic container and adding chlorocresol.

Wing venation is used in identification and can best be seen on the underside of the wing. If necessary some scales may be scraped away from the underside of the wing or a few drops of xylene placed on the wings to make the veins more visible. If the entire wing needs clearing, a bleaching and mounting method is used (see page 20).

Species of butterflies: North America, 700; world, 15,000. Families: North America, 10. Species of moths: North America, 13,000; world, 165,000. Families: North America, ca. 68.

KEY TO COMMON FAMILIES OF LEPIDOPTERA

The wings of some specimens may require bleaching to see the complete venation (see page 20 for procedure). Males are generally recognized by two small, protruding, platelike claspers located laterally at the tip of the abdomen. The claspers may be pressed together with only a vertical longitudinal slit between them. A female's abdomen is usually larger and more pointed at the tip than a male's. Males may also have a patch of coarse scent scales on the dorsal side of the front wing. Keys to larvae are contained in Forbes (1923-1960) and Peterson (1948). Most of the microlepidopteran families are not included in the following key and the reader should refer to a more detailed key such as that in Borror, DeLong, and Triplehorn (1976) for family identification.

1a	Wings present and well developed.. 2
1b	Wings absent or appear as remnants (female) 47
2a(1a)	Coiled proboscis absent; venation and shape of front and hind wings similar; front and hind wings connected by a jugum on the front wing (Fig. 238); wingspread 25 mm or more......... (p. 314) *Hepialidae*
2b	Coiled proboscis usually present (Fig. 233); venation and shape of wings dissimilar; front and hind wings connected by a frenulum (Fig. 237) or an expanded humeral area on the hind wing (Fig. 239A); wingspread variable 3

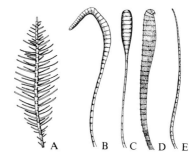

Figure 236 Antennae. A, bipectinate (giant silkworm moths); B, hooked (skippers); C, knobbed (butterflies); D, flattened (sphinx moths); E, filamentous (noctuid moths).

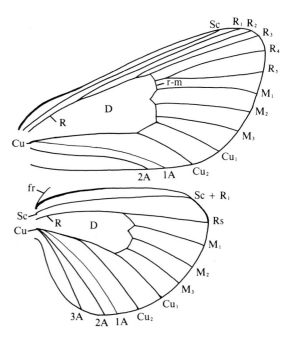

Figure 237 General wing venation. D, discal cell; fr, frenulum.

Order Lepidoptera

3a(2b) Antennae slender and knobbed, hooked, or otherwise swollen at tips (Figs. 236B, C); frenulum absent; diurnal (butterflies and skippers).... 4

3b Antennae variable in shape but usually not knobbed, hooked, or otherwise swollen at tips (Figs. 236A, D, E); frenulum usually present (Fig. 237); mostly nocturnal (moths)..................... 16

4a(3a) Antennae widely separated at bases and usually hooked at tips (Fig. 236B); radius of front wing with 5 simple branches originating on top of discal cell (Fig. 239A); body stout....................... 5

4b Antennae narrowly separated at bases and not hooked at tips; radius of front wing 3- to 5-branched, if 5 branches then some are on stalks (Fig. 241A); body usually slender .. 6

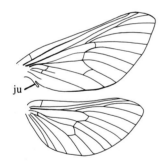

Figure 238 Wings of a hepialid moth (Hepialidae). ju, jugum.

5a(4a) Antennal club extended into a recurved hook (Fig. 236B), or wingspread less than 30 mm; head as wide as or wider than thorax; widely distributed....... (p. 287) **Hesperiidae**

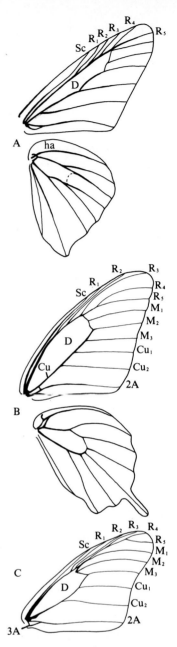

Figure 239 Wings of butterflies. A, skippers (Hesperiidae); B, swallowtails (Papilionidae); C, front wing of milkweed butterflies (Danaidae). D, discal cell; ha, humeral area.

262 Subclass Pterygota

5b Antennal club not extended and does not form a recurved hook; wingspread 40 mm or more; head narrower than thorax; southern and western U.S. .... (p. 286) *Megathymidae*

6a(4b) Cu of front wing appears 4-branched due to closeness of M_2 and M_3 (Fig. 239B); (Papilionidae) 7

6b Cu of front wing appears 3-branched due to closeness of M_3 (Fig. 239C) 8

7a(6a) Hind wing usually with one or more taillike extensions (Fig. 239B); radius of front wing 5-branched (Fig. 239B); widely distributed; (subfamily Papilioninae, swallowtails) .. .. (p. 279) *Papilionidae* (in part)

7b Hind wing without taillike extensions (Fig. 259F); radius of front wing 4-branched; western U.S.; (subfamily Parnassiinae) .... (p. 279) *Papilionidae* (in part)

8a(6b) Labial palpi longer than thorax, very hairy, and extending beaklike to the front (Fig. 258B) .... (p. 279) *Libytheidae*

8b Labial palpi shorter than thorax, and hairiness and forward extension not particularly prominent 9

9a(8b) Front legs usually less than one-half length of other legs, folded, and not used for walking; radius in front wing 5-branched (Figs. 239C; 240A, B)...................... 10

9b Front legs usually about same size as other legs, not folded, and used for walking; radius in front wings 3- or 4-branched (Fig. 241) 14

10a(9a) Front wings with 3A very short (Fig. 239C); antennae without scales; fairly large; orange-brown .... (p. 278) *Danaidae*

10b Front wing lacks 3A; antennae with scales; color and size variable 11

11a(10b) Usually dark brown or grayish with eye spots in wings (Fig. 257); some veins of front wing greatly swollen at base (Fig. 240A) .. (p. 277) *Satyridae*

11b Front wing veins usually not greatly swollen; color variable 12

12a(11b) Usually white with orange and/or black markings; tarsal claws forked; front legs nearly normal in size; M_1 in front wing joins stalk of R (Fig. 241A)..................... .... (p. 281) *Pieridae* (in part)

12b Color variable; tarsal claws absent; front legs greatly reduced in size; M_1 in front wing does not join stalk of R (Fig. 240B) (Nymphalidae) 13

13a(12b) Hv of hind wing extends or is bent forward (Fig. 240B); wings broadly triangular; widely distributed .... (p. 271) *Nymphalidae* (in part)

13b Hv of hind wing bent toward wing base (Fig. 240C); southern U.S. (subfamily Heliconiinae)............. ..... (p. 271) *Nymphalidae* (in part)

Order Lepidoptera   263

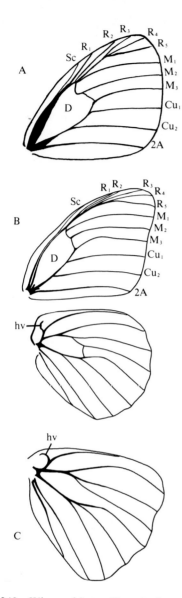

Figure 240 Wings of butterflies. A, front wing of satyrs (Satyridae); B, brushfooted butterflies (Nymphalidae); C, hind wing of heliconians (Nymphalidae). D, discal cell; hv, humeral vein.

14a(9b) White, yellow, or orange patterned with black; M_1 in front wing joins stalk of R beyond discal cell (Fig. 241A)..................... .... (p. 281) *Pieridae* (in part)

14b Not colored as above; M_1 in front wing usually does not join stalk of R beyond discal cell (Fig. 241C).... 15

15a(14b) Small, slender, and usually bluish, coppery-orange, or gray-brown with hairlike tails on hind wings; C not thickened along base of hind wing and hv absent (Fig. 241B)......... .... (p. 284) *Lycaenidae*

15b Similar to some Lycaenidae but colors darker; C thickened along base of hind wing and hv very short (Fig. 241C)..... (p. 283) *Riodinidae*

16a(3b) Much of wing, especially hind wing, lacks scales and is transparent (Fig. 284B); front wings at least 4 times as long as greatest width; resemble bees and wasps; diurnal .... (p. 307) *Sesiidae*

16b Wings completely scaled; if partly transparent then front wings are more triangular, not 4 times longer than greatest width, and the tips of antennae are enlarged.......... 17

17a(16b) Front wing split lengthwise into 2-4 plumelike lobes and hind wing split into 3 lobes (Fig. 289A).......... .... (p. 312) *Pterophoridae*

17b Wings not split or front wings only slightly split 18

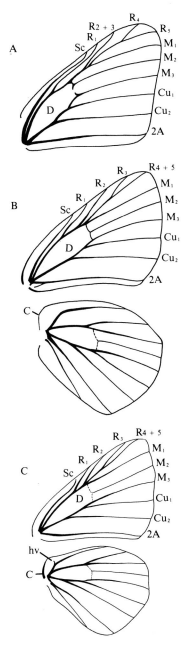

Figure 241 Wings of butterflies. A, front wing of whites, sulfurs, and orangetips (Pieridae); B, blues, coppers, and hairstreaks (Lycaenidae); C, metalmarks (Riodinidae).

18a(17b) Posterior fringe of hind wings narrower than wings; hind wings usually broader than front wings and not spear-shaped 19

18b Posterior fringe of hind wings as wide as or wider than wings (Fig. 284C); hind wings usually no wider than front wings and often spear-shaped (Fig. 247B) (Microlepidoptera) 38

19a(18a) Three anal veins in hind wing (Fig. 242A)................. 20

19b One or 2 anal veins in hind wing .. 24

20a(19a) Sc and Rs of hind wing fused or closely parallel for varying distances beyond discal cell (Fig. 242A)...... .... (p. 310) *Pyralidae*

20b Sc and Rs of hind wing beyond discal cell are widely separated (Fig. 242B) 21

21a(20b) Sc and Rs of hind wing fused beyond middle of anterior margin of discal cell (Fig. 242B) 22

21b Sc and Rs of hind wing separate at base or at most fused along basal half of discal cell (Fig. 242C) 23

22a(21a) Wings dark gray or blackish and sparsely scaled; prothorax often reddish ... (p. 313) *Zygaenidae* (in part)

22b Wings white or yellowish with dense, soft scales .. (p. 313) *Megalopygidae*

Order Lepidoptera 265

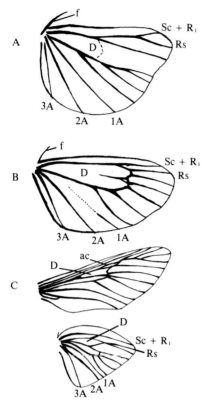

Figure 242 A, hind wing of pyralid moths (Pyralidae); B, hind wing of leaf skeletonizer moths (Zygaenidae); C, carpenterworm moths (Cossidae). ac, accessory cell; D, discal cell.

23a(21b) Front wing with an accessory cell (ac) extending beyond discal cell (Fig. 242C); front wings twice as long as greatest width . . (p. 309) *Cossidae*

23b Front wing without an accessory cell (ac) and length less than twice greatest width (Fig. 243A) . . . . (p. 313) *Limacodidae*

24a(19b) Front wing with 2 (rarely 3) separate and distinct anal veins; wings dark gray or blackish and sparsely scaled; prothorax often reddish . . . . (p. 313) *Zygaenidae* (in part)

24b Front wing with 1 complete anal vein (Fig. 243C), or 1A and 2A fused near ends (Fig. 243B) or connected by a crossvein 25

25a(24b) 1A and 2A of front wings fused near ends (Fig. 243B) or connected by a crossvein; mouthparts usually absent; antennae bipectinate . . . . (p. 305) *Psychidae*

25b Front wing with 1 complete anal vein (Fig. 243C); mouthparts usually present and antennae variable. . . . 26

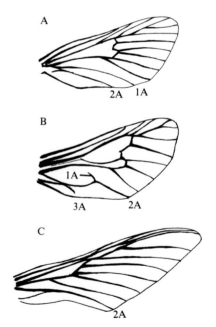

Figure 243 Front wings of: A, slug caterpillar moths (Limacodidae); B, bagworm moths (Psychidae); C, sphinx moths (Sphingidae).

26a(25b) Antennae thickened, tapered at both ends, and sometimes with small hook at ends (Fig. 236D); stout bodied and usually large (>50 mm); front wings long and narrow (Fig. 243C), sometimes partly transparent . . . .
. . . . (p. 290) *Sphingidae*

26b Antennae variable but rarely tapered at both ends 27

27a(26b) M_2 in front wing originates about midway between M_1 and M_3 giving the cubitus a 3-branched appearance (Fig. 244A) 28

27b M_2 in front wing originates closer to M_3 than to M_1 giving a 4-branched appearance to the cubitus (Fig. 246C) 31

28a(27a) Sc and Rs in hind wing enlarged at base, fused to beyond middle of discal cell, then diverging; small and slender . . . .
. . . . (p. 301) *Arctiidae* (in part)

28b Sc and Rs in hind wing not enlarged or fused at base, but may be fused or closely parallel further out for a short distance (Fig. 244A) 29

29a(28b) Sc in hind wing abruptly angled downward at base (Fig. 244A), and Sc and Rs fused or closely parallel for a short distance along discal cell; large wings often marked with slender, wavy lines . . . .
. . . . (p. 303) *Geometridae*

29b Sc in hind wing nearly straight at base and Sc and Rs not fused 30

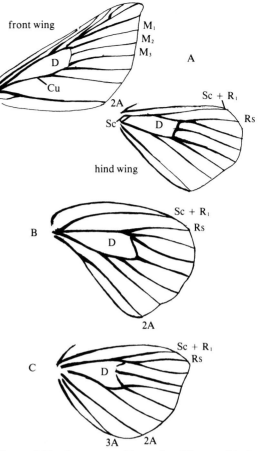

Figure 244 A, geometrid moths (Geometridae); B, hind wing of giant silkworm moths and royal moths (Saturniidae); C, hind wing of notodontid moths (Notodontidae).

30a(29b) Sc and Rs diverge at base of hind wing (Fig. 244B); antennae pectinate (σ); some with transparent "windows" in wings; large (25-150 mm wingspread) . . . . (p. 292) *Saturniidae*

30b Sc and Rs in hind wing closely parallel along most of discal cell length (Fig. 244C); medium-sized, stout-bodied, and hairy; sometimes tufts of hind wing scales project upward when wings folded . . . .
. . . . (p. 296) *Notodontidae*

31a(27b) Cu$_2$ in front wing originates on lower half of discal cell (Fig. 245A); humeral veins present in hind wing (Fig. 245A); stout-bodied and hairy ...... (p. 295) *Lasiocampidae*

31b Cu$_2$ in front wing originates on upper half of discal cell (Fig. 245B); no humeral veins in hind wing (Fig. 245B) 32

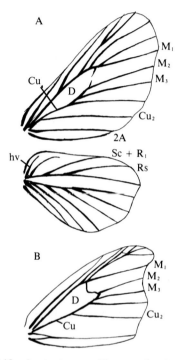

Figure 245 A, tent caterpillar moths (Lasiocampidae); B, front wing of hooktip moths (Drepanidae). D, discal cell; hv, humeral vein.

32a(31b) Tips of front wings usually sickle-shaped (Fig. 245B); frenulum present or absent .... (p. 303) *Drepanidae*

32b Tips of front wings usually not sickle-shaped; frenulum present .. 33

33a(32b) Tips of antennae swollen; usually blackish with white or yellow spots; wingspread about 25 mm .... (p. 300) *Agaristidae*

33b Tips of antennae usually not swollen; color variable 34

34a(33b) Generally dark gray or blackish; sometimes metallic or with bright markings; diurnal; Sc in hind wing absent .... (p. 302) *Ctenuchidae*

34b Colors variable; Sc in hind wing present 35

35a(34b) Ocelli present 36

35b Ocelli absent 37

36a(35a) Sc and Rs in hind wing separated much before middle of discal cell and Sc not swollen at base (Figs. 246A, B); cubitus in hind wing appears 3- or 4-branched (Figs. 246A, B); antennae usually threadlike; often dark gray .. .. (p. 297) *Noctuidae*

36b Sc and Rs in hind wing usually fused to middle of discal cell, or Sc swollen at base (Fig. 246C); cubitus in hind wing appears 4-branched (Fig. 246C); light colors .... (p. 301) *Arctiidae* (in part)

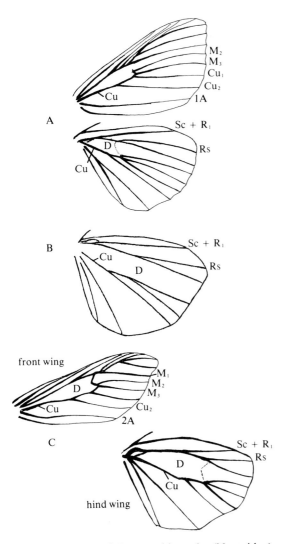

Figure 246 A and B, noctuid moths (Noctuidae): A, cubitus (Cu) of hind wing 4-branched; B, cubitus (Cu) of hind wing 3-branched. C, tiger moths (Arctiidae).

37a(35b) Hind wing with fairly large basal areole (Fig. 247A) .... (p. 295) *Lymantriidae*

37b Hind wing with very small basal areole; often brightly colored with reddish or yellowish, sometimes white .... (p. 301) *Arctiidae* (in part)

38a(18b) Ocelli absent; labial palpi minute or absent; wings very narrow (Fig. 247B) .... (p. 305) *Lyonetiidae*

38b Ocelli present or absent; labial palpi apparent and curved upward or projected forward; wings variable ... 39

39a(38b) Apex of wings pointed; often a hump near base of anterior margin of hind wings (Fig. 247C)................ .... (p. 306) *Gracillariidae*

39b Apex of wings somewhat rounded; no hind wing hump........... 40

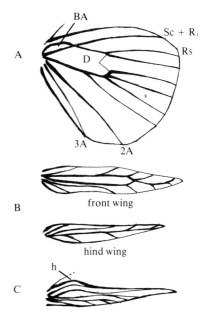

Figure 247 A, hind wing of tussock moths (Lymantriidae); B, lyonetiid moths (Lyonetiidae); C, hind wing of leafblotch miners (Gracillariidae). BA, basal areole; D, discal cell; h, hump.

40a(39b) Maxillary palpi well developed and folded at rest (Fig. 248A) 41

40b Maxillary palpi lacking or project forward at rest without folding... 42

Order Lepidoptera 269

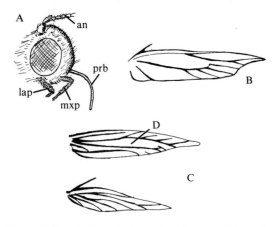

Figure 248 A, lateral view of an incurvariid moth head (Incurvariidae); B, hind wing of a gelechiid moth (Gelechiidae); C, casebearer moths (Coleophoridae). an, antenna; D, discal cell; lap, labial palpus; mxp, maxillary palpus; prb, proboscis.

41a(40a) Minute spines under scales on wing membrane; antennae often long and smooth; rigid piercing ovipositor (female) ..... (p. 304) *Incurvariidae*

41b No spines on wing membrane; antennae usually rough with erect scales on each segment; ovipositor membranous and retracted ....
.... (p. 305) *Tineidae* (in part)

42a(40b) Long stout hairs on face and top of head; antennae rough with erect scales; no ocelli; hind wings very narrow and pointed, venation and anal region of wings generally reduced ..
.... (p. 305) *Tineidae* (in part)

42b Face and head with smooth scales and without stout hairs; antennae moderately smooth; wings variable 43

43a(42b) Outer margin of hind wings concave and wings do not taper evenly to elongated wing apex (Fig. 248B)....
.... (p. 308) *Gelechiidae*

43b Outer margin of hind wings not concave and wings taper evenly to apex (Fig. 248C) 44

44a(43b) Antennae extended forward at rest; no ocelli or maxillary palpi; wings very narrow and pointed (Fig. 248C); discal cell of front wing long, curved downward slightly, and nearly reaching posterior margin (Fig. 248C)....
.... (p. 306) *Coleophoridae*

44b Antennae variable; ocelli present or absent; wings not narrow or pointed; discal cell not unusually long or curved downward 45

45a(44b) Cu_2 in front wing originates on outer fourth of discal cell (Fig. 249A); front wings relatively large, often brightly patterned ....
.... (p. 307) *Yponomeutidae*

45b Cu_2 in front wing originates on basal three-fourths of discal cell (Figs. 249B, C); front wings variable in size and color (Tortricidae) 46

46a(45b) Fringe of long hairs usually on upper side of basal part of Cu in hind wing; M_2, M_3, and Cu_1 of front wings converge distally or R_4 and R_5 are separate (Fig. 249B) (subfamily Olethreutinae) ....
.... (p. 309) *Tortricidae* (in part)

46b Fringe of long hairs absent on upper side of Cu in hind wings; M_2, M_3, and Cu_1 of front wings parallel or divergent distally (Fig. 249C); R_4 and R_5 in front wing generally fused or stalked . . . .
. . . . (p. 309) *Tortricidae* (in part)

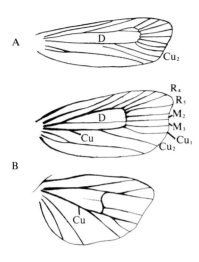

Figure 249 A, front wing of ermine moths (Yponomeutidae); B, leafrolling moths (Tortricidae; Olethreutinae); C, front wing of leafrolling moths (Tortricidae). D, discal cell.

47a(1b) Wingless female in a case or sac constructed of leaves, twigs, or debris (Fig. 281D) . . . .
. . . . (p. 305) *Psychidae*

47b Females not in a case or sac. 48

48a(47b) Body of female slender and scaly or hairy (Fig. 280A); legs long; proboscis present . . . . (p. 303) *Geometridae*

48b Body of female stout and very woolly; legs short; proboscis absent or undeveloped . . . .
. . . . (p. 295) *Lymantriidae*

Abbreviations—Abbreviations used in the description of species are: FW and HW, front and hind wing; DFW, dorsal side of the front wing; VFW, ventral side of the front wing; DHW, dorsal side of the hind wing; VHW, ventral side of the hind wing.

Illustrations—Butterfly wings detached from the body represent a ventral view (p. 272-290).

Species Size—Measurements are of the wingspread.

BUTTERFLIES

Family Nymphalidae— Brushfooted Butterflies

Nymphalidae, the largest (excluding skippers) and one of the most diverse families of butterflies, includes many familiar species. The colors are often orange or brown with black markings, the front legs are short with numerous brushlike hairs, and the discal cell of the hind wing is nearly always open to the outer wing margin. Several species (painted lady, buckeye, California tortoiseshell) are well known for their migratory habits. Larvae are typically brown or black and in later stages covered with branched spines.

Common Species

Fritillaries and Silverspots—Common genus *Speyeria* (Greater Fritillaries) typically orange-brown with large silvery spots on VHW; FW > 25 mm long; larvae brown or black with 6 rows black, branched spines with dorsal prothoracic spine smallest. Genus *Boloria* (Lesser Fritillaries) similar but front wing < 25 mm long.

Boloria bellona (Fabricius) (Fig. 250A). Eastern meadow fritillary—40-45 mm; outer margin FW squared off at tip and not rounded; VHW with bluish bands; northern 1/3 U.S. except Northwest, Appalachian Mts., southern Canada (marshes, meadows).

Boloria selene (Denis and Schiffermüller). Silver-bordered fritillary—37-50 mm; brownish orange or yellowish with black markings; silver spots VFW along margin; northern 1/3 U.S. except Northwest, southwest to KS, Appalachian Mts., most of Canada (bogs, marshes, meadows).

Euptoieta claudia (Cramer) (Fig. 250B). Variegated fritillary—55-60 mm; no silver spots VHW; black zigzag line from middle of anterior margin to middle of posterior margin of both DFW and VFW separates brownish inner area from outer orange band and outermost darker orange, marginal band; southern, central, and southwestern U.S., occasionally north to MN, WY, Manitoba.

Speyeria aphrodite (Fabricius). Aphrodite fritillary—76-82 mm; DFW and DHW yellow orange, marginal border only slightly darkened; like *S. cybele* but smaller; northern 1/2 U.S., southern Canada.

Speyeria cybele (Fabricius) (Fig. 250C). Great spangled fritillary—82-96 mm; brownish orange; DFW and DHW basal 1/3 chocolate brown (♀), light brown margins, heavy black markings; VHW with many silver spots; most of U.S. except Gulf coast states and arid Southwest, southern 1/2 Canada.

Speyeria idalia (Drury). Regal fritillary—75-90 mm; DFW yellowish brown; DHW blue-black with brownish orange at base; 2 rows spots DHW, both creamy white (♀) or only inner row white (♂); eastern U.S. west to Rocky Mts., south to AR (wet meadows).

Figure 250 Brushfooted butterflies (Nymphalidae). A, eastern meadow fritillary, *Boloria bellona*; B, variegated fritillary, *Euptoieta claudia*; C, great spangled fritillary, *Speyeria cybele*.

Speyeria leto (Behr)—75-80 mm; DFW and DHW deep orange with reduced dark markings (♂) or cream color with basal 1/2 dark brown and moderate to heavy black markings (♀); Far West, northeast to ND, southwestern Canada.

Speyeria zerene (Boisduval)—40-70 mm, generally 55-60 mm; coloration highly variable, generally yellow-brown above; spots on underside silvery or not;

variations: (1) ventral side wings pale yellow with silvery spots or (2) VHW lavender, VFW outer 1/4 lavender or (3) broad longitudinal orange band on posterior 1/2 DFW or (4) dorsal side wings dark brown on basal 1/3 and base very hairy; Rocky Mts. westward.

Checkerspots—Typically light and dark markings form a checkered pattern; silvery spots uncommon on VHW but crescents often occur; *Euphydryas* spp. usually blackish brown with bright orange-red, yellow, and cream spots.

Chlosyne nycteis (Doubleday) (Fig. 251A). Nycteis crescent or checkerspot—35-42 mm; DFW and DHW with wide orange-brown band and wide black marginal border; VFW and VHW pale yellowish and orange with greatly reduced dark areas; Rocky Mts. to Atlantic coast, south to GA and TX, southern Canada.

Euphydryas chalcedona (Doubleday) (Fig. 251B). Chalcedon checkerspot—50-75 mm; colors and markings highly variable; commonly brownish black with yellowish spots on DFW and DHW; VFW and VHW orange with yellowish rows of spots and dark markings; Far West.

Euphydryas phaeton (Drury). Baltimore checkerspot—45-65 mm; only checkerspot lacking orange or red-brown band or row of spots across middle of VHW; eastern U.S. south to GA and MO, west to NB, southeastern Canada (near turtlehead plants).

Crescents—Pale white or silvery crescents occur on outer margins of VHW as in some checkerspots; orange or yellowish brown background with dark markings; common genus is *Phyciodes*.

Phyciodes campestris Behr (Fig. 251C). Field crescent—27-35 mm; DFW and DHW brownish black with orange and red-orange spots; VFW and VHW bright yellow-orange with yellowish spots (♂); west of Great Plains, western Canada (moist meadows).

Phyciodes mylitta Edwards. Mylitta crescent—27-38 mm; DFW and DHW with large area deep orange or yellowish brown with black wavy lines; VFW and VHW lighter, VHW with chocolate brown area near outer margin; west of Rocky Mts., southwest Canada (meadows and streams).

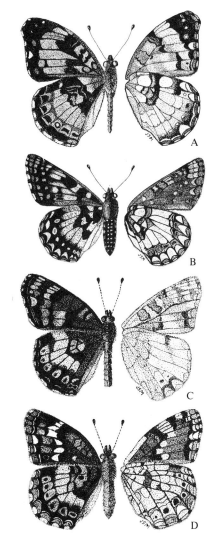

Figure 251 Brushfooted butterflies (Nymphalidae). A, nycteis crescent or checkerspot, *Chlosyne nycteis;* B, chalcedon checkerspot, *Euphydryas chalcedona;* C, field crescent, *Phyciodes campestris;* D, phaon crescent, *P. phaon.*

Phyciodes phaon Edwards (Fig. 251D). Phaon crescent—22-33 mm; DFW and DHW dark brown with rows of orange; VFW with large orange-brown areas; VHW with distinct pattern pale yellow, white and black, or brown; southern U.S. west to southeastern CA, north to KS.

Phyciodes tharos (Drury). Pearl crescent—32-43 mm; DFW and DHW primarily orange-brown with black borders; VHW yellow or cream (summer form) or light gray or brown (spring and fall form); 1 silvery crescent on VHW; Rocky Mts. eastward, southern Canada (open fields, roadsides).

Anglewings—Genus *Polygonia* is distinguished by the concave curvature and deeply indented outer margin of FW, and short taillike projections of HW; DFW and DHW brownish orange with black spots; VFW and VHW mottled brown or gray; VHW with silver "C" or "comma" marking; hibernate as adults; in sunny areas of woods.

Polygonia comma (Harris) (Fig. 252A). Comma or hop merchant—43-50 mm; DFW and DHW brownish yellow or orange with bluish gray to black margin; black diffused over much of DHW in summer form; eastern U.S. south to GA, west to Great Plains, eastern Canada.

Polygonia faunus (Edwards). Green comma—43-48 mm; wings typical color and pattern for genus but DFW and DHW with greenish luster; VFW and VHW with 2 rows submarginal pale green spots; eastern U.S. south to GA, Pacific Northwest, southern 1/2 of Canada.

Polygonia interrogationis (Fabricius) (Fig. 252B). Question mark—62-70 mm; DFW and DHW reddish brown with only slight indentations on outer margin; silvery curved line and dot together form a broad "question mark" on VHW; east of Rocky Mts., southern Canada.

Polygonia satyrus (Edwards). Satyr anglewing—43-50 mm; DFW and DHW bright orange-yellow; VFW and VHW yellow brown with brown bands; "comma" mark on VHW hooked or enlarged at lower end to form "G"; western 1/2 North America (wooded canyons).

Polygonia zephyrus (Edwards). Zephyr anglewing—45-50 mm; DFW and DHW light orange with brown marginal band and pale yellow spots; VFW and VHW gray or gray-brown; Rocky Mts. westward (above 1,500 m).

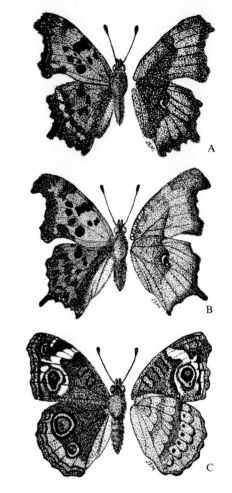

Figure 252 Brushfooted butterflies (Nymphalidae). A, comma or hop merchant, *Polygonia comma;* B, question mark, *P. interrogationis;* C, buckeye, *Precis coenia.*

Buckeyes—Two species occur in North America.

Precis coenia (Hübner) (Fig. 252C). Buckeye—56-63 mm; DFW and DHW grayish brown or darker and eyespots with reddish orange, cream, and blackish rings; DFW with light patch on outer 1/2 and pair of black-bordered orange markings near base; throughout U.S., southern Canada.

Thistle Butterflies—Four species of *Vanessa* occur in continental U.S.; wings are large and triangular; hibernate as adults.

Vanessa annabella (Field). West coast lady—50-58 mm; similar to *V. cardui* but has orange apical spot along with white spots on DFW and blue eyespot centers on VHW; differs from *V. virginiensis* by having 4-5 small eyespots on VHW; Rocky Mts. westward, TX, southwestern Canada (classified by some in the genus *Cynthia*).

Vanessa atalanta (Linnaeus) (Fig. 253A). Red admiral—44-58 mm; DFW dark brown near base, base separated by orange diagonal band from white-spotted, blackish outer 1/2; DHW dark brown with orange marginal band containing 2 bluish spots; VFW with black, pink, brown, white, and blue markings; throughout North America.

Vanessa cardui (Linnaeus) (Fig. 253B). Painted lady—50-58 mm; DFW and DHW brownish orange; heavy dark markings and apical white spots on DFW; VFW brown, pink, and white; VHW brown with row of 4-5 small eyespots; throughout U.S., southern Canada (common migrant in western U.S.; classified by some in the genus *Cynthia*).

Vanessa virginiensis (Drury). Painted beauty—44-58 mm; similar to *V. cardui* but with only 2 large eyespots on VHW; throughout U.S., southern Canada (classified by some in the genus *Cynthia*).

Tortoiseshells—Similar to anglewings but outer margin of FW not deeply concave; wing margin smoothly notched; brown or black with orange, yellow, or black markings; hibernate as adults.

Nymphalis antiopa (Linnaeus) (Fig. 253C). Mourningcloak—68-81 mm; DFW and VFW dark reddish brown with pale yellow outer margins and 1 parallel row submarginal blue patches; throughout North America.

Nymphalis milberti (Godart) (Fig. 253D). Milbert's tortoiseshell—42-44 mm; DFW and DHW dark brown with broad, submarginal orange-yellow band; WV to CA and northward, southern 1/2 of Canada.

Figure 253 Brushfooted butterflies (Nymphalidae). A, red admiral, *Vanessa atalanta;* B, painted lady, *V. cardui;* C, mourningcloak, *Nymphalis antiopa;* D, Milbert's tortoiseshell, *N. milberti.*

Figure 254 Brushfooted butterflies (Nymphalidae). A, viceroy, *Limenitis archippus;* B, white admiral or banded purple, *L. arthemis;* C, Lorquin's admiral, *L. lorquini.*

Nymphalis californica (Boisduval). California tortoiseshell—40-55 mm; DFW and DHW orange with white-margined black spots, brown wing margin and black submarginal band; VFW and VHW dull brown with broad lighter band and row of grayish spots near outer margin; Rocky Mts. westward.

Admirals and Viceroys—Five species (all in genus *Limenitis*) in North America; long antennal club.

Limenitis archippus (Cramer) (Fig. 254A). Viceroy—65-70 mm; DFW and DHW uniform orange or orange-brown with black veins; mimics monarch (Danaidae) but smaller and DHW with long, curved black line from about middle of front margin to hind margin; throughout U.S., southern Canada (near aspen, poplar, willow).

Limenitis arthemis (Drury) (Fig. 254B). White admiral or banded purple—70-80 mm; DFW and DHW purplish black with central row of 8 white patches on FW and 2-4 white spots near apex FW; DHW with 7 white central patches and 7 dark red- orange spots outside of them; bluish crescents along outer margin DFW and DHW; VFW and VHW maroon to reddish brown; northeastern and northcentral U.S., southeastern Canada.

Limenitis astyanax (Fabricius). Redspotted purple—77-85 mm; DFW and DHW bluish black with 2-4 orange-red spots near apex FW; no white band on wings; southern form without orange-red spots, and crescents near margin whitish blue; eastern U.S. west to SD, southwest to AZ, southeastern Canada (open forests, canyons).

Limenitis lorquini (Boisduval) (Fig. 254C). Lorquin's admiral—58-72 mm; DFW and DHW brown with row of white patches; apex DFW orange; northwestern U.S. east to WY and CO, southwestern Canada (near water).

Limenitis weidemeyerii Edwards. Weidemeyer's admiral—75-83 mm; like *L. lorquini* and *L. arthemis* but lacks orange patch on apex DFW and maroon color on VFW; SD and NB form has orange bands on DHW; Rocky Mts. east to Nebraska.

Hackberry Butterflies (genus *Asterocampa*) and Goatweed Butterflies (genus *Anaea*)—*Asterocampa* spp. commonly fly near hackberry trees; *Anaea* spp. have a pointed FW and a tail on HW; both genera sometimes classified as family Apaturidae.

Anaea andria Scudder (Fig. 255A). Goatweed butterfly—55-65 mm; DFW and DHW bright red-orange with brownish margins, yellowish orange wavy band bordered with dark brown (♀); VFW and VHW mottled pale brown; WV to CO and southward.

Asterocampa celtis (Boisduval and LeConte) (Fig. 255B). Hackberry butterfly—45-55 mm; DFW and DHW gray-olive to olive-brown; prominent round and black eyespot near outer margin DFW; eastern U.S. south to FL, west to TX.

Figure 255 Brushfooted butterflies (Nymphalidae). A, goatweed butterfly, *Anaea andria;* B, hackberry butterfly, *Asterocampa celtis.*

Figure 256 Heliconians (Nymphalidae). A, gulf fritillary, *Agraulis vanillae;* B, zebra butterfly, *Heliconius charitonius.*

Heliconians—Most occur in American tropics; FW elongated; larvae feed on passion flower and adults distasteful; sometimes classified as family Heliconiidae.

Agraulis vanillae (Linnaeus) (Fig. 256A). Gulf fritillary—65-70 mm; DFW and DHW bright orange-brown with dark brown and black markings; VHW with large, elongated, silvery spots; southern and southcentral U.S. west to CA.

Heliconius charitonius (Linnaeus) (Fig. 256B). Zebra butterfly—75-85 mm; DFW and DHW black with yellow stripes; SC and Gulf coast states, occasionally KS and CO.

Family Satyridae—Satyrs

These butterflies are typically brown or grayish, the wings have eyespots, and one to three major front wing veins are swollen at the base. The short front legs are not used for walking and the flight of satyrs is erratic and bobbing. Larvae are green with longitudinal stripes and the tip of the abdomen is forked; all feed on grasses or sedges.

Common Species

Cercyonis pegala (Fabricius) (Fig. 257A). Wood nymph—50-69 mm; color and patterns highly variable; DFW and DHW dark brown with broad yellow, ochre, or orange band in DFW that contains 2 eyespots; VHW with up to 6 eyespots; DFW band often absent; most of U.S., southern Canada.

Lethe eurydice (Johannson). Eyed brown—45-50 mm; DFW and DHW grayish brown; DFW with 4 small, nearly round eyespots; DHW with 5-6 eyespots; VFW and VHW with wavy reddish brown lines; DE to CO and northward, eastern Canada (marshes, wet meadows with sedges).

Wood Satyrs

Euptychia cymela (Cramer) (Fig. 257B). Little wood satyr—44-50 mm; DFW and DHW brownish gray with eyespots; 2 large eyespots on VFW; eastern U.S. west to NB and TX, southeastern Canada (woods and adjacent meadows).

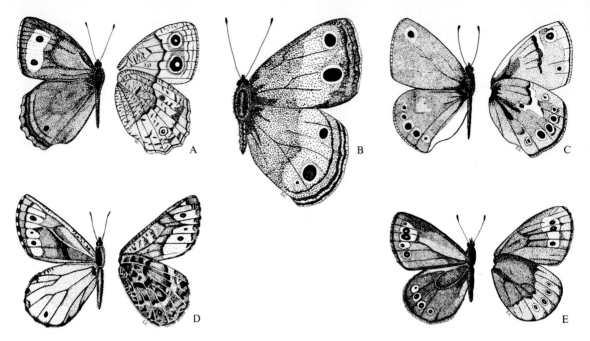

Figure 257 Satyrs (Satyridae). A, wood nymph, *Cercyonis pegala;* B, little wood satyr, *Euptychia cymela;* C, ochre ringlet, *Coenonympha ochracea;* D, chryxus arctic, *Oeneis chryxus;* E, common alpine, *Erebia epipsodea.*

Euptychia hermes (Fabricius). Carolina satyr—37-38 mm; DFW and DHW dark brown with eyespots very small or absent; southern U.S. north to NJ, west to TX (moist woods).

Ringlets

Coenonympha california Westwood. California ringlet—28-45 mm; DFW and DHW grayish white; VFW and VHW darker; CA and OR (foothills, mts.).

Coenonympha ochracea Edwards (Fig. 257C). Ochre ringlet—28-45 mm; DFW and DHW ochre-yellow; VHW with up to 6 black eyespots and wavy band from front to hind margin; Rocky Mts. south to AZ, western 1/3 Canada.

Arctics

Oeneis chryxus (Doubleday) (Fig. 257D). Chryxus arctic—45 mm; DFW and DHW yellow-brown, basal 2/3 FW often grayish brown; VHW mottled gray-black; mts. of western U.S., Dakotas, northern MI, southern 1/2 Canada.

Alpines

Erebia epipsodea Butler (Fig. 257E). Common alpine—45 mm; DFW and DHW very dark brown with row of 3-4 eyespots surrounded by red patches; length of Rocky Mts., eastern OR and WA.

Family Danaidae—Milkweed Butterflies

Danaids are large, orange-brown or chocolate-brown butterflies with black wing margins that are dotted white. The front legs are small and not used in walking. Males have a dark patch of scent scales on each hind wing below vein Cu_2. Adult danaids are distasteful to predators because as larvae they feed on milkweed which contains toxic chemicals. The caterpillars in the U.S. are striped black and yellowish green, and have two or three pairs of long filaments.

The monarch is the best known migratory butterfly, occurring as a migrant or an established species throughout much of the world. The southward migration begins in the fall and much of the northwestern population winters in trees near San Francisco, California, while other groups continue into Mexico. Most of the eastern population flies to pine trees in a mountainous area north of Mexico City although many individuals hibernate in the Gulf coast states. As spring approaches adults disperse northward, begin to lay eggs on milkweed in the southern states, and eventually many die before completing the return trip. The offspring mature and continue into the northern states and Canada.

Common Species

Danaus gilippus (Cramer). Queen—77-83 mm; uniform chocolate brown wings with black veins on dorsal side; larva brownish with dark brown and yellow cross-stripes, yellowish green lateral stripe, and 3 pairs filaments; Gulf coast states, occasionally Mississippi Valley, UT, CA.

Danaus plexippus (Linnaeus) (Figs. 234; 258A). Monarch—87-96 mm; bright orange-brown wings with black veins; 2 rows small white spots on black marginal band of wings, similar viceroy (Nymphalidae) with 1 row spots; larva yellowish green with many narrow black bands, 2 filaments each end of body; chrysalis pale green with gold spots; throughout U.S., southern 1/2 Canada.

Family Libytheidae— Snout Butterflies

Three species of this small family occur in the Americas but only one is established in the U.S. Members have snoutlike, elongated labial palpi and the front wings are extended apically and squared off (Fig. 258B). The front legs of males are reduced but those of females are normal and used for walking. The larvae of the U.S. species have two humped thoracic segments that bear a pair of black tubercles with yellow at their bases.

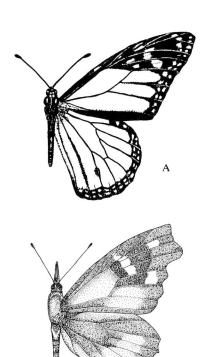

Figure 258 A, monarch, *Danaus plexippus* (Danaidae); B, snout butterfly, *Libytheana bachmanii* (Libytheidae).

Common Species

Libytheana bachmanii (Kirtland) (Fig. 258B). Snout butterfly—34 mm; DFW and VFW dark blackish brown with large orange-brown patches; DFW with subapical white spots; eastern U.S. south to FL, west to Rocky Mts., TX to AZ.

Family Papilionidae—Swallowtails

Swallowtails are the largest butterflies in the U.S. and most are easily recognized by the tail-like projections on the hind wings. The larvae are usually smooth although some have soft

Order Lepidoptera   279

projections or many short hairs. All larvae have an osmeterium, a forked, yellow or orange scent organ on the dorsum of the thorax, that can be turned outward as a means of defense.

The subfamily Parnassiinae is sometimes classified as the family Parnassiidae. The parnassians lack hind wing tails, they are white, pale yellowish, or gray, and they usually have two small reddish spots on the hind wing. Females sometimes will have a hardened structure around the terminal end of the abdomen; this structure originated as a secretion from a male during copulation. Only three species (genus *Parnassius*) inhabit North America and they occur in the mountainous regions of western U.S.

Common Species

Battus philenor (Linnaeus) (Fig. 259A). Pipevine swallowtail—75-113 mm; DFW and DHW black with HW increasingly iridescent blue to greenish toward margins; no orange spots on DFW but present on VHW; eastern U.S. west to NB, south to FL, southwest to southern CA.

Graphium marcellus (Cramer) (Fig. 259B). Zebra swallowtail—93-113 mm; spring forms primarily greenish white with dark bands; summer forms primarily dark brown with greenish white bands, body larger, tails longer; blue and red spots on anal area DHW; eastern U.S. south to FL.

Papilio cresphontes Cramer. Orangedog or giant swallowtail—100-138 mm; DFW and DHW blackish with broad yellow band; VFW and VHW chiefly yellow; dorsal notch near tip of male abdomen; eastern U.S. south to FL, southwest to AZ (larva [orangedog] brown and white; on citrus, prickly ash).

Papilio glaucus Linnaeus (Fig. 259C). Tiger swallowtail—100-163 mm; DFW and DHW primarily yellow or yellow-orange with black bands and blue submarginal spots on DHW; some southern forms dark brownish with yellow and blue marginal spots (♀); North America east of Rocky Mts.

Papilio multicaudatus Kirby. Twotailed swallowtail—100-152 mm; primarily yellow with black markings; HW with 2 tails rather than 1; Rocky Mts. westward, TX.

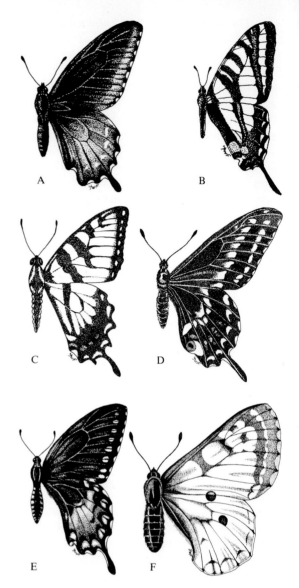

Figure 259 Swallowtails (Papilionidae). A, pipevine swallowtail, *Battus philenor;* B, zebra swallowtail, *Graphium marcellus;* C, tiger swallowtail, *Papilio glaucus;* D, black swallowtail, *Papilio polyxenes asterius;* E, spicebush swallowtail, *Papilio troilus;* F, a parnassian, *Parnassius clodius*.

Papilio palamedes Drury. Palamede swallowtail—113-138 mm; basal half of DFW and DHW dark brown bordered by yellow band; outer 1/3 DHW faint greenish yellow with yellow and black spots; southeastern U.S. (swampy woods).

Papilio polyxenes asterius Stoll (Fig. 259D). Black swallowtail, parsleyworm—69-88 mm; DFW and DHW black with 2 rows yellow spots and blue submarginal band in DHW; east of Rocky Mts., FL, NM, AZ, southcentral and southeastern Canada (larva on carrot, wild carrot, parsley).

Papilio rutulus Lucas. Western swallowtail—90-120 mm; like *P. glaucus* but VFW with continuous submarginal band of yellow; Rocky Mts. westward.

Papilio troilus Linnaeus (Fig. 259E). Spicebush swallowtail—100-125 mm; DFW nearly all dark brown with 2 rows yellow spots; DHW dark brown at base changing to iridescent green or blue toward margin, black pattern present; orange spot near costa of DHW; eastern U.S. south to FL, southwest to TX, southcentral and southeastern Canada (woods and fields).

Parnassians

Parnassius clodius Ménétriés (Fig. 259F)—63-75 mm; solid black antennae; northwestern U.S., Rocky Mts., far-western Canada. *P. phoebus* Fabricius, antennal segments black and white, often red spots in FW.

Family Pieridae—Whites, Sulphurs, and Orangetips

Members of this worldwide family are small- to medium-sized white, yellow, or orange butterflies with black wing margins. The front legs are well developed and the tarsal claws are forked. Individual color variations are common and females often differ from males in pattern and sometimes color (sexual dimorphism). The word "butterfly" probably was derived from the bright, butter-yellow color of European sulphurs. Numerous species are migratory. Larvae are green, smooth and slender, and often have longitudinal stripes. Older larvae have short, fine hairs.

Figure 260 Whites (Pieridae). A, pine butterfly, *Neophasia menapia;* B, southern cabbageworm, *Pieris protodice;* C, imported cabbageworm, *P. rapae*.

Common Species

Whites, Marbles, and Orangetips—3rd segment labial palpi long and hairy; generally whitish wings.

Anthocharis midea Hübner. Falcate orange-tip—40-43 mm; white with apex of DFW orange and curved backward slightly; eastern 1/2 U.S. *A. sara* Lucas similar, apex of FW orange but not curved backward, DHW often greenish, Rocky Mts. westward.

Ascia monuste (Linnaeus). Great southern white—44-57 mm; DFW and DHW white or brownish gray with tip and narrow outer margin FW diffuse blackish gray; VHW straw colored; VFW straw colored and white; Gulf coast states to VA coast.

Neophasia menapia (Felder and Felder) (Fig. 260A). Pine butterfly—37-50 mm; DFW and DHW white (♂) or cream (♀); DFW with oval, white markings in blackish apex and dark and narrow costal band curves downward near middle of costa; Rocky Mts. westward, Black Hills of SD (pines, fir).

Pieris protodice Boisduval and LeConte (Fig. 260B). Southern cabbageworm—32-40 mm; wings with brown and white mottled or checkered pattern; throughout U.S., southernmost central Canada.

Pieris rapae (Linnaeus) (Fig. 260C). Imported cabbageworm—32-45 mm; DFW and DHW white with diffuse dark patch at apex FW; 1 (sometimes 2) dark spot in FW and anterior margin of HW; underside wings pale yellowish; throughout North America (larva velvety green with 3 faint yellow lines; on cabbage, other crucifers).

Sulphurs—Generally yellow or orange wings, 3rd segment of labial palpi short and not hairy; most species are in genus *Colias* and their dark wing borders are solid (♂) or contain light spots or are very narrow (♀).

Colias cesonia Stoll (Fig. 261A). Dog's head or dog face—45-60 mm; front wing pointed at tip; DFW with yellow "dog face" profile including dark eye; southern 1/3 U.S., sporadic north into southern Canada.

Colias eurydice Boisduval. California dog face—45-65 mm; like *C. cesonia* but DHW usually without dark margins; CA.

Colias eurytheme Boisduval (Fig. 261B). Alfalfa butterfly—40-60 mm; hybridizes with *C. philodice;* DFW and VFW yellow with some orange to primarily orange; no adjacent spot next to red ring on VHW as in *C. philodice;* white forms (♀) common; wings do not reflect UV light (♂); most of North America, more common eastern 1/2 (especially in alfalfa fields).

Colias philodice Godart (Fig. 261C). Clouded sulphur—32-48 mm; hybridizes with *C. eurytheme,* both species variable in color and markings; yellow wings with blackish margin; sharply defined red ring with silvery center on VHW, usually small adjacent spot; white forms (♀) common; wings reflect UV light (♂); spring and fall forms smaller with greenish HW; most of North America (especially clover fields, mud puddles).

Eurema lisa Boisduval and LeConte (Fig. 261D). Little sulphur—27-38 mm; dirty yellow or pale yellow wings, blackish margins; southern 1/2 U.S., sporadic to southern Canada (especially mud puddles).

Eurema nicippe (Cramer). Sleepy yellow or sleepy orange—35-45 mm; bright orange wings, wide and irregular dark border; VFW and VHW yellow or orange-yellow, HW sometimes pale tan (♀); eastern U.S. south to FL, southwest to CA, rare in North.

Nathalis iole Boisduval (Fig. 261E). Dainty sulphur—20-32 mm; DFW yellow with black apical border; DHW yellow with narrow costal bar on anterior margin; South and southern Great Plains, west to southern CA, occasionally northcentral states to WY.

Phoebis sennae (Linnaeus) (Fig. 261F). Cloudless sulphur—55-69 mm; DFW and DHW clear yellow without markings (♂) or yellow-orange with marginal black spots and DFW eyespot (♀); white form (♀); southeastern U.S. west to CA, strays to NY.

Figure 261 Sulphurs (Pieridae). A, dog's head or dog face, *Colias cesonia;* B, alfalfa butterfly, *C. eurytheme;* C, clouded sulphur, *C. philodice;* D, little sulphur, *Eurema lisa;* E, dainty sulphur, *Nathalis iole;* F, cloudless sulphur, *Phoebis sennae.*

Family Riodinidae (= Nemeobiidae)—Metalmarks

Metalmarks are moderately small butterflies, found primarily in the New World tropics, that usually have metalliclike spots or lines on their wings. Tropical species have a tremendous diversity of color, ornamentation, and form. Riodinids are similar to the Lycaenidae but the costa is thickened at the base of the hind wing and the subcosta has a spur near the base. The antennae are unusually long and each supports a slender, flattened club. At rest metalmarks typically hold their wings out to the side or at a 45° angle. Larvae are stout, have large heads, and bear raised spots with clumped setae.

Common Species

Apodemia mormo (Felder and Felder)—28 mm; DFW and DHW brown and orange with many squarish white spots; VFW and VHW similar but lighter color; no metallic marks; arid regions WY to TX and westward.

Apodemia palmeri (Edwards)—18-20 mm; DFW and DHW dark brown with orange margins; VFW and VHW pale orange; deserts of Southwest (flies around mesquite).

Calephelis virginiensis (Guérin-Méneville) (Fig. 262)—16-20 mm; DFW and DHW rusty orange-brown with lines of metallic spots; wing fringes not checkered; southeastern U.S.

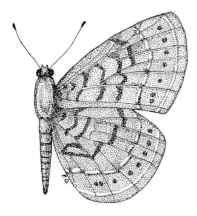

Figure 262 A metalmark, *Calephelis virginiensis* (Riodinidae).

Family Lycaenidae—Blues, Coppers, and Hairstreaks

Members of this large family are small, slender, and often iridescent. Males of most species have reduced front legs but the females' legs are normal and used for walking. Flight is rapid and often darting. Larvae are slug-shaped, broad and flattened, and covered with short hairs. Larvae of some species secrete honeydew and are tended by ants; other immatures are predators of aphids.

Common Species

Blues—Usually some shade of blue, some grayish (♀) or brown (♀); genus *Philotes* has outer fringes of wings checkered, VHW gray with submarginal row orange spots.

Brephidium exilis (Boisduval). Western pygmy blue—11-17 mm; DFW and DHW brown shading to blue at base; VFW and VHW grayish white at base shading to brown distally; VHW with tiny gold and black circles in submarginal area; TX to CA, strays north to NB and OR (dry disturbed areas, washes; some males = 6 mm and are world's smallest butterflies).

Celastrina argiolus pseudargiolus (Boisduval and LeConte). Spring azure—20-32 mm; DFW and DHW pale violet-blue with whitish margin (♂) or DFW with broad dark margin and DHW blue washed with white (♀); VFW and VHW gray-white, small markings often crescentlike; eastern U.S., southeastern Canada (flies March-April).

Everes amyntula (Boisduval). Western tailed blue—like *E. comyntas* but VHW chalk white and not gray; Rocky Mts. westward, northernmost U.S., western 1/2 Canada.

Everes comyntas (Godart) (Fig. 263A). Eastern tailed blue—23-28 mm; DFW and DHW pinkish blue (♂) or dark brown to gray (♀ summer forms); HW with short tail and 2 orange spots near base; VHW gray; east of Rocky Mts., sporadic westward.

Glaucopsyche lygdamus (Doubleday). Silvery blue—23-30 mm; DFW and DHW shining silvery blue (♂) or duller with diffuse dark margins (♀); VFW and VHW light brown with no markings except single row black dots on outer 1/3; northcentral U.S. south to GA.

Hemiargus isola (Reakirt). Reakirt's blue—19-28 mm; DFW and DHW blue with narrow brown margin (♂) or outer 1/2 brown (♀); VFW and VHW gray, area just beyond center of FW with row of black spots ringed with white; west of Mississippi River, MI, MN, southwestern Canada.

Plebejus acmon (Westwood). Acmon blue—18-25 mm; DFW and DHW blue (♂) or dark brown (♀); DHW and VHW with orange submarginal band; Far West, east to MN and KS.

Plebejus melissa (Edwards) (Fig. 263B). Melissa blue—23-30 mm; ♂: DFW and DHW clear blue with white margin and narrow black submarginal band, and VFW and VHW gray with black spots and submarginal row orange spots; ♀: DFW and DHW bluish dark brown with submarginal row orange spots on DHW (north central and northeastern U.S.), or dark brown with prominent orange submarginal band on DFW and DHW; wings with white margins; western U.S.; northeastern population called Karner's blue.

Plebejus saepiolus Boisduval. Saepiolus blue—23-31 mm; color highly variable; DFW and DHW iridescent blue often with silvery cast and small spot at tip of discal cell (♂); or brown or blue and brown (♀); Rocky Mts. westward, Great Lakes states, ME, southern Canada.

Figure 263 Blues and coppers (Lycaenidae). A, eastern tailed blue, *Everes comyntas;* B, Melissa blue, *Plebejus melissa;* C, purplish copper, *Lycaena helloides;* D, great copper, *L. xanthoides dione.*

Coppers—Variable coloration, many orange-red or coppery; stout bodied.

Lycaena helloides (Boisduval) (Fig. 263C). Purplish copper—28-33 mm; DFW and DHW light to dark purplish orange-brown (♂), scattered spots; orange crescents on outer margin DHW; VFW and VHW light orange-brown; faint wavy red line near outer margin VHW; western U.S. southeast to KS, northeast to MI, British Columbia to Quebec.

Lycaena phlaeas americana Harris. American copper—25-27 mm; DFW spotted and metallic copper color with grayish or dark brown border; DHW grayish or dark brown with coppery and spotted outer margin; VFW pale copper, spotted, gray outer margin; VHW spotted, gray, with row of coppery spots near outer margin; primarily northeastern U.S.; south to NC and AR, west to KS and ND.

Lycaena xanthoides dione (Scudder) (Fig. 263D). Great copper—38-43 mm; DFW and DHW shiny gray-brown; DHW with orange marginal crescents; VFW light gray with black spots; VHW light gray with orange marginal patch; Midwest, central Canada (prairies with bitter dock). *L. xanthoides xanthoides* (Boisduval), 29-31 mm, similar but with diagonal and spotted orange band on DFW, OR and CA.

Hairstreaks—Most have 1 to several filamentous tails on HW; usually dark gray or brownish; in meadows and roadsides early spring. Elfins (genus *Callophrys*) lack tails; DFW and DHW gray, brown, or orange-brown; VFW and VHW lighter brown, gray, green, or blue; outer margin wings slightly scalloped; tip of HW slightly pointed.

Atlides halesus Cramer (Fig. 264A). Great blue hairstreak—34-35 mm; DFW and DHW brilliant blue-green with dark brown margins; 1 long and 1 short tail on each HW; southern 1/2 U.S.

Order Lepidoptera

Figure 264 Hairstreaks (Lycaenidae). A, great blue hairstreak, *Atlides halesus;* B, *Satyrium behrii;* C, banded hairstreak, *Satyrium calanus falacer;* D, gray hairstreak, *Strymon melinus.*

Calycopis cecrops (Fabricius). Redbanded hairstreak—25-26 mm; DFW and DHW brownish black with blue-gray patch on anal angle HW; VFW and VHW dark brown with bright red band bordered on outside by white and black lines; South, north to NJ and OH, west to TX.

Harkenclenus titus (Fabricius). Coral hairstreak—32-33 mm; DFW and DHW uniformly graybrown; VHW with submarginal band of coral-red spots; most of U.S. except Gulf coast and extreme Southwest.

Satyrium behrii (Edwards) (Fig. 264B)—26-28 mm; DFW and DHW orange with wide blackish margin on FW and narrow margin on HW; VFW and VHW grayish brown with marginal dots and small orange patch near outer margin HW; Rocky Mts. westward.

Satyrium calanus falacer (Godart) (Fig. 264C). Banded hairstreak—25-27 mm; DFW and DHW grayish black without markings except gray scent patch near middle of anterior margin FW (♂); VFW with 2 dark lines and inner line not broken into spots; blue and red patches at anal angle HW; eastern U.S. south to FL, southwest to TX, southeastern Canada.

Strymon melinus (Hübner) (Fig. 264D). Gray hairstreak—27-28 mm; DFW and DHW gray to dark gray; VFW and VHW light gray; bright orange patch on DHW and VHW; throughout U.S., southern Canada.

Family Megathymidae— Giant Skippers

These large, stout, fast-flying skippers occur in the western and southern U.S. The antennae are clubbed and not hooked as are those of many Hesperiidae (skippers). Larvae of the genus *Agathymus* feed in stems of various *Agave* plants, and those in the genus *Megathymus* bore into roots of *Yucca* species.

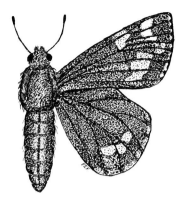

Figure 265 A giant skipper, *Megathymus yuccae* (Megathymidae).

Common Species

Megathymus yuccae (Boisduval and LeConte) (Fig. 265)—44-82 mm; DFW and DHW dark brown to black with orange-yellow, yellow, or white spots; discal cell spot separate from closest spot on DFW; NC to CA.

Family Hesperiidae—Skippers

Skippers are a very large family of small- to medium-sized butterflies with stout bodies and generally dark brown and/or orange wings. There are over 3,000 species worldwide (mostly subtropical) and about 240 in the U.S. Features separating skippers from other families of butterflies are (1) the antennal club is usually hooklike (Fig. 236B); (2) the antennae are widely separated at the base; and (3) the five branches of the radial (R) vein in the front wing arise from the discal cell and are not stalked (Fig. 239A). Unlike other butterflies, most skippers have a darting (skipping) flight with little fluttering. Many species hold the front and hind wings at different angles from each other while at rest, a good field identification characteristic. The larva has a large head, a constricted "neck," and the smooth body is tapered to the front and back. Silken tubes or leaf shelters are formed for feeding or as a retreat.

Skippers are most easily caught while resting and then placed in a fully charged killing jar rather than attempting to stun them by pinching the thorax. Wings should be spread while the specimen is fresh.

Common Species

Branded Skippers—Majority of species; small and orange, some brown; usually patch (= brand or stigma) of specialized scent scales on DFW (♂); middle tibiae spined; common genus *Hesperia* characterized by arched stigma on DFW with silvery line in middle (♂).

Amblyscirtes vialis (Edwards) (Fig. 266A). Roadside skipper—22-26 mm; DFW and DHW dark brown; VFW and VHW dark brown at base, outer 1/2 VHW and apex of VFW violet-gray; spots absent and outer margin noticeably checkered on VHW; throughout U.S., southern Canada (streams, ravines, woods).

Ancyloxypha numitor (Fabricius). Least skipper—20-26 mm; DFW completely black or dark brown with basal 1/2 orange-washed; DHW orange with broad black margin; VFW black with orange costal margin and apex; VHW orange, gold or unmarked cream-orange; eastern U.S. west to Rocky Mts., southwest to TX, southeastern and southcentral Canada.

Atalopedes campestris (Boisduval) (Fig. 266B). Field skipper—24-36 mm; DFW and DHW orange with large central brown patch (stigma, ♂) or orange to dark brown but always with clear spot under discal cell in center of FW (♀); VHW olive to dark yellow with pale yellow spots; southern 1/2 U.S., sporadic to NB and NY.

Copaeodes minima (Edwards). Southern skipperling—14-20 mm; DFW and DHW bright orange; VFW and VHW with white streak from base to outer margin; southern U.S.

Euphyes vestris (Boisduval). Dun skipper—24-30 mm; both sides wings dark brown and without spots (♂) or very small light spots on DFW ( ); CA form with brownish orange wings and yellow spots on DFW; throughout U.S., southern Canada.

Hesperia leonardus Harris. Leonardus skipper—22-36 mm; DFW orange basally with wide, dark brown outer margin (♀); VFW and VHW brick red or reddish brown with white or pale yellow spots in a loose band; eastern U.S. west to MN.

Figure 266 Skippers (Hesperiidae). A, roadside skipper, *Amblyscirtes vialis;* B, field skipper, *Atalopedes campestris;* C, *Ochlodes sylvanoides;* D, Peck's skipper, *Polites coras;* E, tawnyedged skipper, *Polites themistocles;* F, little glassy wing, *Pompeius verna.*

Lerodea eufala (Edwards). Eufala skipper—22-28 mm; DFW and DHW gray-brown, unmarked except for 3-5 white spots on FW; VFW and VHW lighter brown with gray dusting on HW and apex of FW; southern 1/2 of U.S., strays to MN.

Ochlodes sylvanoides (Boisduval) (Fig. 266C)—20-28 mm; DFW and DHW reddish orange with wide brown marginal border; black elongated spot (stigma) in center of DFW and often with gray central line (♂); DFW with orange and brown patches (♀); VFW and VHW yellowish and unmarked to dark brown with band of square yellow spots; Rocky Mts. westward.

Panoquina ocola (Edwards). Ocola skipper—28-34 mm; DFW and DHW dull brown; DFW with 3-4 whitish spots but no spot in discal cell; VHW dull brown without conspicuous markings; southeastern U.S. west to TX, north to NJ.

Poanes hobomok (Harris). Hobomok skipper—26-30 mm; DFW and DHW yellow-orange with broad blackish border, occasionally dark brown with pale yellow or whitish spots (♀); VFW orange and VHW yellowish, both with purplish brown border; eastern U.S. west to MN and KS, eastern Canada.

Polites coras (Cramer) (Fig. 266D). Peck's skipper—20-26 mm; DFW and DHW orange and dark brown, broad grayish brown patch from outside of stigma to dark border of DFW (♂); VHW reddish or orange-brown with large, squarish, yellow spots which may fuse; eastern U.S. (rare near southern limits), southwest to AZ, northwest to OR.

Polites themistocles (Latreille) (Fig. 266E). Tawny-edged skipper—20-26 mm; DFW and DHW dull brown, FW discal cell and costal area bright orange; black stigma on DFW broad and S-shaped (♂); DFW and DHW darker with less orange (♀); most of U.S., southern Canada.

Pompeius verna (Edwards) (Fig. 266F). Little glassy wing—24-32 mm; both sides wings dark brown and without orange; white-glassy, squarish spots on DFW, largest spot below end of discal cell; eastern U.S. west to NB and TX.

Pyrginae Skippers—Generally dark brown, some checkered black and white; medium to large; no stigma on FW (♂); no spines on front and middle tibiae. Common genus *Erynnis* (dusky wings) recognized by large size, dark with some whitish clear spots on FW, costal margin often folded (♂); wings out to side when resting.

Achalarus lyciades (Geyer). Hoary edge (Fig. 267A)—38-46 mm; DFW and DHW brown with orange to gold glassy spots on FW; VHW with broad and diffuse whitish area on outer 1/2; eastern U.S. south to FL, west to MN and TX.

Epargyreus clarus (Cramer) (Fig. 267B). Silverspotted skipper—44-60 mm; DFW and DHW dark brown with large yellow-orange spots on DFW; VHW dark brown with large silvery patch; throughout U.S., southern Canada.

Erynnis juvenalis (Fabricius). Juvenal's dusky wing—32-43 mm; both sides wings dark brown (♂) or light brown (♀); DFW with several small whitish-clear spots below apex and center of wing, dark markings scattered on wing and also form band near outer margin; long, curved, whitish hairs above DFW surface and directed toward outer margin (♂); costal area of FW folded (♂); VHW with 2 subapical pale brown spots; eastern U.S. south to FL, west to WY and TX.

Pholisora catullus (Fabricius). Common sooty wing—22-30 mm; DFW with 2 to many minute white dots; DHW with submarginal row small white spots or none; VHW uniformly sooty brown; throughout U.S. except FL, southern Canada.

Pyrgus communis (Grote) (Fig. 267C). Checkered skipper—20-32 mm; DFW and DHW grayish (♂) or blackish (♀) and checkered with white; VHW white with pale greenish brown bands; most of U.S. and Canada, uncommon in New England and eastern Canada.

Thorybes pylades (Scudder) (Fig. 267D). Northern cloudy wing—30-46 mm; DFW medium to dark brown with triangular whitish-clear spots on DFW; costal fold on FW (♂); wing fringes checkered white and brown; VHW grayish brown with finely outlined wavy lines; eastern U.S. south to FL, west to CA, southern Canada.

Urbanus proteus (Linnaeus). Longtailed skipper—38-50 mm; green head, thorax, and wing base; hind wing with long tail; CT to FL, Gulf coast states, sporadic in AZ and CA.

Figure 267 Skippers (Hesperiidae). A, hoary edge, *Achalarus lyciades;* B, silverspotted skipper, *Epargyreus clarus;* C, checkered skipper, *Pyrgus communis;* D, northern cloudy wing, *Thorybes pylades.*

MOTHS

Family Sphingidae—Sphinx Moths

Sphinx moths are medium to large moths with front wings that are longer and narrower than the hind wings. The antennae frequently are thickened in the middle and usually curved or hooked at the tips (Fig. 236D). The abdomen often is pointed and extends beyond the hind wings. These moths typically hover over flowers beginning at dusk and thrust their long proboscis into a flower to siphon up the nectar. Their large size and rapid speed and wingbeat have led to names such as hummingbird moths and hawk moths. The larvae are large and commonly have a hornlike spine on the anal segment and often oblique stripes on the sides (Fig. 269B). The pupa often has a long handlelike appendage consisting of the adult's proboscis that extends beneath the body (Fig. 269C). There are 115 species of sphinx moths in America north of Mexico.

Common Species

Agrius cingulatus (Fabricius). Sweetpotato hornworm—105-110 mm; HW and abdomen with pink bands; larva on sweet potato; southern U.S., strays to Nova Scotia, MI, KS.

Ceratomia catalpae (Boisduval). Catalpa sphinx—63-97 mm; yellowish brown with gray shading; 4-5 dark streaks on FW, otherwise markings indistinct on DFW and DHW or lacking on VFW and VHW; larva: greenish or dorsal surface black, sides yellow with vertical and irregular black stripes, black horn, on catalpa trees; eastern U.S. south to FL, west to KS.

Ceratomia undulosa (Walker). Waved sphinx—80-110 mm; FW grayish with distinct wavy lines, 5 dark streaks point toward outer margin, and distinct alternating white and dark spots on outer margin; larva: grayish green or green, 7 pairs lateral oblique bands, head with 2 bands, on ash; eastern U.S. south to FL, west to TX, central and southeastern Canada.

Eumorpha achemon (Drury) (Fig. 268A). Achemon sphinx—75-100 mm; DFW pale yellow gray-brown; dark brown patches on DFW and sides of thorax; HW pink with dark spots on brown border; larva: green, tan, pink or brownish red, 6-8 pale oblique and narrow blotches on sides, horn replaced by black spot on older larvae, on grape and Virginia creeper; eastern U.S. south to FL, southwest to CA.

Hemaris thysbe (Fabricius) (Fig. 268B). Hummingbird moth or common clearwing—42-48 mm; FW and HW with large clear areas; bumble bee appearance; ventral surface thorax yellow and without dark bands; ventral surface abdomen red-brown to dark brown; larva: greenish, granular surface, no oblique lines, horn present; eastern U.S. south to FL, west to TX and Great Plains, southern Canada (diurnal).

Hyles lineata (Fabricius) (Fig. 268C). Whitelined sphinx—50-90 mm; brown; white stripes on head and thorax; FW with white stripe from posterior margin to tip, veins and margins white; HW with pinkish white band; larva: usually green, head and horn yellowish or orange, 2 black lateral lines (occasional black form with yellow lateral and middorsal stripes); throughout U.S., southern Canada.

Figure 268 Sphinx moths (Sphingidae). A, Achemon sphinx, *Eumorpha achemon;* B, hummingbird moth, *Hemaris thysbe;* C, whitelined sphinx, *Hyles lineata;* D, blinded sphinx, *Paonias excaecatus;* E, great ash sphinx, *Sphinx chersis.*

Manduca quinquemaculata (Haworth) (Fig. 269). Tomato hornworm—100-125 mm; brownish gray wings; FW with wavy, black line near and parallel to outer margin; usually with 5 pairs orange-yellow spots on sides of abdomen; larva: light or dark green, 8 oblique lateral white stripes curved backward to form a "V", red or black horn, primarily on tomatoes, potatoes; most of U.S., southeastern Canada.

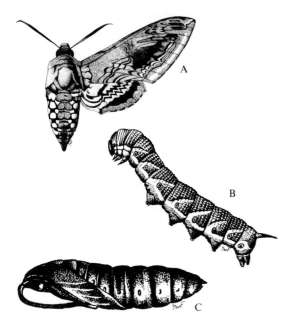

Figure 269 Tomato hornworm, *Manduca quinquemaculata*. A, adult; B, larva; C, pupa.

Manduca sexta (Linnaeus). Tobacco hornworm—like *M. quinquemaculata* but wings more brown with gray shades; FW with narrow white spots on outer margin; usually 6 pairs orange-yellow spots on abdomen; larva: yellow green with 7 oblique lateral white stripes which do not curve backward, red horn, on tomatoes, tobacco, potatoes; eastern U.S. south to FL, west to MN.

Pachysphinx modesta (Harris). Modest sphinx—100-130 mm; FW dark gray, basal 1/3 pale yellow-gray; HW with broad and diffuse reddish brown band, and diffuse triangular marking with apex pointing toward HW base; larva: whitish green, 7 pairs lateral oblique white lines, head pink with 2 white bands, body surface with pale outgrowths, on willow, poplar; most of U.S., southern Canada.

Paonias excaecatus (J.E. Smith) (Fig. 268D). Blinded sphinx—55-90 mm; DFW mottled grayish brown; DHW pink at base with bluish-black eyespot; VFW pink at base; FW outer margin scalloped and inner parts of scallops margined with white; larva: pale yellowish green, granular, 7 pairs oblique yellow stripes with red blotch on dorsal end; eastern U.S., mts. of western U.S., southern Canada.

Sphinx chersis (Hübner) (Fig. 268E). Great ash sphinx—90-120 mm; dark to pale blue-gray; 4 dark dashes on FW, remaining surface indistinctly marked; 2 dorsolateral dark lines on thorax which diverge toward abdomen; larva: pale bluish green, paired lateral yellow bands on head, pale oblique lateral stripes, 1st 3 body segments darkest, on ash, aspen, privet, others; throughout U.S., southern Canada.

Family Saturniidae—Giant Silkworm Moths and Royal Moths

These moths are medium to very large insects that usually have feathery (pectinate) antennae and clear "windows" or eyespots on the wings. The proboscis is rudimentary or absent and the adults probably never feed. There are 65 species in America north of Mexico. Some subtropical species are giant sizes (Atlas moth wingspread is 250 mm) or exhibit very long hind wing tails. Larvae are brownish to pale green and are armed with spines or tubercles. Large, brownish cocoons are spun from coarse silk; the silk has no commercial value. The royal moths, considered here as a subfamily, have sometimes been classified as a separate family, Citheroniidae.

Common Species

Subfamily Saturniinae—Colored apical wing patch on most species.

Actias luna (Linnaeus). Luna moth—75-92 mm; pale green; clear eyespots; long HW tail; larva: like *Antheraea polyphemus* but with yellow lateral stripe and no oblique stripes on sides, on hickory, walnut, sweet gum, persimmon; eastern U.S. south to FL, west to NB and TX, southeastern Canada.

Antheraea polyphemus (Cramer) (Fig. 270A). Polyphemus moth—100-125 mm; usually reddish or light brown; large yellow-margined eyespot on FW; blue- and black-margined eyespot on HW; larva: green, pale yellow oblique stripes, orange tubercles on red spots, on common deciduous trees; cocoon: brown, broadly oval, hanging on branch (Fig. 270B); throughout North America.

Callosamia promethea (Drury) (Fig. 270C). Promethea moth—64-100 mm; DFW and DHW reddish brown (♀) or brownish to black (♂); pale brown with yellow outer wing margins; FW apex elongated; ♂ DFW with lighter patch and dark eyespot; VFW and VHW dark maroon (♂); diurnal (♂); larva: bluish green, 2 pairs orange-red spines on thorax, 1 yellow spine near tip of abdomen, 3 rows small black tubercles on sides, on spicebush, sassafras, tulip tree, wild cherry; east of Great Plains, south to FL, southeastern Canada.

Hyalophora cecropia (Linnaeus) (Fig. 270D). Cecropia moth—100-155 mm; body dark pinkish brown with white bands; wings reddish brown, reddish crescent on FW and white crescent on HW; white band on wings with reddish outer border; wing margins grayish; larva: dark bluish green, 2 pairs orange-red tubercles on thorax, 15 yellow tubercles on back, 2 rows blue lateral tubercles, on common deciduous trees (Fig. 270E); cocoon: gray-brown, attached lengthwise to twig (Fig. 270F); east of Rocky Mts., UT, WA.

Hyalophora euryalus (Boisduval). Ceanothus silk moth—87-127 mm; body and wings reddish brown; narrow white band separates basal 2/3 wing from outer 1/3, outer 1/3 with pale brown outer wing margin; crescent spots in center of wing long and angular; larva: dorsal tubercles yellow and slender, those on abdomen pointed, on buckbush, other plants; Pacific coast northeast to MT.

Samia cynthia (Drury). Cynthia moth—75-127 mm; wings greenish brown with white markings; large white crescent near center of both FW and HW is bordered by yellow on lower side; black apical spot on FW; white spots on abdomen; larva: green with black spots, bluish dorsal and lateral tubercles, yellow head, on ailanthus (tree-of-heaven), other trees and shrubs; cities of Atlantic coast states, straying to IN.

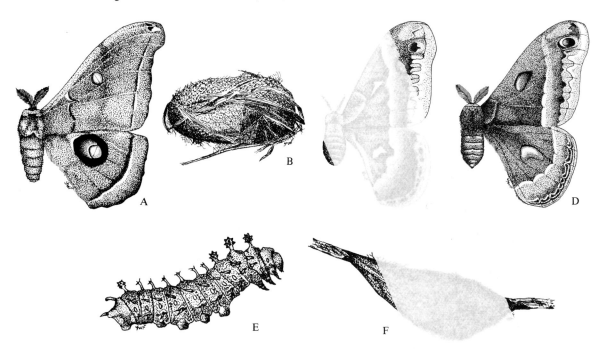

Figure 270 Giant silkworm moths (Saturniidae). A and B, polyphemus moth, *Antheraea polyphemus*: A, adult; B, cocoon; C, promethea moth, *Callosamia promethea*; D-F, cecropia moth, *Hyalophora cecropia*: D, adult; E, larva; F, cocoon.

Subfamily Citheroniinae—Royal Moths

Anisota stigma (Fabricius). Spiny oakworm—38-64 mm; bright orange- or red-brown wings; many speckles; small clear dot on each FW; larva: brownish, rows of short spines, 1 pair longer spines on prothorax, orange head, on oaks; eastern U.S. south to FL.

Anisota virginiensis (Drury)—38-62 mm; reddish brown with nearly clear central area of FW (♂), or yellow body with brownish yellow wings and wide pink outer margin of FW (♀); larva: dark green with white speckling, 2 purplish-red stripes, 3 lateral rows short black spines, 1 pair long spines on prothorax, on oaks; eastern U.S. south to Carolinas, west to KS, southeastern and southcentral Canada.

Citheronia regalis (Fabricius). Regal moth, hickory horned devil—90-150 mm; brown or reddish brown body with yellowish bands; FW dark gray or olive-colored with yellow patches and reddish brown veins; HW brown or orange-brown with yellowish streaks; larva: brown and green with colorful markings, long black-tipped red horns on thorax (nonstinging), on hickory, walnut, other trees; eastern U.S. south to FL, southwest to TX.

Eacles imperialis (Drury) (Fig. 271A). Imperial moth—88-140 mm; dark yellow; numerous purplish gray speckles and irregular wide bands on wings; larva: brown or green, densely clothed with hairs; east of Great Plains, southeastern Canada.

Sphingicampa bicolor (Harris). Honey locust moth—45-68 mm; highly variable coloration; DFW dark gray to light speckled brown, often with pair of white dots; DHW pinkish with lighter outer margin; larva: green, spined, orange stripe on side, very long light-colored spine cluster on thorax, on honey locust, Kentucky coffee tree; eastern U.S. west to NB and TX.

Subfamily Hemileucinae—Larvae with stinging spines.

Automeris io (Fabricius) (Fig. 271B). Io moth—50-76 mm; FW bright yellow (♂) or reddish brown (♀); HW yellow with reddish orange submarginal band and large black eyespot; larva: green or brown, pink and white lateral stripe, many clusters of stinging hairs, on many plants; east of Rocky Mts., southcentral and southeastern Canada.

Coloradia pandora Blake. Pandora moth—50-100 mm; dark gray body; grayish brown wings with small dark spot in discal cell; HW pinkish in far western forms; narrow band beyond FW spot; terminal spines on front tibia at least 1 mm long; larva: brown to yellowish green, short and stout branched spines, on pine needles; Rocky Mts. westward.

Hemileuca maia (Drury) (Fig. 271C). Buck moth—50-73 mm; wings blackish brown with reddish tinge and white median band from anterior to posterior margin; white median wing band on DFW 1/2 width of DHW median band; wings very slightly transparent; larva: brownish black, yellow dots, tubercles with branched stinging spines, on oak, willow; eastern U.S. south to FL, southwest to TX (diurnal). *H. nevadensis* Stretch similar but in western 1/2 U.S. east to IL, differs in having broader wing band and band width equal on DFW and DHW.

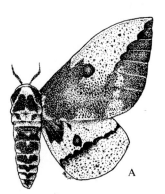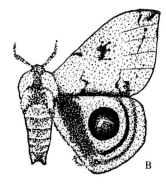

Figure 271 Giant silkworm moths (Saturniidae). A, imperial moth, *Eacles imperialis;* B, io moth, *Automeris io;* C, buck moth, *Hemileuca maia.*

Family Lasiocampidae—Tent Caterpillar Moths and Allies

Lasiocampids are medium-sized, typically brown or gray moths with stout, hairy bodies. The hind wing lacks a frenulum but has humeral veins. The six *Malacosoma* species (tent caterpillars) in the U.S. are gregarious defoliators of trees and *M. disstria* is an important forest pest.

Common Species

Malacosoma americanum (Fabricius) (Fig. 272). Eastern tent caterpillar—25-45 mm; yellowish brown with 2 whitish parallel stripes from anterior to posterior margin on DFW; outer wing margins checkered; larva: black with whitish yellow median stripe, bluish spots and colored lines and markings on sides, slightly hairy, gregarious and spins tentlike silken nest in tree crotch, on apple, cherry, crab apple, others; eggs: hard and elongated silvery-gray mass which surrounds a tree twig; Rocky Mts. eastward, southern Canada (silken nest toward end of branch in summer probably is fall webworm [Arctiidae]).

Malacosoma californicum (Packard). Western tent caterpillar—20-40 mm; highly variable color ranging from straw color to medium brown; 1 band running from anterior to posterior margin on DFW is straw-colored bordered by dark lines on pale specimens to dark brown bordered by light lines on darkest specimens; FW on darker form often lighter than HW; larva: highly variable, usually orange and black, may have white continuous or broken median-dorsal stripe or not, prefers aspen; primarily western U.S., also northeastern U.S., southern Canada.

Malacosoma disstria Hübner. Forest tent caterpillar—20-40 mm; pale to slightly translucent yellow-brown; 2 dark parallel stripes on DFW from anterior to posterior margins, occasionally broad dark band replaces stripes; larva: dark brown with median row of keyhole- or diamond-shaped spots, lateral bluish and reddish brown stripes, spins silken mat (but not tent) on limb or trunk, prefers aspen, birch, maple; throughout U.S. except NV and AZ, southern Canada.

Family Lymantriidae (= Liparidae)—Tussock Moths

Tussock moths are medium-sized, usually brown, gray, or whitish insects. The antennae of males are feathery and some females are wingless or very weak fliers. Females of some species cover their eggs with their abdominal hairs or secretions that harden. The larvae have conspicuous tufts of hairs and some species have irritating hairs. The species listed below are pests of forest and shade trees.

Common Species

Dasychira plagiata (Walker). Pine tussock moth—25-37 mm; gray-brown with light and dark stripes on FW; larva gray-brown, dorsum with 4 tufts, 1st tuft with 2 groups long black hairs toward front and 3 toward rear, on pine needles, especially jack pine; northeastern and Great Lakes states, southeastern Canada.

Lymantria dispar (Linnaeus) (Fig. 273C). Gypsy moth—36-38 mm (♂), 38-50 mm (♀); dark brown wings with blackish bands across FW (♂); dull white wings with black-spotted outer wing margins, zigzag submarginal lines, broad "V" near discal cell DFW (♀); thick abdomen with yellowish brown

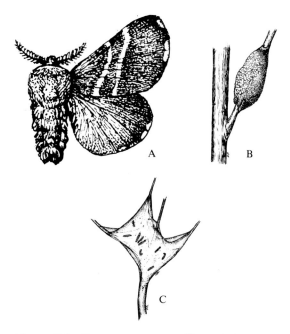

Figure 272 Eastern tent caterpillar, *Malacosoma americanum* (Lasiocampidae). A, adult; B, egg case; C, silken tent.

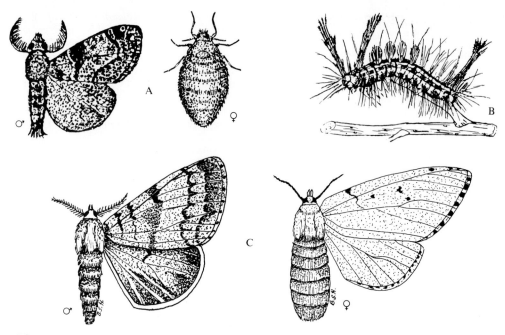

Figure 273 Tussock moths (Lymantriidae). A and B, whitemarked tussock moth, *Orgyia leucostigma:* A, male and wingless female; B, larva. C, gypsy moth, *Lymantria dispar,* male and female.

hairs (♀); larva: double row of 5 pairs blue spots and double row of 6 pairs red spots on dorsum, body hairy, on oaks, other forest trees; primarily northeastern U.S.

Orgyia leucostigma (J.W. Smith) (Figs. 273A, B). Whitemarked tussock moth—29-31 mm; ash gray; FW with dark wavy bands and white spot near anal angle (♂); ♀ wingless, 12 mm long, grayish or light brown, hairy; larva: head and dorsum of thorax reddish, pale yellow abdomen, 2 very long upright tufts of black hairs on prothorax and 1 near tip of abdomen, thick whitish tufts on 1st 4 abdominal segments followed by 2 reddish dots, on deciduous trees; northern 2/3 U.S., southern Canada.

Orgyia vetusta (Boisduval). Western tussock moth—20-25 mm; brown with gray markings (♂); ♀ wingless, 12-15 mm long, gray; larva: gray, many colored spots, 4 median and 1 posterior white tufts on dorsum, 2 very long black tufts on prothorax and 1 near tip of abdomen, on fruit trees, other trees; CA to British Columbia.

Family Notodontidae— Notodontid Moths

Moths in this family typically are brown or brownish yellow and some species have protruding tufts of scales on the hind wing margin that project backward when the wings are folded. Notodontids are similar to the Noctuidae but in the former the cubitus of the front wing appears 3-branched because vein M_2 arises near the middle of the discal cell apex. With noctuids M_2 arises closer to M_3 and thus the cubitus appears 4-branched (Fig. 246A).

Some larvae have prominent tubercles on the dorsum and often the anal prolegs are rudimentary or spinelike. Species of *Datana*, sometimes called the handmaid moths, are gregarious as larvae and when disturbed they characteristically raise both ends of the body so that only the prolegs support them. Species in other genera, often called the prominents, may be solitary or gregarious and feed exposed, within tents (e.g., genus *Ichthyura*), or inside folded leaves.

Common Species

Datana integerrima Grote and Robinson. Walnut caterpillar—40-58 mm; chestnut to dull brown wings; FW darker on anterior 1/2 and crossed by 3 dark, wide bands; larva: black with yellow longitudinal stripes, white hairs, gregarious, on walnut, butternut, hickory, pecan; eastern U.S., southeastern Canada.

Datana ministra (Drury) (Fig. 274A). Yellownecked caterpillar—40-58 mm; FW light brown with irregular darker brown lines from anterior to posterior margin, outer margin scalloped; HW pale grayish yellow; larva: black head, orange-yellow prothorax, yellowish and black longitudinal stripes, long white hairs, gregarious, many deciduous trees; eastern U.S. south to FL, west to CO and NM, southeastern and southcentral Canada. *D. perspicua* Grote and Robinson, the sumac datana, is similar as an adult but 40-50 mm larva has yellow and black stripes, on sumac leaves, throughout most of U.S., southernmost Canada.

Heterocampa manteo (Doubleday). Variable oakleaf caterpillar—37-42 mm; brownish gray wings, DFW mottled, darker triangular patch near apex (♂); ♀ wings ash gray with few markings; larva: yellowish green, broad dark band along dorsum, on oaks, other deciduous trees; eastern U.S. south to FL, west to TX, CO, MN, southeastern Canada.

Ichthyura inclusa Hübner. Poplar tentmaker—22-28 mm; brownish gray; DFW with 3 whitish lines from anterior to posterior margin and 1 oblique line, submarginal row dark dots, and orange-brown patch toward apex; DHW with oblique, wavy, whitish band; thorax with crest of dark hairs; larva: dark with 4 yellow lines on dorsum, black tubercle on 1st and 8th abdominal segment, white hairs, gregarious, makes tents from 1+ leaves of poplar, willow; eastern U.S. west to CO, southeastern Canada.

Schizura concinna (J.E. Smith) (Figs. 274B, C). Redhumped caterpillar—25-35 mm; FW grayish brown with reddish brown shading along posterior margin and wing base, faint curved row of dots in center; HW translucent grayish white with dark spot at anal angle; larva: head and hump on 1st abdominal segment red, body with black and yellowish lines, posterior end held up, gregarious, on deciduous trees; throughout U.S., southern 1/2 Canada.

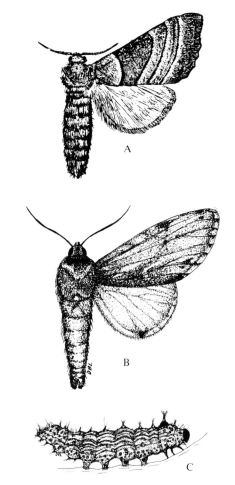

Figure 274 Notodontid moths (Notodontidae). A, yellownecked caterpillar, *Datana ministra;* B and C, redhumped caterpillar, *Schizura concinna*: B, adult; C, larva.

Family Noctuidae—Noctuids or Owlet Moths and Underwings

Noctuids are extremely common, usually medium-sized, dull-colored moths that frequent lights throughout the year. The family is the largest in the order, containing about 2,700 species in the U.S. and Canada. The antennae are typically hairlike (Fig. 236E) and the thorax is tufted on some species. Noctuids are similar to the Notodontidae but the cubitus in the front wing of noctuids appears 4-branched because

Figure 275 Noctuids or owlet moths (Noctuidae). A, bollworm, corn earnworm or tomato fruitworm, *Heliothis zea;* B, fall armyworm, *Spodoptera frugiperda;* C, black cutworm, *Agrotis ipsilon;* D, spotted cutworm, *Amathes c-nigrum.*

vein M_2 originates closer to M_3 than to M_1 (Fig. 246A). This family is well known for its paired auditory organs located at the base of the abdomen. These organs detect the high frequency sounds of bats and enable the moth to evade the predaceous bats which are using echo-location to hunt food.

The larvae are smooth, nearly hairless, mottled or striped, and usually possess five pairs of prolegs (loopers have fewer). These caterpillars are nocturnal and most are foliage feeders. Others bore into stalks or fruit and some live in the soil feeding on roots. Many species are important crop pests.

Common Species

Erebus odora (Linnaeus). Black witch—100-150 mm; medium or dark brown; prominent brownish white, scalloped band from anterior to posterior margin of wings; blackish eyespot on basal 1/3 DFW and 2 dark, oval spots on anal angle DHW; southern U.S., strays into northern U.S. east of Rocky Mts.

Heliothis zea (Boddie) (Fig. 275A). Bollworm, corn earworm, tomato fruitworm—33-45 mm; DFW light brown, pale yellowish brown, or grayish brown with numerous speckles and darker irregular band toward outer margin; HW brownish white with brown band adjacent to outer margin; larva: yellowish, light green, pinkish, brown, or dark brown, alternating dark and light longitudinal stripes, in cotton bolls, ears of corn, tomato fruit; throughout U.S., southeastern Canada. *H. virescens* (Fabricius), tobacco budworm, similar, larva feeds on tobacco and cotton buds in southern U.S.

Armyworms—Larvae commonly migrate in groups at night to new feeding areas; many important pests are in genus *Spodoptera*.

Pseudaletia unipuncta (Haworth). Armyworm—32-45 mm; light brown or grayish with very fine dark spots scattered on wings; small white spot on DFW; larva: greenish to dark gray, 3 yellowish longitudinal stripes on dorsum, broader dark yellow stripe on sides; prefers cereals and corn, on other crops; throughout U.S., southwestern Canada.

Spodoptera exigua (Hübner). Beet armyworm—21-32 mm; DFW mottled grayish brown, zigzag lines, blackish dots on outer margin; HW brownish white, brown submarginal border; larva: olive or pale green dorsally, pale yellowish ventrally, dark dorsal stripe, yellow subdorsal stripe on each side, prefers sugarbeets; most of U.S. except Northwest and Northeast.

Spodoptera frugiperda (J.E. Smith) (Fig. 275B). Fall armyworm—25-38 mm; like *S. exigua* but DFW with rectangular pale marking near upper center and at apex; larva: green, light tan, or dark brown, 3 yellowish white longitudinal lines on dorsum, wider dark stripe laterally with wavy yellowish and pink-blotched stripe below, late summer and fall, on corn, sorghums, grasses, other crops; southern U.S., CA, strays north to southcentral and southeastern Canada.

Spodoptera ornithogalli (Guenée). Yellowstriped armyworm—25-45 mm; FW brownish gray and heavily marked with dark brown, gray, and light tan; DFW with long, oblique light band near center, wide irregular light band near outer margin; HW silvery gray; larva: paired dark blotches on dorsum of each segment, fine yellow stripes laterally and pinkish below stripes, on cotton, alfalfa, other crops; southern U.S., southern CA, strays north to New England.

Cutworms—Larvae soft, fat, may roll up body when disturbed, usually nocturnal, in shallow holes in soil near base of plant or under stones during day; cut off small, young plants at soil surface often leaving most of plant to wilt on ground, some climb to tops of fruit trees to feed on leaves, others chew into fruit or roots; adults very similar in appearance and only a few of many common species listed below.

Agrotis ipsilon (Hufnagel) (Fig. 275C). Black cutworm—38-58 mm; DFW dull brownish or reddish gray with dull silvery patch on outer 1/3 and at base to lesser extent; HW and abdomen silvery gray; larva: dull brown to blackish, broken pale line on dorsum, 2 faint lines laterally, surface shiny and greasy appearing, burrows in soil; throughout U.S., southern Canada.

Agrotis orthogonia Morrison. Pale western cutworm—33-38 mm; DFW mottled brownish gray; larva: greenish gray, unmarked, brown head and prothorax, blackish spiracles, under soil feeding in late spring; adults come to lights Aug.-Sept.; western U.S.

Amathes c-nigrum (Linnaeus) (Fig. 275D). Spotted cutworm—38-45 mm; DFW dark gray at base shading to grayish brown at outer margin, numerous stripes, whitish triangular patch nearly halfway out along anterior margin; HW brownish white; abdomen banded brown and whitish; larva: dorsal elongated and pointed black dashes on each segment of posterior 1/2 body, stripe through spiracles; northern 1/2 U.S., southern 1/2 Canada.

Peridroma saucia (Hübner) (Figs. 276A, B). Variegated cutworm—38-50 mm; DFW grayish brown and mottled, 2 irregular circles near center toward anterior margin; HW whitish brown darkening toward margin; thorax with dorsal "X" marking and tuft posterior to center of "X"; larva: grayish or dull brown mottled dorsally with darker brown, pale yellow middorsal spot on most segments, often blackish "W" near last segment; throughout North America.

Dagger Moths—Genus *Acronicta*; some with daggerlike mark near anal angle of FW.

Acronicta americana (Harris). American dagger moth—50-70 mm; pale yellow-gray with brown dots on wing margins; DFW with 2 light zigzag lines in outer 1/4; thorax pale grayish; abdomen brownish gray; larva: yellowish white body and fine hairs, very long and black hair clusters on abdominal segments 1, 3, and 8, on hardwoods; eastern U.S. west to Great Plains, southeastern and southcentral Canada.

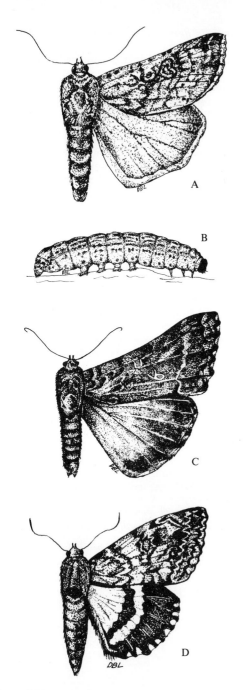

Figure 276 Noctuids or owlet moths (Noctuidae). A and B, variegated cutworm, *Peridroma saucia:* A, adult; B, larva. C, cabbage looper, *Trichoplusia ni;* D, an underwing, *Catocala* sp.

Loopers—Larvae with only 3 pairs prolegs as with loopers in family Geometridae.

Alabama argillacea (Hübner). Cotton leafworm—25-38 mm; DFW medium brown and lightly speckled, dark spot near center; apex FW slightly pointed and curved; spines at tip of proboscis can injure ripe fruit; larva: greenish, dark dorsal stripe, 4 black spots form square on dorsum of each segment, on cotton primarily; southern 1/3 U.S., sporadic adults north to southeastern Canada.

Anagrapha falcifera (Kirby). Celery looper—33-45 mm; DFW purplish brown except for light brown band on outer margin, teardrop-shaped whitish spot in center with whitish line curving to base of posterior margin; larva: like *Trichoplusia ni,* on celery, sugar beets, lettuce, succulent weeds; throughout U.S., southern Canada.

Autographa californica (Speyer). Alfalfa looper—25-45 mm; similar to *Trichoplusia ni* but DFW more gray, silvery white spot in center of wing elongated and angular; larva: dark green, 1 dorsal and 2 lateral dark lines, general feeder; western U.S., southwestern Canada.

Trichoplusia ni (Hübner) (Fig. 276C). Cabbage looper—25-38 mm; DFW brownish and mottled, silvery-white spot resembling number "8" near center; HW pale brown at base darkening to margin; larva: greenish, white lateral line along body, humps middle of body when moving, on cabbage, other vegetables; throughout U.S., southern Canada.

Underwings—ca. 100 species of large moths (ca. 35-85 mm wingspread) in genus *Catocala* (Fig. 276D); HW usually brightly colored with concentric bands of pink, orange, white, or brown, but HW concealed at rest; FW often grayish and camouflaged like tree bark on which insect rests during day; forest inhabitants; collected at lights or with bait (sugaring); species' comparison requires color illustrations such as in Holland (1903) and Sargent (1976).

Family Agaristidae— Forester Moths

This family is classified by some taxonomists as a subfamily of Noctuidae. Members are commonly black with yellowish or whitish wing patches and the antennae are slightly clubbed.

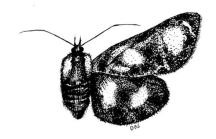

Figure 277 Eightspotted forester, *Alypia octomaculata* (Agaristidae).

Common Species
Alypia octomaculata (Fabricius) (Fig. 277). Eightspotted forester—25-35 mm; black with 2 yellowish patches on each FW and 2 whitish patches on HW; sides of thorax yellowish white; eastern U.S. west into CO (diurnal; larvae on Virginia creeper and grape).

Family Arctiidae—Tiger Moths
These small- to medium-sized moths are generally spotted or banded conspicuously and hold their wings rooflike while at rest. Most larvae have a dense coat of hairs which are shed and mixed with silk when the cocoon is made.

Common Species

Subfamily Lithosiinae—Footman Moths; usually small, slender, often dull-colored; larvae feed on lichens.
Hypoprepia miniata (Kirby). Striped footman moth—30-35 mm; FW bright pink with 3 longitudinal gray stripes; basal 1/2 HW pinkish or yellowish, outer 1/2 grayish black; eastern U.S. west to MN, southeastern and southcentral Canada.

Subfamily Arctiinae—Tiger Moths; most species occur in this group; genus *Apantesis* characterized by white, yellow, or red stripes on black FW, and HW yellowish or pink with black spots (Fig. 278A); larvae in genus *Halisidota* with hair tufts and similar to larvae in genus *Orgyia* of family Lymantriidae.

Apantesis virgo (Linnaeus) (Fig. 278A). Virgin tiger moth—47-75 mm; FW and thorax black with white irregular stripes; HW and abdomen pinkish with black spots; eastern North America west to Rocky Mts., southwest to AZ. *A. arge* (Drury) smaller, HW whitish with brown spots, throughout U.S. (except Northwest), southeastern Canada.

Estigmene acrea (Drury) (Fig. 278B). Saltmarsh caterpillar—45-62 mm; FW whitish with many dark spots; HW white (♀) or yellowish (♂), some spots; thorax white; abdomen yellowish or pinkish; larva: yellowish brown, dense long brown hairs, common in fall; east of Rocky Mts. Larva of yellow woollybear, *Diacrisia virginica* (Fabricius), similar but more whitish with black head, throughout U.S., southern Canada.

Euchaetias egle Drury. Milkweed tiger moth—32-38 mm; wings white; abdomen yellowish brown with median row black dots; larva: many yellowish tufts on abdomen, many long black and white tufts each end, on milkweed in late summer; eastern U.S. south to KY, southeastern Canada.

Halisidota caryae (Harris). Hickory tussock moth—33-58 mm; FW light or yellowish brown with 4-5 crossbands of silvery white spots; HW translucent, yellowish, unspotted; larva: grayish white hairs, black tufts on 1st 8 abdominal segments, 2 pairs long black hair clusters each end, on hickory, walnut, other trees; most of U.S. except Southeast and Northwest, southcentral and southeastern Canada.

Halisidota tessellaris (J.E. Smith) (Fig. 278C). Pale tussock moth—33-50 mm; FW translucent dirty yellow with 4 crossbands of white, translucent rectangular spots; HW translucent yellowish white; larva: dense tufts yellowish to gray hairs, 2 pairs long black hair clusters on thorax and 1 pair on tip of abdomen, on deciduous trees; eastern U.S. south to FL, west to TX and Dakotas.

Hyphantria cunea (Drury). Fall webworm—20-42 mm; FW and HW white with dark spots; front legs orange or red at base; abdomen yellow with black lateral dots; all-white specimens in northern U.S. and Canada; larva: pale yellow or greenish with yellow lateral stripe, long gray hairs in tufts, gregarious larvae in large web toward end of branch, many deciduous trees; throughout U.S., southern 1/2 Canada.

Figure 278 Tiger moths (Arctiidae). A, virgin tiger moth, *Apantesis virgo;* B, saltmarsh caterpillar, *Estigmene acrea;* C, pale tussock moth, *Halisidota tessellaris;* D, banded woollybear, *Isia isabella.*

Isia isabella (J.E. Smith) (Fig. 278D). Banded woollybear—37-50 mm; wings dull yellowish brown, very few spots; abdomen with 3 longitudinal rows dark spots; larva: black at ends, brown in middle, very fuzzy, common in fall; throughout U.S., southern 1/2 Canada.

Family Ctenuchidae— Ctenuchas and Wasp Moths

These moths are diurnal and some are wasplike in appearance due to their slender bodies, narrow and clear or partly clear wings, and rapid flight. The genus *Ctenucha* consists of moderate-sized blackish, blue-black, or dark brown moths, usually with orange or yellow heads. The family is often classified as Amatidae or occasionally as a subfamily of Arctiidae.

Common Species

Ctenucha virginica (Charpentier). Virginia ctenucha—35-42 mm; wings brownish black; head orange or yellow; body metallic blue-black; northeastern U.S.

Lycomorpha pholus (Drury). Lichen moth—25-28 mm; wings narrow; basal 1/2 FW yellow-orange, outer 1/2 blackish; basal 1/3 HW yellowish, outer 2/3 blackish; body blackish; throughout U.S., southern Canada.

Scepsis fulvicollis (Hübner) (Fig. 279A). Yellow-collared scape moth—25-32 mm; brownish black wings; FW central area lighter; HW central area nearly translucent; prothorax dark yellow; eastern U.S. west to Rocky Mts., southwest to AZ.

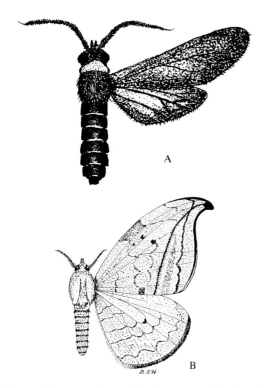

Figure 279 A, yellowcollared scape moth, *Scepsis fulvicollis* (Ctenuchidae); B, a hooktip moth, *Drepana arcuata* (Drepanidae).

Family Drepanidae— Hooktip Moths

The apex of each front wing is usually curved in this family. *Drepana arcuata* Walker (Fig. 279B) is a very pale brownish species with dark brown markings, 28-33 mm wingspread, and occurs in the Atlantic coast states.

Family Geometridae— Geometrid Moths

This large family (over 1,200 species) consists of slender-bodied, often small and delicate moths. The wings are relatively broad, sometimes greenish, and may have wavy lines. The front wings have a cubitus that appears to be 3-branched and in the hind wing the subcosta abruptly bends downward at its base (Fig. 244A). The females of some species are wingless.

Larvae are called inchworms, measuring worms, spanworms, and loopers. The slender caterpillar lacks prolegs in the middle of the body (Fig. 280C); it moves by drawing its posterior end up to the thoracic legs forming a loop, and then extending the body forward. A larva may extend its body outward to resemble a twig for concealment. Many species feed on deciduous trees, shrubs, and conifers, and some attack all three. There are numerous common species, a few of which are described below.

Common Species

Alsophila pometaria (Harris) (Fig. 280A). Fall cankerworm—22-35 mm (σ); ♀ wingless, 10-13 mm body length; pale brown; DFW with 2 whitish jagged cross-stripes, whitish spot on anterior margin interior and adjacent to upper end of outer cross-stripe; larva: pale green to brownish green, broad middorsal brown stripe, 3 pairs prolegs, on apple, elm, other trees; eastern U.S. south to TN, northwest to MT, southcentral and southeastern Canada (adults in fall).

Cingilla catenaria (Drury) (Fig. 280B). Chainspotted geometer—28-47 mm; grayish white wings with row of spots on outer margins and a parallel row linked in chain fashion on outer 1/3 of wing; head and part of thorax orange-yellow; body white; larva: straw-colored with black spots, spots on sides chainlike; northern 1/2 U.S. excluding Far West, southern Canada.

Ennomos magnarius Guenée. Notchwing geometer—35-50 mm; outer margin of wings with many notches giving tattered appearance; wings reddish yellow shading to brown toward outer margin, many small brown spots; northern 1/2 U.S., southern Canada.

Ennomos subsignarius (Hübner). Elm spanworm—28-37 mm; powdery white; larva: dull black with reddish brown head, on many hardwoods; eastern U.S., southeastern Canada.

Haematopis grataria (Fabricius). Chickweed geometer—20-25 mm; reddish yellow; wing margin and 2 crossbands pink; larva on chickweed; eastern U.S. southwest to TX.

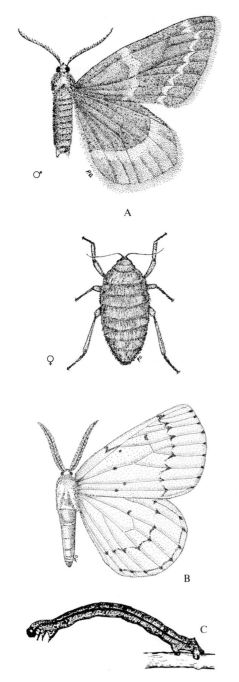

Paleacrita vernata (Peck). Spring cankerworm—21-32 mm (♂); ♀ wingless, 10-13 mm body length; similar to fall cankerworm but DFW without whitish spot on anterior margin; ♀ with median dorsal stripe and 2 transverse rows reddish spines on each abdominal segment; larva: yellowish brown or yellowish green, striped, 2 pairs prolegs, on apple, elm, other trees; eastern U.S. southwest to TX, also CO, CA (adults in spring).

Sabulodes caberata Guenée. Omnivorous looper—40-42 mm; dull yellow or brown; FW with oblique crossband from apex to middle of posterior margin; FW incurved at apex; HW with faint oblique crossband; larva: yellow, brown, or green stripes, webs leaves together, on many plants; CA.

Family Incurvariidae

The three subfamilies in this group are sometimes classified as families. The subfamily Prodoxinae (or family Prodoxidae) contains four species of yucca moths (genus *Tegeticula*) that have a symbiotic relationship with yucca plants. The white female moth collects yucca pollen, flies to another yucca flower and oviposits into its ovary, and then rubs the pollen on the flower's stigma. With the necessary cross-pollination completed, the seeds develop with some serving as food for the moth larvae. Larvae of the bogus yucca moths (genus *Prodoxus*) feed on yucca and agave fruit and stems, but do not pollinate these plants. *Adela* moths have very long antennae.

Common Species

Paraclemensia acerifoliella (Fitch). Maple leafcutter—8-13 mm; head orange-yellow; FW steel blue with black fringes; HW translucent pale brown and fringed; thorax steel blue; legs whitish; abdomen brown; larva: whitish, longitudinal stripe, carries oval case of leaves giving turtlelike appearance; on sugar and red maples, beech, birch, elm; eastern U.S. south to VA and west to IL, southeastern Canada.

Figure 280 Geometrid moths (Geometridae). A, fall cankerworm, *Alsophila pometaria,* male and female; B, chainspotted geometer, *Cingilla catenaria*; C, a geometrid larva.

Family Tineidae—Clothes Moths or Tineid Moths

The moths of this moderately large family are generally small, narrow-winged, and tan, brown, or grayish. The maxillary palpi are usually large and folded and the labial palpi are typically short (large in genus *Acrolophus*). Most larvae are scavengers or feed on fungi but three species are of economic concern because the larvae attack wool clothing, fur, hair, skins, feathers, silk, and other animal products. The webbing and casemaking clothes moths are the main pests, and the carpet moth is occasionally damaging.

Common Species

Acrolophus popeanellus (Clemens)—25-38 mm; brown with some blackish spots; long palpi arch over head and thorax; throughout U.S.

Tinea pellionella (Linnaeus) (Figs. 281A, B). Casemaking clothes moth—10-16 mm; FW pale brown or grayish with 3 dark spots; HW straw-colored; wings fringed; larva: whitish, builds elongated silken case open at both ends; throughout North America.

Tineola bisselliella (Hummel). Webbing clothes moth—10-16 mm; straw-colored; wings fringed and unspotted; larva: straw or pale greenish colored, no case; throughout North America.

Trichophaga tapetzella (Linnaeus). Carpet moth—12-24 mm; basal 1/3 FW blackish, outer 2/3 grayish white; larva: forms long silken galleries in carpet, upholstery, blankets (primarily eastern 1/2 U.S.).

Family Psychidae—Bagworm Moths

The larvae of this family construct a large silken bag with bits of leaves and twigs. The bags enclose the larvae and are carried while they feed on leaves and are eventually fastened to twigs prior to pupation. Adult males are winged and leave the bag. The wingless and legless adult females resemble the larvae, usually remain in the bag, and mate and lay eggs inside the bag.

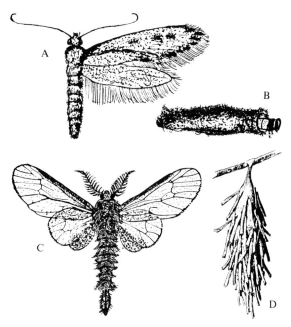

Figure 281 A and B, casemaking clothes moth, *Tinea pellionella* (Tineidae): A, adult; B, larva and case. C and D, bagworm, *Thyridopteryx ephemeraeformis* (Psychidae): C, adult male; D, bag of larva.

Common Species

Thyridopteryx ephemeraeformis (Haworth) (Figs. 281C, D). Bagworm—20-28 mm (♂); ♂: body blackish, thick, and hairy, antennae feathery, wings with large clear areas; ♀ yellowish white; larva: dark brown, on arborvitae, red cedar, other conifers, deciduous trees; eastern U.S. south to FL, southwest to TX.

Family Lyonetiidae

These tiny moths have very narrow, broadly fringed wings and many larvae first mine and later skeletonize tree leaves. Most species are in the genus *Bucculatrix* and the minute, longitudinally ribbed cocoons are often common on trees and shrubs (Fig. 282B). Larvae of *B. thurberiella* Busck, the cotton leafperforator, eat small holes in cotton leaves in Arizona and California. *B. albertiella* Busck larvae mine and skeletonize live oak in California and form white, ribbed cocoons on their trunks.

Common Species

Bucculatrix ainsliella Murtfeldt. Oak skeletonizer—7-8 mm; silvery white sheen; long tuft of hairs on head; FW with dark costal band curving and widening downward toward outer margin; dark semicircular patch on posterior wing margin; larva mines oak leaves; cocoon: 3 mm, white, elongated and ridged, on trunks and branches of trees fall to spring; eastern U.S., southern Canada.

Bucculatrix canadensisella Chambers (Figs. 282A, B). Birch skeletonizer—8-9 mm; body brown and white; FW with diagonal white bars; larva: forms wavy mines in leaf, later spins small web on underside of leaf and feeds externally as skeletonizer, on birch; eastern U.S., southern Canada.

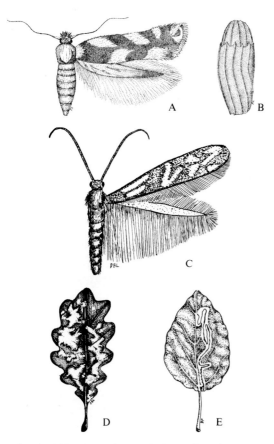

Figure 282 A and B, birch skeletonizer, *Bucculatrix canadensisella* (Lyonetiidae): A, adult; B, cocoon. C-E, leafblotch miners (Gracillariidae): C, *Lithocolletis hamadryadella*; D, *L. hamadryadella* larval mining on oak; E, *Phyllocnistis populiella* larval mining on poplar.

Family Gracillariidae—Leafblotch Miners

This is the largest leaf-mining family of moths (over 200 species). The minute adults have very narrow wings and often are covered by silver- or gold-colored scales. At rest, the front of the body is raised and the wing tips touch the surface on which the moth rests. The mines of larvae are typically blotchy, although some are winding, and more mature larvae may fold over leaves and skeletonize them. *Lithocolletis* species are common miners of deciduous trees; *L. hamadryadella* (Clemens) forms blotches on oaks (Fig. 282D) in eastern U.S. and southeastern Canada. Adults have a 6 mm wingspread, bronze-banded front wings, and silvery hind wings (Fig. 282C). Nearly square mines in basswood leaves are produced by *L. lucetiella* Clemens and *L. hamameliella* Busck mines the leaves of witch hazel. *Phyllocnistis liquidambarisella* Chambers mines sweet gum leaves, and *P. populiella* Chambers forms winding mines on one side of the midrib of trembling aspen and other poplars. Feces often form a median dark line in the mine (Fig. 282E). *Gracillaria negundella* Chambers, the boxelder leafroller, mines the leaves of boxelder and later vacates the mine to roll the leaf tip under and form a shelter for feeding. *G. sassafrasella* Chambers mines sassafras leaves but later abandons the mine to turn the tips under for further feeding. *G. syringella* (Fabricius), the lilac leafminer, feeds inside leaves of lilac, privet, euonymus, and ash. *Marmara pomonella* Busck, the apple fruitminer, produces a white, winding mine on the skin of apples grown in the Far West.

Family Coleophoridae—Casebearer Moths

Adults are plain moths with sharply pointed wings, an oblique discal cell in the front wing, and a wingspread of 12 mm or less. The larvae begin as leaf miners, construct a case as second instars, and either continue as miners or feed externally. All species in the U.S. are in the genus *Coleophora*. *C. fuscedinella* Zeller, the birch casebearer, attacks birch in northeastern U.S. and southeastern Canada. *C. laricella* (Hübner), the larch casebearer, defoliates larch in eastern North America. *C. laticornella* Clemens, the pecan cigar casebearer, is enclosed in a 6 mm long brown, cigarlike case. The larvae feed on pecan, walnut, and hickory from New England to Florida and west to Texas. *C. malivorella* Riley, the pistol casebearer, forms a case curved at the posterior end and carries it in an upright position while feeding on leaves of apple and other fruit trees in the eastern U.S. (Fig. 283A). *C. serratella* (Linnaeus), the cigar casebearer, has an 8 mm long cigarlike case and feeds on apple leaves in the northern U.S. and southern Canada (Fig. 283B).

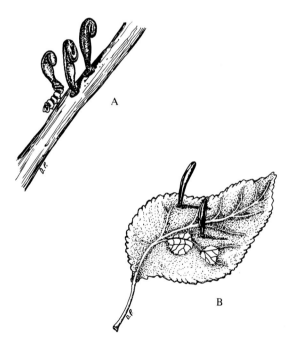

Figure 283 Casebearer moths (Coleophoridae). A, pistol casebearer, *Coleophora malivorella*; B, cigar casebearer, *C. serratella*.

Family Yponomeutidae— Ermine Moths

The front wings of these small moths are often brightly patterned. Species in the genus *Yponomeuta* have white wings with black dots. Several *Argyresthia* species feed on leaves of cedar and cyprus. Adults have a wingspread of about 7-12 mm, fold their wings close to the body, extend their front legs forward, and slant their hind legs upward while at rest.

Common Species

Atteva punctella (Cramer) (Fig. 284A). Ailanthus webworm—25-30 mm; FW yellow crossed with 4 grayish blue bands each surrounding a row of yellow spots; larva spins a silken web on ailanthus leaves; southern U.S., ranges north to NY and MI.

Family Sesiidae—Clearwing Moths

Many moths in this family resemble bees and wasps. One or both pairs of wings are transparent over much of their area, the front wings are narrow, the body is slender and often long, and the moths fly during the day. Larvae bore into most parts of plants and many are serious pests.

Common Species

Melittia satyriniformis (Hübner) (Fig. 284B). Squash vine borer—21-35 mm; FW, thorax, and 1st abdominal segment dark metallic green; HW clear; hind legs heavily covered with orange-red hairs; abdomen orange; larva in squash stems; most of North America except Pacific coast states.

Pennisetia marginata (Harris). Raspberry crown borer—20-35 mm; head brown with yellow posterior collar; thorax dark brown, yellow markings; wings transparent with brown borders; legs yellow and black; abdomen dark brown with golden color on posterior part of each segment, tip with yellow and black tufts (♂) or tip all yellow (♀); larva in canes of raspberries, blackberries; most of North America.

Podosesia syringiae (Harris). Ash borer, lilac borer—26-38 mm; blackish or reddish brown; FW opaque with long clear streaks between veins, fringe dark brown; HW clear, yellowish brown veins, fringe darkens toward wing base; larva in ash, lilac, privet, other trees and shrubs; eastern 1/2 U.S. and Canada.

Synanthedon exitiosa (Say). Peachtree borer—18-32 mm; ♂ : FW and HW transparent with black margins, blue-black abdomen with narrow yellow crossbands; ♀ : FW dark brown, HW clear, blue-black abdomen with orange-red crossband in middle; larva in peach tree trunks below ground, other stone fruit trees; throughout U.S., southern Canada.

Synanthedon scitula (Harris). Dogwood borer—14-20 mm; blue-black; wings clear with blue-black margin; legs yellow-banded; abdominal segments 2-4 yellow-striped; larva in dogwood, pecan, oaks, other trees; eastern 1/2 U.S., southeastern Canada.

Family Gelechiidae— Gelechiid Moths

The nearly 600 North American species are small moths with long and upcurved labial palpi and narrow hind wings usually with a curved outer margin (Fig. 248B). Larvae are leaf miners, leaf folders, gall formers, and internal feeders of buds, seeds, and roots. Members of the genus *Gnorimoschema* cause elongated, spindle-shaped galls on goldenrod stems (round galls are caused by a fly in the family Tephritidae).

Common Species

Anarsia lineatella Zeller. Peach twig borer—15-18 mm; gray with white and dark scales; 2 black spots on FW costa; larva in tips of peach twigs; throughout U.S., southern Canada.

Dichomeris marginella (Denis and Schiffermüller). Juniper webworm—14-16 mm; brownish or pinkish brown; FW with white anterior and posterior margins; larvae web together juniper needles; most of U.S., southern Canada.

Pectinophora gossypiella (Saunders). Pink bollworm—15-20 mm; dark brown or grayish brown; FW with several poorly defined spots, blackish tip; wings fringed; legs ringed with black; larva: pinkish with darker crossbands, yellow head, in cotton buds and bolls; South, Southwest, and CA.

Phthorimaea operculella (Zeller). Potato tuberworm—12-16 mm; head cream white; body and wings powdery gray or silvery; pale yellowish brown regions sometimes between wing veins; tiny dark specks on FW; larva in potato tuber, stem, and leaves, other solanaceous plants; chiefly southern 1/2 U.S., strays northward.

Sitotroga cerealella (Olivier) (Fig. 284C). Angoumois grain moth—8-15 mm; pale yellow brown or pale grayish brown; FW sometimes with some darker markings; larva: white, yellow head, in corn, wheat, other grains in fields (fall) or storage (all year); throughout North America.

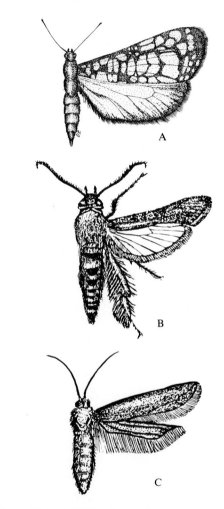

Figure 284 A, Ailanthus webworm, *Atteva punctella* (Yponomeutidae); B, squash vine borer, *Melittia satyriniformis* (Sesiidae); C, Angoumois grain moth, *Sitotroga cerealella* (Gelechiidae).

Family Cossidae—
Carpenterworm Moths

Adults resemble sphinx moths in body size and wing shape. The larvae bore into trees.

Common Species

Prionoxystus robiniae (Peck) (Fig. 285). Carpenterworm—50-85 mm; mottled gray-brown body and wings (♀) or FW mottled dark gray and HW dark yellow or orange-red on outer 1/2 (♂); HW much smaller than FW; ♂ ca. 2/3 size of ♀; larva bores into deciduous trees; throughout U.S. except FL to southern TX, southernmost Canada.

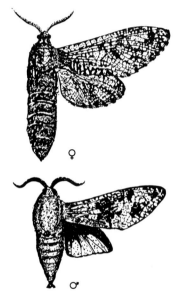

Figure 285 Carpenterworm, *Prionoxystus robiniae* (Cossidae).

Family Tortricidae—
Leafrolling Moths

These small moths have mottled or banded front wings that are somewhat squared-off at the tip (Fig. 286B) and held rooflike over the body. When the wings are folded, the whole insect often has a bell-shaped appearance (Fig. 286A). Many larvae are leaf rollers, miners, tyers, or fruit borers, and are important pests. The spruce budworm of eastern North America, *Choristoneura fumiferana* (Clemens), and the western spruce budworm of western North America, *C. occidentalis* Freeman, are serious defoliators of spruce, fir, and other conifers.

Members in the Olethreutinae subfamily often have a fringe of long hairs along the basal part of the cubitus in the hind wing. There are over 700 species in this subfamily in North America and many are important pests of forest and fruit trees. This group is often classified as the family Olethreutidae. The "Mexican jumping beans" which are sold as novelties contain the larvae of *Laspeyresia saltitans* (Westwood); after eating the seed's contents, the larva throws itself against the thin wall causing the seed to jump.

Common Species

Archips argyrospilus (Walker). Fruittree leafroller—16-25 mm; pale yellow, reddish brown, or orange-red; FW mottled, white spot near middle of anterior margin and other spots in places; HW dark with dirty white fringes; larva: light green, rolls leaves with webbing, on apple and other fruit trees, shade trees; throughout U.S., southern Canada.

Argyrotaenia velutinana (Walker) (Fig. 286A). Redbanded leafroller—12-16 mm; brownish with broad reddish band on FW; larva on apple trees, other deciduous trees, conifers; eastern U.S. west to IA and TX, southeastern and southcentral Canada.

Choristoneura rosaceana (Harris). Obliquebanded leafroller—18-20 mm; reddish brown; FW with 3 dark brown oblique bands and many fine irregular dark lines; larva on deciduous trees, conifers, shrubs; most of U.S., southern Canada.

Cydia pomonella (Linnaeus) (Fig. 286B). Codling moth—15-22 mm; FW gray or brownish with dark crosslines and a large coppery patch near apex; HW pale brown with fringed border; larva: light with dark head, most common larva in apple and pear fruit, English walnut; throughout apple growing area of U.S. and Canada (recently changed from genus *Laspeyresia*).

Melissopus latiferreanus (Walsingham). Filbertworm—12-15 mm; pale or dull bronze or brownish orange; 2 coppery areas at tip of FW; most of U.S.

Rhyacionia frustrana (Comstock). Nantucket pine tip moth—9-15 mm; gray body; FW with irregular brick-red and blackish brown patches separated by bands of gray scales; larva on most pines; east of Rocky Mts.

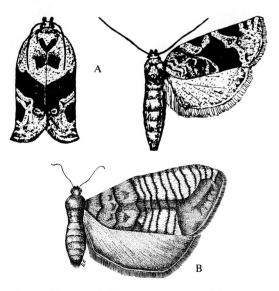

Figure 286 Leafrolling moths (Tortricidae). A, redbanded leafroller, *Argyrotaenia velutinana*; B, codling moth, *Cydia pomonella*.

Family Pyralidae—Pyralid Moths or Snout and Grass Moths

Pyralids in the U.S. are small- to medium-sized moths that are often dull-colored but may also exhibit bright colors. All have a pair of auditory organs (tympana) at the base of the abdomen on the ventral side and some have long, extended labial palpi. Veins Sc and Rs of the hind wing are closely parallel beyond the discal cell and fork near the margin giving a stalked appearance (Fig. 242A). The family is very large (ca. 1,300 species) and only a few of the subfamilies are mentioned below.

Most larvae conceal themselves when feeding. They are leaf miners, stem borers of terrestrial and aquatic plants, scavengers, carnivores, stored grain feeders, or inhabitors of bee and wasp nests. Some species are fully aquatic as larvae and in one case the adult, wingless female remains underwater except for emerging briefly to mate.

Common Species

Subfamily Scopariinae (Fig. 287A)—Triangular FW mottled gray to brown (resembles FW of Noctuidae) with tufts of scales; HW pale and translucent; sometimes in large numbers on tree trunks during day; cool or forested regions; common genera are *Scoparia* and *Eudonia*.

Subfamily Nymphulinae—Often pale; usually distinct and complex pattern of transverse wing bands; some with raised metallic spots on outer margin of HW; common genera have (1) aquatic larvae that live in standing water, make leaf cases, and feed on leaves or (2) live on rocks in rapidly flowing streams, spin webs, and feed on algae. *Nymphula ekthlipsis* (Grote) (adult, Fig. 287B) larva forms leaf case in slow rivers; *Parapoynx badiusalis* (Walker) larva forms leaf case on pondweed in lakes and slow streams; *Synclita obliteralis* (Walker) larva forms leaf case on waterlilies, duckweed, pondweed, aquatic plants in greenhouses; all 3 spp. occur east of Rocky Mts.

Subfamily Glaphyriinae
Hellula rogatalis (Hulst). Cabbage webworm—16-21 mm; FW mottled light to dark brown with ca. 3 wavy white lines from anterior to posterior margin; DFW with kidney-shaped bluish black spot on outer edge discal cell; HW whitish, light brown at margins, white marginal fringe; larva: grayish yellow, 5 purplish longitudinal stripes, body tapers at each end, spins webbing on cabbage, radish, beet; southern U.S. west to southern CA.

Subfamily Evergestinae
Evergestis rimosalis (Guenée). Cross-striped cabbageworm—22-31 mm; FW straw-colored, mottled with olive- or violet-brown, clear-yellowish apical patch divided by dark vein, outer margin yellowish and dark-checkered; HW translucent, pale straw color, outer margin yellowish with brown submargin; larva: bluish gray, black tubercles, yellow longitudinal band over spiracles, on cabbage, other crucifers; southern 2/3 U.S., WA.

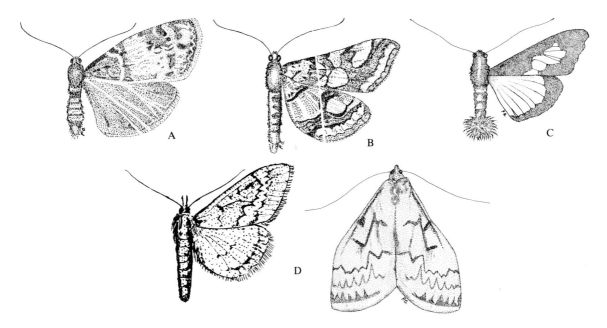

Figure 287 Pyralid moths or snout moths (Pyralidae). A, *Scoparia* sp.; B, *Nymphula ekthlipsis;* C, pickleworm, *Diaphania nitidalis;* D, European corn borer, *Ostrinia nubilalis*.

Subfamily Pyraustinae

Achyra rantalis (Guenée). Garden webworm—20-21 mm; yellow to yellow-orange wings with grayish markings; larva: pale green or dark yellow, spotted, forms webs on leaves, general feeder; most of U.S., southern Canada.

Desmia funeralis (Hübner). Grape leaffolder—22-24 mm; wings brownish black; silvery iridescence; white margins; FW with 2 oval white spots; HW with 1 (♂) or 2 (♀) white spots; larva rolls part of grape leaf into pencil-sized cylinder; eastern U.S. west to Great Plains, CA, British Columbia, eastern Canada.

Diaphania hyalinata (Linnaeus). Melonworm—43-46 mm; wings shiny white with brownish black band around anterior and outer margins; prothorax dark, remaining thorax and abdomen white; tip of abdomen with large dark tuft of long scales; larva: greenish with 2 longitudinal white stripes, on foliage of cucumbers, cantaloupe, squash, pumpkin; eastern U.S. south to FL, southeastern Canada.

Diaphania nitidalis (Stoll) (Fig. 287C). Pickleworm—25-29 mm; body yellowish brown; wings with wide yellowish brown margin; FW with large translucent yellowish white patch in center; HW with basal 2/3 translucent yellowish white; tip of abdomen with large tuft of long scales; larva: greenish white, bores into cucumber, cantaloupe, squash; eastern U.S. south to FL, west to TX and NB.

Loxostege cerealis (Zeller). Alfalfa webworm—similar to *Achyra rantalis;* row of spots near apical margin VHW, primarily western U.S. *L. sticticalis* (Linnaeus), the beet webworm, is 25 mm, purplish brown mottled with dark and light markings, VHW with dark line near outer margin; larva on sugar beets, other plants; primarily western U.S.

Ostrinia nubilalis (Hübner) (Fig. 287D). European corn borer—25-27 mm; pale yellowish brown with irregular wavy lines and bands (♀), or darker yellowish brown with greenish brown markings; larva: light pinkish brown with small brown spots, bores in corn stalks and ears, other crops; east of Rocky Mts., southeastern Canada.

Subfamily Pyralinae

Pyralis farinalis (Linnaeus) (Fig. 288A). Meal moth—25 mm; FW light brown in middle bordered by white lines, dark brown at base and outer margin; HW gray with 2 whitish irregular lines; larva: grayish, in silken tube, infests cereals, flour, and meal of granaries, warehouses, homes; throughout North America.

Order Lepidoptera   311

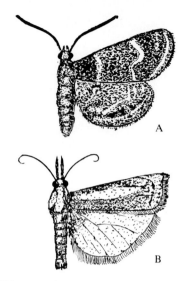

Figure 288 Pyralid moths or snout moths and grass moths (Pyralidae). A, meal moth, *Pyralis farinalis;* B, bluegrass webworm, *Crambus teterrellus.*

Subfamily Crambinae—Grass Moths (Fig. 288B); most species are in genus *Crambus* and are pale yellowish brown or whitish, 12-25 mm, wings held close to body, long labial palpi form snout, adults common in lawns (e.g., *C. teterrellus* [Zincken], the bluegrass webworm [Fig. 288B]; larvae: some called sod webworms, construct webs and bore into roots, crowns and stems of grasses; throughout U.S., southern Canada.

Diatraea saccharalis (Fabricius). Sugarcane borer—25-27 mm; straw-colored; FW with black dots in V-shape; larva: yellowish white with brown spots, bores in sugarcane, corn, sorghums, rice; Gulf coast states.

Subfamily Galleriinae

Galleria mellonella (Linnaeus). Greater wax moth—28-30 mm; wings pale brown or gray with black markings; wing tips dull dark gray; FW scalloped (♂); larva feeds on wax comb of bee hives; throughout U.S., southern Canada.

Subfamily Phycitinae—FW narrow and HW broad; *Cactoblastis cactorum* (Berg), the cactus moth, from Argentina introduced into Australia, larvae bore into prickly pear cactus and help control this weed of grazing lands.

Anagasta kuehniella (Zeller). Mediterranean flour moth—24-26 mm; dark gray; FW with black zigzag markings, sparsely speckled; HW dull white, margins darker; larva: in silken tube, infests cereal, flour, meal in warehouses, granaries, homes; throughout North America.

Dioryctria zimmermani (Grote). Zimmerman pine moth—25-37 mm; gray; FW mottled with gray shades and red zigzag lines; larva: pink or greenish, small black dots, bores into pines; most of U.S., southern Canada.

Plodia interpunctella (Hübner). Indian meal moth—20 mm; pale gray, apical 2/3 FW dark brown or coppery; larva: variable colors, spins silken web over cereals, meal, nuts, dried fruit; throughout North America.

Family Pterophoridae— Plume Moths

The wings of these slender-bodied, long-legged moths are split into plumelike divisions. The front wing usually has two divisions and the hind wing three. The wings are folded close together and held horizontally at right angles to the body while at rest. Larvae are stem borers and leaf rollers.

Common Species

Adaina ambrosiae (Murtfeldt). Ragweed plume moth—14-20 mm; FW white with black spotting, 1st division with black basal area and brown streak through middle; HW gray; larva on ragweed; throughout U.S.

Platyptilia carduidactyla (Riley). Artichoke plume moth—20-27 mm; straw-colored to brown; FW with light spot on anterior margin near apex; HW gray; larva: yellowish with black head, thorax, and legs, on artichoke, thistles; most of U.S., southern Canada.

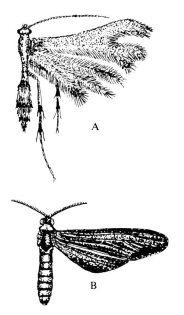

Figure 289 A, grape plume moth, *Pterophorus periscelidactylus* (Pterophoridae); B, grapeleaf skeletonizer, *Harrisina americana* (Zygaenidae).

Pterophorus periscelidactylus Fitch (Fig. 289A). Grape plume moth—17-20 mm; FW yellowish brown with whitish linear markings; HW with 1st 2 divisions dark brown, 3rd division whitish with dark apex; larva on grape leaves; eastern U.S.

Family Zygaenidae— Leaf Skeletonizer Moths

These small moths are gray or black and often have a red, orange, or yellow prothorax and other bright markings. Larvae skeletonize leaves of grape and Virginia creeper. The family is sometimes classified as Pyromorphidae.

Common Species

Harrisina americana (Gúerin-Ménéville) (Fig. 289B). Grapeleaf skeletonizer—18-25 mm; iridescent bluish or greenish black; yellow or orange prothorax; larva: whitish with black hair tufts, gregarious, on grapes; eastern 1/2 U.S., southeastern Canada.

Family Megalopygidae— Flannel Moths

The flannel or woolly texture of the wings in this family derives from the fine, crinkly hairs that are mixed with a dense covering of scales. Adults are brown and usually small- to medium-sized. Larvae are sluglike, have seven pairs of prolegs, and possess stinging hairs.

Common Species

Megalopyge crispata (Packard) (Fig. 290A). Crinkled flannel moth—26-28 mm; cream-colored; FW with black wavy lines in basal 2/3, brownish spots or bands, brownish curled hair; larva: body oval, long brown hairs form a median ridge then slope off rooflike on sides, on oak, elm, apple, blackberry, raspberry; eastern U.S.

Megalopyge opercularis (J.E. Smith) (Fig. 290B). Puss caterpillar—24-26 mm; yellowish brown; wings with brownish spots and white wavy hairs; larva: very dense covering of extra-long yellow hairs and gray or reddish brown hairs, hairs at posterior end taillike, on deciduous trees and shrubs; southern U.S. west to TX.

Family Limacodidae— Slug Caterpillar Moths

Adults are small- or medium-sized, stout-bodied moths with a dense covering of wing scales. The moths are generally yellowish or brownish and some are marked with spots of other colors. Larvae are sluglike, moving with a creeping motion due to the short thoracic legs and lack of prolegs. The caterpillars may be naked or bear hairs or stinging spines.

Common Species

Sibine stimulea (Clemens) (Fig. 290C). Saddleback caterpillar—20-38 mm; velvety reddish brown; FW with 1 white dot near apex and base; larva: green dorsal patch covers most of back, large oval brown spot in center, conspicuous stinging spines, on cherry, oak, pawpaw, rose, corn; southeastern U.S. northeast to New England.

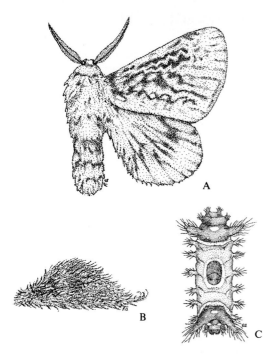

Figure 290 A and B, flannel moths (Megalopygidae). A, crinkled flannel moth, *Megalopyge crispata;* B, puss caterpillar, *M. opercularis.* C, saddleback caterpillar, *Sibine stimulea* (Limacodidae).

Family Hepialidae— Hepialid Moths

This small, primitive family contains medium- to large-sized moths, some resembling sphinx moths, that are generally gray or brown with silvery white wing spots. The jugum in the front wing is slender and fingerlike. Larvae bore into roots and stems. The smaller species (25-50 mm wingspread) are in the genus *Hepialus* and the larger species belong to the genus *Sthenopis.* None of the species is common and only a few *Hepialus* species come to lights.

GENERAL REFERENCES

Baker, W.L. 1972. Eastern Forest Insects. U.S. Dept. Agric. For. Serv. Misc. Publ. 1175. 642 pp.

Chu, H.F. 1949. *How to Know the Immature Insects.* Wm. C. Brown Company Publishers, Dubuque, Iowa. 234 pp.

Collins, M.M., and Weast, R.D. 1961. *Wild Silk Moths of the United States: Saturniinae.* Entomological Reprint Specialists, Los Angeles, Calif. 138 pp.

Ebner, J.A. 1970. *The Butterflies of Wisconsin.* Popular Sci. Hnbk., Milwaukee Pub. Museum, Milwaukee, Wis. 205 pp.

Ehrlich, P.R., and Ehrlich, A.H. 1961. *How to Know the Butterflies.* Wm. C. Brown Company Publishers, Dubuque, Iowa. 262 pp.

Emmel, T.C. 1975. *Butterflies: Their World, Their Life Cycle, Their Behavior.* Alfred A. Knopf, N.Y. 260 pp.

Emmel, T.C., and Emmel, J.F. 1973. *The Butterflies of Southern California.* Natural History Museum Los Angeles Co., Sci. Series, Los Angeles, Calif. 148 pp. (also useful for western states).

Field, W.D.; dos Passos, C.F.; and Masters, J.H. 1974. A Bibliography of the Catalogs, Lists, Faunal and Other Papers on the Butterflies of North America North of Mexico Arranged by State and Province (Lepidoptera:Rhopalocera). Smiths. Contr. Zool. 157. 104 pp.

Forbes, W.T. 1923-1960. Lepidoptera of New York and Neighboring States. Mem. Cornell Univ. Agric. Expt. Sta. Part I. Primitive Forms, Microlepidoptera; Mem. 68 (1923), 729 pp. Part II. Geometridae, Sphingidae, Notodontidae, Lymantriidae; Mem. 274 (1948), 263 pp. Part III. Noctuidae; Mem. 329 (1954), 433 pp. Part IV. Agaristidae Through Nymphalidae Including Butterflies; Mem. 371 (1960), 188 pp.

Harris, L., Jr. 1972. *Butterflies of Georgia.* Univ. Oklahoma Press, Norman, Okla. 326 pp.

Holland, W.J. 1903. *The Moth Book.* Dover, N.Y. 479 pp. (1968 reprint).

———. 1931. *The Butterfly Book.* (rev. ed.). Doubleday, Garden City, N.Y. 424 pp.

Howe, W.H. 1975. *The Butterflies of North America.* Doubleday, Garden City, N.Y. 633 pp. (most completely illustrated single book).

Kimball, C.P. 1965. *The Lepidoptera of Florida.* Florida Dept. Agric. 363 pp.

Klots, A.B. 1951. *A Field Guide to the Butterflies of North America East of the Great Plains.* Houghton Mifflin, Boston. 349 pp.

———. 1958. *The World of Butterflies and Moths.* McGraw-Hill, N.Y. 207 pp.

Lewis, H.L. 1973. *Butterflies of the World.* Follett, Chicago. 312 pp.

Mitchell, R.T., and Zim, H.S. 1964. *Butterflies and Moths.* Golden Press, N.Y. 160 pp. (Golden Nature Guide).

Peterson, A. 1948 *Larvae of Insects. Part 1. Lepidoptera and Plant Infesting Hymenoptera.* J.W. Edwards, Ann Arbor, Mich. 315 pp.

Sargent, T.D. 1976. *Legion of Night: The Underwing Moths.* Univ. Massachusetts Press, Amherst, Mass. 222 pp.

The Moths of America North of Mexico. 1971 and continuing. E. Classey and R.B.D. Publ., London. Fascicles exist on: Pyralidae (in part) by E.G. Munroe; Apatelodidae, Lasiocampidae, and Bombycidae by J.G. Franclemont; Saturniidae by D.C. Ferguson; Sphingidae and Oecophoridae by R.W. Hodges.

Tietz, H.M. 1973. *An Index to the Described Life Histories, Early Stages and Hosts of the Macrolepidoptera of the Continental United States and Canada.* E. Classey, Middlesex, England. 1,042 pp.

Urquhart, F.A. 1960. *The Monarch Butterfly.* Univ. Toronto Press, Toronto. 361 pp.

Villard, P. 1969. *Moths and How to Rear Them.* Funk and Wagnalls, N.Y. 242 pp.

Watson, A., and Whalley, P.E.S. 1975. *The Dictionary of Butterflies and Moths in Colour.* Michael Joseph, London. 296 pp.

ORDER DIPTERA[31]
Flies

Flies are characterized by a single pair of membranous wings attached to an enlarged mesothorax. The hind wings on the metathorax are reduced to a pair of knoblike appendages, the halteres, which help to balance the insect (Figs. 292; 304A). A few species, especially those that are parasites as adults, are wingless. Although a few insects in other orders have only two wings (e.g., certain mayflies and beetles, male scales), all lack halteres. When the word "fly" is used with nondipterous insects, it is part of the descriptive word (e.g., caddisfly, dragonfly); with dipterans "fly" is separated from the descriptive word (e.g., flesh fly).

Flies range in size from less than 1 mm in length (midges or gnats) to 75 mm (tropical robber flies). The compound eyes are often large; in many groups the expanded eyes of males meet on top of the head whereas they do not in females. Most adult flies have sucking mouthparts that are modified for piercing, lapping, or sponging; the mouthparts are nonfunctional in some species. Food consists chiefly of nectar but numerous species feed on fruit juices, plant sap, liquids from decaying organic materials, and blood. Some adults capture and feed on other insects. Eggs are usually laid singly on or near larval food. Many females use an ovipositor to push their eggs into a substrate. The eggs may hatch immediately or remain inactive for months through severe weather conditions. The eggs of many parasitic flies hatch just before being released and the larvae are deposited on the host.

Fly larvae are usually legless and have either a distinct (Figs. 306; 307A, C, D) or greatly reduced head (Fig. 330C). The latter type are called maggots. Larvae molt 3-8 times, the number decreasing in the more advanced groups (e.g., suborder Cyclorrhapha). Larvae of some species (e.g., a few gall midges [Cecidomyiidae]) produce daughter larvae without fertilization for several generations. Fly larvae prefer moist habitats such as fresh or brackish water, decaying plant and animal materials (fungi, humus, fruit, flesh, feces), or inside plants and animals. Larvae may be scavengers, leaf miners, gall formers, or fruit, stem, and root borers. Larvae that parasitize animals feed on blood, lymphatic liquids, and tissues. Some larvae are predators of other insects. A pupa may have its appendages on the outside and visible, or the pupa may be concealed inside a hardened larval skin (puparium) (Fig. 330D).

---

31. Diptera: *di,* two; *ptera,* wings.

Dipterans are the major disease-carrying order of insects. Through their blood-sucking activity some transmit malaria, African sleeping sickness (tsetse flies in the family Glossinidae), encephalitis, yellow fever, or filariasis. Flies that feed on feces may contaminate food and water with bacteria and viruses causing typhoid fever, dysentery, and other intestinal ailments. Some species are simply pests because of their biting behavior. Numerous species are agricultural pests (e.g., fruit flies). However, the vast majority of flies are of little agricultural or medical concern. Many adults are useful pollinators. Larvae are important in breaking down and redistributing organic materials and serve as an essential element in food chains.

The Diptera are divided into either two or three suborders. Most members in the suborder Nematocera are slender, long-legged, and mosquitolike in appearance. The antennae have six or more segments (Fig. 294A). Larvae have a distinct head, most live in water or wet soil, and they molt into pupae with external and visible appendages. The suborder Brachycera typically contains medium- to large-sized, somewhat stout-bodied flies. The antennae usually are 3-segmented; the third segment may be ringed or bear a style (Figs. 294B-D). Larvae have a small head and most are parasitic or predaceous on other insects. The pupa is concealed and the adult emerges through a "T"-shaped slit on the end of the puparium. The Cyclorrhapha are classified as a subgroup of the Brachycera or as a third suborder. These flies have 3-segmented and aristate antennae (Fig. 294E). A frontal suture in the majority of the species (muscoid flies) marks the break in the adult's head where a retractable, bladderlike sac, the ptilinum, was once everted (Fig. 291A). The ptilinum is used to push out a circular opening from one end of the puparium, thus allowing the adult to emerge. The muscoid flies are generally divided into those with and without calypteres (Fig. 293). Larvae lack a distinct head and are generally parasites or scavengers.

Dipterans are collected by sweeping a variety of habitats. By moving rapidly through vegetation and making wide, shallow net sweeps, one will capture fast-flying species. Weak-flying specimens are obtained by slow, deep sweeping near the collector's feet. Leaf litter may be sifted or placed in a Berlese funnel. Traps baited with a fermenting sugar solution (e.g., molasses, fruit juice, beer, and yeast), fermenting fruit, decomposing fungi, decaying meat, dung, crushed snails, or rotten eggs will attract specific groups of flies. Dry ice may be placed in a sealed plastic bag and the slow release of carbon dioxide through the bag will attract black flies and mosquitoes. In all cases the trap should allow the flies to enter a chamber and then fly upward into a holding chamber (Figs. 12B, C). A sticky, yellow panel (wood, plastic, etc.) or a tray of water with a yellow panel in the bottom (detergent is added to the water to trap the flies) will attract certain groups. Malaise traps (Fig. 13) are particularly effective in capturing Diptera. Lights of various types also attract nocturnal dipterans. Rearing adults from the larval or pupal stage is an important means of obtaining aquatic, parasitic, fungus- and debris-inhabiting, and gall-forming species. The galls, fungi, or leaf litter may be placed in a plastic bag to insure adequate humidity for rearing the insects. Adults are pinned or, if small, glued to points or placed on slides. Specimens on points are usually mounted on their sides. Soft-bodied adults (primarily wingless types) and larvae are preserved in 70% alcohol.

Species: North America, 18,000; world, 90,000.
Families: North America, 106.

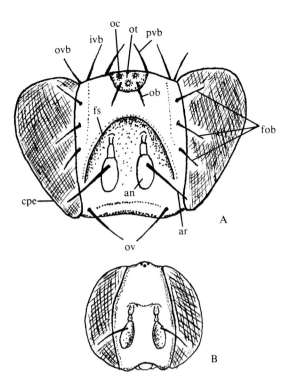
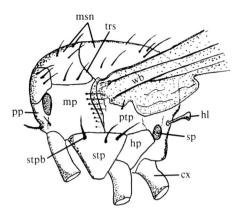

Figure 292 General structure and bristle arrangement of a fly thorax (Muscidae). cx, coxa; hl, halter; hp, hypopleuron; mp, mesopleuron; msn, mesonotum; pp, propleuron; ptp, pteropleuron; sp, spiracle; stp, sternopleuron; stpb, sternopleural bristle; trs, transverse suture; wb, wing base.

Figure 291 General bristle arrangement on a fly head (anterior view). A, frontal suture present; B, frontal suture absent. an, antenna; ar, arista; cpe, compound eye; fob, fronto-orbital bristles; fs, frontal suture; ivb, inner vertical bristle; ob, ocellar bristle; oc, ocellus; ot, ocellar triangle; ov, oral vibrissae; ovb, outer vertical bristle; pvb, postvertical bristle.

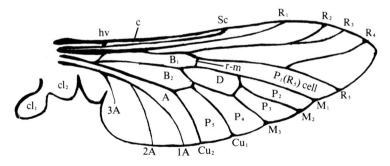

Figure 293 General wing venation. A, anal cell; B_1 and B_2, first and second basal cells; cl_1 and cl_2, lower and upper calypteres; D, discal cell; hv, humeral crossvein; P, posterior cell; r-m, radio-medial crossvein.

Order Diptera 317

KEY TO COMMON FAMILIES OF DIPTERA

1a	Wings absent or very small; coxae of each segment widely separated; external parasites of birds or mammals . . . . . . (p. 350) *Hippoboscidae* (in part)	3a(2b)	Antennae with 6 or more freely joined segments (Fig. 294A); antennae sometimes long and very plumose 4
1b	Wings present and well developed; coxae usually very close or touching; usually not parasites of birds or mammals, but if so body is very flattened (Fig. 331A) and coxae are widely separated 2	3b	Antennae with 5 or fewer segments; third segment often weakly ringed (Figs. 294B, C) or bearing a style (Fig. 294D) or large bristle (arista) (Fig. 294E) 17
		4a(3a)	"V"-shaped suture between prothorax and mesothorax (Fig. 304B); legs long, slender, loosely jointed (Figs. 304A, D) 5
2a(1b)	Body very flattened (Fig. 331A); coxae widely separated; anterior veins of wing broad and crowded forward leaving posterior veins narrow or absent . . . . (p. 350) *Hippoboscidae* (in part)	4b	No "V"-shaped suture between prothorax and mesothorax; legs variable 7
		5a(4a)	Ocelli present . . . . (p. 326) *Trichoceridae*
2b	Above combination of characteristics not present 3	5b	Ocelli absent 6

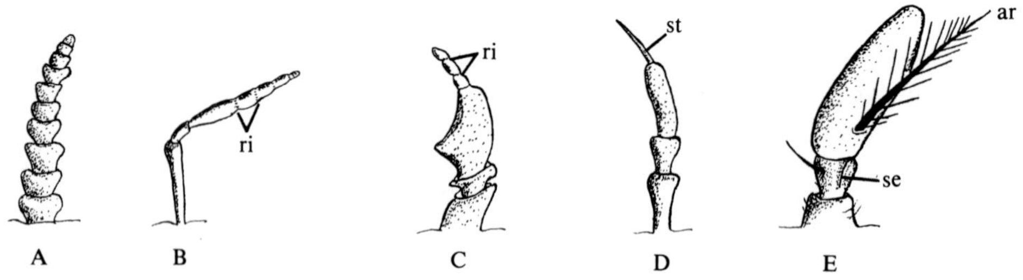

Figure 294 Antennae. A, March flies (Bibionidae); B, soldier flies (Stratiomyidae); C, horse flies (Tabanidae); D, robber flies (Asilidae); E, blow fly (Calliphoridae). ar, arista; ri, rings, se, seam; st, style.

6a(5b)	Two anal veins reach wing margin (Fig. 295A) ...... (p. 326) *Tipulidae*	9b	Antennae usually longer than thorax and insert between compound eyes at middle or higher; pulvilli very thin or absent. 10
6b	One anal vein reaches wing margin (Fig. 295B). . (p. 327) *Ptychopteridae*	10a(9b)	Compound eyes curved inward and meet above bases of antennae (Fig. 296A) ...... (p. 333) *Sciaridae*
		10b	Compound eyes do not curve inward and meet above bases of antennae .. .. (p. 333) *Mycetophilidae*

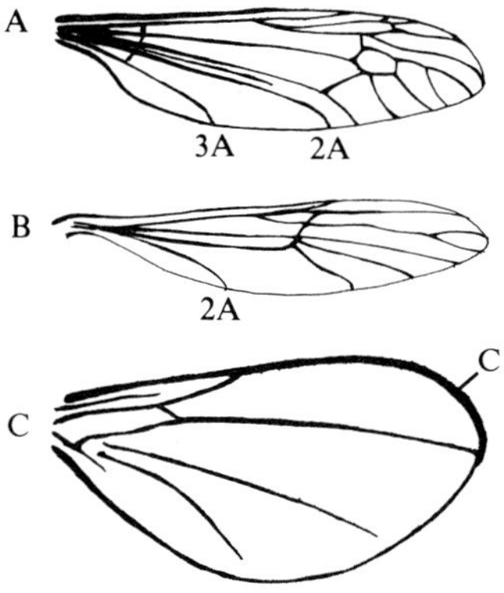

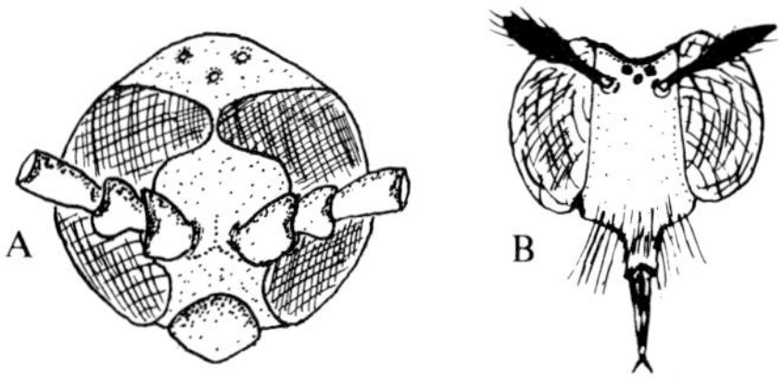

Figure 295 Wings. A, crane flies (Tipulidae); B, phantom crane flies (Ptychopteridae); C, gall midges (Cecidomyiidae).

Figure 296 A, head and eyes of a darkwinged fungus gnat (Sciaridae); B, concave dorsal area of a robber fly head (Asilidae).

7a(4b)	Wings short, broadly oval, pointed at apex, and held rooflike over body when at rest (Fig. 305B); wings densely hairy .. (p. 327) *Psychodidae*	11a(8b)	Costa ending before tip of wing .. 12
7b	Wings usually long and relatively narrow, or if broad then not pointed at apex; wings held flat over body when at rest; wings not densely hairy but may have sparse hair or scales on wing margins or veins. 8	11b	Costa continuing around wing tip although often more narrow on posterior margin (Fig. 295C) 14
8a(7b)	Ocelli present 9	12a(11a)	Wings broad, their anterior veins heavy and posterior veins weak (Fig. 311A); antennae short, about length of head; body stout; biting flies. .... (p. 332) *Simuliidae*
8b	Ocelli absent. 11		
9a(8a)	Antennae usually shorter than thorax and insert under compound eyes; pulvilli present (Fig. 298) .... (p. 332) *Bibionidae*	12b	Wings long and narrow, their posterior veins often well developed; antennae much longer than head; body often slender 13

Order Diptera

13a(12b) Proboscis absent; posterior part of head flattened; metanotum with median groove or ridge; tarsi of front legs of some species very long making front legs longest (Fig. 310B) . . . . (p. 331) *Chironomidae*

13b Proboscis present; posterior part of head rounded; metanotum without a median groove or keel; tarsi of front legs not unusually long, hind legs longest . . . (p. 331) *Ceratopogonidae*

14a(11b) Venation usually reduced to less than 7 longitudinal veins reaching wing margin (Fig. 295C); antennae usually very long, 10-36 segments. . . . . (p. 333) *Cecidomyiidae*

14b Venation with 9 or more veins reaching wing margin; antennae variable 15

15a(14b) Wing margin and veins without scales or hairs. . . . . (p. 328) *Dixidae*

15b Wing margin and usually veins with scales 16

16a(15b) Proboscis very long; scales or hairs on wing margin and veins. . . . . (p. 328) *Culicidae*

16b Proboscis short, extending only slightly beyond head; scales primarily on wing margin . . . . (p. 328) *Chaoboridae*

17a(3b) Antennae 4-segmented, the fourth elongated and often thickened; 1 ocellus; only 1-2 veins reach posterior margin behind the wing apex, others curving forward (Fig. 316C); large flies . . . . (p. 337) *Mydidae*

17b Antennae 3-segmented, the third with rings (Figs. 294B, C), an elongated style (Fig. 294D), or a bristle (arista) (Fig. 294E) 18

18a(17b) Third antennal segment with rings (Figs. 294B, C) 19

18b Third antennal segment with an elongated style or arista (Figs. 294D, E) 20

19a(18a) Calypteres small or vestigial; discal cell small, its dimensions nearly equal (Fig. 297A). . . . . (p. 334) *Stratiomyidae*

19b Calypteres large and conspicuous (Fig. 293); discal cell at least twice as long as wide (Fig. 297B) . . . . (p. 335) *Tabanidae*

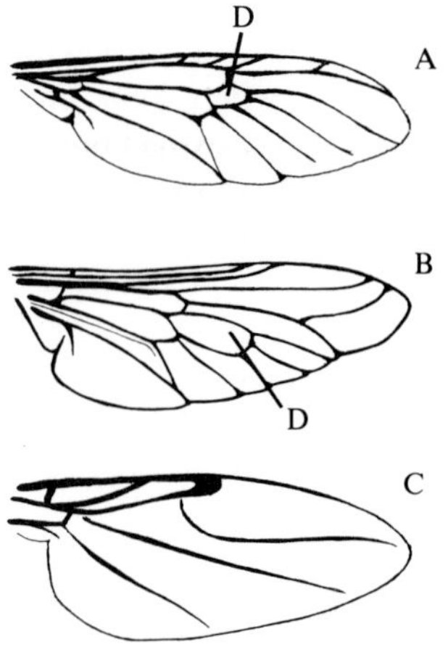

Figure 297 Wings. A, soldier flies (Stratiomyidae); B, deer and horse flies (Tabanidae); C, humpbacked flies (Phoridae). D, discal cell.

20a(18b) Tarsi with 3 pads at tip (Fig. 298A); style on tip of antennae longer than 3 antennal segments combined . . . . (p. 336) *Rhagionidae*

20b Tarsi with 1 pad under each claw and a bristle between claws (Fig. 298B); antennae usually with an arista. 21

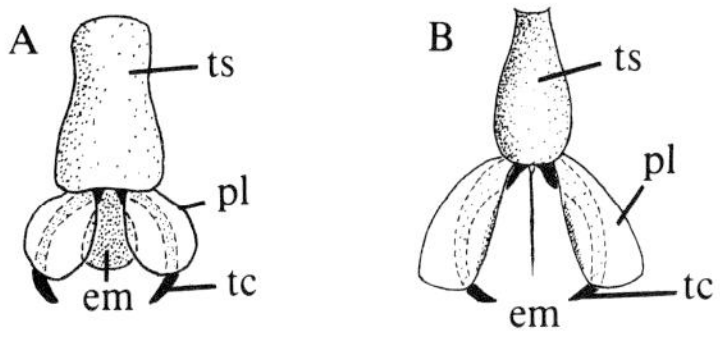

Figure 298 Tarsi. A, 3 pads; B, 2 pads and a central bristle. em, empodium; pl, pulvillus; tc, tarsal claw; ts, last tarsal segment.

21a(20b) Wings with 2 strong, longitudinal veins anteriorly and 4-5 weak veins posteriorly (Fig. 297C); antennae appear 1-segmented; humpbacked, very small . . . . (p. 340) *Phoridae*

21b Wing veins similar in width; antennae not 1-segmented; appearance and size variable 22

22a(21b) Wings pointed at apex (Fig. 299A); crossveins absent except near base (Fig. 299A); third antennal segment rounded and with arista at tip . . . . (p. 340) *Lonchopteridae*

22b Wings rounded at apex; crossveins usually present beyond wing base; antennae variable 23

23a(22b) Vein above radio-medial crossvein (r-m) is 2-branched (R_4 and R_5) giving radius a 4-branched appearance (Fig. 299B) 24

23b Vein above radio-medial crossvein (r-m) unbranched (R_{4+5}) giving radius a 3-branched appearance (Fig. 299C) 27

24a(23a) Top of head strongly concave and eyes bulging when viewed from front (Fig. 296B); body elongate to very elongate, sometimes stout and hairy resembling a bumble bee (Fig. 317B); face usually very hairy . . . . (p. 337) *Asilidae*

24b Top of head not concave; eyes and body variable; face not unusually hairy. 25

25a(24b) Wing with 5 posterior cells (Fig. 293); antennae usually with a sharp style on apex of third segment; abdomen long and tapering . . . . (p. 336) *Therevidae*

25b Wing with 4 or fewer posterior cells (Figs. 299B, C); antennae and abdomen variable. 26

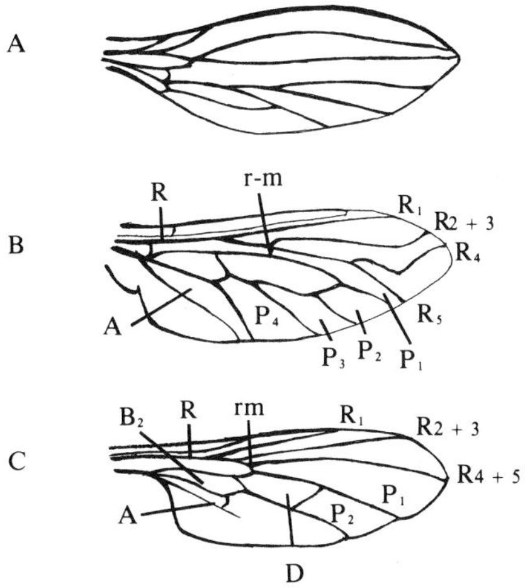

Figure 299 Wings. A, lonchopterid flies (Lonchopteridae); B, bee flies (Bombyliidae); C, dance flies (Empididae). A, anal cell.

26a(25b)   Anal cell open to posterior wing margin (Fig. 299B), or if closed then closure near wing margin and apex of cell acute; body often hairy and stout .... (p. 338) *Bombyliidae*

26b   Anal cell absent or closed far from posterior wing margin (Fig. 299C), or if closed very near wing margin then apex of cell not acute; body rarely hairy or stout .... (p. 339) *Empididae* (in part)

27a(23b)   Anal cell longer than second basal cell, apex usually pointed and closed near wing margin (Fig. 300A); no frontal suture (Figs. 291B; 301A) 28

27b   Anal cell absent or short, closed far from posterior wing margin (Fig. 299C); frontal suture present (Fig. 291A) or absent (Figs. 291B; 301A) 30

28a(27a)   Proboscis longer than head, thin, often folded (Fig. 322A); face with distinct grooves beneath antennae .... (p. 342) *Conopidae*

28b   Proboscis very short, shorter than head; face without grooves 29

29a(28b)   First posterior cell (cell P_1) closed at or near wing margin and usually stalked at tip (Fig. 300A); spurious vein usually present between 3rd and 4th longitudinal veins (Fig. 300A); head with distinct cheeks (area below eyes); eyes not unusually large .... (p. 341) *Syrphidae*

29b   First posterior cell always open at wing margin; spurious vein absent; head without cheeks; eyes greatly enlarged and encompassing almost all of head (Fig. 320B) .... (p. 340) *Pipunculidae*

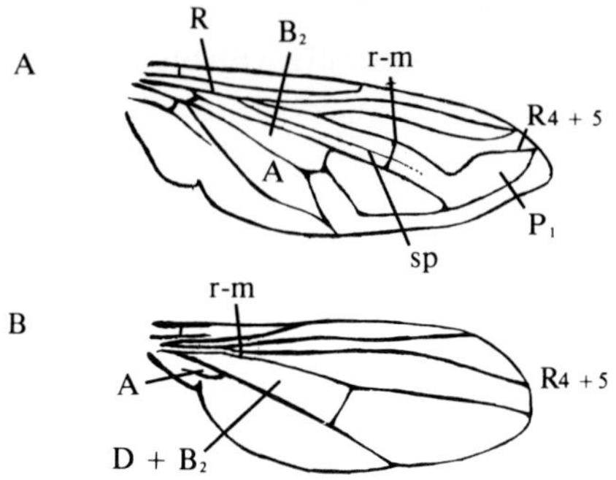

Figure 300 Wings. A, flower flies (Syrphidae); B, longlegged flies (Dolichopodidae). sp., spurious vein.

30a(27b)   Frontal suture absent (Fig. 301A) 31

30b   Frontal suture present (Fig. 291A) 32

31a(30a)   R-M crossvein in basal fourth of wing (Fig. 300B); discal cell and second basal cell united into 1 cell (Fig. 300B); male genitalia often folded forward under abdomen (Fig. 319C); body usually metallic blue or green .... (p. 340) *Dolichopodidae*

31b   R-M crossvein beyond basal fourth of wing (Fig. 299C); discal cell and second basal cell usually separate (Fig. 299C); male genitalia not folded under abdomen; not metallic .... (p. 339) *Empididae* (in part)

32a(30b) Mouthparts vestigial or absent and mouth opening small; body hairy (but not bristly) and beelike 33

32b Mouthparts well developed and evident, mouth opening normal; body usually with bristles 35

33a(32a) P_1 cell (Figs. 293; 335C) widens toward wing apex............... .... (p. 354) *Gasterophilidae*

33b P_1 cell narrows toward wing apex or is closed 34

34a(33b) Scutellum very short but postscutellum enlarged (Fig. 301B) .... (p. 353) *Oestridae*

34b Scutellum extends much beyond metanotum and postscutellum not enlarged (Fig. 301C)............. .... (p. 353) *Cuterebridae*

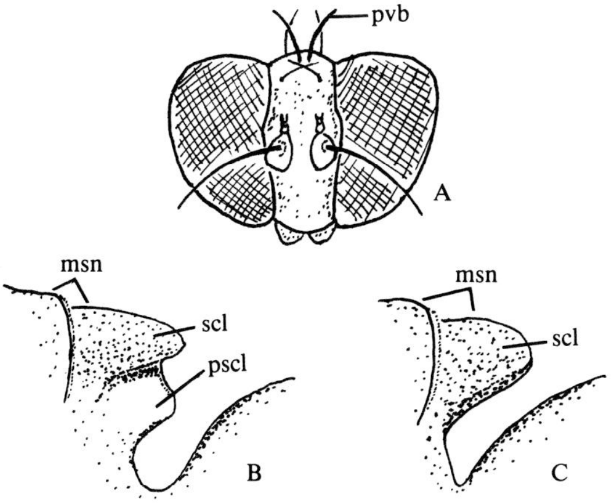

Figure 301 A, head without a frontal suture and with postvertical bristles bent outward; B, scutellum (lateral view) of bot and warble flies (Oestridae); C, scutellum (lateral view) of rabbit bots and rodent bots (Cuterebridae). msn, mesonotum; pscl, postscutellum; pvb, postvertical bristles; scl, scutellum.

35a(32b) Second antennal segment without a longitudinal seam on outer side; innermost (lower) calypter small or absent; thorax usually without a complete transverse suture 36

35b Second antennal segment with a longitudinal seam on outer side (Fig. 294E); innermost (lower) calypter usually large (Fig. 293); thorax usually with a transverse suture (Figs. 292; 303)............... 51

36a(35a) Sc does not extend to costa (Fig. 302A); anal cell present or absent 37

36b Sc extends to costa or ends very near costa; anal cell present 43

37a(36a) Tip of Sc bent at right angle toward costa (Fig. 302A); anal cell usually with pointed tip posteriorly (Fig. 302A); wings usually banded or spotted...................... ...... (p. 343) *Tephritidae* (in part)

37b Not with above combination of characteristics. 38

38a(37b) Costa broken near end of Sc or R_1 and also near humeral crossvein (Fig. 302B) 39

38b Costa only broken near end of Sc or R_1 41

39a(38a) Anal cell absent (Fig. 302B); if postvertical bristles present (on top of head behind ocelli) they are bent outward (Fig. 301A)............ .... (p. 345) *Ephydridae*

Order Diptera 323

39b   Anal cell present (Fig. 293); postvertical bristles bent inward (Fig. 291A).................. 40

40a(39b)   At least 1 pair fronto-orbital bristles bent inward; arista not plumose.............. (p. 345) *Milichiidae*

40b   Fronto-orbital bristles not bent inward (Fig. 291A); arista plumose .... (p. 346) *Drosophilidae*

41a(38b)   Anal cell absent; large, often shining, slightly raised triangular area surrounding ocelli and extending onto front of face; postvertical bristles (on top of head behind ocelli) bent inward (Fig. 291A) or occasionally parallel .... (p. 346) *Chloropidae*

41b   Anal cell present; ocelli not surrounded by large, shining, raised triangular area; postvertical bristles variable 42

42a(41b)   Postvertical bristles bent inward (Fig. 291A)................... .... (p. 347) *Anthomyzidae*

42b   Postvertical bristles bent outward (Fig. 301A)... (p. 347) *Agromyzidae*

43a(36b)   Head spherical, body cylindrical, abdomen often narrowed at base; palpi vestigial .... (p. 344) *Sepsidae*

43b   Body shape variable; palpi usually well developed 44

44a(43b)   Thorax flattened dorsally; conspicuous bristles on legs and abdomen; on seaweed, other vegetation near seashore... (p. 344) *Coelopidae*

44b   Thorax rounded dorsally, or if somewhat flattened then legs are not bristly; not on seaweed 45

45a(44b)   Oral vibrissae present (Fig. 291A); small spines on costa .... (p. 347) *Heleomyzidae*

45b   Oral vibrissae absent; spines rarely on costa 46

46a(45b)   Postvertical bristles bent inward (Fig. 291A); 2A vein short, not reaching posterior wing margin .... (p. 345) *Lauxaniidae*

46b   Postvertical bristles bent outward, parallel, or absent; 2A vein reaching wing margin, or extended as a fold reaching margin 47

47a(46b)   Antennae generally projecting forward; face somewhat concave on lower side (Fig. 324B); femora with bristles, middle femur with bristle near middle of anterior side; bristles near apex of tibiae; brownish to grayish, wings occasionally patterned ... . (p. 344) *Sciomyzidae*

47b   Not with above combination of characteristics................ 48

48a(47b)   Anal cell with a pointed tip posteriorly (Fig. 302A); wings commonly spotted or banded 49

48b   Anal cell without a pointed tip posteriorly; wing color variable .. 50

49a(48a) Tip of Sc slightly curved toward costa and usually reaching costa; costa not broken near end of Sc . . . . (p. 343) *Otitidae* (in part)

49b Tip of Sc sharply bent toward costa at nearly a right angle, usually not reaching costa (Fig. 302A); costa broken near end of Sc . . . . (p. 343) *Tephritidae* (in part)

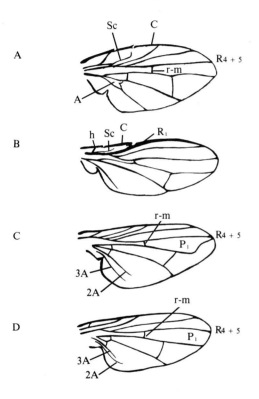

Figure 302 Wings. A, fruit flies (Tephritidae); B, shore flies (Ephydridae); C, house flies and stable flies (Muscidae); D, *Fannia* spp. (Muscidae). A, anal cell.

50a(48b) Costa not broken near end of Sc; anterior side of anal cell less than one-fourth length of posterior side of discal cell; tip of Sc slightly curved toward costa and usually reaching costa . . . . . (p. 343) *Otitidae* (in part)

50b Costa broken near end of Sc (Fig. 302A); anal cell length variable; tip of Sc sharply bent toward costa at nearly a right angle, usually not reaching costa (Fig. 302A) . . . . (p. 343) *Tephritidae* (in part)

51a(35b) Hypopleura usually without strong bristles (Fig. 292), with only weak and scattered hairs or hairs absent; if hypopleural bristles present then pteropleural bristles absent or the P_1 cell not narrowing toward wing tip 52

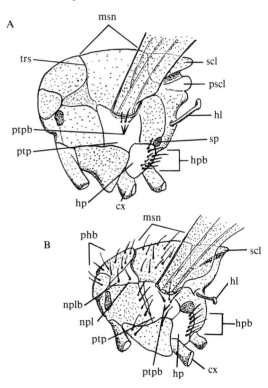

Figure 303 Lateral view of thorax. A, tachina flies (Tachinidae); B, blow flies (Calliphoridae). cx, coxa; hl, halter; hp, hypopleuron; hpb, hypopleural bristles; msn, mesonotum; npl, notopleuron; nplb, notopleural bristle; phb, posthumeral bristles; pscl, postscutellum; ptp, pteropleuron; ptpb, pteropleural bristles; scl, scutellum (mesoscutellum); sp, spiracle; trs, transverse suture.

Order Diptera   325

51b	Hypopleura and pteropleura with strong bristles (Fig. 303); P_1 cell narrowing or closed toward wing tip 53	
52a(51a)	2A vein not reaching posterior wing margin (Figs. 302C, D); usually more than 1 sternopleural bristle (Fig. 292) . . . . (p. 349) *Muscidae*	
52b	2A vein reaching wing margin as a vein or fold; 1 sternopleural bristle . . . . (p. 348) *Anthomyiidae*	
53a(51b)	Postscutellum developed into convex bump (Fig. 303A); strong bristles usually on abdominal tergites . . . . (p. 352) *Tachinidae*	
53b	Postscutellum not developed; abdominal bristles not common 54	
54a(53b)	Arista of antennae usually plumose to tip (Fig. 294E); 2 bristles on notopleuron (Fig. 303B); body often metallic . . . . (p. 351) *Calliphoridae*	
54b	Arista of antennae usually plumose only on basal half; 3 or 4 bristles on notopleuron; body not metallic; often with gray and black stripes on thorax and checkerboard pattern on abdomen . . . (p. 351) *Sarcophagidae*	

SUBORDER NEMATOCERA

Family Trichoceridae— Trichocerids or Winter Crane Flies

Trichocerids closely resemble small crane flies (Tipulidae) but differ by having ocelli. Adults in the genus *Trichocera* often swarm in late fall, on mild winter days, or in early spring. Most species occur in the northern half of the U.S. Larvae live in decaying organic materials.

Family Tipulidae—Crane Flies

Crane flies are very long-legged insects that usually resemble quite large mosquitoes; however, crane flies do not bite. Tipulids are the largest family of dipterans with about 12,000 species worldwide and nearly 1,500 species in North America. Adults are especially common in damp areas where there is abundant vegetation such as woods, streams, and lakes. Many species are attracted to lights. Larvae (Fig. 304C) typically occur in moist soil and decaying organic material in wooded areas; some are aquatic. Larvae of *Tipula simplex* Doane, the range crane fly, feed on leaves of grasses and grains in California in early spring and are sometimes pests of grain crops.

Common Species

Nephrotoma ferruginea (Fabricius)—15-16 mm; brownish yellow; thorax with rusty red stripes, ends of transverse suture dark; abdominal tergites with row of dark spots; east of Rocky Mts.

Pedicia albivitta Walker—24-35 mm; whitish gray; very large, dark brown, triangular marking on wings, posterior edge of marking reaches wing margin; femora yellow with black tips; eastern U.S., southcentral and southeastern Canada.

Tipula abdominalis (Say) (Fig. 304A)—25-38 mm; thorax patterned black and gray dorsally, sides striped gray and brownish black; wings clouded with gray, 3 brown areas along anterior margin; abdomen dark orange with black sides; eastern U.S. south to FL (late summer).

Tipula acuta Doane—15-25 mm; gray; dark brown middorsal line on abdomen; oblique white mark and small dark spots near wing tips; Pacific Coast states (late summer, fall).

Tipula trivittata trivittata Say—14-22 mm; thoracic stripe pale in center; abdomen with 3 brown stripes dorsally, gray lateral margins; eastern U.S.

with black and white; 1st tarsal segment distinctly swollen; most of North America except Pacific coast states. *B. occidentalis* Aldrich in Pacific coast states.

Family Psychodidae—Moth Flies

These mothlike, tiny flies are usually less than 4 mm long, very hairy, and many hold their wings rooflike over the body. They occur in damp, shady areas and in or near drains and sewers; some are attracted to lights. *Telmatoscopus albipunctatus* (Williston), sometimes flies out of drains in houses in the eastern half of the U.S. southwest to AZ. Larvae breed in moist, decaying organic matter (including seaweed), moss, mud, and water. Larvae are 3-10 mm long and have faint rings between each segment. Most moth flies do not bite; those that do, often called sand flies (genus *Phlebotomus* [Fig. 305A]), occur chiefly in the southern states. Tropical sand flies transmit various diseases.

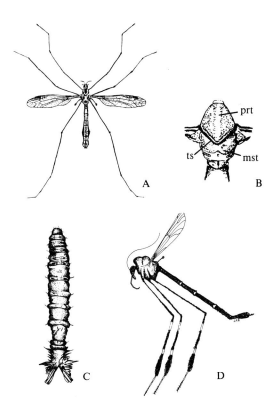

Figure 304 A-C, crane flies (Tipulidae): A, *Tipula abdominalis;* B, V-shaped thoracic suture; C, crane fly larva. D, a phantom crane fly, *Bittacomorpha clavipes* (Ptychopteridae). mst, mesothorax; prt, prothorax; ts, thoracic suture.

Family Ptychopteridae— Phantom Crane Flies

These dipterans resemble crane flies (Tipulidae) but have only one anal vein (crane flies have two) that reaches the wing margin (Fig. 295B). The aquatic larvae occur in decaying vegetation along margins of marshes, ponds, and streams (especially in shoreline iron seepages). Larvae are recognized by the very long breathing tube on the tip of the abdomen and their small heads.

Common Species

Bittacomorpha clavipes (Fabricius) (Fig. 304D)—10-15 mm; grayish and dark brown; legs long, banded

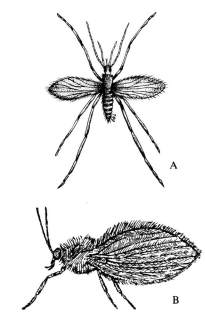

Figure 305 Moth flies (Psychodidae). A, a sand fly, *Phlebotomus* sp.; B, *Psychoda alternata*.

Common Species

Psychoda alternata Say (Fig. 305B)—2 mm; yellowish or pale tan; wings faintly mottled black and white; antennae 15-segmented, segments 13-15 fused, segment 15 very small; throughout North America.

Family Dixidae—Dixid Midges

Dixids are slender, long-legged, mosquitolike insects but differ partly from mosquitoes by lacking wing or body scales. Adults do not bite. Dixids occur on rocks near the water's surface and around vegetation where moisture is abundant; they often swarm at dusk. Larvae are aquatic and resemble mosquito larvae. However, dixids form a broad "U" shape (Fig. 306A) when alive and swim by opening and closing the "U." Most North American species are in the genus *Dixa*.

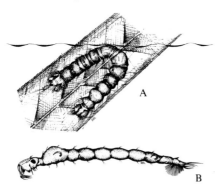

Figure 306 A, a dixid midge larva, *Dixa* sp. (Dixidae); B, a phantom midge larva, *Chaoborus flavicans* (Chaoboridae).

Family Chaoboridae—Phantom Midges

These nonbiting midges resemble mosquitoes but have a very short proboscis and the wing scales are primarily on the margin. Adults occur near water and larvae are aquatic. Larvae of the genus *Chaoborus* are nearly transparent (hence the family name "phantom") and lack a posterior breathing tube. Larvae in the genus *Mochlonyx* are darker and have a breathing tube near the tip of the abdomen projecting at nearly a right angle to the body. Larvae are predaceous on mosquito larvae, other insects, and minute crustaceans. The antennae are modified to capture and hold prey.

Common Species

Chaoborus flavicans (Meigen) (Fig. 306B)—4.7-7.0 mm; dark longitudinal bands on mesonotum; wings clear; northern 1/2 U.S.

Family Culicidae—Mosquitoes

The presence of scales on the wings and body and the long proboscis help to identify mosquitoes. Males have prominent plumose antennae. Females feed on nectar as well as blood; the majority of species feed on animals other than humans. A bloodmeal is usually required for egg production. The female may deposit on or near water several clusters of 50-200 eggs each over her lifetime of a few weeks. Males feed on flower nectar and fruit juices and in doing so they act as pollinators of various wildflowers. There are 148 species known in North America.

Mosquito larvae and pupae (Figs. 307; 308) are aquatic and occur in a variety of habitats, depending on the particular species. Most larvae breathe at the water's surface by using a tube or a pair of spiracles at the posterior end of the body; tracheal gills also assist in breathing. Algae, protozoans, and organic debris serve as food for larvae. They typically mature in 4-10 days and molt four times. Pupae (Fig. 308) are active swimmers that rise to the water's surface to breathe through two small tubes on the thorax.

Female mosquitoes are major vectors of diseases to humans and other animals. *Aedes* species transmit yellow fever (virus) and certain types of encephalitis (virus). Malaria (protozoan) is transmitted by *Anopheles* species. *Culex* species transmit several kinds of encephalitis, filariasis (filarial worms) including dog heartworm, and bird malaria.

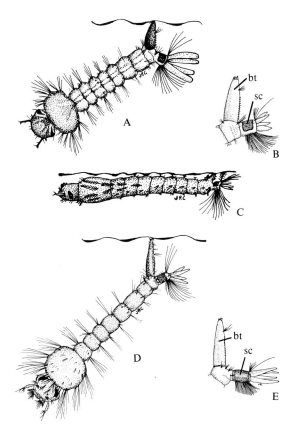

Common Genera

The four major genera are described as follows:

Aedes—adult with bristles just behind mesothoracic spiracle; female with tip of abdomen pointed and often white or yellowish markings on thorax. Larva with short and stout breathing tube (Fig. 307A) and 1 pair hair tufts on tube; anal segment partly covered with a sclerotized plate (Fig. 307B); common in flooded areas, woodland pools, salt marshes.

Anopheles—adult with wings spotted, maxillary palpi as long as central proboscis; body angles distinctly upward at rest. Larva lacks breathing tube, lies parallel to water's surface (Fig. 307C).

Culex—adult without bristles just behind mesothoracic spiracle; female with tip of abdomen blunt, thorax without markings. Larva with long and slender breathing tube and several pairs hair tufts on tube (Fig. 307D).

Psorophora—adult large; hind tibiae with long erect scales; bristles just in front of mesothoracic spiracles. Larva like *Aedes* but anal segment with a complete sclerotized ring (Fig. 307E); common in flooded areas, woodland pools, salt marshes.

Figure 307 Mosquito larvae (Culicidae). A, *Aedes* sp.; B, anal segment of *Aedes* sp.; C, *Anopheles* sp.; D, *Culex* sp.; E, *Psorophora* sp. anal segment. bt, breathing tube; sc, sclerotized area.

Common Species

Aedes aegypti (Linnaeus) (Fig. 309A). Yellowfever mosquito—wing length 2.5-3.0 mm; body blackish; head with narrow light stripes; thorax with 4 silvery lines, outer pair curved; abdomen and legs banded with white; southern 2/3 U.S. (prefers artificial containers for breeding).

Aedes dorsalis (Meigen)—wing length 4.0-4.5 mm; yellowish gray; mouth parts and legs speckled white; head with pale patch; western and northern U.S., southern Canada. *A. nigromaculis* (Ludlow), wing length 3-4 mm, blackish, reddish yellow thorax, white rings on tarsi, western 1/2 North America (in overflow water, rain pools).

Aedes sollicitans (Walker) (Fig. 309B). Saltmarsh mosquito—wing length 3.5-4.5 mm; thorax yellow-brown; abdomen with white or yellow crossbands and middorsal pale yellowish line; wings with white or brown speckles; 1st segment hind tarsus with yellow median ring; Atlantic and Gulf coast states, parts of Midwest (breeds in marshes flooded with salt water; disperses great distances). *A. squamiger* (Coquillett), California saltmarsh mosquito, wing length 4.5-5.0 mm, brownish gray, ringed tarsi, salt

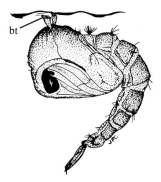

Figure 308 Mosquito pupa (Culicidae). bt, breathing tube.

Order Diptera

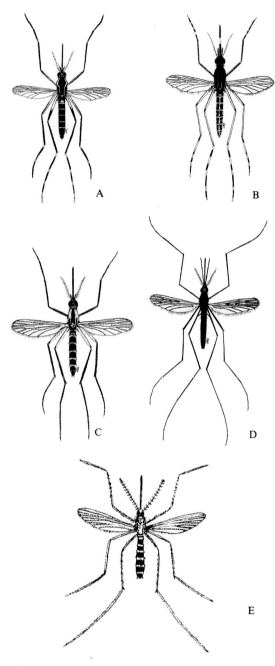

Figure 309 Mosquitoes (Culicidae). A, yellowfever mosquito, *Aedes aegypti;* B, saltmarsh mosquito, *Aedes sollicitans;* C, floodwater mosquito, *Aedes sticticus;* D, common malaria mosquito, *Anopheles quadrimaculatus;* E, northern house mosquito, *Culex pipiens.*

marshes along Pacific coast. *A. taeniorhynchus* (Wiedemann), wing length 3 mm, black, abdomen with white crossbands but not middorsal line, Atlantic and Gulf coast states, southeastern U.S., southern CA coast.

Aedes sticticus (Meigen) (Fig. 309C). Floodwater mosquito—wing length 3-4 mm; head pale yellowish; thorax with 2 narrow middorsal yellow-brown lines; white patch on 1st abdominal segment, other segments dark with narrow white basal band on each, ends of bands broaden into triangle; throughout North America. *A. communis* (De Geer), like *A. sticticus,* 4.5-5.0 mm, high elevations in western states, Canada (June, July; one of snow mosquito group, breeds in snow pools).

Aedes vexans (Meigen)—wing length 3.5-4.0 mm; brown; white rings on base of tarsus; abdomen black with broad white rings; most of North America (in woods, near temporary flooded areas).

Anopheles punctipennis (Say)—wing length 4 mm; brown; wings with white or yellowish spot (with large dark spot on each side) on outer 1/3 of anterior margin; throughout U.S., southern Canada.

Anopheles quadrimaculatus Say (Fig. 309D). Common malaria mosquito—wing length 4.5 mm; dark brown; head with pale scales on top; wings with 4 dark spots near center; mouthparts, thorax, and tarsi with dark scales; eastern 1/2 U.S. (especially Southeast) southwest to TX, southeastern Canada.

Culex pipiens Linnaeus (Fig. 309E). Northern house mosquito—wing length 3.5-4.0 mm; mouthparts, wings, and tarsi dark; thorax pale brown or grayish; dorsum of abdomen with white bands at base of each segment, spots on sides of bands; northern 3/4 U.S. (prefers artificial containers for breeding; enters houses). *C. quinquefasciatus* Say, southern house mosquito, similar, posterior margin of each white abdominal band rounded or bands not joined to side spots, southeastern U.S. west to UT and CA.

Culex tarsalis Coquillett—4 mm; pale brown to blackish; proboscis with white ring near middle; yellow-brown scales on thorax; white bands on tarsi; dorsum of abdomen with yellowish white bands at base of each segment; primarily western U.S., east to MI, SC and FL, southwestern Canada (enters houses).

Family Ceratopogonidae—Biting Midges

These tiny biting flies are usually less than 3 mm long. They resemble stout-appearing midges (Chironomidae) but ceratopogonids have a well-developed proboscis and broader wings. Adults suck blood from other insects, reptiles, humans, and other mammals, although they also feed on flower nectar. These insects are also called no-seeums and punkies. Some species transmit diseases. Most larvae develop in moist or wet sand, mud, and decaying vegetation of salt and freshwater marshes, ponds, and streams; a few occur under bark of rotting trees.

Common Species

Culicoides variipennis variipennis (Coquillett) (Fig. 310A)—2.5 mm; bluish gray; thorax with brown spots; wing with brown and gray streaks; antennae 15-segmented; northern and eastern U.S., most of Canada (in forests; 4 other subspecies occur near salt marshes, other alkaline water in U.S.).

Family Chironomidae—Midges

Midges are small, mosquitolike dipterans that often occur in large swarms and are commonly attracted to lights. Adults in the subfamily Chironominae have long front tarsi, the front legs are held up in the air when resting, and the antennae of males are highly plumose. The thorax is large and may extend slightly over the small head. Midge adults do not have a piercing mouthpart like mosquitoes and do not bite. They typically live 5-10 days. Most larvae live in fresh water (some are intertidal or live in wet moss) and occupy habitats from streams to the mud of deep lakes. Some larvae feed on algae and other plant materials; others are filter-feeders or are predaceous. Many immatures build silk-lined tubes of sand, mud, and debris. Larvae of the tribe Chironomini are generally called bloodworms because of the red, hemoglobin-containing blood. Midge larvae are 12-segmented; they have a pair of prolegs on both the first thoracic and last abdominal segments, and often tubular gills near the posterior end (Fig. 310C).

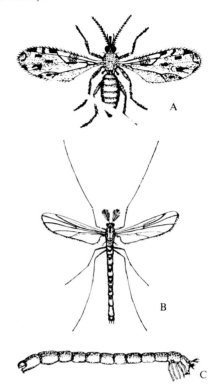

Figure 310 A, a biting midge, *Culicoides variipennis variipennis* (Ceratopogonidae); B and C, midges (Chironomidae): B, *Chironomus plumosus;* C, a midge larva.

Common Species

Ablabesmyia monilis (Linnaeus)—wing length 2-3 mm; head and thorax black-brown; tibia with one light band; abdominal tergites 2-5 each with basal 1/2-2/3 brown; most of North America.

Chironomus attenuatus Walker—wing length 3.5-4.0 mm; light brown to pale green; thoracic markings and abdominal bands dark brown; throughout North America.

Chironomus plumosus (Linnaeus) (Fig. 310B)—wing length 6.0-7.5 mm; pale brown, often with green tinge; thoracic markings and abdominal bands light to dark brown; abdomen green; northern 1/2 U.S., eastern U.S., southern Canada.

Paratendipes albimanus (Meigen)—wing length 2 mm; shiny black, base of abdomen brownish; base of tarsus white, apex brown; middle and hind tibiae and tarsi pale yellow; most of U.S. except southernmost states.

Family Simuliidae—Black Flies

These small, humpbacked dark flies (sometimes called buffalo gnats) have broad wings and short legs. Both sexes feed on nectar but only females are bloodsucking. They are infamous, especially in the northern states and Canada, for their biting in the late spring and early summer. Various species attack humans, livestock, birds, and other animals. Some subtropical species transmit filarial worms to humans; in the U.S. protozoan parasites are transmitted to birds. Black fly larvae (Fig. 311B) are aquatic and attach themselves with a posterior sucker to rocks, sticks, and vegetation in streams. Pupae have prominent, thoracic tracheal gills for breathing (Fig. 311C).

Common Species
Cnephia pecuarum (Riley) (Fig. 311A). Southern buffalo gnat—4 mm; black; base of abdomen light; eastern 1/2 U.S.

Simulium venustum Say—wing length 2.5-3.0 mm; shiny black; thorax and legs yellowish (♀) or legs yellow and white (♂); antennae 11-segmented (♀); eastern 1/2 U.S., throughout Canada.

Simulium vittatum Zetterstedt—wing length 2-3 mm; dark gray or black; mesothorax with 5 dark longitudinal stripes; basal 1/2 tarsi yellow; 11-segmented antennae (♀); throughout North America.

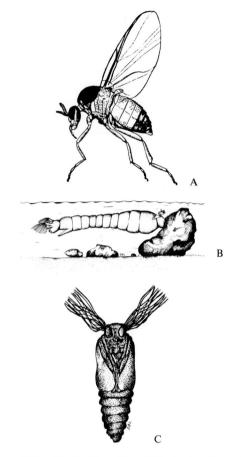

Figure 311 Black flies (Simuliidae). A, *Cnephia pecuarum;* B, a black fly larva; C, a black fly pupa.

Family Bibionidae—March Flies

March flies are usually dark, hairy flies that appear from early spring to summer. They often fly "aimlessly" in swarms. The antennae are short, the tibiae have prominent apical spurs, and the wings have a dark spot on the anterior margin. The thorax is red or yellow in some species and many males have very large eyes. Larvae feed on decaying organic matter. Great swarms of mating *Plecia nearctica* Hardy (often called the lovebug) may occur in May or September in the Gulf coast states.

Common Species

Bibio albipennis Say (Fig. 312A)—winglength 7.5-10.0 mm; shiny black; dense gray or golden pubescence; veins and wing spot yellow-brown; inner spur on front tibia 1/2 length of outer; throughout North America.

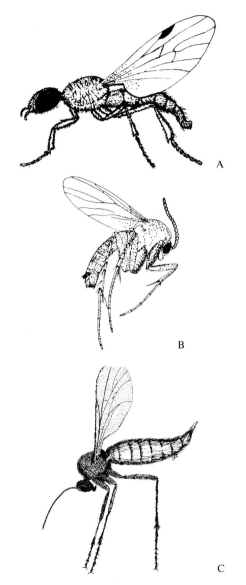

Figure 312 A, a march fly, *Bibio albipennis* (Bibionidae); B, a fungus gnat (Mycetophilidae); C, a darkwinged fungus gnat, *Bradysia* sp. (Sciaridae).

Family Mycetophilidae— Fungus Gnats

These long-legged gnats have very elongated coxae (Fig. 312B). Adults are common in dark places around damp, decaying vegetation. Larvae live in and feed on fungi and other plant materials; some are predators. This family is large with over 600 species in North America.

Family Sciaridae— Darkwinged Fungus Gnats

These small gnats resemble fungus gnats (Mycetophilidae) but have shorter coxae and the kidney-shaped eyes usually meet above the antennae (Fig. 296A). Adults and larvae occur in moist, shaded areas. Larvae feed on fungi and a few species damage mushrooms, potatoes, and roots of greenhouse plants. A *Bradysia* species is illustrated (Fig. 312C).

Family Cecidomyiidae— Gall Midges

Gall midges are small to minute (6 mm or less), slender dipterans with long legs and antennae. Seven or fewer veins reach the wing margin. This large family contains over 1,200 North American species. The majority of the species cause plant galls and each species usually causes a characteristic gall. The pinecone willow gall on the tips of willow branches (Fig. 313B) is caused by *Rhabdophaga strobiloides* (Osten Sacken). *Caryomyia* species cause flat, globular (and sometimes fuzzy), or pointed galls on hickory leaves (Figs. 313C, D). These hickory galls lack an opening; if one is present the gall is caused by a phylloxeran (Homoptera). Yellow blisters with red margins on maple leaves are caused by *Cecidomyia ocellaris* Osten Sacken. Larvae of other species feed on foliage, stems, or seeds, prey on other insects, or are scavengers. Larvae of some gall midges produce daughter larvae (paedogenesis) for several generations.

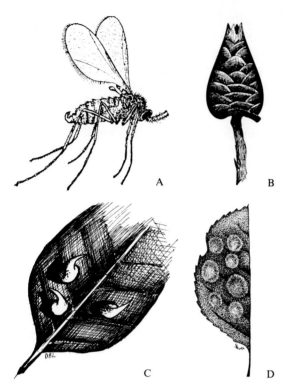

Figure 313 Gall midges (Cecidomyiidae). A, sorghum midge, *Contarinia sorghicola;* B, pinecone willow gall caused by *Rhabdophaga strobiloides;* C and D, galls on hickory leaves caused by *Caryomyia* spp.

Common Species

Contarinia sorghicola (Coquillett) (Fig. 313A). Sorghum midge—2 mm; orange-red; head yellow; antennae and legs brown; thorax with black spot; long ovipositor (♀); southern U.S. west to NM (pink or orange larvae in sorghum and Sudan grass seeds).

Dasineura leguminicola (Lintner). Clover seed midge—1.5 mm; thorax dark brown to black; abdomen red; wings smoky; ovipositor very long (♀); OR to VA, southern Canada (larvae in clover heads).

Mayetiola destructor (Say). Hessian fly—2-4 mm; dark brown or blackish; abdomen reddish orange (♀); wheat-growing regions of North America (larva white with greenish stripe; in wheat, barley, and rye stems causing them to break).

SUBORDER BRACHYCERA

Family Stratiomyidae— Soldier Flies

These small- to medium-sized flies are generally dark-colored although some have bright green or yellow markings. In the genus *Stratiomys* the abdomen is very broad and flat and the elongated, third antennal segment is ringed. Other species are more wasplike in appearance. The antennae of many species are held closely parallel except for the apical halves which abruptly angle away from each other. Adults frequent flowers and occur primarily in wooded areas or meadows near water. Most larvae are terrestrial and often occur in decaying vegetation and wood; some larvae are aquatic.

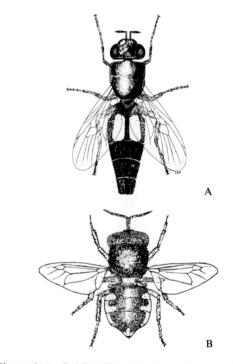

Figure 314 Soldier flies (Stratiomyidae). A, black soldier fly, *Hermetia illucens;* B, *Stratiomys meigenii.*

Common Species

Hermetia illucens (Linnaeus) (Fig. 314A). Black soldier fly—15-20 mm; black; 2 light elongated patches at base of abdomen, apex sometimes reddish; most of U.S. except Northwest.

Stratiomys maculosa Loew—14 mm; black; yellow markings on head; abdomen with curved yellow markings on sides and tip, segments 3-5 each with yellow, triangular, middorsal marking; CA to WA and UT, British Columbia.

Stratiomys meigenii Wiedemann (Fig. 314B)—12-14 mm; bluish black; thorax with gray pubescence on sides; curved yellow stripes on sides of abdomen, tip of abdomen with middorsal yellow line; eastern U.S. southwest to TX, southeastern Canada.

Family Tabanidae— Deer and Horse Flies

Tabanids are medium to large (usually 6-25 mm long), biting flies with large heads. The third antennal segment is elongated and ringed (Fig. 294C). Veins R_4 and R_5 reach the wing margin on each side of the wing tip (Fig. 293). The eyes of males touch each other but are separated in females; the eyes are often iridescent or colored. Most females are bloodsucking and inflict a painful bite on horses, cattle, humans, and other animals. Males (and females to some extent) feed on nectar or pollen. These flies are rapid fliers and are common near marshes and bogs, ocean beaches, lakes, streams, and the edges of woods. Larvae are predaceous and typically occur in water, damp soil, or rotten logs. The immatures are tapered at both ends and the first seven abdominal segments bear a pair of dorsal, lateral, and ventral prolegs (Fig. 315E).

The larger, common tabanids (horse flies) are in the genus *Tabanus*. The third antennal segment of these flies has a basal toothlike projection (Fig. 294C) and apical spurs are absent on the hind tibiae. The smaller tabanids (deer flies) are generally in the genus *Chrysops*. The third antennal segments of these species lack projections but the hind tibiae have apical spurs. Most deer flies are black or brown with dark bands on the wings. Some *Chrysops* species are among the many arthropods that transmit the bacterial diseases anthrax and tularemia to mammals in North America.

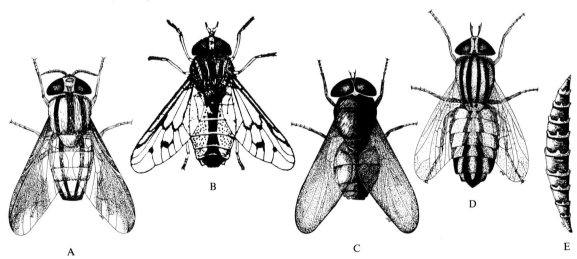

Figure 315 Deer and horse flies (Tabanidae). A, a deer fly, *Chrysops vittatus;* B, a horse fly, *Hybomitra lasiophthalma;* C, black horse fly, *Tabanus atratus;* D, striped horse fly, *Tabanus lineola;* E, horse fly larva.

Common Species

Chrysops aestuans Wulp—10 mm; black raised area between eyes on lower front of face; dorsum blackish with yellow markings on midline and posterior margins of each segment except 1st; wing with posterior basal cell clear; northern 1/2 U.S., OK, Canada.

Chrysops callidus Osten Sacken—10.0-12.5 mm; head black with yellow raised area between eyes on lower front of face; green or golden eyes; thorax striped black and yellow; abdomen with yellow lateral markings and black "V"-shaped middorsal markings; wings with dark smoky markings; eastern 1/2 U.S., southern Canada.

Chrysops noctifer Osten Sacken—9 mm; shiny black; 2 orange spots at base of abdomen; wings with dark smoky markings, both basal cells dark; ID and OR south to NM. *C. proclivis* Osten Sacken similar, wing with posterior basal cell clear, CA and CO northward.

Chrysops vittatus Wiedemann (Fig. 315A)—10 mm; thorax yellowish green with 3 dark longitudinal stripes; abdomen yellow-brown with 4 dark brown longitudinal stripes, tip of abdomen broad and blunt; eastern 1/2 U.S., southeastern Canada.

Hybomitra lasiophthalma (Macquart) (Fig. 315B)—13-15 mm; thorax black with narrow gray stripes; wings clear with brown markings; abdomen reddish on sides; Rocky Mts. eastward, Canada.

Tabanus atratus Fabricius (Fig. 315C). Black horse fly—20-28 mm; black; thorax with whitish pubescence; abdomen with bluish white tinge; wings dark brown or blackish; most of U.S. *T. punctifer* Osten Sacken similar, but more distinct gray-white or yellowish thoracic pubescence, dark spot in center of blackish wings, western 1/2 U.S., southwestern Canada.

Tabanus lineola Fabricius (Fig. 315D). Striped horse fly—20-25 mm; brown with reddish tinge; eyes purple with 3 green stripes; thorax and abdomen with pale yellowish or reddish yellow stripes; conspicuous white median stripe on abdomen; eastern U.S. southwest to TX.

Tabanus quinquevittatus Wiedemann—13 mm; yellowish brown; brilliant green eyes; abdominal segments with 2 lighter patches on each side of a central line; eastern 1/2 U.S., southeastern Canada (common on lake beaches). *T. nigrovittatus* Macquart similar, darker stripes, East Coast and Gulf Coast (salt marshes and beaches).

Tabanus sulcifrons Macquart—20-25 mm; dark gray or dark reddish brown; whitish or yellow middorsal triangles on abdomen; eastern 1/2 U.S.

Family Rhagionidae—Snipe Flies

These long-legged flies are generally 8-15 mm in length and the relatively long abdomen tapers posteriorly. Most species are dark but some have yellow bands or spots. Adults and larvae are predaceous on small insects although adults in the genus *Symphoromyia* bite and feed on the blood of humans and other vertebrates. Larvae in one genus are aquatic; most larvae occur in decaying plant material.

Common Species

Rhagio mystaceus (Macquart) (Fig. 316A)—6-8 mm; blackish with light thoracic stripes and abdominal crossbands; eastern 1/2 North America.

Symphoromyia atripes Bigot—5-8 mm; black; legs reddish, 1 spur on hind tibia; CA to CO and northward, western Canada (bites humans). *S. hirta* Johnson similar, 6.0-7.5 mm, yellow markings, eastern 1/2 U.S.

Family Therevidae—Stiletto Flies

The slender, tapered abdomen of many stiletto flies gives them a resemblance to robber flies (Asilidae). However, the top of the head is flat or convex and not concave as with robber flies. Adults may be plant feeders although larvae appear to be predators. Adults occur on blossoms, foliage, and in open fields and sandy beaches.

Common Species

Psilocephala aldrichi Coquillett (Fig. 316B)—6.0-7.5 mm; black and silver; 2 silvery longitudinal stripes on thorax (♀) or abdomen silver dorsally and relatively stout (♂); tibia and 1st tarsal segment of middle and hind legs light except at apex; CA to CO and northward.

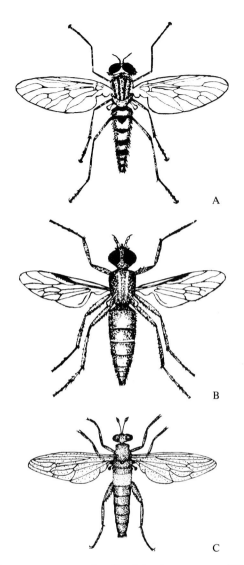

Figure 316 A, a snipe fly, *Rhagio mystaceus* (Rhagionidae); B, a stiletto fly, *Psilocephala aldrichi* (Therevidae); C, a mydas fly, *Mydas clavatus* (Mydidae).

Psilocephala haemorrhoidalis (Macquart)—6.0-7.5 mm; black with white fine pubescence; antennal segments 1-3 with black bristles; thorax with middorsal black stripe and faint white stripe on each side, blue-gray pubescence; stem of halteres yellow and knob black; bases of tibiae and tarsi yellow; eastern U.S. southwest to TX.

Family Mydidae—Mydas Flies

These large, elongated flies have conspicuous 4-segmented antennae with the last segment swollen. Adults and larvae appear to be predaceous. Larvae occur in decaying wood.

Common Species

Mydas clavatus (Drury) (Fig. 316C)—25 mm; black; yellow or orange band on 2nd abdominal segment; red and yellow markings on hind legs; throughout U.S., Ontario.

Family Asilidae—Robber Flies

Robber flies typically have a long abdomen that tapers posteriorly, a bearded face (and often a hairy body), and a concavity on top of the head. The family is large with about 850 species in North America. Some are hairy and stout and resemble yellow and black bumble bees; these are in the genus *Laphria* (Fig. 317B). Asilids are predaceous and will attack other flies, bees and wasps, grasshoppers, etc. A piercing mouthpart is used to suck the prey's blood. Larvae live in the soil or in wood and may feed on beetle larvae and other immature insects.

Common Species

Efferia pogonias (Wiedemann) (Fig. 317A)—13-20 mm; black with gray-brown to white pollenlike covering; abdomen whitish and segments 3-6 with pair of black spots (♀), or segments 1-2 and posterior margins of 3-5 whitish and segments 6-7 silvery (♂); tibiae orange with black tips; most of U.S.

Laphria flavicollis Say (Fig. 317B)—13-19 mm; black; yellow hairs on face and scutellum; abdomen with fine black hair; most of U.S. except Southwest, southern Canada.

Leptogaster flavipes (Loew) (Fig. 317C)—9-13 mm; black; dorsum of thorax dull red-brown; bristle between tarsal claws; pale markings on abdomen; eastern 1/2 U.S.

Order Diptera   337

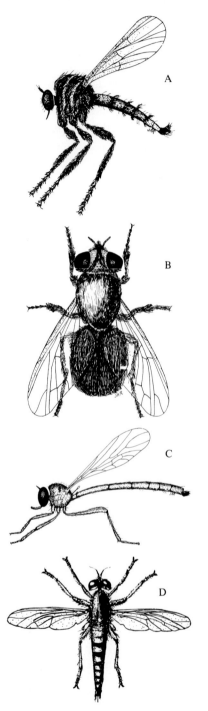

Figure 317 Robber flies (Asilidae). A, *Efferia pogonias;* B, *Laphria flavicollis;* C, *Leptogaster flavipes;* D, *Promachus vertebratus.*

Promachus vertebratus (Say) (Fig. 317D)—28-35 mm; thorax black, thickly covered with gray pubescence; abdomen pale gray with black crossband at base of each segment; tibiae and tarsi orange-yellow; Midwest.

Family Bombyliidae—Bee Files

These stout-bodied flies are usually densely hairy and the long, slender wings are often patterned or spotted. Many species have a very long and slender proboscis. There are over 750 species in North America. Adults commonly hover over flowers, grass, or open ground; they will also land in open areas with their wings outspread. Adults feed on nectar and pollen. Larvae are parasites on other insect larvae and pupae and grasshopper eggs.

Common Species

Anthrax analis Say—7-11 mm; thorax black or brownish; wings dark brown at base, color extends from point where anal veins reach posterior margin along an irregular oblique line to point where R_1 reaches anterior wing margin; sides of 1st abdominal segment with black, white, or mixed hairs; throughout U.S., southern 1/2 Canada.

Anthrax tigrinus (De Geer)—7-11 mm; reddish brown with mixture of black hairs; wings heavily patterned with dark brown interconnected markings; side of 1st abdominal tergum with white hairs in upper 1/2 and black hairs in lower 1/2; eastern U.S. southwest to TX.

Bombylius lancifer Osten Sacken—11 mm; like *B. major;* legs reddish; marking on anterior margin of wing brown shading to gray; Far West.

Bombylius major Linnaeus (Fig. 318A)—7-12 mm; black; long black and brown hairs and gray pollenlike dust on face; thorax with dense yellowish hairs; whitish tuft in front of wings, brown tuft below wings; anterior 1/2 wings dark brown; abdomen with dense yellowish hairs, black tuft on sides of 3rd segment; most of U.S., southern Canada.

Sparnopolius lherminierii (Macquart) (Fig. 318B)—9-10 mm; black with thick golden yellow hairs; eastern 1/2 U.S., MT.

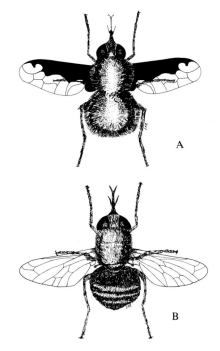

Figure 318 Bee flies (Bombyliidae). A, *Bombylius major;* B, *Sparnopolius lherminierii.*

Family Empididae—Dance Flies

These small- to medium-sized flies frequently have a large thorax and a long, tapering abdomen. The male genitalia are generally conspicuous as a large lobe on the tip of the abdomen (Fig. 319A). Some adults occur in swarms, "dancing" up and down over forest streams and other damp areas with vegetation. Other species occur on flowers and tree trunks. The family is large with over 700 North American species. Males of some species capture a prey and carry it to attract a female. Males of other species surround their prey in a silken balloon or frothy secretion before carrying it. Larvae live in soil, decaying vegetation and wood, or in water; they probably are precadeous.

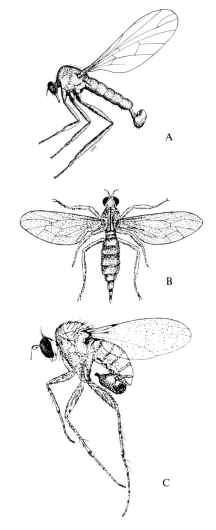

Figure 319 A and B, dance flies (Empididae): A, a male dance fly; B, *Rhamphomyia rava.* C, a long-legged fly, *Dolichopus* sp. (Dolichopodidae).

Common Species

Rhamphomyia rava Loew (Fig. 319B)—9 mm; head, thorax, and abdomen weakly marked with brown; eyes dark reddish brown; wings and legs yellowish brown; eastern U.S., NV.

Family Dolichopodidae—Longlegged Flies

These metallic green, coppery, or blue flies are slender and usually less than 10 mm long. The legs are long and ornamented on many males, the antennae bear an arista, and the male genitalia are often large and folded forward under the abdomen. Dolichopodids are a large family with over 1,230 North American species. These flies frequent lightly shaded areas near water, swamps, and meadows; they are more common in the cooler and mountainous regions of North America. Adults and larvae are predaceous. Larvae are chiefly aquatic but some occur in decaying vegetation and under bark. The genus *Dolichopus* contains the largest and most common species (Fig. 319C).

SUBORDER CYCLORRHAPHA

Family Lonchopteridae—Lonchopterid Flies

These slender flies are 2-5 mm long. The wings are pointed apically and have no crossveins except at the base (Fig. 299A). The third antennal segment is rounded and bears an arista. Lonchopterids occur in grassy, moist, or shady areas. All four North American species are in the genus *Lonchoptera*.

Family Phoridae—Humpbacked Flies

Phorids are tiny, black or sometimes yellowish flies with a humped back and a low, small head. The costal vein extends only about halfway along the anterior wing margin. Strong veins occur in the costal area but all others are weak (Fig. 297C). The hind femora are flattened. Adults occur near decaying vegetation. Larvae live in fungi or decomposing plant material, in nests of termites and ants, and some are parasites of other insects, snails, and, rarely, humans.

Common Species

Megaselia rufipes (Meigen) (Fig. 320A)—2.5-3.0 mm; black or dull to dark brown; legs yellowish or brown; abdomen with long flat hairs on lateral and posterior margins of each segment; throughout North America.

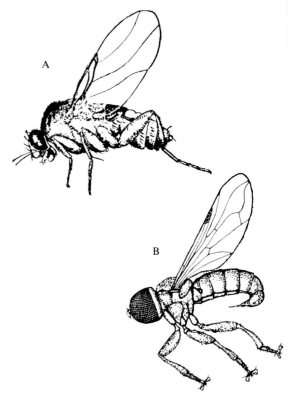

Figure 320 A, a humpbacked fly, *Megaselia rufipes* (Phoridae); B, a bigheaded fly, *Pipunculus* sp. (Pipunculidae).

Family Pipunculidae—Bigheaded Flies

These small flies have a very large head composed chiefly of the compound eyes. Adults commonly hover but also land on flowers. Larvae are parasitic primarily on leafhoppers and planthoppers (order Homoptera). Adults in the most common genus, *Pipunculus* (Fig. 320B), have a darkened area (stigma) in the outer third of the anterior wing margin. Species in the other common genus, *Tomosvaryella*, lack a stigma.

Common Species

Tomosvaryella subvirescens (Loew)—2.5 mm; greenish black; thorax with brownish pollenlike covering; tip of femur and base of tibia yellow, rest of leg black; halteres brown with yellow knob; ovipositor sharp, as long as hind tibia (♀); throughout North America.

Family Syrphidae—Flower Flies

Syrphid flies are medium- to large-sized, common flies that frequently hover and dart rapidly over flowers. These flies are usually recognized by the presence of a spurious vein in the wings (Fig. 300A). The wings are often spread open at rest. Many species are dark brown with yellow bands. Some common syrphids closely resemble bees and wasps in color, shape, and the presence of dense hair. Syrphids are important pollinators. There are about 950 North American species. Larvae vary considerably in appearance but some are sluglike, somewhat wrinkled, and pointed at one end. These types are predators of aphids, other homopterans, and thrips. Others live in ant, termite, or bee nests, and in decomposing vegetation. A few injure bulbs and other plant parts. The rattailed maggots, genus *Eristalis* (Fig. 321C) and related genera, have a long, posterior breathing tube and live in stagnant, polluted water.

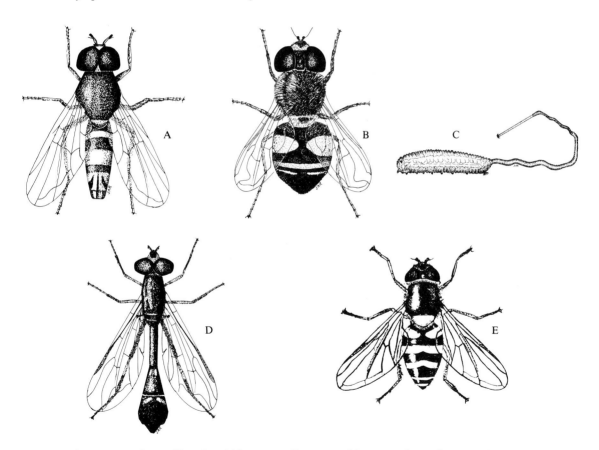

Figure 321 Flower flies (Syrphidae). A, *Allograpta obliqua;* B, drone fly, *Eristalis tenax;* C, a rattailed maggot, *Eristalis* sp.; D, *Pseudodorus clavatus;* E, *Syrphus ribesii.*

Order Diptera

Common Species

Allograpta obliqua Say (Fig. 321A)—6-7 mm; thorax greenish; abdomen dark brown with 2 yellow-orange crossbands in basal 1/2, tip with longitudinal bands; throughout U.S., southern Canada.

Didea fuscipes Loew—12-13 mm; black; side of face and scutellum yellowish brown; abdomen yellow-banded like *Syrphus ribesii;* 4th vein ($R_4 + 5$) with broad dip but not deep as in *Eristalis tenax;* most of U.S., southern Canada.

Eristalis tenax (Linnaeus) (Fig. 321B). Drone fly—15 mm; brown to black; eyes with vertical stripe of tiny dark brown hairs; scutellum yellow; thorax hairy; abdomen with lighter patches as shown; 4th vein (R_{4+5}) with deep dip (Fig. 300A); resembles large honey bee; throughout North America.

Eupeodes volucris Osten Sacken—7-10 mm; shiny green and black; face white with black stripe; yellow spots on abdominal segments 2-5; western 1/2 U.S., LA, British Columbia to Ontario.

Metasyrphus americanus (Wiedemann)—9-10 mm; metallic green; side of face black, dark brown median stripe on face; 3 yellow abdominal crossbands, 2nd and 3rd not reaching margins; throughout U.S., southern Canada.

Paragus bicolor (Fabricius)—4-6 mm; stout-bodied; shiny green to blackish; abdomen red; southern CA to VA and northward, Canada. *P. tibialis* (Fallén) similar, 3-5 mm, tip of abdomen red or yellow, throughout U.S.

Pseudodorus clavatus (Fabricius) (Fig. 321D)—9-11 mm; slender, wasplike; greenish blue and reddish brown; last 2 abdominal segments expanded, each with 2 spots; most of U.S.

Scaeva pyrastri (Linnaeus)—11-14 mm; face yellow; eyes and antennae reddish brown; thorax metallic blue-black, covered with dense, fine hairs; abdomen shiny black with 3 pairs of arched spots; Rocky Mts. westward, southeast to AR, western 1/3 Canada.

Syrphus opinator Osten Sacken—9-11 mm; yellow face; eyes without hairs; 3 abdominal crossbands, 1st broken in middle but reaching sides, 2nd and 3rd unbroken and reaching sides; western 1/2 U.S., southwestern Canada.

Syrphus ribesii (Linnaeus) (Fig. 321E)—8-12 mm; blackish; eyes without hairs; 4-5 pale yellow crossbands on abdomen, 1st band broken and 3rd constricted in middle; 4th vein ($R 4 + 5$) without dip; throughout North America. *S. torvus* Osten Sacken similar, eyes hairy.

Family Conopidae—Thickheaded Flies

The head of these brown flies is broader than the thorax and the proboscis is long and slender. Because the abdomen of many species is long and often slender at its base, these species resemble small threadwaisted wasps (Sphecidae). Adults frequent flowers; larvae are parasites of adult bees and wasps. Species in the genus *Physocephala* (Fig. 322A) have a greatly constricted abdomen basally, an r-m crossvein that meets the discal cell much beyond the middle of the cell, and hind femora that are thickened in the basal half. Species in the genus

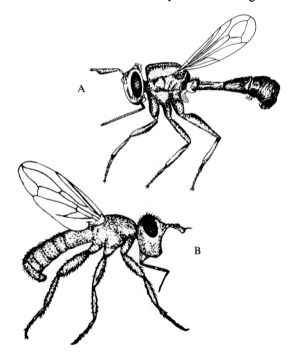

Figure 322 Thickheaded flies (Conopidae). A, *Physocephala* sp.; B, *Myopa* sp.

Physoconops are similar except that the r-m crossvein is at or before the middle of the discal cell. Members in the genus *Myopa* (Fig. 322B) lack a greatly constricted abdomen, the proboscis is elbowed near the middle and base, and the area below the eye is as wide as the eye is high.

Family Otitidae—Otitid Flies

The wings of Otitids usually have black, brown, or yellowish markings (the family has been called picture-winged flies for this reason). The body may be metallic or shiny. Otitids closely resemble fruit flies (Tephritidae) but the subcosta is not abruptly bent upward in the former. Adults occur in meadows and other grassy, moist areas. Larvae live in decayed plant materials; some are pests, feeding on onions, sugar beets, and other crops.

Common Species

Delphinia picta (Fabricius) (Fig. 323A)—7-8 mm; reddish brown; scutellum yellow; wings reddish brown and opaque white; eastern 1/2 U.S.

Tetanops myopaeformis (Röder). Sugarbeet root maggot—6 mm; shiny black; smoky patch along basal 1/3 anterior margin; Rocky Mts. westward, southwestern Canada (in sugar beet fields).

Family Tephritidae—Fruit Flies

These small- to medium-sized flies have banded or spotted wings and varied body colors. Fruit flies resemble the Otitidae but the former have the apex of the subcosta more sharply angled (Fig. 302A). Adults occur on foliage, flowers, and fruit. Many species slowly move their wings up and down while walking or resting. Larvae feed chiefly in fruit, nuts, flowers, stems, and roots (some forming galls); a few are leaf miners. Larvae in the genus *Eurosta* cause round galls in stems of goldenrod (Fig. 323C) and other plants. Many larvae are serious fruit pests throughout the world.

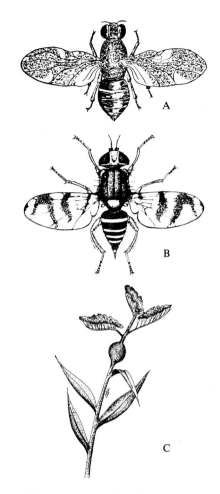

Figure 323 A, an otitid fly, *Delphinia picta* (Otitidae); B and C, fruit flies (Tephritidae): B, cherry fruit fly, *Rhagoletis cingulata*; C, gall in goldenrod stem caused by *Eurosta* sp.

Common Species

Eurosta solidaginis (Fitch)—8.5-9.0 mm; reddish brown; 2 bristles on dark scutellum; wings broad, yellow speckles, bristles on 3rd major vein; ovipositor conical and stout; northern 2/3 U.S., Canada.

Rhagoletis cingulata (Loew) (Fig. 323B). Cherry fruit fly, cherry maggot—2-5 mm; blackish; thorax with yellow-white lines and margins; tibiae and tarsi yellow; abdomen with white bands; eastern U.S. (larva in cherries, other fruit). *R. fausta* (Osten Sacken), black cherry fruit fly, similar, 5-6 mm, ab-

Order Diptera

domen all black, northern 1/2 U.S. (larva more common in sour cherries). *R. indifferens* Curran, western cherry fruit fly, similar, Far West.

Rhagoletis pomonella (Walsh). Apple maggot—5-6 mm; shiny black; head orange with greenish eyes; thorax with yellow-white lines on sides; scutellum and crosslines on abdomen yellow-white; legs yellowish; wings black-banded; eastern 1/2 U.S., southeastern Canada (larva in apples, other fruit).

Family Coelopidae— Seaweed Flies

These flies have a distinctly flattened thorax and a bristly body and legs. They are common on seaweeds (kelp) that have washed up on beaches. Adults also occur on flowers. Larvae live in the decomposing seaweed. All five North American species are in the genus *Coelopa*. *C. vanduzeei* Cresson is 5-6 mm long and occurs on the Pacific Coast. *C. frigida* (Fabricius) is on the Atlantic Coast from Rhode Island northward.

Family Sepsidae— Black Scavenger Flies

Sepsids are typically shining black, or sometimes purple, flies. The head is round and the abdomen narrows basally; some species are antlike in appearance. A dark spot may occur near the tip of the wing along the anterior margin. Adults and larvae frequent manure, carrion, and decaying vegetation, but adults also are collected in grassy meadows and woods.

Common Species

Sepsis punctum (Fabricius) (Fig. 324A)—4 mm; purplish black; dull spot on anterior wing margin near tip; tubercle on front femora (♂); abdomen with several bristles; throughout U.S.

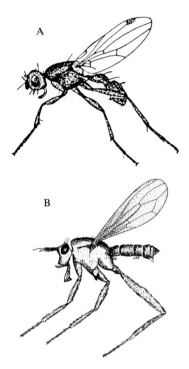

Figure 324 A, a black scavenger fly, *Sepsis punctum* (Sepsidae); B, a marsh fly, *Sepedon fuscipennis* (Sciomyzidae).

Family Sciomyzidae—Marsh Flies

Marsh flies are common, small- to medium-sized yellowish brown or grayish flies. They occur around marshes, ponds, lakes, bogs, or wooded areas. The wings may be spotted or otherwise marked. The antennae extend forward noticeably (Fig. 324B). Larvae are aquatic or terrestrial and prey on freshwater or terrestrial snails, slugs, fingernail clams, or snail eggs.

Common Species

Sepedon fuscipennis Loew (Fig. 324B)—6-7 mm; reddish brown; median stripe on front of face, central depression of face 3x wider than lateral depression; 2nd antennal segment longer than 3rd; scutellum with 2 bristles; hind femora extending much beyond abdomen; eastern 1/2 U.S. northwest to OR, Canada.

Family Lauxaniidae—Lauxaniid Flies

These small flies (usually 2-7 mm long) are typically stout-bodied and vary in color, although some common ones are yellowish. The wings may be spotted; the subcosta reaches the anterior wing margin but the anal vein does not reach the posterior wing margin. The postvertical bristles are bent inward (Fig. 291A) and the oral vibrissae are absent. Adults occur in moist and shady areas, meadows, and on bushes. Larvae live in decaying plant material; a few are leaf miners.

Common Species

Minettia lupulina (Fabricius) (Fig. 325A)—6.5-7.0 mm; stout-bodied; grayish black; thorax dark gray on sides, middorsal gray line; scutellum dark gray; wings long, slightly yellowish-opaque; most of North America.

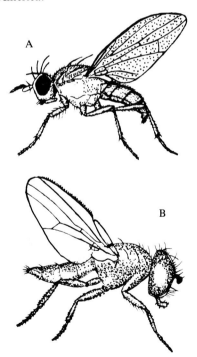

Figure 325 A, a lauxaniid fly, *Minettia lupulina* (Lauxaniidae); B, a milichiid fly, *Pholeomyia indecora* (Milichiidae).

Family Milichiidae—Milichiid Flies

The small flies in this family are black or sometimes the abdomen is silvery. Adults occur in sunny, open areas on logs, fences, and in grass. Larvae live in decaying plant material, carrion, and manure.

Common Species

Pholeomyia indecora (Loew) (Fig. 325B)—3 mm; thorax black, grayish or brownish, heavy covering of pollenlike dusting; abdomen blackish brown; wings tinted brownish; mesopleuron with a row of 3 bristles along posterior margin; throughout U.S. (primarily eastern), Canada.

Family Ephydridae—Shore Flies

Shore flies are small (many are less than 5 mm long) and dark colored. The face sometimes bulges outward anteriorly. The anterior wing margin has two breaks (Fig. 302B) and the anal cell is absent. Adults occur along streams, ponds, marshes, alkaline lakes, and the seashore. They often appear in great swarms and many will walk on the water's surface feeding on algae. Larvae live in fresh or brackish water, salt water, and even the highly saline water of the Great Salt Lake in Utah. Common species whose larvae live in brackish water or salt water are in the genus *Ephydra*; larvae have a long anal tube with two slender breathing tubes inside (Fig. 326A). One California species lives in crude petroleum pools and a few species (genus *Hydrellia*) live in stems or mine leaves of aquatic plants. Shore flies are an important food item for waterfowl.

Common Species

Ephydra cinerea Jones (Fig. 326A)—2.5-3.5 mm; opaque gray; front of face shiny green; tibiae and tarsi yellowish; CA to UT, south to TX (on saline and alkaline lakes, Great Salt Lake).

Hydrellia griseola (Fallen)—2.0-2.5 mm; dull greenish brown; front of head white, yellow, or brown; thorax and abdomen metallic blue-green on sides; femora blue-green, tibiae dark gray; most of U.S., southern Canada (mines leaves of irrigated cereals).

Ochthera mantis (De Geer) (Fig. 326B)—4-5 mm; brownish black; face very broad, yellowish brown; front femur very enlarged, tibia with apical spur 1/3 its length, front tarsi black; most of U.S., southern Canada.

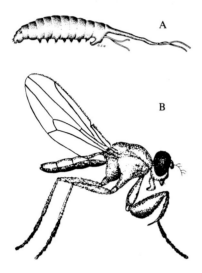

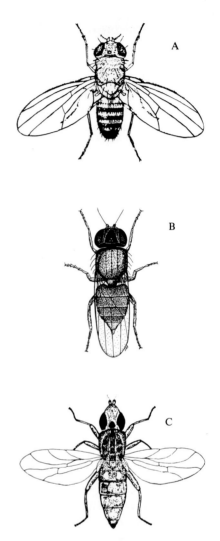

Figure 326 Shore flies (Ephydridae). A, *Ephydra cinerea;* larva B, *Ochthera mantis.*

Family Drosophilidae— Vinegar Flies

The vinegar flies are 3-4 mm long and generally yellowish. Adults are very common around ripe and fermenting fruit and also occur on decaying vegetation. Larvae live in decomposing fungi and fruit, feeding on yeasts that grow in the fruit. The larvae of a few species are external parasites or predators of insects. Adults in the common genus *Drosophila* have a plumose arista on the antennae (Fig. 294E), the postvertical bristles are large, and the face is ridged.

Common Species

Drosophila melanogaster Meigen (Fig. 327A)— 3 mm; yellowish brown; dark crossbands on abdomen; eyes red; throughout U.S., southern Canada (species used commonly in genetics research).

Figure 327 A, a vinegar fly, *Drosophila melanogaster;* B and C, chloropid flies (Chloropidae): B, eye gnat, *Hippelates pusio;* C, wheat stem maggot, *Meromyza americana.*

Family Chloropidae— Chloropid Flies

These tiny (1.5-2.5 mm long), stout flies have a large, triangular, raised area on top of the head extending down toward the antennae. Some species are black and yellow and others are all black or gray. Adults are most common in grass and other low vegetation. Species in the genus *Hippelates* (Fig. 327B) are attracted to blood and animal secretions. These gnats crawl into the eyes of humans and other animals and are severe nuisances. They do not bite but can transmit pinkeye (bacterial conjunctivitis). *Hippelates* species have a distinct, shining black spur on the apex of the hind tibia. Some chloropid larvae breed in grass stems and a few are plant pests; others live in manure and soil with high organic content (e.g., *Hippelates* larvae).

Common Species

Hippelates pusio Loew (Fig. 327B). Eye gnat—1.5 mm; black or dark gray; legs and tip of abdomen pale yellow; throughout U.S.

Meromyza americana Fitch (Fig. 327C). Wheat stem maggot—1.5 mm; yellowish green; 3 dark dorsal stripes on thorax; hind femora large; west of Rocky Mts.

Oscinella frit (Linnaeus). Frit fly—1-2 mm; black; halteres yellow; legs may have yellow markings; throughout U.S. except FL (on grasses, cereals; larvae inside stems).

Family Agromyzidae— Leafminer Flies

The tiny to small black or yellowish flies in this family are common on vegetation but easily overlooked because of their small size. Larvae are chiefly leaf miners and often make a narrow, winding mine that widens toward one end as the larva grows. Some miners are crop pests. Many kinds of plants are mined and it is frequently easiest to identify the fly species by its characteristic mine and the particular host plant. For example, *Ophiomyia simplex* (Loew) larvae mine asparagus stems throughout the U.S. (Fig. 328A). *Phytomyza* species mine leaves of chrysanthemum, holly, columbine, and other plants.

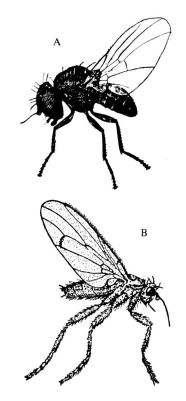

Figure 328 A, a leafminer fly, *Ophiomyia simplex* (Agromyzidae); B, a heleomyzid fly, *Aecothea fenestralis* (Heleomyzidae).

Family Anthomyzidae— Anthomyzid Flies

Anthomyzids are small, slender, slightly elongated flies that occur on grasses in wet areas. The postvertical bristles (Fig. 291A) are bent inward or absent, and at least one pair of fronto-orbital bristles point upward. Larvae breed in sedges and marsh grasses.

Family Heleomyzidae— Heleomyzid Flies

These brownish flies are generally 4-6 mm long and have distinct bristles on the anterior wing margin. These flies resemble the marsh flies (Sciomyzidae) but differ by having smaller, nonprojecting antennae and oral vibrissae

(Fig. 291A). Adults occur in humid, shady areas. Larvae live in fungi and decaying animal and plant material including bird nests and rodent burrows.

Common Species
Aecothea fenestralis (Fallén) (Fig. 328B)—5 mm; brown; legs and wings yellowish, dark crossveins on wings; northern 1/2 U.S. west to Dakotas, British Columbia (also common in caves).

Family Anthomyiidae— Anthomyiid Flies

Anthomyiids are medium-sized (average length 6 mm), dark-colored flies that often resemble houseflies (Muscidae). Anthomyiids differ from muscids by having the anal vein (or its fold) reaching the posterior wing margin. One subfamily, Scatophaginae, contains some common yellowish, slender, and densely hairy species quite unlike other anthomyiids (Fig. 329B). Adults and larvae in this subfamily are often predators of other insects. Scatophagines have sometimes been classified as a separate family, Scatophagidae, in other texts. Larvae of numerous anthomyiids feed on roots and other parts of plants; many are crop pests. Adults of some species are common on kelp and seaweed along the Pacific coastline and the aquatic larvae are predaceous. There are over 560 North American species of anthomyiids.

Common Species
Fucellia rufitibia Stein—3 mm; slender; gray; tibiae reddish; Pacific Coast (on kelp piled on beaches).

Hylemya brassicae (Wiedemann) (Fig. 329A). Cabbage maggot—5 mm; resembles small, slender housefly; light or dark gray; thorax with black stripes; hind femur with ventral clump short bristles at base; northern 1/2 U.S., southeastern U.S., southern 1/2 Canada (larvae in roots of cabbage, cauliflower, radish, others). *H. antiqua* (Meigen), onion maggot, similar, 3-5 mm, dull gray, northern 1/2 U.S., southern Canada (larvae in onions).

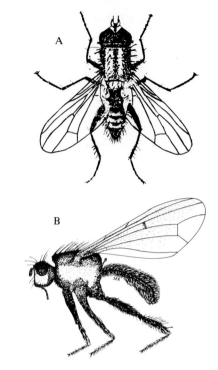

Figure 329 Anthomyiid flies (Anthomyiidae). A, cabbage maggot, *Hylemya brassicae;* B, *Scatophaga stercoraria.*

Hylemya platura (Meigen). Seedcorn maggot— 5 mm; resembles *H. brassicae;* hind femur with row of short ventral bristles nearly to tip (♂); throughout North America (larvae in seeds or stems of corn, beans, peas, melons, others).

Pegomya hyoscyami (Panzer). Spinach leafminer— 5 mm; gray; white face; northern 1/2 U.S., southern Canada (larvae mine spinach leaves).

Scatophaga stercoraria (Linnaeus) (Fig. 329B)— 7-9 mm; slender; very dense yellow or yellowish brown hair; arista with several long hairs on basal 1/2, longest is 1/2 width of 3rd antennal segment; 5th abdominal segment with posterior corners elongated and 2 apical processes between them; most of North America (often on cow manure in pastures). *S. furcata* (Say) similar, arista bare, front femora dark but covered with yellowish pollenlike dusting, middle and hind femora light.

Family Muscidae—House Flies Stable Flies, and Allies

The muscids are a large family (over 700 North American species) of typically dark-colored, medium-sized flies. They resemble some species of the Anthomyiidae but wing vein 2A does not meet the wing margin in muscids (Figs. 302C, D). Muscids may appear similar to some tachina flies (Tachinidae) or flesh flies (Sarcophagidae) but members of the latter two families have strong bristles on the hypopleura and pteropleura (Fig. 303) whereas muscids have only weak hairs or the pleura are bare (Fig. 292). Many muscids are major pests through their biting activities (both males and females) or transmission of diseases. The house fly, *Musca domestica* Linnaeus, can be a vector of dysentery, typhoid fever, anthrax, and other diseases because of its association with excrement that contains disease organisms. Muscid larvae breed primarily in decaying plant material or manure.

Members in the genus *Fannia* are common and resemble small house flies. The wings of *Fannia* are characterized by a parallel-sided P_1 cell and an upward-curved 3A vein (Fig. 302D). Adults generally hover or zigzag through the air and typically occur under sheds (especially near poultry and other animals) or in entrances to buildings.

Common Species

Fannia canicularis (Linnaeus) (Fig. 330A). Little house fly—3-5 mm; like smaller, more slender house fly; dark gray; 3 darker longitudinal stripes on thorax; wings as in Fig. 302D; sides of abdomen pale yellow; throughout North America.

Fannia scalaris (Fabricius). Latrine fly—4-6 mm; like *F. canicularis;* thorax and abdomen dark gray to bluish black; 4 darker longitudinal stripes on thorax; middorsal abdominal stripe forms triangular markings; middle tibia with tubercle (♂); most of North America.

Haematobia irritans (Linnaeus). Horn fly—4-5 mm; similar to *S. calcitrans* but body smaller, more slender; long conspicuous proboscis with black and yel-

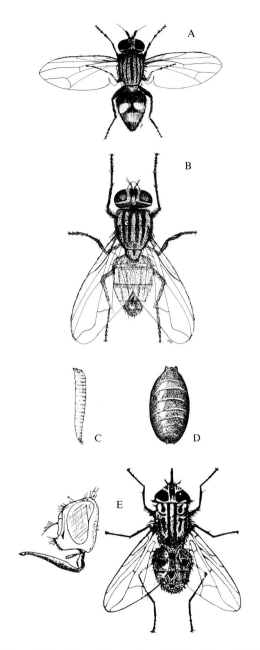

Figure 330 House flies and stable flies (Muscidae). A, little house fly, *Fannia canicularis;* B-D, house fly, *Musca domestica:* B, adult; C, larva; D, pupa. E, stable fly, *Stomoxys calcitrans.*

low palpi on each side; wings widespread at rest; most of U.S., southern Canada (bites cattle on their backs, clusters around base of horns; also on horses, sheep).

Musca autumnalis De Geer. Face fly—8-10 mm; nearly identical to *M. domestica* but darker gray; eyes nearly meet (♂), silver stripe around eyes (♀); abdomen black on sides (♀) or brownish orange (♂); most of U.S. except southernmost states, southern Canada (on face of cattle; do not bite but suck blood from wounds).

Musca domestica Linnaeus (Figs. 330B-D). House fly—5-9 mm; resembles *S. calcitrans;* proboscis with fleshy apical lobe; outer thoracic stripes complete; R_5 cell nearly closed at wing tip; abdomen with lighter patch on each side near base, not spotted; throughout North America (larva yellow-white, 7-11 mm long, anterior end pointed with 2 black mouth hooks, posterior end blunt with 2 distinct dark spiracles).

Muscina stabulans (Fallén). False stable fly—7-10 mm; similar to *S. calcitrans* but stouter, usually larger; scutellum yellowish at tip; legs reddish brown; abdomen blotchy black and gray; most of North America (enters houses).

Ophyra leucostoma (Wiedemann)—3.5-6.0 mm; shiny black or blue-black; calypteres (Fig. 293) dark; hind tibiae curved, ventral surface with fine hairs; similar to *Fannia* spp. but stouter and vein 3A not curved upward; throughout U.S., southern Canada (around garbage dumps, urban areas).

Stomoxys calcitrans (Linnaeus) (Fig. 330E). Stable fly—6-7 mm; gray, may have greenish yellow luster; proboscis protrudes forward, very slender at tip; outer thoracic stripes broken; R_5 cell widely open at wing tip; dark spots on abdomen; most of North America (bites; resembles *M. domestica;* on lake beaches, seashores, near barns).

Family Hippoboscidae— Louse Flies

Adult louse flies are external, bloodsucking parasites of birds and, to a lesser extent, mammals. Winged adults are very flat, leathery brown, 4-6 mm long, and found on birds. Hippoboscids are the only common flies on living birds. Wingless adults are generally found on mammals such as deer and goats. Larvae of some species develop in a uteruslike pouch of the female, feeding on her glandular secretions. The larvae are ready to pupate when deposited on the host.

Common Species

Icosta americana (Leach) (Fig. 331A)—7.0-8.5 mm; dark brown; palpi no longer than 1/2 height of head; thorax with middorsal line and transverse groove forming a cross, sides grayish; abdomen lighter above; throughout North America (on grouse, quail, hawks, owls). *I. albipennis* (Say) similar, 4-5-6.0 mm, eastern 1/2 U.S. northwest to Wa, southern Canada (on shore birds, doves).

Melophagus ovinus (Linnaeus) (Fig. 331B). Sheep ked—5-6 mm; reddish brown; wings reduced to knobs; long bristly hairs; most of U.S., southern Canada (on sheep).

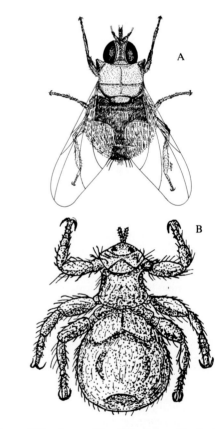

Figure 331 Louse flies (Hippoboscidae). A, *Icosta americana;* B, sheep ked, *Melophagus ovinus*.

Pseudolynchia canariensis (Macquart)—wing length 4.5-7.5 mm; metallic greenish; anal area of wings without hairs; most of U.S. (on pigeons; may enter houses in winter).

Family Calliphoridae—Blow Flies

Blow flies are very common, metallic blue, green, or blackish flies that are usually somewhat larger than house flies and occasionally much more common. Some blow files are similar to flesh flies (Sarcophagidae) but lack the typically gray and black striped thorax. Blow flies also have the arista plumose to the tip (Fig. 294E) whereas it is bare or plumose only on the basal half in flesh flies. Adults that frequent carrion and manure for food or oviposition can spread diseases such as dysentery. Larvae of most species are scavengers and feed on carrion and excrement. Adults of some species deposit eggs in wounds, sores, or nostrils of animals (also humans). The larvae develop in the tissue and cause severe irritation and weakening of the host. The invasion of animal tissue (myiasis) is well documented in the Calliphoridae. The screwworm, *Cochliomyia hominivorax* (Coquerel), was once a major parasite of cattle in the southern third of the U.S., but massive releases of sterilized males into the native population have reduced the pest to minor importance.

Common Species

Calliphora vicina Robineau-Desvoidy—10-11 mm; eyes red; side of face orange; thorax blue-gray; abdomen metallic blue; sclerite at base of costa yellow or orange; throughout North America. *C. vomitoria* (Linnaeus) similar, 10-14 mm, side of face black.

Lucilia illustris (Meigen)—6-9 mm; shiny metallic green, sometimes ranging to dark blue; palpi yellow; like *P. sericata* but slightly smaller, sclerite at base of costa with setae; most of North America.

Phaenicia sericata (Meigen) (Fig. 332A)—6.0-9.5 mm; shiny, metallic yellowish green or bluish green, sometimes with coppery tinge; no markings; often silvery dusting on tip of abdomen; sclerite at base of costa without setae; throughout U.S., southern Canada.

Phormia regina (Meigen). Black blow fly—6-11 mm; robust; bluish black, greenish black, or blue-green; dark thoracic stripes; mesothoracic spiracle with bright orange hairs; like *Phaenicia sericata* but darker, slightly larger; most of North America.

Pollenia rudis (Fabricius) (Fig. 332B). Cluster fly—8.5-10.0 mm; thorax black with dense golden hairs; abdomen checkered black and silver; wings overlap at rest; throughout U.S., southern Canada (on shrubs, flowers, cluster in houses in winter).

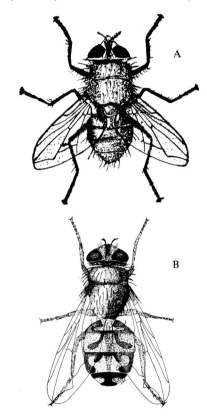

Figure 332 Blow flies (Calliphoridae). A, *Phaenicia sericata;* B, cluster fly, *Pollenia rudis*.

Family Sarcophagidae—Flesh Flies

Flesh flies are medium-sized flies that often have distinct, black and gray, longitudinal thoracic stripes and a checkered abdomen. The arista are bare or plumose to the basal half (rarely two-thirds) whereas the similar Muscidae have fully plumose arista. Females of most species deposit larvae rather than eggs on carrion. Larvae of numerous species in the largest and most common genus, *Sarcophaga*, are parasites of grasshoppers and, to a lesser extent, beetles. Some small species live in ground burrows of wasps and bees, feeding on the stored food or the hymenopteran larvae.

Common Species

Helicobia rapax (Walker)—3-8 mm; gray thorax with 3-5 black stripes; abdomen checkered black and silvery; throughout U.S., southern Canada.

Sarcophaga haemorrhoidalis (Fallén) (Fig. 333A)—10-14 mm; gray thorax with 3-5 distinct black stripes; tip of abdomen reddish; throughout U.S., southeastern Canada.

Senotainia rubriventris Macquart—6-7 mm; gray thorax with indistinct dark stripes; abdomen bright red except at base; throughout U.S., southern Canada.

Wohlfahrtia vigil (Walker)—11-12 mm; black; palpi and 2nd antennal segment yellow; gray thorax with 3 black stripes; abdomen blackish with 3-4 rows gray patches; northern 1/2 U.S., NM, most of Canada.

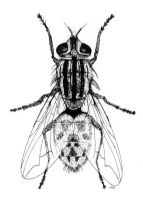

Figure 333 A flesh fly, *Sarcophaga haemorrhoidalis* (Sarcophagidae).

Family Tachinidae—Tachina Flies

This family consists of medium- to large-sized flies that often have numerous bristles on the posterior end of the abdomen or sometimes covering the entire abdomen. Identifying features are (1) postscutellum appears as a prominent lobe beneath the scutellum (Fig. 303A), (2) hypopleural and pteropleural bristles are present (Fig. 303A), and (3) the abdominal terga often overlap the sterna. Tachina flies comprise the second largest family of flies (Tipulidae is largest) with over 1,300 species in North America. Adults feed on nectar, honeydew from homopterans, and other liquids. These species are major parasites of other insects; the larvae commonly feed on larvae of moths, butterflies, sawflies (Hymenoptera), beetles, and other insects. Females deposit eggs or larvae on the host's body and the larvae bore inside to feed; some females lay eggs on a leaf to be eaten by the host along with the leaf.

Common Species

Archytas apicifer (Walker) (Fig. 334A)—11-15 mm; thorax gray with 5 darker indistinct stripes; abdomen shiny metallic blue, 4th segment yellowish on sides and tip; throughout U.S., southern Canada. *A. californiae* (Walker) similar, 10-11 mm, side of face with white hairs, thorax brownish yellow with pollenlike covering, calypteres white, abdomen black.

Bombyliopsis abrupta (Wiedemann) (Fig. 334B)—12-13 mm; very stout body; head and thorax greenish gray, thorax with pale brown margins; abdomen orange-yellow and extremely bristly; most of U.S., southern Canada.

Compsilura concinnata (Meigen)—7-8 mm; black; head white with dorsal black stripe; palpi orange; thorax whitish with 4 black stripes; MN to NJ, CA to British Columbia, southeastern Canada (reared from > 200 spp. lepidopteran larvae).

Paradejeania rutilioides (Jaennicke)—13.5-19.0 mm; body very bristly; head black and brown; thorax black and yellow; abdomen yellow with middorsal black spots, or with middorsal stripe and tip black; southwestern U.S., CA.

Spoggosia claripennis (Macquart) (Fig. 334C)—5-9 mm; black with silvery gray or metallic bluish sheen; head and thorax with gray hairs; dark thoracic stripes; scutellum brown; calypteres whitish and prominent; sides of 2nd abdominal segment brown; throughout U.S., southern Canada.

Winthemia rufopicta (Bigot)—8-12 mm; black with grayish pollenlike covering; greatly narrowed anteriorly (♂); 5 black stripes on mesonotum; scutellum red; hind tibiae with dense hairs of equal length (♂); abdomen red on sides, reddish yellow at tip, broad black median stripe; most of U.S., southern Canada.

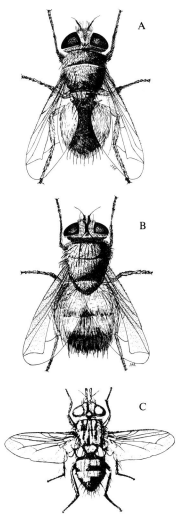

Figure 334 Tachina flies (Tachinidae). A, *Archytas apicifer;* B, *Bombyliopsis abrupta;* C, *Spoggosia claripennis.*

Family Cuterebridae— Rabbit Bots, Rodent Bots

These parasitic, stout-bodied and hairy flies resemble bumble bees. Adults occur in open woodlands and in canyons but are not common. The large, stout larvae (bots) of most species parasitize rabbits and rodents (especially squirrels and chipmunks) by penetrating beneath the skin. The larvae likely to be encountered are in the genus *Cuterebra*. An interesting subtropical species parasitizes animals indirectly by grasping a mosquito, fly, or tick, and laying its eggs on the captive's body. When the carrier lands on birds and mammals (including humans) to feed, the eggs drop off and the larvae hatch and penetrate the host.

Family Oestridae— Bot and Warble Flies

These large, swift flies resemble bees as do bot flies in two other families. The R_5 cell narrows toward the wing apex or the cell is closed, and the postscutellum is large (Fig. 301B). The larvae (bots) (Fig. 335B) are internal parasites of mammals and a warble is a skin swelling caused by a larva. Adults occur around their hosts and although adults do not bite they often annoy animals. The ox warble flies, genus *Hypoderma,* are pests that lay eggs on the legs of cattle. The larvae penetrate the skin and move up to the back where they develop in warbles. Mature larvae exit from the back and pupate in the ground. Adults are not commonly caught and specimens are best obtained by rearing pupae or squeezing larvae from the skin of cattle and preserving them.

Common Species

Hypoderma lineatum (Villers) (Fig. 335A). Common cattle grub—13-14 mm; black; thorax with white longitudinal stripes, white tufts on side of prothorax; posterior part of mesonotum yellowish; reddish hairs on tip of abdomen; throughout U.S., southern Canada. *H. bovis* (Linnaeus), northern cattle grub, similar, 17-18 mm, posterior part of mesonotum black, causes cattle to run when fly is ovipositing.

Oestrus ovis Linnaeus. Sheep bot fly—12-14 mm; head and thorax dull yellowish gray; round black dots on dorsum; abdomen shiny gray; throughout U.S., southcentral Canada (larvae in nasal passages and sinuses of sheep).

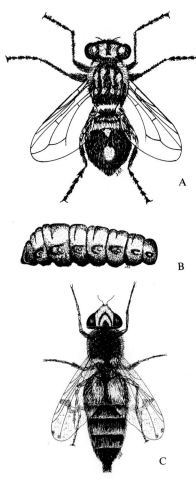

Figure 335 A and B, bot and warble flies (Oestridae): A, common cattle grub, *Hypoderma lineatum;* B, an oestrid larva (bot). C, horse bot fly, *Gasterophilus intestinalis* (Gasterophilidae).

Family Gasterophilidae— Horse Bots

These parasitic flies are usually 9-14 mm long and resemble hairy honey bees. The mouthparts are very small and the flies do not feed. They occur in pastures and near horse barns but are not very common. Eggs are laid on the legs, shoulders, lips, or jaws of horses and usually taken into the mouth when the horse licks itself. The larvae (bots) hatch from the sudden warmth and make their way into the stomach or other areas where they feed on tissues and liquids of the host. The mature larvae leave with the feces and pupate in the ground.

Common Species

Gasterophilus intestinalis (De Geer) (Fig. 335C). Horse bot fly—12-17 mm; yellowish brown; hairy; wings with faint gray median crossband and 2 spots on apex; faint spots and bands on abdomen; throughout U.S., southern Canada. *G. haemorrhoidalis* (Linnaeus), nose bot fly, similar, 15 mm, yellowish white hairs, end of abdomen orange-red, northern 1/2 U.S. south to VA. *G. nasalis* (Linnaeus), throat bot fly, 9 mm, reddish brown thorax, dark crossband around middle of abdomen, most of North America.

GENERAL REFERENCES

Carpenter, S.J., and LaCasse, W.J. 1955. *Mosquitoes of North America (North of Mexico).* Univ. California Press, Berkeley. 360 pp.

Cole, F.R. 1969. *The Flies of Western North America.* Univ. California Press, Berkeley. 693 pp.

Curran, C.H. 1934. *The Families and Genera of North American Diptera.* H. Tripp, Woodhaven, N.Y. 515 pp. (1965 reprint).

Diptera of Connecticut. 1942-1964. Guide to the Insects of Connecticut. Part VI. The Diptera, or True Flies of Connecticut. Bulletins of the Conn. State Geol. and Nat. Hist. Survey 64, 68, 69, 75, 80, 87, 92, 93, 97.

Felt, E.P. 1940. *Plant Galls and Gall Makers.* Comstock Publ., Ithaca, N.Y. 364 pp.

Greenberg, B. 1971. *Flies and Disease.* Vol. 1. *Ecology, Classification and Biotic Associations.* 856 pp. Vol. 2. 1973. *Biology and Disease Transmission.* 447 pp. Princeton Univ. Press, Princeton, New Jersey.

Hall, D.G. 1948. *The Blow Flies of North America.* Thomas Say Foundation. Publ. 4. 477 pp.

James, M.T., and Harwood, R.F. 1969. *Herms's Medical Entomology, 6th ed.* Macmillan, N.Y. 484 pp.

Oldroyd, H. 1964. *The Natural History of Flies.* Weidenfeld and Nicholson, London. 324 pp.

Peterson, A. 1951. *Larvae of Insects. Part II. Coleoptera, Diptera, Neuroptera, Siphonaptera, Mecoptera, Trichoptera.* Edwards Bros., Ann Arbor, Mich. 519 pp.

Stone, A.; Sabrosky, C.W.; Wirth, W.W.; Foote, R.H.; and Coulson, J.R. (eds.). 1965. *A Catalog of the Diptera of America North of Mexico.* USDA Agric. Handbook 276. 1,696 pp. (contains references on taxonomy of all families, genera, and species).

Wirth, W.W., and Stone, A. 1956. Aquatic Diptera. In *Aquatic Insects of California,* R.L. Usinger (ed.). Univ. California Press, Berkeley. Pp. 372-482.

Manual to the Families and Genera of Nearctic Diptera. Biosystematics Research Institute; Agriculture Canada, Ottawa, Ontario. In preparation.

ORDER SIPHONAPTERA[32]
Fleas

Adult fleas are bloodsucking ectoparasites of mammals and to a lesser extent, birds. Flea larvae generally are not found on the host animal and instead ordinarily utilize organic debris as a food source. The larvae undergo a holometabolous type of development.

Adults range in length from 1-10 mm although most are less than 5 mm. Wings are absent and the body is laterally compressed giving a distinct flattened appearance (Fig. 336). The antennae are short, 3-segmented structures which usually fit into grooves on the head. The piercing-sucking mouthparts are enclosed within a beak. Short, stiff bristles of the hard cuticle point posteriorly, enabling adults to move rapidly between the hairs or feathers of a host and preventing their falling off or being readily captured by the host. One or more combs (ctenidia) of broad, flattened teeth (spines) may be present on the head, thorax, or both. The coxae are long and the tarsi are five-segmented. Long, powerful legs enable some fleas to jump 200 times their body's length.

Adult fleas can live several weeks without a blood meal. Their life span ranges from a few weeks to two years or more. Most fleas require blood to reproduce and the majority of the species infest mammals such as rabbits, rats, mice, squirrels, and other rodents. Birds, bats, pigs, dogs, cats, bears, and humans are also hosts.

Fleas respond to warmth and most roam freely over the host's body and sometimes from one host to another. Eggs may be deposited by the female while on or off the host. If the eggs are on the animal, they soon fall to the ground, floor, or the animal's bedding or nest.

Flea larvae are small, whitish, legless individuals with two short, curved posterior spines and chewing mouthparts. They live in the nests of their hosts in dust and other accumulated debris, and feed on the dried excrement of adult fleas and other organisms, dried blood, and dead mites. A few suck blood from the host. Developmental time averages two weeks to several months depending on the environmental conditions. The mature larva spins a silken cocoon prior to pupating.

Fleas may be extremely irritating to the host, causing skin inflammation and itching. The infestation of a house and its occupants with cat fleas several weeks after a pet or stray cat (or dog) has departed is a common occurrence. Although most of the adults were on the cat, the eggs and larvae remained in the house and the adults that eventually developed found

---

32. Siphonaptera: *siphon,* tube; *aptera,* wingless.

only humans available as a food source. Generally cat fleas will survive only a few weeks on human blood and will not reproduce.

Many species of fleas transmit diseases to their hosts and some of these species are discussed later under the appropriate family. For further information on diseases a medical entomology textbook such as *Herms's Medical Entomology* (James and Harwood 1969) should be consulted.

Fleas may be collected on the host or in the immediate living quarters of the host. The cat flea (and less commonly the dog flea) is frequently encountered on cats and dogs during the summer and fall. Hens' nests commonly contain the western chicken flea and hog pens may harbor the human flea. The majority of fleas are found on wild rodents and rabbits which may be difficult to collect, so it is often advisable to live-trap or locate the nests or burrows of the animals. Fleas from a live animal may be collected by running a comb downward against the fur. If the animal has died not more than a few hours earlier, it may be placed in a sealed plastic bag or jar and searched for fleas at a later date. To collect fleas from a dead animal, suspend the host inside a closed jar with a few drops of chloroform. After several minutes remove and brush the host. Preserve all fleas in 70% alcohol. If fleas are to be mounted on a slide, clearing first with 10% sodium or potassium hydroxide produces the most useful specimen.

Species and subspecies: North America, 275; world, 2,100. Families: North America, 7.

KEY TO FAMILIES OF SIPHONAPTERA

The major external structures used to identify fleas are shown in Figure 336. Antepygidal bristles are large bristles situated just in front of the sensilium.

1a	Three combined thoracic terga shorter than first abdominal tergum; spinelike bristles on inner side of hind coxae near apex *(Echidnophaga, Tunga)* .... (p. 357) *Pulicidae* (in part)
1b	Three combined thoracic terga longer than first abdominal tergum (Fig. 336) 2
2a(1b)	One row of bristles on abdominal terga 2-7 (Fig. 336); each side of sensilium (Fig. 336) with 14 pits....... .... (p. 357) *Pulicidae* (in part)
2b	Usually 2 rows of bristles on abdominal terga 2-7 (Fig. 339B); 14-16 pits on each side of sensilium.. 3
3a(2b)	Outer dorsal apical bristle of front femur shorter than inner bristle; on large carnivores .... (p. 360) *Vermipsyllidae*
3b	Outer dorsal apical bristle of front femur longer than inner bristle 4
4a(3b)	Forehead (frons) with a tubercle sunk in a groove; no combs .... (p. 360) *Rhopalopsyllidae*
4b	Without above combination of characters.................... 5
5a(4b)	Posterior margin of metanotum without small spines............. .... (p. 358) *Hystrichopsyllidae*

5b	Posterior margin of metanotum with small spines 6		8a(6b)	Genal comb present or absent; arch of tentorium visible in front of eye .. .... (p. 359) *Leptopsyllidae* (in part)
6a(5b)	Interantennal suture present (Fig. 338B)................. 7		8b	Genal comb absent; no arch in front of eye ..... (p. 359) *Ceratophyllidae*
6b	Interantennal suture very weak or absent 8			
7a(6a)	Genal comb consists of 2 broad lobes on each side of head; on bats .... (p. 360) *Ischnopsyllidae*			
7b	Genal comb usually present but not shaped as above; not on bats .... (p. 359) *Leptopsyllidae* (in part)			

Family Pulicidae

Most of the species that attack man and domestic animals are found in this small family. The abdominal terga have one transverse row of bristles and the three thoracic terga combined are longer than the first abdominal tergum (Fig. 336). Two groups, including the sticktight flea *(Echidnophaga)* and the chigoe flea *(Tunga),* have the combined three thoracic terga shorter than the first abdominal tergum.

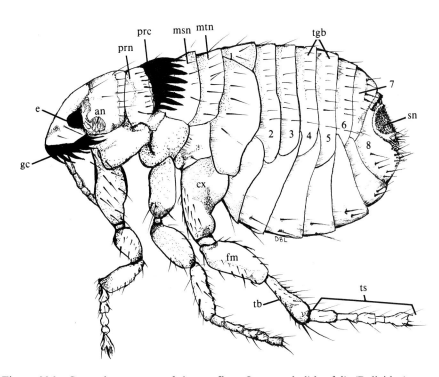

Figure 336 General structures of the cat flea, *Ctenocephalides felis* (Pulicidae). an, antennae; cx, coxa; e, eye; fm, femur; gc, genal comb; msn, mesonotum; mtn, metanotum; prc, pronotal comb; prn, pronotum; sn, sensillium; tb, tibia; tgb, tergal bristles; ts, tarsus; 1-8, abdominal terga.

The oriental rat flea, *Xenopsylla cheopis* (Rothschild), is the major vector of bubonic plague (Black Death), a disease caused by bacteria. Plague is most commonly transmitted by regurgitating the bacteria into the wound while biting the host. Murine (endemic) typhus, a rickettsial disease of man, is most often transmitted by scratching infected feces of the oriental rat flea into the skin. The dog flea, *Ctenocephalides canis* (Curtis), is an intermediate host for the dog tapeworm. The tapeworm develops in the flea larva after the larva ingests a tapeworm egg. When a dog bites an adult flea, the tapeworm enters the dog's digestive system. Rodent tapeworms are transmitted by the human flea, *Pulex irritans* Linnaeus, the dog flea, and the oriental rat flea. The chigoe flea, *Tunga penetrans* (Linnaeus), is a subtropical species (placed in the family Tungidae by some authors) that ranges along the Gulf of Mexico in the southern U.S. Females burrow into the skin of man (especially the feet) and other animals causing painful sores.

Common Species

Cediopsylla inaequalis (Baker) (Fig. 337A)—0.9-2.0 mm; forehead angulate in front; 5-6 stout, straight and blunt genal teeth; 7-12 bristles evenly spaced on margin of clasper (♂); western North America (on jack rabbits and cottontails).

Cediopsylla simplex (Baker)—Similar to *C. inaequalis* but clasper bristles not evenly spaced and usually 8 genal teeth; Great Plains to Atlantic, southeastern Canada (on cottontails, occasionally foxes and rodents).

Ctenocephalides canis (Curtis) (Fig. 337B). Dog flea—1.5-2.0 mm; forehead high and well-rounded; genal comb with 7-8 sharp teeth and 1st tooth may be much shorter than the 2nd; throughout U.S. and southern Canada (primarily on dogs, also cats, humans).

Ctenocephalides felis (Bouché) (Fig. 336). Cat flea—1.5-2.5 mm; low and sloping forehead; genal comb with 7-8 sharp teeth and 1st (most anterior) tooth about equal in length to the 2nd; throughout U.S. and southern Canada (most common species on both cats and dogs in western U.S., also on coyotes, foxes, rabbits, rats, humans).

Echidnophaga gallinacea (Westwood) (Fig. 337C). Sticktight flea—1 mm; 3 combined thoracic terga shorter than the 1st abdominal tergum; head angular and not rounded; sticks tightly to host; primarily southern U.S. (on poultry, other birds, mammals).

Family Hystrichopsyllidae

This family is the largest in the order, containing over 100 North American species. The majority are parasites of small rodents and shrews, and a few feed on small carnivores. Adults have an interantennal suture, and the posterior margin of the metanotum lacks spines.

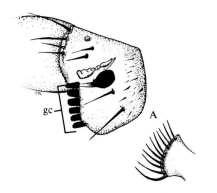
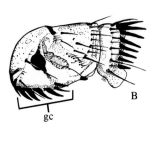
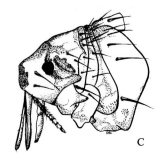

Figure 337 Pulicidae. A, head and clasper on tip of abdomen of male *Cediopsylla inaequalis*; B, head of dog flea, *Ctenocephalides canis*; C, head of sticktight flea, *Echidnophaga gallinacea*. gc, genal comb.

Common Species

Ctenophthalmus pseudagyrtes Baker—2.9-3.3 mm; genal comb with 3 teeth; preantennal region with an upper row of 5 bristles and a lower row of 3 longer bristles; east of Rocky Mts. (on moles, shrews, mice, ground squirrels, other rodents).

Doratopsylla blarinae C. Fox—2.0-2.5 mm; genal comb with 4 teeth, and last tooth nearly horizontal and pressed to genal border; northeastern U.S. and southeastern Canada (on shrews).

Epitedia wenmanni (Rothschild) (Fig. 338A)— 1.8-2.6 mm; genal comb consists of 2 overlapping teeth; 9th abdominal sternite with 9-12 short, black spines (♂); throughout North America (prefers deermice [especially *Peromyscus* in eastern U.S.], other small rodents).

Meringis parkeri Jordan—2.0-2.3 mm; genal comb with 2 overlapping teeth; apex of 9th abdominal sternite rounded (♂); arid regions of western U.S. (on kangaroo rats, other desert rodents).

Stenoponia americana (Baker)—4.0-5.0 mm; 12-14 teeth on genal comb; comb of 1st abdominal tergite similar in size to pronotal comb; most abundant in fall and winter; eastern North America (prefers mice, especially *Peromyscus*).

Family Leptopsyllidae

Most of the species of this family are parasitic an Old World rats *(Rattus)* and mice *(Mus)* and to a lesser extent on native mice *(Peromyscus, Microtus)* and other small mammals. The interantennal suture may be present (Fig. 328B) or absent, a row of spines occurs on the posterior margin of the mesonotum, and an arch occurs in front of the eyes of some species. *Leptopsylla segnis* (Schönherr) (Fig. 338B) is an imported species confined primarily to port cities. Most of the native species are in the genus *Peromyscopsylla*.

Common Species

Leptopsylla segnis (Schönherr) (Fig. 338B). European mouse flea—1.0-1.75 mm; genal comb with 4 teeth; imported species in coastal areas of U.S. (on commensal rats and mice, especially the housemouse).

Peromyscopsylla hesperomys (Baker)—2.5 mm; genal comb with 2 teeth; throughout northern U.S., southern Canada (on mice [prefers *Peromyscus]).*

Family Ceratophyllidae

Most species of this large family are parasites on rodents although a few attack birds and small carnivores. Adults have two or more rows of bristles on each abdominal tergite and lack both a genal comb and interantennal suture.

The California ground squirrel flea of the western U.S. and Mexico, *Diamanus montanus* (Baker), and Franklin's ground squirrel flea of central U.S. and Canada, *Opisocrostis bruneri* (Baker), transmit sylvatic plague to rodents and occasionally to man. The disease is an endemic strain of the bacterium that causes bubonic plague and is generally confined to the Far West.

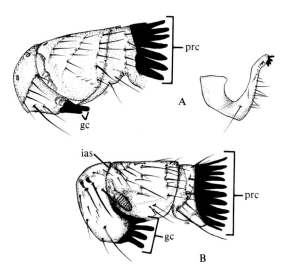

Figure 338 A, head and ninth sternite of *Epitedia wenmanni* (Hystrichopsyllidae); B, head of European mouse flea, *Leptopsylla segnis* (Leptopsyllidae). gc, genal comb; ias, interantennal suture; prc, pronotal comb.

Order Siphonaptera

Common Species

Ceratophyllus niger C. Fox (Fig. 339A). Western chicken flea—3.0-3.3 mm; dark color; 2 rows of bristles on preantennal region with 3-6 bristles in upper row and 3 much longer bristles in lower; pronotal comb with 26-28 teeth; western North America (on poultry, other birds).

Foxella ignota (Baker). Pocket gopher flea—2.7-3.0 mm; eyes absent; pronotal comb with 10-11 teeth; 2 (♂) or 3 (♀) antepygidal bristles; central and western U.S., southwestern Canada (on pocket gophers).

Orchopeas howardii (Baker). Gray squirrel flea—2.0-2.5 mm; preantennal region with 1 distinct row of bristles; mesonotum and metanotum with 2 rows of bristles; throughout North America (primarily eastern) (on gray squirrel, other rodents).

Orchopeas leucopus (Baker) (Fig. 339B)—2.0-2.5 mm; 3 postantennal bristles in addition to a row of 5 bristles along posterior lateral margin of head; pronotal comb with 9-10 teeth; 3 rows of bristles on mesonotum and metanotum; throughout North America (primarily eastern) (on mice, especially *Peromyscus* and *Microtus*).

Family Rhopalopsyllidae

The three species found in the U.S. are restricted to the Gulf coast states. *Polygenis gwyni* (C. Fox), a parasite of opossums, rats, and dogs, has the broadest distribution and ranges from southeastern U.S. to Texas.

Family Ischnopsyllidae—Bat Fleas

Members of this family are normally parasitic on bats and are readily recognized by the hard comblike flaps on each side of the mouth opening. *Myodopsylla insignis* is a common species on *Myotis* bats of eastern North America.

Family Vermipsyllidae—Carnivore Fleas

The five species in North America (genus *Chaetopsylla*) are parasites of large carnivores including foxes, coyotes, wolves, and bears.

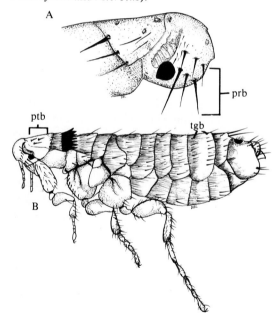

Figure 339 Ceratophyllidae. A, head of western chicken flea, *Ceratophyllus niger;* B, *Orchopeas leucopus.* plb, posterior lateral bristles of head; prb, preantennal bristles; ptb, postantennal bristles; tgb, tergal bristles.

GENERAL REFERENCES

Fox, I. 1940. *Fleas of Eastern United States.* Hafner Publ. Co., N.Y. 191 pp. (1968 reprint).

Holland, G.P. 1949. The Siphonaptera of Canada. Canada Dept. Agric. Pub. 817, Tech. Bull. 70. 306 pp.

Holland, G.P. 1964. Evolution, Classification, and Host-relationships of Siphonaptera. Ann. Rev. Entomol. 9:123-146.

Hopkins, G.H.E., and Rothschild, M. 1953-71. An Illustrated Catalogue of the Rothschild Collection of Fleas (Siphonaptera) in the British Museum (Natural History). London: British Museum. Vol. 1-5.

———. 1964. Evolution, Classification, and Host-relations of Siphonaptera. Ann. Rev. Entomol. 9:123-146.

Hubbard, C.A. 1947. *Fleas of Western North America.* Hafner Publ. Co., N.Y. 533 pp. (1968 reprint).

James, M.T., and Harwood, R.F. 1969. *Herms's Medical Entomology,* 6th ed. Macmillan, N.Y. 484 pp.

Lewis, R.E. 1972-75. Notes on the Geographic Distribution and Host Preferences in the Order Siphonaptera. Parts 1-6. J. Med. Ent. 9:511-520, 10:255-260, 11:147-167, 11:403-413, 11:525-540, 11:658-676.

ORDER HYMENOPTERA[33]
Ants, Bees, Sawflies, Wasps, and Allies

The Hymenoptera range from some of the smallest known insects (0.2 mm) to large specimens 75 mm or more in length. Most species have four membranous wings with relatively few veins. The hind wings are smaller than the front wings and are attached to the front wings by a row of tiny hooks on the anterior margin. Most species have a constriction between the thorax and abdomen (Figs. 356; 367). Common exceptions to the above characteristics are ants (wingless workers), velvet ants (wingless females), and sawflies (thorax broadly connected to the abdomen). Mouthparts are used for chewing (e.g., leaves, pollen, other insects), or for both chewing and sucking (e.g., nectar, insect body fluids). The female's ovipositor is used to deposit eggs and is often very elongated and modified to sting, pierce, or saw (Figs. 355C; 356). Adults are the principal insect pollinators of flowering plants.

Parthenogenic reproduction (i.e., without fertilization) is especially common in the Hymenoptera; unfertilized eggs develop into males and fertilized ones become females. Some species have two generations which are markedly different from each other (i.e., all females or both males and females, host and behavior differences) or the males are quite different in appearance from females. A few hymenopterans apparently produce only females (e.g., larch sawfly). The eggs of certain parasitic wasps will divide into additional embryos (polyembryony) at the beginning of embryogenesis and each new individual (2-1,000) develops into an adult.

Larvae are either maggotlike or caterpillarlike and undergo a holometabolus type of development. They may be parasites, predators, plant feeders, or scavengers. Adults and larvae exert a continuous population control over other insects and spiders. Solitary wasps and bees are hymenopterans that live independently and do not depend on other members of a colony to share in the raising of young or the maintaining of a nest. A solitary wasp may build a nest in the ground, in wood and plant stems, or it may construct a mud nest on various objects. The female provisions the cells of the nest with paralyzed insects or spiders for her larvae's food and often seals the cells when finished. Other solitary wasps deposit eggs in or on a host's eggs, larvae, or pupae. If the wasp larva feeds externally on a host larva, the latter is first stung by the adult female to paralyze it before laying the egg. Sawflies are solitary hymenopterans with larvae that feed on leaves, stems, and wood. Most bees are solitary in behavior and gather pollen and nectar for larval food rather than insects and spiders.

Ants, social wasps, and social bees live in colonies where castes and a division of labor occur. The workers are sterile females that continuously feed the young and maintain and defend the nest. The queen only lays eggs and the few males, alive only for a short time, mate with the new queens. For further information refer to the families Formicidae, Vespidae, and Apidae.

Hymenopterans are divided into two suborders: Symphyta and Apocrita. Members of the Symphyta (sawflies and horntails) have the abdomen broadly joined to the thorax rather than constricted or stalked (Figs. 352A, B; 354A). The well-developed ovipositor of females is used to insert eggs into host plants. The name "sawflies" is derived from the sawlike appearance of the ovipositor of most females. The larvae of all but one family are phytophagous; the majority feed externally on foliage although some are stem or wood borers or leaf miners. Most larvae resemble caterpillars in the order Lepidoptera but differ by having six or

---

33. Hymenoptera: *hymeno*, membrane; *ptera*, wings.

more pairs of prolegs and the prolegs lack crochets (Fig. 352C). Lepidopterous larvae have five or fewer pairs of prolegs and all bear crochets. There are nearly 1,000 species of sawflies and horntails in North America and about 7,500 in the world.

The majority of hymenopterans are in the suborder Apocrita. The base of the abdomen is constricted and often stalked (Figs. 356; 367; 370A), and the ovipositor is modified to pierce or sting. Larvae are legless. There are numerous superfamilies in the suborder. For example, the chalcidoids are tiny wasps belonging to 18 families in the superfamily Chalcidoidea. The antennae are elbowed and there are few or no wing veins. Adults feed primarily on plant exudations and insect honeydew. Females puncture insect hosts with their ovipositors to insert eggs and/or sometimes feed on blood from the wounds. Chalcidoid larvae are primarily internal parasites of the eggs and larvae of Lepidoptera, Homoptera, and some Diptera and Coleoptera. Some species are hyperparasites (i.e., parasites of other parasites). Entomologists have used these wasps as parasites in the biological control of scale insects and mealybugs (Homoptera) and other plant pests.[34]

The superfamily Apoidea includes about 3,500 species of bees in North America and 20,000 species worldwide. Bees have branched body hairs and the first segment of the hind tarsus is flattened in most cases. These features are lacking in wasps. Bees also feed their young honey and pollen rather than insects and other animal material fed by wasps. The mouthparts of bees are noticeably elongated to suck up nectar. About 90% of the bees are solitary, building individual nests in the ground or in stems or wood. The honey bee and bumble bees are examples of social species. Cuckoo bees (used here to mean species in several families) lay eggs in nests of other bee species and the foster parents raise the young. Bees are the most important pollinators of plants. Pollen sticks to the body hairs and is periodically transferred to the hind tibiae or to hairs on the ventral side of the abdomen for transport.

Hymenopterans may be obtained by using common insect collecting nethods such as net sweeping and operating lights at night. Malaise traps work very well. Parasitic forms are commonly acquired by sweeping with a net or collecting and rearing large numbers of potential hosts to obtain the adults that emerge from the parasitized insects. Galls, mud nests, and small paper nests may be collected and placed in a plastic bag to await adult emergence. Sugar baits will attract ants. Ants are picked up with forceps or stunned first by touching them with a small brush dipped in alcohol. Some solitary wasps and bees may be induced to build nests in bored-out doweling and pithy stems, or in plastic straws. These tubes are grouped side by side in an open container such as a milk carton or inserted individually into holes drilled in a block of wood. The container or block is placed outdoors in a horizontal position.

Adults are pinned and larvae are placed in 70% alcohol. Minute wasps are placed in alcohol, glued on their sides to points, or placed on microscope slides.

Species: North America, 17,100; world, 110,000. Families: North America, 74.

---

34. The term "wasps" generally refers to all species in the suborder except ants, velvet ants, and bees. Some taxonomists use the term mostly for families with stinging species (e.g., Vespidae, Pompilidae, Sphecidae, and a few others).

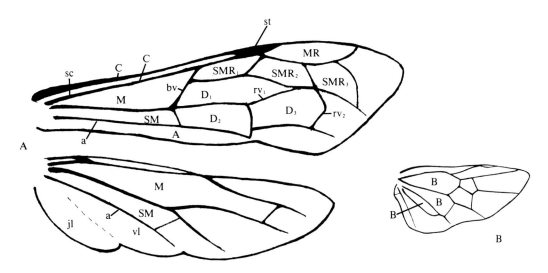

Figure 340 General wing venation. A, wasp; B, hind wing of a sawfly. a, anal vein; A, anal cell; B, basal cell; bv, basal vein; c, costa; C, costal cell; D_{1-3}, first, second, and third discoidal cells; jl, jugal lobe; M, median or basal cell; MR, marginal cell; rv_{1-2}, first and second recurrent veins; sc, subcosta; SM, submedian or basal cell; SMR, submarginal cell; st, stigma; vl, vannal lobe.

KEY TO COMMON FAMILIES OF HYMENOPTERA

1a   Base of abdomen broadly joined to thorax (Fig. 352B); hind wings commonly with 3 closed basal cells (Fig. 340B); trochanters 2-segmented (Fig. 344A) (Suborder Symphyta). . 2

1b   Base of abdomen gradually or abruptly narrowed, often joined to thorax by a slender stalk with or without dorsal nodes (Figs. 343; 367); hind wings with 2 or fewer closed basal cells M, SM, (Fig. 340A); trochanters 1- or 2-segmented (Suborder Apocrita)............ 7

2a(1a)   Front tibia with 1 apical spur 3

2b   Front tibia with 2 apical spurs (Fig. 341C).................. 4

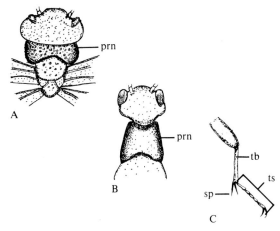

Figure 341 A, pronotum of a horntail (Siricidae); B, pronotum of a stem sawfly (Cephidae); C, front tibia with two apical spurs. prn, pronotum; sp, spur; tb, tibia; ts, tarsus.

3a(2a)   Pronotum constricted in center and wider than long (Fig. 341A); dorsal spine or plate on apex of abdomen (Fig. 354A); ovipositor long, stout (♀) (Fig. 354A)............ .... (p. 372) Siricidae

Order Hymenoptera   363

3b	Pronotum only slightly constricted in center and about as wide as long (Fig. 341B); no dorsal spine or plate on apex of abdomen; ovipositor not long or stout . . . . . (p. 372) *Cephidae*
4a(2b)	Antennae 3-segmented, 3rd segment very long . . . . (p. 370) *Argidae*
4b	Antennae with 4 or more segments 5
5a(4b)	Antennae clubbed (Fig. 342C); large, resemble bumble bees . . . . (p. 370) *Cimbicidae*
5b	Antennae not clubbed; do not resemble bumble bees 6
6a(5b)	Antennae 7- to 12-segmented but usually 9-segmented, generally filiform . . . . . . (p. 371) *Tenthredinidae*
6b	Antennae with 13 or more segments and usually serrate (♀) or pectinate (♂) (Fig. 342A) . . . . (p. 370) *Diprionidae*

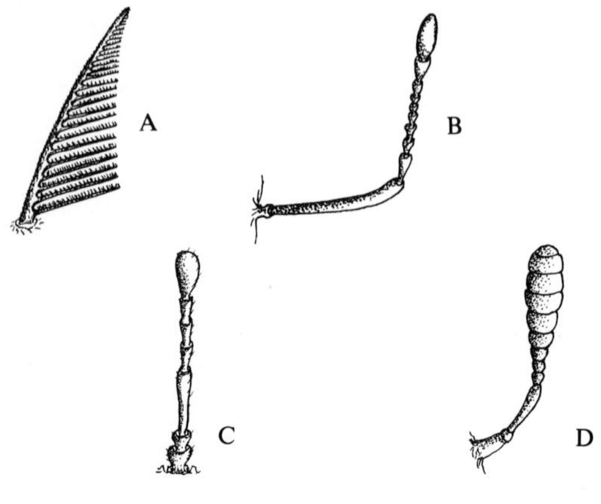

Figure 342 Antennae. A, male conifer sawfly (Diprionidae); B, ant (Formicidae); C, cimbicid sawfly (Cimbicidae); D, Masaridae.

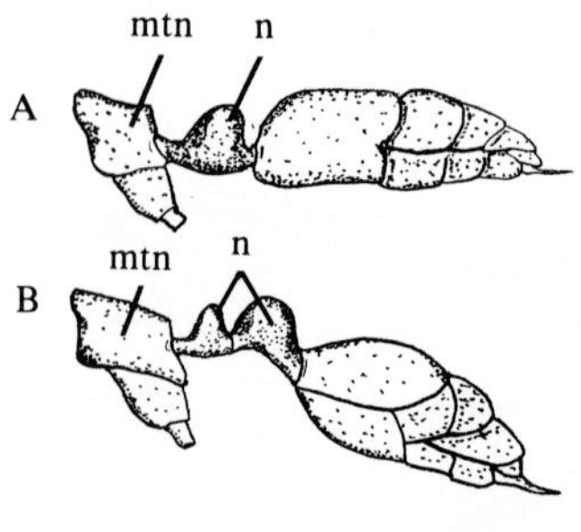

Figure 343 Abdomen of ants (Formicidae). A, one node; B, two nodes. mtn, metanotum; n, node.

7a(1b)	First or first 2 abdominal segments narrowed and bearing 1-2 nodes or humps (Fig. 343); antennae usually elbowed, first segment long (Fig. 342B); winged or wingless . . . . (p. 380) *Formicidae*
7b	First or first 2 abdominal segments narrow or broad and do not bear a node or hump; antennae not elbowed; winged or wingless 8
8a(7b)	Wingless (♀); body usually very hairy and bright-colored (Fig. 362B); narrow band of dense hair on side of second abdominal tergite . . . . (p. 379) *Mutillidae* (in part)
8b	Winged; body hairy or bare 9
9a(8b)	Hind leg with 2 trochanters between femur and coxa (Fig. 344A) 10
9b	Hind leg with 1 trochanter 21
10a(9a)	Front wings without closed cells (Fig. 345A); generally small, parasitic wasps 11
10b	Front wings with 1 or more closed cells (Fig. 345B) 15

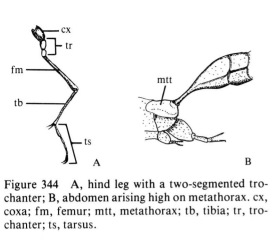

Figure 344 A, hind leg with a two-segmented trochanter; B, abdomen arising high on metathorax. cx, coxa; fm, femur; mtt, metathorax; tb, tibia; tr, trochanter; ts, tarsus.

11a(10a) Tarsi 3-segmented; antennae 3- to 8-segmented; 1 mm or less in length ...... (p. 374) *Trichogrammatidae*

11b Tarsi 4- or 5-segmented; antennal segmentation variable; usually 1 mm or longer. 12

12a(11b) Hind femora greatly enlarged (Fig. 358C), often toothed ventrally; thorax large. ... (p. 376) *Chalcididae*

12b Hind femora and thorax not as above 13

13a(12b) Hind tibiae with 2 apical spurs; pronotum wide and nearly square; black or yellowish insects; not metallic .... (p. 375) *Eurytomidae*

13b Hind tibiae with 1 apical spur; pronotum variable in shape; often metallic 14

14a(13b) Antennae 4- to 8-segmented; apical spur of front tibia short and straight .... (p. 375) *Aphelenidae*

14b Antennae usually 13-segmented; apical spur of front tibia long and curved ...... (p. 375) *Pteromalidae*

15a(10b) Costal cell present (Fig. 340A); ovipositor long or very short 16

15b Costal cell absent (Fig. 345C); ovipositor often very long 19

16a(15a) Stigma of front wing absent (Fig. 359A); abdomen usually oval and shiny, arises on thorax between bases of hind coxae or very near .... (p. 376) *Cynipidae*

16b Stigma of front wing present (Fig. 340A); abdomen oval or elongated, arises on thorax much above bases of hind coxae (Fig. 344B) 17

17a(16b) Abdomen oval and compressed (especially females), held on end of slender stalk much above thorax (Fig. 340A); abdomen oval or not protruding ... (p. 377) *Evaniidae*

17b Abdomen elongated and held much above thorax; ovipositor often as long as body 18

18a(17b) One recurrent vein or none (Fig. 360B); antennae relatively short .... (p. 377) *Gasteruptiidae*

18b Two recurrent veins (Fig. 340A); antennae relatively long .... (p. 378) *Aulacidae*

Order Hymenoptera 365

19a(15b)	Costa and stigma extremely broad (Fig. 345C).................. (p. 372) *Braconidae* (in part)		22a(21a)	Abdomen extremely long (about 40 mm) and slender (Fig. 360C); antennae long, 14 or more segments; black, shiny . . . . (p. 378) *Pelecinidae*
19b	Costa and stigma not extremely broad (Figs. 345B, D).......... 20		22b	Not as above................. 23

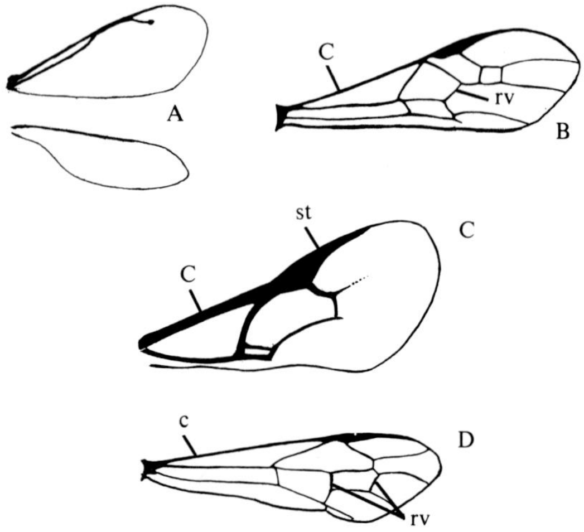

23a(22b) Hind wing with a distinct lobe at base (Fig. 346A); antennae 13-segmented; abdomen often concave ventrally; metallic green, blue or purplish; dense, distinct punctures on body; beelike, 6-12 mm .... (p. 379) *Chrysididae*

23b Hind wing without a lobe at base; antennae 12-segmented at most; abdomen, color, and sculpturing not as above; 5 mm or less 24

Figure 345 Wings. A, Pteromalidae; B and C, front wings of braconids (Braconidae); D, front wing of ichneumons (Ichneumonidae). c, costa; rv, recurrent vein; st, stigma.

20a(19b) One recurrent vein (Fig. 345B) or none; abdomen usually not greatly elongated...................... .... (p. 372) *Braconidae* (in part)

20b Two recurrent veins (Fig. 345D), or if only one then abdomen is 3 times as long as remaining body .... (p. 373) *Ichneumonidae*

21a(9b) Hind wings without closed cells (Fig. 346A)................... 22

21b Hind wings with 1 or more closed cells (Figs. 346B, C) 25

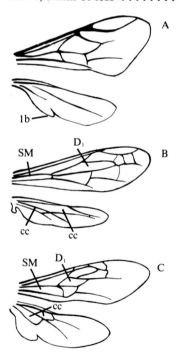

Figure 346 Wings. A, cuckoo wasps (Chrysididae); B, hornets and yellowjackets (Vespidae); C, Scoliidae. cc, a closed cell; D_1, first discoidal cell; lb, lobe; SM, submedian cell.

24a(23b) Antennae 11- to 12-segmented, or 7-9 segments if apical club is unsegmented; nearly veinless front wings usually with costal vein arising from stigma; abdomen usually without dorsal curved process . . . . (p. 378) *Scelionidae*

24b Antennae 9- to 10-segmented; no costal or stigmal veins, often without any veins; abdomen of some with dorsal, forward-curving process . . . . (p. 378) *Platygastridae*

25a(21b) First segments of hind tarsi slender, cylindrical and not flattened, usually shorter than other tarsal segments combined; body hairs unbranched; abdomen often stalked at base (Figs. 365A, 367D) 26

25b First segments of hind tarsi usually as long as or longer than other tarsal segments combined, generally broadened, flattened or thickened; some thoracic hairs branched (Fig. 347A); abdomen not stalked; bees 33

26a(25a) Pronotum with a rounded lobe on each side, lobe does not reach the tegula (Fig. 347C); viewed from above, collarlike pronotum has a straight posterior margin . . . . (p. 385) *Sphecidae*

26b Pronotum without a rounded lobe on each side, pronotum sometimes reaching the tegula (Fig. 347B); viewed from above, posterior margin of pronotum curves forward. 27

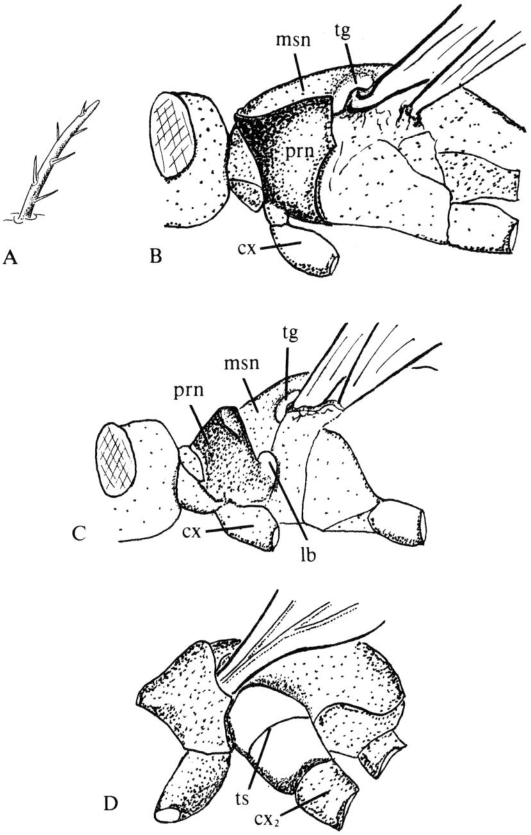

Figure 347 A, branched thoracic hair of a bee; B and C, lateral view of pronotum: B, a vespid (Vespidae); C, a sphecid (Sphecidae). D, lateral view of thorax of a spider wasp (Pompilidae). cx, coxa; lb, lobe; msn, mesonotum; prn, pronotum; tg, tegula; ts, transverse suture.

27a(26b) Wings folded once lengthwise when at rest; first discoidal cell of front wing longer than submedian cell (Fig. 346B) 28

27b Wings usually not folded lengthwise when at rest; first discoidal cell of front wing shorter than submedian cell (Fig. 346C) 29

Order Hymenoptera

28a(27a)	Tarsal claws toothed .... (p. 382) *Vespidae*		31a(30b)	Antennae gradually clubbed (Fig. 342D); two submarginal cells; western U.S. .... (p. 384) *Masaridae*
28b	Tarsal claws not toothed .... (p. 384) *Eumenidae*		31b	Antennae not clubbed; 2-3 submarginal cells; widely distributed 32
29a(27b)	Hind femora long, when extended backward usually reaching to or beyond tip of abdomen; side of mesothorax with a transverse suture between base of wing and middle coxa (Fig. 347D) .... (p. 385) *Pompilidae*		32a(31b)	Mesosternum with a small plate slightly overlapping each middle coxa (Fig. 348B); slightly to moderately hairy but without narrow band of dense hair on side of second abdominal tergite .... (p. 379) *Tiphiidae*
29b	Hind femora usually do not reach tip of abdomen; side of mesothorax without a transverse suture 30		32b	Mesosternum without a small plate slightly overlapping each middle coxa; very hairy with dense narrow band of hair on side of second abdominal tergite; males .... (p. 379) *Mutillidae* (in part)
30a(29b)	Mesosternum and metasternum together form a large plate divided by a transverse suture, the suture meeting the middle coxae (Fig. 348A); fine longitudinal wrinkles on wings beyond closed cells .... (p. 380) *Scoliidae*		33a(25b)	Hind wing with a jugal lobe equal to or longer than submedian cell (Figs. 349A, B) 34
30b	Mesosternum and metasternum do not form a plate although a small plate may slightly overlap each middle coxa (Fig. 348B); wing membrane not wrinkled beyond closed cells 31		33b	Hind wing with a jugal lobe shorter than submedian cell (Fig. 349C) or absent (Fig. 349D) 38
			34a(33a)	Basal vein of front wing distinctly arched, usually 3 submarginal cells (Fig. 349B); glossa pointed (Fig. 350A) ..... (p. 388) *Halictidae*
			34b	Basal vein straight or only slightly arched (Fig. 349A), 2-3 submarginal cells; tip of glossa variable 35
			35a(34b)	Two submarginal cells in front wing 36
			35b	Three submarginal cells in front wing (Fig. 349A) 37

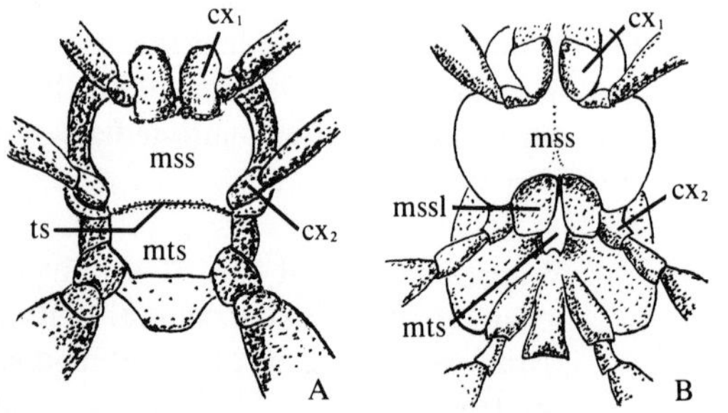

Figure 348 Ventral view of the thorax. A, mesosternum and metasternum joined to form a plate (Scoliidae); B, sterna not joined. cx, coxa; mss, mesosternum; mssl, lobe or plate of mesosternum; mts, metasternum; ts, transverse suture.

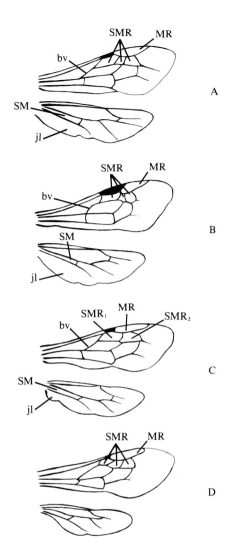

Figure 349 Wings. A, andrenid bees (Andrenidae); B, halictid and sweat bees (Halictidae); C, leafcutting bees (Megachilidae); D, bumble bees and honey bees (Apidae). bv, basal vein; jl, jugal lobe; MR, marginal cell; SM, submedian cell; SMR, submarginal cell.

36a(35a) Glossa pointed (Fig. 350A); 2 subantennal sutures below base of each antenna (Fig. 351A); tip of marginal cell in front wing usually squared-off; submarginal cells equal in length ...... (p. 388) *Andrenidae* (in part)

36b Glossa usually bilobed or squared-off (Fig. 350B); 1 subantennal suture below base of each antenna; tip of marginal cell pointed; second (outer) submarginal cell shorter than first .... (p. 387) *Colletidae* (in part)

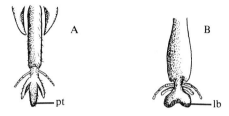

Figure 350 Proboscis. A, pointed glossa (Andrenidae, Halictidae); B, bilobed glossa (Colletidae). lb, lobe; pt, point.

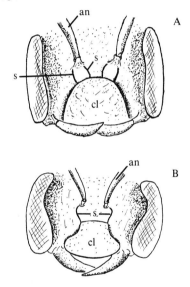

Figure 351 Subantennal sutures. A, two sutures (Andrenidae); B, one suture (Megachilidae). an, antenna; cl, clypeus; s, suture.

37a(35b) Glossa pointed (Fig. 350A); 2 subantennal sutures below base of each antenna (Fig. 351A) .... (p. 388) *Andrenidae* (in part)

37b Glossa bilobed or squared-off (Fig. 350B); 1 subantennal suture be-

low base of each antenna ....
.... (p. 387) *Colletidae* (in part)

38a(33b) Two submarginal cells in front wing (Fig. 349C); 1 subantennal suture extending downward from outer margin of each antennal base (Fig. 351B); underside of abdomen often densely covered with hairs and pollen (females) ....
.... (p. 389) *Megachilidae*

38b Three submarginal cells (Fig. 349D); 1 subantennal suture extending from center of each antennal base; abdomen not as above 39

39a(38b) Hind tibiae without apical spurs, or if spurs present then hind wing lacks a jugal lobe (Fig. 349D) ....
.... (p. 390) *Apidae*

39b Hind tibiae with apical spurs; hind wing with a jugal lobe. ....
.... (p. 389) *Anthophoridae*

SUBORDER SYMPHYTA
Sawflies and Horntails

Family Argidae—Argid Sawflies

These dark sawflies are less than 16 mm long and are recognized by their greatly elongated third antennal segment. This segment is U- or Y-shaped in males of a few species. Larvae feed chiefly on shrub or tree leaves.

Common Species

Arge pectoralis (Leach). Birch sawfly—9-11 mm; orange with black antennae, head, and legs (♀); black with dorsum of thorax red (♂); wings blackish; northeastern and northcentral U.S., southeastern Canada (larvae: head red-yellow, body yellowish with 6 rows dorsal black spots and 3 rows lateral spots; on birch, willow).

Family Cimbicidae— Cimbicid Sawflies

The cimbicids are large (18-25 mm long), stout insects with clubbed antennae. A few resemble bumble bees. There are only six species in North America.

Common Species

Cimbex americana Leach (Fig. 352A). Elm sawfly—18-28 mm; dark blue; 3-4 yellowish spots on sides of abdomen (♀) or distinct white spot near thorax (♂); wings smoky; most of U.S. and Canada, uncommon in southeastern U.S. (larvae: pale green or yellow, black dorsal stripe, feed in coiled position on elm, willow, other trees).

Family Diprionidae— Conifer Sawflies

Members of this family are usually less than 12 mm long and stout-bodied. The antennae have 13 or more segments and are serrate (♀) or pectinate (♂). Larvae of many species in the genus *Neodiprion* are serious defoliators of conifers (especially pines).

Common Species

Neodiprion lecontei (Fitch) (Fig. 352B). Redheaded pine sawfly—5-10 mm; head and thorax pale reddish brown and abdomen black (♀) or body all black (♂); ♂ shorter and more slender than ♀; eastern 1/2 U.S., southeastern Canada (larvae: yellowish white with reddish head, 6 longitudinal rows black spots).

Neodiprion sertifer (Geoffroy) (Fig. 352C). European pine sawfly—10-11 mm; yellow-brown to dark brown; dark wing venation; northeastern U.S. west to WI, southeastern Canada (adults in Sept.-Oct.; larvae: gray-green or dull yellowish, light dorsal and dark lateral stripe, head black, cluster on tips of pine needles).

Neodiprion taedae linearis Ross. Loblolly pine sawfly—8-10 mm; orange and black (♀) or entirely black (♂); SC to TX, north to IL (larvae: dull green, black stripes, on loblolly and shortleaf pines).

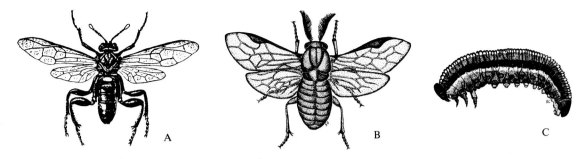

Figure 352 A, elm sawfly, *Cimbex americana* (Cimbicidae); B and C, conifer sawflies (Diprionidae): B, redheaded pine sawfly, *Neodiprion lecontei;* C, European pine sawfly larva, *N. sertifer*.

Family Tenthredinidae—Sawflies

Adults in this family usually have 9-segmented antennae (range: 9-12) and front tibiae with two spurs (Fig. 341C). This is the largest and most common sawfly family with about 750 species in North America and 6,000 worldwide. Adults frequent a variety of vegetation. Some larvae are gall makers and leaf miners. *Fenusa ulmi* Sundevall, the elm leafminer, occurs in the eastern U.S. and Canada, and *F. pusilla* (Lepeletier), the birch leafminer (Fig. 353A), ranges throughout the northern states and eastern Canada. Members of the genus *Pontania* form pinkish or yellow-green, round galls on willow leaves.

Common Species

Ametastegia glabrata (Fallén). Dock sawfly—10-12 mm; bluish black; legs reddish brown; northern U.S., Canada (larvae: bright green with white tubercles, on dock, sheep sorrel, others).

Caliroa cerasi (Linnaeus) (Fig. 353B). Pearslug—7-9 mm; shiny black; wings smoky; throughout North America (larvae: sluglike and slimy, dull green, anterior end broad, posterior end narrow, chiefly on pear and cherry leaves).

Endelomyia aethiops (Fabricius) (Fig. 353C). Roseslug—5-6 mm; black; wings smoky with brown veins; most of U.S., southern Canada (larvae on roses).

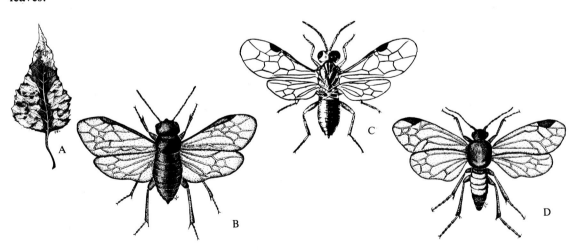

Figure 353 Sawflies (Tenthredinidae). A, birch leafminer damage, *Fenusa pusilla;* B, pearslug, *Caliroa cerasi;* C, roseslug, *Endelomyia aethiops;* D, larch sawfly, *Pristiphora erichsonii*.

Monophadnoides geniculatus (Hartig). Raspberry sawfly—5-6 mm; blackish, sometimes with abdomen partly red; throughout North America (in spring; larvae: dull green with white spiny tubercles, on raspberry, blackberry).

Pristiphora erichsonii (Hartig) (Fig. 353D)—Larch sawfly—10-11 mm; black with broad orange band on abdomen; most northern states, most of Canada (larvae: dull green, black head, on larch). *P. abbreviata* (Hartig), California pear sawfly, 5-6 mm, black with yellow prothoracic markings, CA to WA, NY to CT (larvae: bright green, on pears).

Family Siricidae—Horntails

The long cylindrical bodies of these large wasps have a spinelike projection on the last abdominal segment (Fig. 354A). The female has a long, stout ovipositor. Larvae bore into limbs and trunks of dying or dead trees.

Common Species

Sirex areolatus (Cresson)—25-35 mm; metallic blue-black; wings darkened (♀) or only slightly darkened at apex (♂); abdominal segments 3-7 yellowish orange (♂); Far West.

Sirex cyaneus Fabricius. Blue horntail—20 mm; metallic blue-black; legs reddish brown except at base; thorax metallic blue or greenish (♂); wings clouded near apex or slightly yellowish; throughout North America.

Tremex columba (Linnaeus) (Fig. 354A). Pigeon tremex—18-50 mm; brown or blackish; head and thorax reddish brown and black; abdomen with yellow crossbands; wings smoky yellow; North America except Far West.

Family Cephidae—Stem Sawflies

These sawflies are slender, 9-13 mm long, and the larvae bore into stems of berry plants, twigs of woody plants, and grasses.

Common Species

Cephus cinctus Norton (Fig. 354B). Wheat stem sawfly—8-11 mm; shiny black; abdomen with yellow crossbands; femora yellow; costa and stigma of wings yellow; west of Mississippi River, central Canada (larvae: in wheat stems, other grasses). *C. pygmaeus* (Linnaeus), European wheat stem sawfly, similar, eastern U.S.

SUBORDER APOCRITA

Family Braconidae—Braconids

Braconids are often dark although many range in color to bright red and yellow. They are usually 2-15 mm long, and the wings contain one recurrent vein or none (Figs. 345B, C). The family is very large with about 2,000 species in North America and 10,000 worldwide. Adults feed on plant discharges, honeydew from insects, and sometimes wound liquids from a host. Braconid larvae are common internal or external parasites primarily on the larvae of Lepidoptera and Coleoptera, but also attack

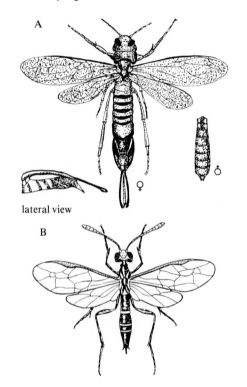

Figure 354 A pigeon tremex, *Tremex columba* (Siricidae); B, wheat stem sawfly, *Cephus cinctus* (Cephidae).

Diptera and Homoptera. Many larvae spin silken cocoons for pupation on the outside of the host. The female adult injects one or more eggs into the host or, if her larva feeds externally, she first paralyzes the host before ovipositing on it. Polyembryony occurs with some braconid species.

Common Species

Apanteles congregatus (Say)—2-3 mm; black; legs partly yellow; throughout North America (larvae: form cocoons on outside of hornworms [Sphingidae]).

Apanteles glomeratus (Linnaeus) (Fig. 355A)—2.5-3.0 mm; black with yellow and pale red markings; throughout North America (larvae: in cabbage butterfly larvae [Pieridae]).

Chelonus texanus Cresson (Fig. 355B)—5 mm; dull black; 2 yellowish white spots near base of abdomen; throughout U.S.

Lysiphlebus testaceipes (Cresson)—2 mm; head and thorax black; legs yellow-brown; abdomen brownish except for pale yellow second segment; most of U.S., Canada.

Macrocentrus ancylivorus Rohwer (Fig. 355C)—4.5 mm; slender; yellow-brown; eastern U.S., OR, CA (nocturnal).

Family Ichneumonidae—Ichneumons

These common, slender insects range from about 3-40 mm in length. Two recurrent veins in the front wing (Fig. 345D) separate this family from some similar Braconidae, the latter having only one or none. The antennae have 16 or more segments and are at least half the body length. Many females bear a long, nonretractable ovipositor (Fig. 356A) and will try to sting by poking only when handled. Others have short ovipositors and may sting readily. Ichneumons are one of the largest insect families with about 8,000 species estimated for North America. Ichneumon larvae are major parasites of other immature insects and are generally internal feeders. They attack Lepidoptera primarily but also parasitize many larvae of Coleoptera and Hymenoptera.

Common Species

Arotes amoenus Cresson—9.5-15.5 mm; black; whitish narrow crossband on posterior margins of most abdominal tergites; front and middle legs mostly ivory; hind leg with trochanters, tip of femur, basal 2/3 of tibia and all of tarsus white; eastern U.S.

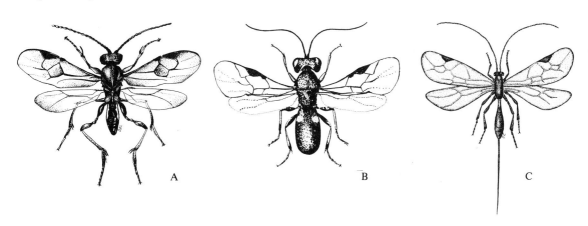

Figure 355 Braconids (Braconidae). A, *Apanteles glomeratus;* B, *Chelonus texanus;* C, *Macrocentrus ancylivorus.*

Order Hymenoptera 373

Ceratogastra ornata (Say)—12-14 mm; blackish; yellowish marks on face, mouthparts and thorax; legs mostly yellow-brown; wings brownish with yellowish tint; yellowish bands on each abdominal segment; eastern 1/2 U.S.

Coccygomimus pedalis (Cresson) (Fig. 356B)—5-13 mm; black; legs yellowish red; hind tibiae and tarsi blackish; wings clear (eastern U.S.) or smoky (western U.S.); northern 1/2 U.S., mts. of other areas, southern Canada.

Cryptus albitarsis (Cresson)—front wing length 4-12 mm; black; face, sides of pronotum, part of front and middle coxae, and upper side of front and middle tibiae white; abdomen red; eastern 1/2 U.S., southern Canada.

Diplazon laetatorius (Fabricius)—5-8 mm; mostly black; antennae, legs, and abdomen reddish; hind tibiae with brown and whitish rings; most of North America.

Enicospilus merdarius (Gravenhorst)—14-24 mm; pale yellowish brown or dark yellow; most of U.S., southern Canada.

Gelis tenellus (Say)—5-8 mm; brown or blackish; front wings each with 2 dark crossbands; most of U.S., southern Canada.

Itoplectis conquisitor (Say)—front wing length 3.5-12.5 mm; black with scattered white markings; hind tibia with white band; distinct narrow white crossband on posterior margin of each abdominal tergite; 4th and 5th tergites shiny (♂); east of Rocky Mts., southern Canada.

Megarhyssa macrurus macrurus (Fabricius) (Fig. 356C)—front wing length 7-29 mm, body length 22-40 mm; light brown; head yellow; thorax with yellow markings; sides of abdomen with row of V-shaped yellow marks bordered with black; eastern U.S. west to SD and western TX (bores into wood with ovipositor to parasitize larvae of the pigeon tremex [Siricidae]). *M. macrurus icterosticta* Michener similar, Southwest.

Therion morio (Fabricius)—22-26 mm; black; antennae, basal 2/3 hind tibiae, and tarsi yellowish; eastern 1/2 U.S., southern Canada.

Trogus pennator (Fabricius)—12-22 mm; reddish or reddish brown; posterior region of mesonotum (scutellum) strongly raised; anterior 1/2 of 1st abdominal segment very slender and higher than broad; abdomen broadly rounded at tip and not pointed; ovipositor very short; eastern 1/2 U.S., Far West (comes to lights at night).

Family Trichogrammatidae—Minute Egg Parasites

These tiny wasps (0.3-1.0 mm long) are parasites of insect eggs. The tarsi are 3-segmented and there are rows of tiny hairs on the wing surface.

Common Species

Trichogramma minutum Riley (Fig. 357A). Minute egg parasite—0.3 mm; yellowish; eyes pink; abdomen brown; throughout North America.

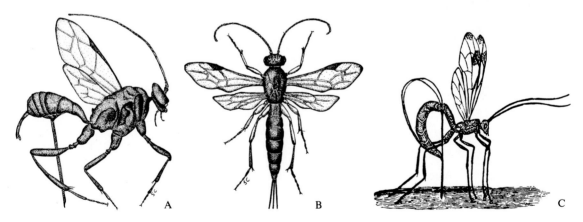

Figure 356 Ichneumons (Ichneumonidae). A, an ichneumon; B, *Coccygomimus pedalis;* C, *Megarhyssa macrurus macrurus.*

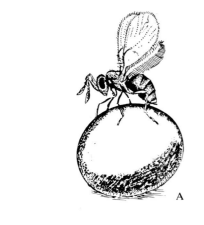

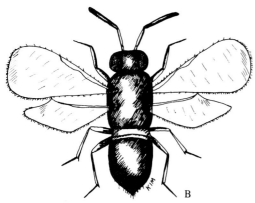

Figure 357 A, minute egg parasite, *Trichogramma minutum* (Trichogrammatidae); B and C, aphelinid wasps (Aphelinidae): B, *Aphelinus mali*; C, *Aspidiotiphagus citrinus citrinus*.

Family Aphelinidae (Eulophidae)

Aphelinids are small (0.7-3.0 mm long), black or metallic insects with 4-segmented tarsi. They are common parasites of other insects and some are hyperparasites.

Common Species

Aphelinus mali (Haldeman) (Fig. 357B)—1.0-1.5 mm; dark brown to blackish; base of abdomen and part of antennae and legs yellow; throughout North America (aphid parasite; leaves a darkened aphid shell glued to plant).

Aphytis proclia (Walker)—2 mm; yellowish; each of 1st 5 abdominal segments with darker broken crossband near middle; throughout U.S.

Aspidiotiphagus citrinus citrinus (Craw) (Fig. 357C)—0.7 mm; pale brownish yellow; ocelli red, compound eyes black; middle of wings with smoky band; crossband on middle of abdomen; most of U.S. (scale parasite).

Family Pteromalidae

Pteromalids are commonly collected wasps, generally 2-4 mm long, and black, metallic green or sometimes bronze. The tarsi are 5-segmented and the apical spur of the front tibia is curved. Larvae are parasites of many insects including crop pests.

Common Species

Dibrachys cavus (Walker)—1.0-1.5 mm; head and thorax shiny green; abdomen dark brown to black; most of U.S. except Southwest, eastern Canada, British Columbia.

Pteromalus puparum (Linnaeus) (Fig. 358A)—1-4 mm; green; antennae yellow, brown, and blackish; legs bronze and green or yellowish; abdomen golden dorsally (♂); throughout North America.

Family Eurytomidae—Eurytomids, Jointworms, Seed Chalcids

These small wasps are dull black, the thorax is rough, and often the body is hairy. This group resembles the Pteromalidae but the pronotum of eurytomids is more square and the abdomen more compressed. Larvae are insect parasites and also plant feeders, the latter feeding in seeds, stems, or galls.

Common Species

Harmolita grandis (Riley) (Fig. 358B). Wheat strawworm—2.5-4.0 mm; black and wingless (smaller form) or winged (larger form); wheat growing areas of North America. *H. tritici* (Fitch), wheat jointworm, similar, 5 mm, winged.

Family Chalcididae—Chalcids

Chalcids are 2-7 mm long and have distinctly enlarged and toothed hind femora. At rest their wings are held out to the sides. Larvae are parasites and hyperparasites of other insects.

Common Species

Brachymeria ovata (Say) (Fig. 358C)—3.5-7.0 mm; black and yellow; tip of hind femora pale yellow or white; throughout North America.

Family Cynipidae— Cynipids or Gall Wasps

These small wasps are generally 6-8 mm long and have 11- to 16-segmented antennae. The family is large with about 750 species in North America. The great majority of larvae cause plants to form galls which serve as food and shelter; some live in the galls made by other species of gall wasps. A few larvae parasitize other insects. Most species cause oak galls of diverse but characteristic shapes (Fig. 359C). Species are often best identified by their gall's shape and the specific host. Some oak galls have a high tannin content and were once used in making ink. Most galls are not injurious to plants.

Adults in the subfamily of gall makers are typically black or brown, the abdomen is oval and compressed laterally, and the second abdominal tergite forms half or more of the abdomen (Fig. 359A). Many species have two distinct generations in a year. One generation consists of females that reproduce without fertilization and their larvae produce a certain type of gall. The second generation has males and females and their young may cause quite different galls.

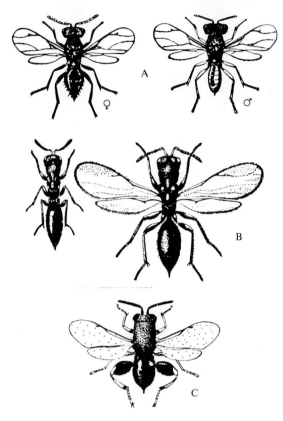

Figure 358 A, a pteromalid wasp, *Pteromalus puparum* (Pteromalidae); B, wheat strawworm, *Harmolita grandis* (Eurytomidae); C, a chalcid, *Brachymeria ovata* (Chalcididae).

Common Species

Andricus californicus (Bassett)—3-5 mm; brown or reddish brown; Pacific coast states (forms oak apples 40-100 mm diameter).

Diplolepis rosae (Linnaeus) (Figs. 359A, B). Mossy-rose gall wasp—4 mm; black; wings brownish; eastern U.S. southwest to KS (forms mossy rose gall on rose stems).

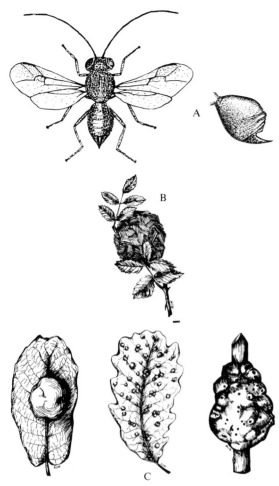

Figure 359 Gall wasps (Cynipidae). A, mossyrose gall wasp, *Diplolepis rosae,* and lateral view of its abdomen; B, mossyrose gall; C, galls on oak.

Family Evaniidae—Ensign Wasps

Ensign wasps are 10-15 mm long, black, and have a small, oval abdomen attached high up on the thorax by a slender stalk. Larvae are parasites of cockroach eggs and thus the adults are more common in the eastern and southern states where cockroaches are more abundant.

Common Species

Evania appendigaster (Linnaeus) (Fig. 360A)—10-12 mm; black; fine pubescence; AZ to FL, Atlantic coast states north to NY (often in and around buildings).

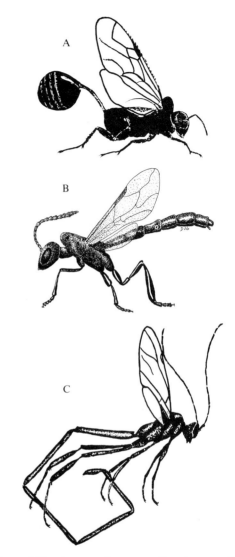

Figure 360 A, an ensign wasp, *Evania appendigaster* (Evaniidae); B, a gasteruptiid wasp, *Gasteruption* sp. (Gasteruptiidae); C, a pelecinid wasp, *Pelecinus polyturator* (Pelecinidae).

Family Gasteruptiidae

Members of this family are black, 13-20 mm long, and the female's ovipositor is often as long as her body. These insects resemble the Ichneumonidae but the head of the Gasteruptiidae is attached to a slender neck, the wings have a costal cell, and the abdomen inserts high up on the thorax (Fig. 360B). Adults occur on

Order Hymenoptera 377

flowers including wild carrot and wild parsnip. Larvae are parasites of bees and solitary wasps. Most species occur in the northern and mountainous regions of the U.S. and much of Canada. *Gasteruption* is the only genus in North America.

Family Aulacidae

Aulacids resemble the Gasteruptiidae but the former have two recurrent veins instead of one in the front wings. Adults occur around logs infested by wood-boring insects which serve as hosts for aulacid larvae. Most species occur in the northern and mountainous regions of the U.S. and much of Canada. There are two genera: *Pristaulacus* (tarsal claws with 2^+ teeth) and *Aulacus* (tarsal claws without teeth).

Family Pelecinidae— Pelecinid Wasps

The only North American species in this family is *Pelecinus polyturator* (Drury) which occurs in the eastern half of the U.S. The nonstinging female is shiny black, 50-75 mm long, and has an extremely long and slender abdomen (Fig. 360C). Males are rarely collected. Larvae parasitize June beetle larvae (Scarabaeidae).

Family Scelionidae— Scelionid Wasps

These minute, black wasps are usually 2 mm long or less. The antennae are elbowed, the wing venation is nearly absent, and the flattened abdomen may have sharp lateral margins. Larvae are parasites of insect or spider eggs. Adult females may ride on top of the larger host until the latter deposits her eggs.

Common Species

Eumicrosoma beneficum Gahan (Fig. 361A)— 0.7-1.0 mm; head and thorax shiny black; antennae reddish yellow with 4-segmented club (♀) or dark brown without a club (♂); abdomen all reddish yellow (♀) or apical 1/2 dark brown ( ); VA to Midwest (range of its host, the chinch bug [Lygaeidae]).

Family Platygastridae

Platygastrids resemble the Scelionidae but the antennae are 9- or 10-segmented in some platygastrids or they have an unsegmented club. Some species are parasites of gall midge larvae (Cecidomyiidae). The Hessian fly is parasitized by species in the genus *Platygaster* (size 1.0-1.5 mm) (Fig. 361B).

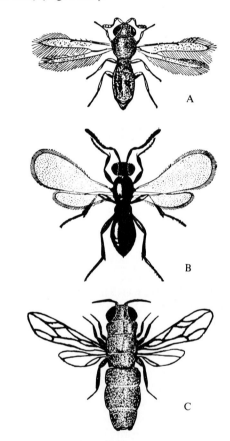

Figure 361 A, a scelionid wasp, *Eumicrosoma beneficum* (Scelionidae); B, a platygastrid wasp, *Platygaster* sp. (Platygastridae); C, a cuckoo wasp, *Chrysis coerulans* (Chrysididae).

Family Chrysididae— Cuckoo Wasps

Cuckoo wasps are generally 6-12 mm long and colored brilliant metallic shades of green or blue (rarely purple or red). Adults are roughly sculptured and the abdomen is concave ventrally. These wasps do not sting and when disturbed often roll up into a ball. Cuckoo wasps are external parasites of larval mud daubers (Sphecidae), potter wasps (Eumenidae) and other hymenopterans. The name "cuckoo" was derived from the parasitic cuckoo birds who lay eggs in exposed nests of bird hosts as the wasps do in nests of their hosts.

Common Species

Chrysis coerulans Fabricius (Fig. 361C)—7-11 mm; brilliant metallic blue or blue-green; tip of abdomen with 4 evenly spaced teeth; throughout U.S., southern Canada.

Parnopes edwardsii (Cresson)—6-12 mm; brilliant green to dark blue; Pacific coast states, ID, British Columbia.

Family Tiphiidae—Tiphiid Wasps

The most common species in this family have a mesosternum with two posterior lobes separated by a deep indentation. Many males also have an upward-curving spine on the tip of the abdomen (Fig. 362A). The species are often blackish and 10 to over 25 mm long. Their young parasitize the larvae of scarab beetles and other insects.

Common Species

Myzinum quinquecinctum (Fabricius) (Fig. 362A)—16-23 mm; shiny black with bright yellow markings on body; grayish hairs; antennae and legs yellowish brown (♀) or antennae grayish black and legs black at base and yellowish brown at tip (♂); upturned black spine at tip of abdomen (♂); most of U.S. except Northwest, southeastern Canada.

Tiphia intermedia Malloch—6-13 mm; blackish; clypeus extended ventrally into 2 teeth; transverse ridge on pronotum; inside of hind tibia prominently ridged (♂); many punctures; eastern 1/2 U.S.

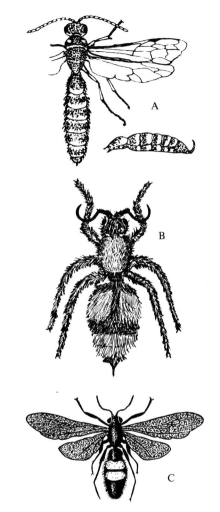

Figure 362 A, a male tiphiid wasp, *Myzinum quinquecinctum* (Tiphiidae), and lateral view of abdomen; B, a female velvet ant, *Dasymutilla occidentalis* (Mutillidae); C, a scoliid wasp, *Scolia bicincta* (Scoliidae).

Family Mutillidae—Velvet Ants

The wingless females of this family resemble ants (Formicidae) and are 6-20 mm long. However, velvet ants lack a node on the narrowed, basal part of the abdomen that is present on ants (Fig. 343). Velvet ants are covered with dense, brightly colored hairs and a narrow band of very dense hair often occurs on each side of the second abdominal segment. Males are winged and larger than females. Female mutillids give a painful sting. Larvae are external parasites of larval and pupal bees, vespid and sphecid wasps, and a few ants. Velvet ants are most common in the South and Southwest.

Common Species
Dasymutilla occidentalis (Linnaeus) (Fig. 362B)—15-30 mm; head and legs blackish; thorax, abdomen, and hairs orange-red; CT to FL, west to TX. *D. bioculata* (Cresson) similar, 12-25 mm, western 2/3 U.S. and Canada.

Dasymutilla sackenii (Cresson)—12-13 mm; black; dense white hairs; OR, CA, NV, AZ. *D. gloriosa* (Saussure) similar, 13-16 mm, white hairs very long, CA and UT south to TX.

Family Scoliidae
These medium to large wasps are commonly black with one or more yellow abdominal bands. Scoliids resemble male velvet ants (Mutillidae) but scoliids have longitudinal wrinkles on the outer third of the wings. Adults frequent flowers and larvae parasitize June beetle larvae (Scarabaeidae).

Common Species
Campsomeris plumipes (Drury)—12-15 mm; black with yellow markings; 1 recurrent vein in front wing; east of Rocky Mts.

Scolia bicincta Fabricius (Fig. 362C)—21-25 mm; black with bluish or purplish sheen; wings dark, 2 recurrent veins; abdomen with two broad golden yellow bands; eastern 1/2 U.S.

Scolia dubia dubia Say—12-18 mm; hairy; black; wings black, 2 recurrent veins; red and yellow markings on abdomen; eastern 1/2 U.S. southwest to AZ.

Family Formicidae—Ants
Ants are common and abundant winged or wingless insects that are characterized by one or two narrow, basal, abdominal segments (pedicel) that bear a node or projection (Figs. 343; 363). The antennae are usually elbowed and the compound eyes are often small or sometimes absent. Ants that sting (females) do so by use of a modified ovipositor. Many species do not sting and instead may spray formic acid or other chemicals from anal glands as a means of defense. There are about 660 species of ants in North America and 10,000 worldwide.

Ants are social insects that live in colonies generally on the ground or in decomposing wood, although colonies occur in many other sites. There are usually three castes—workers, females (queens), and males. Workers are sterile females that may occur in varying forms in a given colony. The head or the whole body of some workers may be larger or different in shape from those of others in the same colony. Workers care for the queen and her eggs and larvae, build and repair the nest, and defend the colony from invaders. Males function primarily to mate with the females and die soon afterwards. Males typically are intermediate in size between the workers and the very large queens. The queens mate only once but lay eggs throughout most of their long lives (often several years).

Swarms of winged males and females typically occur in the spring and fall. Mating occurs in the air or on the ground. The queen pulls her wings off, burrows into a site to lay eggs, and takes care of the first generation of tiny maggotlike larvae. Larvae are fed regurgitated food or partly chewed plant or animal material. These resulting workers then tend the queen's succeeding broods. Some species of ants form new colonies when the queen or another fertilized female leaves the nest along with some workers. With other species workers may invade and take over a nest of an alien ant species or carry off pupae to enslave the adults that emerge.

Some ants feed on flower nectar, plant secretions, and insect honeydew. Aphids and other homopterans are tended, protected, and sometimes transported by ants who feed on the homopteran's honeydew. This relationship is a standard example of symbiosis. Numerous common species also gather and feed on seeds, leaves and blossoms, dead insects and other animals, or are predators of other insects. The leafcutting ants feed on fungi in their nest by culturing the fungi on macerated leaves enriched by the ants' feces. Exchange of food between individuals is common. The army ants of the New World tropics and driver ants of Africa move in great hordes, devouring most arthropods (and occasionally small vertebrate animals) in their paths.

Common Species

Subfamily Dorylinae—includes tropical army and driver ants; a few species of these highly predaceous ants occur in southern U.S.; females sting.

Labidus coecus (Latreille)—3-10 mm; bright reddish brown or brown; eyes tiny or absent; 2-segmented pedicel; each tarsal claw with tooth near middle; TX and LA north to OK and AR.

Subfamily Myrmicinae—largest subfamily; pedicel 2-segmented; females sting.

Monomorium pharaonis (Linnaeus) (Fig. 363A). Pharaoh ant—2 mm; light or yellowish brown to reddish; head, thorax, and pedicel with punctures; throughout North America (in soil or houses). *M. minimum* (Buckley), little black ant, similar, 1.5-2.0 mm, shiny black or dark brown.

Pheidole bicarinata vinelandica Forel—1.5-3.0 mm; yellowish brown to dark brown; head very large, middorsal groove; antennae with 3-segmented club; eastern 1/2 U.S. southwest to AZ (on beaches, deserts, mts.).

Pogonomyrmex barbatus (F. Smith). Red harvester ant—6-13 mm; reddish brown; some forms have head, thorax, and legs black, or abdomen partly or entirely brown, or pedicel surface roughened; southwestern U.S. (clears vegetation to form circle 1-8 m in diameter with nest in center).

Pogonomyrmex occidentalis (Cresson) (Fig. 363B). Western harvester ant—6-7 mm; reddish brown, similar to *P. barbatus*; western U.S. except Northwest (mound at nest entrance). *P. californicus* (Buckley), California harvester ant, similar, 5.0-7.5 mm, TX to NV and CA, nest in sandy areas, fan-shaped low mounds on one side of entrance.

Solenopsis invicta Buren. Red imported fire ant—3-6 mm; reddish or reddish brown; abdomen dark brown and sometimes with yellow crossband at base; mandibles with 4 distinct teeth and not curved inward; antennae 10-segmented with 2-segmented club; NC to FL, west to TX (stinging pest).

Solenopsis xyloni McCook (Fig. 363C). Southern fire ant—1-6 mm; reddish or yellowish; abdomen with dark dense hairs; antennae 10-segmented with small 2-segmented club; mandibles with 3 distinct teeth and not curved inward; SC and FL to CA (nest is irregular mound of loose soil). *S. molesta* (Say), thief ant, similar, 1-2 mm, yellowish or bronze, antennal club large, throughout North America (enters houses).

Tetramorium caespitum (Linnaeus). Pavement ant—2.5-3.0 mm; light to dark brown or blackish; hairy; head and prothorax with narrow parallel grooves; area above hind coxae with pair of tubercles or short spines; most of U.S., chiefly Atlantic Coast (in lawns, pavement edges, houses).

Subfamily Dolichoderinae—pedicel 1-segmented; anal opening slitlike, located ventrally at tip of abdomen; females stingless.

Iridomyrmex humilis (Mayr) (Fig. 363D). Argentine ant—1-2 mm; light to dark brown; antennae long and not clubbed; southern U.S., CA (in soil, pavement, under stones, houses; musty odor if crushed).

Tapinoma sessile (Say). Odorous house ant—2.0-2.5 mm; dark reddish brown to black; anterior part of abdomen extends over pedicel; most of North America (enters houses; musty odor if crushed).

Subfamily Formicinae—2nd largest subfamily; pedicel 1-segmented and dorsal node thin; circular anal opening at tip of abdomen; females stingless, some inject formic acid.

Acanthomyops claviger (Roger). Smaller yellow ant—3-4 mm; shiny pale yellow or yellowish red; hairy; node on pedicel long and pointed; most of U.S., chiefly eastern and central (lemonlike odor if crushed). *A. interjectus* (Mayr), larger yellow ant, similar, 4.0-4.5 mm, nests in soil.

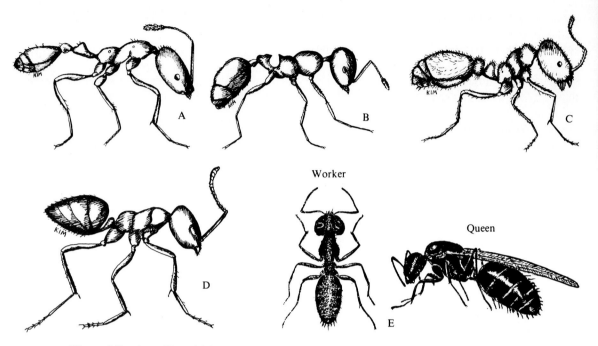

Figure 363 Ants (Formicidae). A, pharaoh ant, *Monomorium pharaonis;* B, western harvester ant, *Pogonomyrmex occidentalis;* C, southern fire ant, *Solenopsis xyloni;* D, Argentine ant, *Iridomyrmex humilis;* E, black carpenter ant, *Camponotus pennsylvanicus.*

Camponotus abdominalis floridanus (Buckley). Florida carpenter ant—5.5-10.0 mm; head reddish, thorax and pedicel yellowish or yellowish red; abdomen blackish; yellowish hairs on body; southeastern U.S., especially FL (often in houses). *C. castaneus* (Latreille) similar, 7-10 mm, yellowish or yellowish red, front of face ridged, head length = or longer than width, southern U.S. north to New England, west to TX and IA.

Camponotus pennsylvanicus (De Geer) (Fig. 363E). Black carpenter ant—6-12 mm; black; thorax, legs, and pedicel may be light brown, reddish brown or dark brown in some forms; pale dorsal hairs on abdominal segments; eastern 1/2 U.S., southeastern Canada (nests in tree trunks, buildings, other wood; does not eat wood; bites but does not sting).

Formica fusca Linnaeus. Silky ant—3-5 mm; generally evenly dark brown or black, also reddish or reddish brown; primarily northern 1/2 U.S., Canada (in fields, forests; nest is small, hard dome of soil). *F. exsectoides* Forel, Allegheny mound ant, 3-6 mm, reddish brown, thorax roughened, anal area red, eastern U.S., southeastern Canada (builds mounds 0.6-1 m high and over 1 m across in coniferous forests).

Family Vespidae— Hornets, Yellowjackets

These common social wasps are generally black with yellow or white markings. The wings at rest are folded lengthwise and parallel to the body. The first discoidal cell on the front wing is longer than the submedian cell (Fig. 346B). The tarsal claws are not toothed.

Adults build paper nests in the ground, in hollow trees, on trees, or under building overhangs. Wood and leaves are chewed into pulp and bits are plastered together to form a paper nest of hexagonal cells. The caste system consists of workers (sterile females), queens, and males (to fertilize the queens). Only young, fertilized queens survive past late fall and they leave the nest to hibernate elsewhere. In the spring the queen builds a comb in which to lay eggs, rears the first generation, and the resulting adult workers assume the nest building and caring of the young. Adults feed on nectar, ripe fruit, and insect pieces, and are commonly attracted to food at picnics and garbage dumps. These insects inflict a painful sting and several species of yellowjackets are pests in picnic-recreational areas. Larvae are fed chewed insects, meat scraps, and nectar by workers. The mature larva seals its cell with silk before pupating.

Common Species

Subfamily Vespinae—Hornets and Yellowjackets; hornets are stout-bodied, mostly blackish, usually have aerial nests; yellowjackets are yellow and black hornets nesting in the ground or in hollow logs.

Vespula maculata (Linnaeus) (Fig. 364A). Baldfaced hornet—12-19 mm; black; face and thorax with whitish or pale yellow markings; abdomen with white markings on tip; throughout North America (builds large, rounded, closed paper nest suspended from tree branch or overhang (Fig. 364B).

Vespula pensylvanica (Saussure) (Fig. 364C). Western yellowjacket—12.0-17.5 mm; black and yellow; underside of 1st antennal segment yellow; upper 3/4 of eye encircled by yellowish line; western 1/2 U.S. *V. maculifrons* (Buysson), eastern yellowjacket, similar, 1st segment of antennae mostly black, east-

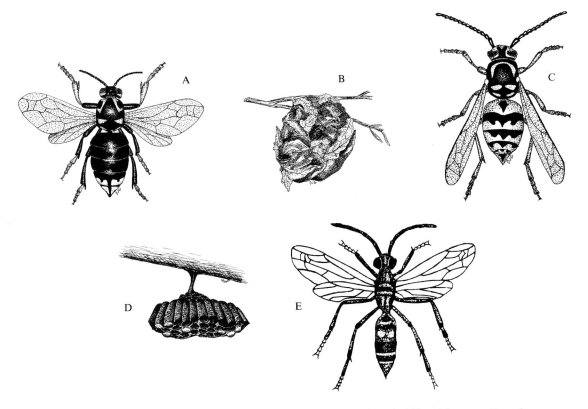

Figure 364 Hornets and yellowjackets (Vespidae). A and B, baldfaced hornet, *Vespula maculata,* and paper nest; C, western yellowjacket, *V. pensylvanica;* D, paper nest of a *Polistes* wasp; E, *Polistes fuscatus fuscatus.*

ern 1/2 North America. *V. vulgaris* (Linnaeus) similar, broad dark stripe down middle of face, throughout U.S., southern Canada.

Subfamily Polistinae—slender, spindle-shaped abdomen, long legs; circular paper nest not covered, cells open downward, attached by stalk to bush or overhang (Fig. 364D); all North American spp. in genus *Polistes*, sometimes called paper wasps.

Polistes exclamans Viereck—16-20 mm; yellow band behind ocelli; mesonotum reddish; black, reddish, and yellowish markings on abdomen; western 1/2 and southern U.S. *P. apachus* Saussure, 16-20 mm, bright reddish brown with large yellow markings, KS and TX west to southern CA.

Polistes fuscatus fuscatus (Fabricius) (Fig. 364E)— 17-22 mm; head and thorax black with brown markings; abdomen blackish with yellow margins on basal segments and reddish yellow spots on sides; eastern U.S. west to NV, southcentral Canada. *P. fuscatus aurifer* Saussure, golden paper wasp, 9-15 mm, black and yellow, abdomen yellow with base of 1st 2-3 segments black, Far West, CO.

Family Eumenidae—Potter Wasps

The tarsal claws of these wasps are toothed, the abdomen sometimes is stalked at its base, and the mandibles are elongate and often crossed. Species in the genus *Eumenes* make small jug-like or vaselike mud nests attached to twigs and provisioned with caterpillars and other insects (Fig. 365B). The family is classified as a subfamily of the Vespidae in many texts although eumenids are solitary wasps and not social as are the vespids.

Common Species

Eumenes fraternus Say (Fig. 365A)—19 mm; black with scattered yellow markings; eastern 1/2 U.S. southwest to TX.

Monobia quadridens (Linnaeus) (Fig. 365C)—20-21 mm; black with yellow markings; eastern U.S. southwest to NM (nest is mud-partitioned wood cavity).

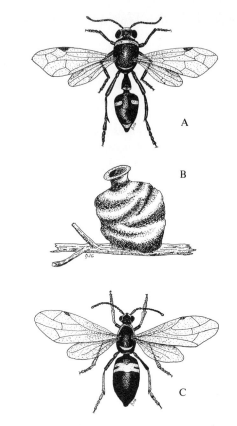

Figure 365 Potter wasps (Eumenidae). A,*Eumenes fraternus*; B, mud nest of a *Eumenes* sp.; C, *Monobia quadridens*.

Family Masaridae

The antennae of these black and yellow solitary wasps are clubbed (Fig. 342D), the wings are not folded, and the front wings contain two submarginal cells. The species are 10-20 mm in length and occur in western and southwestern U.S. Nests are made of sand or mud adhered to twigs or rocks, and provisioned with nectar and pollen. The family is classified as a subfamily of the Vespidae in many texts.

Common Species

Pseudomasaris vespoides (Cresson)—15-22 mm; black with extensive yellow markings; antennae and most of face yellow; transverse ridge between antennae (♀) or last dorsal abdominal segment with 6 tubercles (♂); wings yellowish; western 1/2 U.S.

Family Pompilidae—Spider Wasps

Spider wasps are usually 10-25 mm long and black with dark- or amber-colored wings although some have brightly colored markings. These wasps are characterized by a transverse groove on the side of the mesothorax (Fig. 347D). The antennae of females are often curled. Individuals are very active and often are seen darting about on flowers or running on the ground (especially on sand). Adults provision their nests in the ground with spiders. Members of the genus *Pepsis* are large (20-40 mm long), blue-black wasps often with bright red or orange wings. Adults sting and paralyze tarantulas in the Southwest and their larvae feed on these spiders.

Common Species

Anoplius americanus (Palisot de Beauvois) (Fig. 366)—12-14 mm; dull black; dorsal part of 1st and 2nd abdominal segments brick red; wings smoky; eastern U.S. southwest to TX. *A. marginalis* (Banks) similar, 16-18 mm, dorsal part of 1st, anterior 2/3 of 2nd, and anterior 1/2 of 3rd abdominal segments orange-red, Rocky Mts. eastward, AZ.

Pompilus luctuosus Cresson—10-12 mm; metallic blue-black; wings darkened with violet iridescence; legs very long; northern 1/3 U.S., mts. and coast of western states, Canada.

Figure 366 A spider wasp, *Anoplius americanus* (Pompilidae).

Family Sphecidae—Cicada Killers, Mud Daubers, and Sand Wasps

Members of this family are characterized by a narrow, collarlike pronotum that has a posteriorly projecting lobe on each side (Fig. 347C). The lobes usually do not reach the tegulae. The family is large (about 1,200 species in North America) and many species are common. The sphecids vary from stout-bodied to slender forms; many of the latter have a long, cylindrical stalk (petiole) connecting the abdomen to the thorax (Fig. 367D). These wasps are solitary, each usually forming a separate nest in the ground, in natural openings, or by building mud cells. Adults provision their nests with paralyzed insects or spiders.

Common Species

Subfamily Larrinae—2 major groups: (1) stout-bodied (Fig. 367A), 10-20 mm, brownish, front wing with 3 submarginal cells (outer cell pointed); (2) slender, 13-25 mm, black, front wing with 1 submarginal cell, inner margin of eyes notched (includes organpipe mud daubers [Fig. 367B, genus *Trypoxylon*] which build long, parallel mud tubes).

Trypoxylon clavatum Say (Fig. 367B)—12-13 mm; shiny black; white pubescence; wings darkened; hind tarsi pale; hind trochanters spined (♂); east of Rocky Mts. except New England. *T. politum* Say similar, 17-18 mm, dark wings with violet tones, hind tibiae white, hind trochanters without spines (♂), Atlantic coast states, southern U.S.

Subfamily Pemphredoninae—Aphid wasps; 8-15 mm, usually black, abdomen may be on short stalk (Fig. 367C).

Subfamily Sphecinae—Threadwaisted wasps; most 10-20 mm long, abdomen with long and slender stalk (petiole); some build mud nest (mud daubers).

Chalybion californicum (Saussure)—12-18 mm; metallic blue, blue-green, or blackish; wings bluish; throughout U.S., Canada (mud dauber; uses nests of *Sceliphron caementarium* and provisions with spiders.

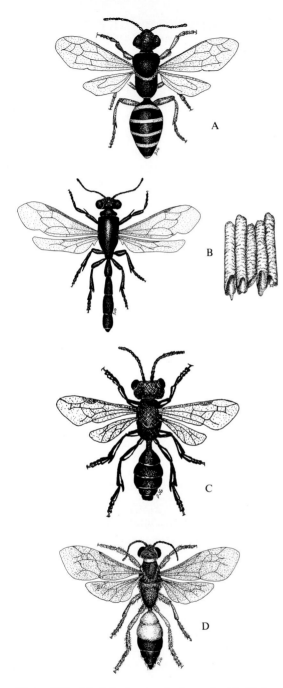

Podalonia violaceipennis (Lepeletier)—12-25 mm; metallic blue-black; basal 1/2 abdomen reddish; eastern U.S. west to CA. *P. luctuosa* Smith entirely metallic blue-black, 12-15 mm, western U.S., northernmost states, Canada.

Prionyx atratus (Lepeletier)—15-17 mm; pubescent; black, including wings; throughout U.S., southern Canada.

Sceliphron caementarium (Drury) (see frontispiece)—25-30 mm; blackish brown with yellow markings on thorax, petiole, and base of abdomen; legs yellow-banded; wings yellowish brown; throughout U.S., southern Canada (mud dauber; builds nest of mud cells 25-35 mm long in open buildings and provisions nest with spiders).

Sphex ichneumoneus (Linnaeus) (Fig. 367D)—15-22 mm; black; base of abdomen red; throughout U.S., southern Canada.

Subfamily Nyssoninae—several forms including: (1) sand wasps, 20-25 mm, stout-bodied, pale greenish markings, distinct long and triangular labrum extends downward in front of mandibles; (2) 10-15 mm, yellow and black markings, 2 apical spurs on middle tibiae.

Bembix americana spinolae Lepeletier (Fig. 368A)—13-16 mm; stout-bodied; black; short white pubescence; greenish yellow markings on front of face, prothorax, tibiae and tarsi; broken crossbands on abdomen greenish yellow; eastern 1/2 U.S. southwest to AZ, southern Canada. *B. americana comata* Parker similar, 10-17 mm, longer and denser pubescence, abdominal bands whitish (♂) or tip of abdomen spotted (♀), Far West (on sand dunes).

Bembix pallidipicta Smith—18-19 mm; black; dense pale gray pubescence on head, thorax, and base of abdomen; greenish white dorsal bands on 1st 5 abdominal segments; most of U.S. (on sand dunes).

Microbembex monodonta (Say)—8-14 mm; black; wings darkened; abdomen yellow or greenish white, often with 1-5 black bands; east of Rocky Mts., southern Canada.

Sphecius speciosus (Drury) (Fig. 368B). Cicada killer—25-38 mm; reddish brown to blackish; abdomen with yellow crossbands; east of Rocky Mts.

Figure 367 Mud daubers and other sphecids (Sphecidae). A, a stout-bodied sphecid; B, an organpipe mud dauber, *Trypoxylon clavatum,* and mud nest; C, an aphid wasp; D, *Sphex ichneumoneus.*

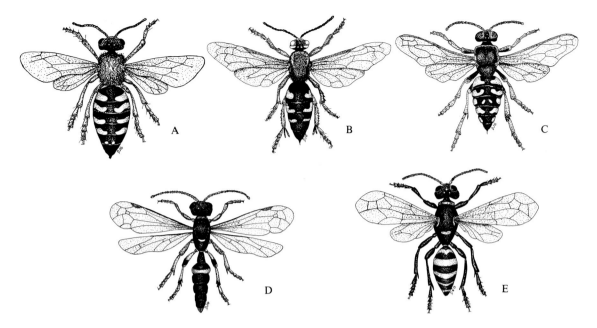

Figure 368 Cicada killers, sand wasps, and other sphecids (Sphecidae). A, a sand wasp, *Bembix americana spinolae;* B, cicada killer, *Sphecius speciosus;* C, *Stictia carolina;* D, *Cerceris clypeata;* E, *Philanthus ventilabris.*

Stictia carolina (Fabricius) (Fig. 368C)—25 mm; black with yellow markings; NM to FL, north to IL and NJ (often around horses to capture flies).

Subfamily Philanthinae—12-18 mm, black with yellow markings, abdomen without a petiole but with a distinct constriction between abdominal segments 1-2.

Cerceris clypeata Dahlbom (Fig. 368D)—17-20 mm; brownish black; light bands on legs; wings shaded; light band near middle of abdomen; Mississippi Valley eastward (provisions nest with weevils).

Philanthus ventilabris Fabricius (Fig. 368E)—12-13 mm; black; legs yellow; wings faintly yellowish; yellowish crossband near apex of each abdominal segment; most of U.S., southern Canada (provisions nest with bees).

Subfamily Crabroninae—6-20 mm, black with yellow markings, common species have head large and squarish, apex of marginal cell squared-off in front wing, only 1 submarginal cell which may lack the vein separating it from the discoidal cell below.

Family Colletidae—Colletid Bees

These bees have a short proboscis that is bilobed or squared-off at the tip. The most common species (genus *Hylaeus*) in North America have two submarginal cells in the front wing (Fig. 349C), are 10 mm or less, and are black with yellow or whitish markings on the face. *Hylaeus* species are not very hairy and resemble small wasps. The remaining species have three submarginal cells (Fig. 349D), are larger and more hairy, and most are in the genus *Colletes*. The family is also called plasterer, yellow-faced, or silk bees. They burrow in the ground or in large, pithy plant stems.

Common Species

Hylaeus modestus Say (Fig. 369A)—5-7 mm; black; face with sides and lower 1/2 of front yellow (♂) or with triangular yellow patch on sides (♀); base of tibiae yellow; tarsi yellow (♂) or reddish brown (♀); posterior margins of abdominal segments brownish; Atlantic coast states, CO, NM, eastern Canada.

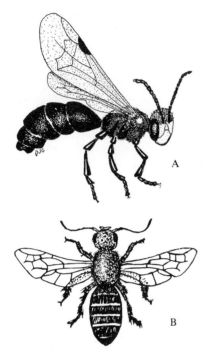

Figure 369 A, a colletid bee, *Hylaeus modestus* (Colletidae); B, a halictid bee, *Agapostemon virescens* (Halictidae).

Family Halictidae— Halictid Bees, Sweat Bees

Most halictids are 5-15 mm long and generally dark brown, metallic green, or a combination of both colors. Some resemble the Andrenidae but halictids have a strongly arched basal vein (Fig. 349B) and only one suture running vertically below the base of each antenna. Most halictids nest in the ground and many are social with nests quite close together. Some cuckoo bee species in this family lay their eggs in nests of other bees. Certain small, black species in the genus *Dialictus* are attracted to perspiration and have acquired the name sweat bees.

Common Species

Agapostemon virescens (Fabricius) (Fig. 369B)— 11-13 mm; head and thorax brilliant metallic green; sides with whitish hairs; legs with brown hairs; abdomen black (♀) or black with yellow bands (♂); most of U.S.

Halictus farinosus Smith—11-12 mm; black; long dense hairs on head and thorax; legs yellow; each abdominal segment with crossband of hairs on posterior margin; western U.S., British Columbia.

Halictus ligatus Say—7-10 mm; black; yellowish white pubescence; clypeus and mandibles yellow (♂); lower side of face with a tubercle (♀); each abdominal segment with crossband of hairs on posterior margin; throughout U.S., southern Canada.

Nomia melanderi Cockerell. Alkali bee—12-13 mm; black; pale brown hairs and black bristles on anterior part mesothorax; 4 light green apical crossbands on abdominal segments 2-4; western U.S. (nests in alkali soil and is an important, commercially managed alfalfa pollinator).

Family Andrenidae— Andrenid Bees

These dark-colored or reddish brown bees range from several to 20 mm in length. The glossa of the proboscis is pointed (Fig. 350A). Two faint antennal sutures run vertically below the base of each antenna (Fig. 351A). These sutures are difficult to see due to the dense hairs on the face. Females have facial patches of velvetlike hair near each compound eye. Andrenids, also called mining bees, are a large family of almost 1,300 species in North America. They tunnel into the ground to build nests which may be grouped into colonies. The common genus, *Andrena* (Fig. 370A), contains hundreds of species with yellowish, reddish brown, or whitish markings. There are usually three submarginal cells in the front wing and the marginal cell is squared-off apically (Fig. 349A). The genus *Perdita* (Fig. 370B) contains small, usually brightly colored or spotted species with sparse hair and two submarginal cells. *Perdita* spp. occur primarily in the western half of the U.S. and secondarily in the Southeast.

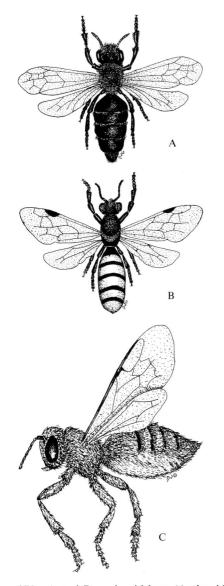

Figure 370 A and B, andrenid bees (Andrenidae): A, *Andrena* sp.; B, *Perdita* sp. C, a leafcutting bee, *Megachile latimanus* (Megachilidae).

Family Megachilidae— Leafcutting Bees

These bees are 10-20 mm long, the jugal lobe of the hind wing is shorter than the submedian cell (Fig. 349C), and there are two submarginal cells of about equal size in the front wing (Fig. 349C). Most species carry pollen on dense hairs located ventrally on the abdomen rather than on the hind legs. Nests are made in the ground, old branch cavities and other openings, and are frequently lined with circular pieces of leaves which the bees have cut. Large semicircles cut from plant leaves (e.g., roses) indicate the presence of these bees. Some species act as cuckoo bees, laying their eggs in nests of other megachilids.

Common Species

Megachile latimanus Say (Fig. 370C)—12-15 mm; black; dense, pale brownish yellow pubescense; abdominal segments 2-4 with faint white bands on posterior margin, pale red hairs on ventral side of abdomen; eastern U.S. northwest to WY, southern Canada. *M. pacifica* Panzer 13-14 mm, black, pale abdominal bands, whitish yellow dense hairs on ventral side of abdomen, eastern U.S., northern 1/2 western U.S. (commercially managed alfalfa pollinator).

Osmia lignaria Say—8-12 mm; shiny, dark bluish green; long whitish hairs; eyes, antennae and legs black; eastern U.S. southwest to TX, southeastern and southcentral Canada.

Family Anthophoridae— Carpenter Bees, Nomadine Bees, and Digger Bees

These bees possess apical spurs on the hind tibiae and a jugal lobe on the hind wing. In many texts the family has been classified as a subfamily of the Apidae. Carpenter bees are small to large, robust bees that nest in wood or plant stems. The large carpenter bees (genus *Xylocopa*) are about 25 mm long, blackish, partly hairy, and resemble bumble bees except that the dorsum of the abdomen is bare and shiny (Fig. 371B). These species nest in wood. The small carpenter bees (genus *Ceratina*) are 6-10 mm long, dark blue-green, and nest in dry stems. These bees have an arched basal vein in the front wing like the Halictidae but differ by having a short jugal lobe in the hind wing.

Most nomadine bees are 8-10 mm long, wasplike, relatively hairless cuckoo bees that

Order Hymenoptera

lay their eggs in nests of other bees. These dark reddish or yellow-spotted parasites have no pollen-collecting structures on the hind legs. The digger bees are large (10-20 mm long), stout, and very hairy (Fig. 371A). They resemble bumble bees and carpenter bees but the posterior margin of the first submarginal cell is longer than that of the second cell. Digger bees nest in the ground.

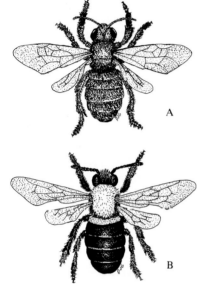

Figure 371   Carpenter bees and digger bees (Anthophoridae). A, a digger bee, *Anthophora occidentalis*; B, carpenter bee, *Xylocopa virginica*.

Common Species

Anthophora occidentalis Cresson (Fig. 371A)—16-17 mm; black; short, dense, yellow pubescence; side of face yellow (♂); apex of wing shaded; tip of abdomen brownish; most of western 1/2 U.S. *A. abrupta* Say similar, eastern 1/2 U.S.

Xylocopa virginica (Linnaeus) (Fig. 371B). Carpenter bee—24-25 mm; black; resembles bumble bee; abdomen bare; eastern U.S. southwest to TX.

Family Apidae— Bumble Bees and Honey Bees

Bumble bees (subfamily Bombinae) are large, robust, hairy bees that are generally black and yellow or sometimes black and orange. There is no jugal lobe in the hind wings (Fig. 349D). Most of the common species are in the genus *Bombus*. Females in the genus *Psithyrus* lay their eggs in nests of *Bombus*. The outer surface of the hind tibiae of *Psithyrus* females is convex and hairy but concave or flat with few hairs in *Bombus* species. Bumble bees nest socially in the ground in openings such as bird or mice nests or under matted vegetation. A colony lasts one season and the large, fertilized queen overwinters. She rears the first brood of workers (sterile females) who then take over the colony's maintenance. Larvae are fed pollen and nectar. Males and new queens are produced in late summer.

Honey bees (subfamily Apinae) are represented by a single introduced species in North America, *Apis mellifera* Linnaeus. There are various strains that differ in color and behavior. Three castes exist: workers, males (drones), and queens (Fig. 372). Workers are sterile females who maintain the colony. They have glands that produce wax to build wax combs, each comb consisting of many 6-sided cells for housing larvae and storing honey. Drones, produced temporarily to fertilize new queens, are expelled or killed by workers. A new queen mates in flight and the old queen leaves the nest with a large number of workers to establish a new colony. Females are produced from fertilized eggs and males from unfertilized eggs, typical of most hymenopteran reproduction. Female larvae become sterile adult workers partly because they are fed primarily pollen instead of royal jelly (a hormone-containing secretion from workers) throughout their developmental period.

Honey bees are very valuable pollinators, the value of this activity far surpassing that of their honey and beeswax. In order to increase

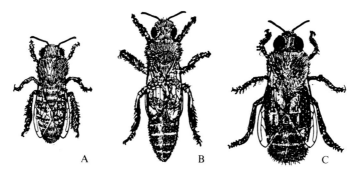

Figure 372 Honey bee, *Apis mellifera* (Apidae). A, worker; B, queen; C, drone.

the harvest, hives are often placed in fruit orchards or other crop sites when the crop's flowers are blooming. A bee that finds a source of nectar and pollen at some distance from the hive returns to the hive and imparts information about the location and food quality to other workers. The returning worker's nectar sample or flower odor, coupled with a particular type of behavior (a specific dance), form a "language" which stimulates and enables the other bees to locate the food source.

Common Species

Apis mellifera Linnaeus (Fig. 372). Honey bee—11-15 mm; generally golden brown, also blackish or grayish; dense, fine hairs on thorax; abdomen often banded with lighter or darker color; throughout North America.

Bombus fervidus (Fabricius)—8-23 mm; black; most of thorax and dorsum of 1st 4 abdominal segments yellow; throughout U.S., southern Canada.

Bombus griseocollis De Geer (Fig. 373A)—13-25 mm; black; thorax yellow; 1st abdominal tergite pale to grayish yellow, anterior 1/2 2nd tergite pale yellow to dull reddish brown, remainder of abdomen black; wings deep brown, darker toward tip; U.S. except Southwest.

Bombus pennsylvanicus De Geer (Fig. 373B)—12-23 mm; black; base of abdomen yellow and remainder black (♀) or most of abdomen yellow dorsally and tip reddish brown (♂); throughout U.S., southern Canada.

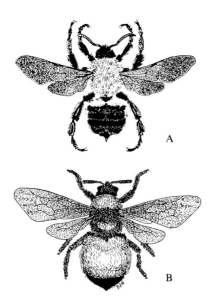

Figure 373 Bumble bees (Apidae). A, *Bombus griseocollis*; B, *Bombus pennsylvanicus*.

GENERAL REFERENCES

Bohart, R.M., and Menke, A.S. 1976. *Sphecid Wasps of the World*. Univ. California Press, Berkeley. 695 pp.

Clausen, C.P. 1940. *Entomophagous Insects*. McGraw-Hill, N.Y. 688 pp.

Creighton, W.S. 1950. The Ants of North America. Harvard Univ., Mus. Comp. Zool. Bull. 104. 585 pp.

Evans, H.E. 1963. *Wasp Farm*. Natural History Press, Garden City, N.Y. 178 pp.

Evans, H.E., and Eberhard, M.J.W. 1970. *The Wasps*. Univ. Michigan Press, Ann Arbor. 265 pp.

Felt, E.P. 1940. *Plant Galls and Gall Makers.* Comstock Publ., Ithaca, N.Y. 364 pp.

Frisch, K. von. 1967. *The Dance Language and Orientation of Bees.* Belknap Press of Harvard Univ. Press, Cambridge, Mass. 566 pp.

Krombein, K.V. 1967. *Trap-Nesting Wasps and Bees.* Smithsonian Press, Washington, D.C. 576 pp.

Michener, C.D. 1974. *The Social Behavior of Bees. A Comparative Study.* Belknap Press of Harvard Univ. Press, Cambridge, Mass. 404 pp.

Mitchell, T.B. 1960-1962. Bees of the Eastern United States. Vol. 1, Tech. Bull. 141, N.C. Agric. Expt. Sta., 1960. 538 pp. Vol. 2., *op. cit.,* Bull. 152, 1962. 557 pp.

Morse, R.A. 1975. *Bees and Beekeeping.* Cornell Univ. Press, Ithaca, N.Y. 320 pp.

Muesebeck, C.F.W.; Krombein, K.V.; Townes, H.K.; *et al.* 1951-1967. Hymenoptera of America North of Mexico; Synoptic Catalog. USDA Agric. Monog. 2, 1951. 1,420 pp. First Supplement to USDA Agric. Monog. 2, 1958. 305 pp. Second Supplement to USDA Agric. Monog. 2, 1967. 584 pp. (Catalog and supplements contain references to the taxonomy and biology of all families, genera, and species. A new catalog is in preparation.)

Peterson, A. 1948. *Larvae of Insects. Part 1. Lepidoptera and Plant-Infesting Hymenoptera.* Edwards Bros., Ann Arbor, Mich. 315 pp.

Spradbery, J.P. 1973. *Wasps. An Account of the Biology and Natural History of Social and Solitary Wasps.* Univ. Washington Press, Seattle. 416 pp.

Wilson, E.O. 1971. *The Insect Societies.* Harvard Univ. Press, Cambridge, Mass. 548 pp.

Index and Glossary

A

ABDOMEN: The third main division of the body, posterior to the head and thorax (Fig. 28).
Abedus, 139
  indentatus, 140
Ablabesmyia monilis, 331
Acalymma trivittatum, 230, 231
  vittatum, 230
Acanalonia bivittata, 165, 166
Acanaloniidae, 158, 165, 166
Acanthocephala femorata, 152, 153
  terminalis, 153
Acanthomyops claviger, 381
  interjectus, 381
Acanthoscelides obtectus, 232
ACCESSORY CELL: A closed cell in the front wing of Lepidoptera (Fig. 242C).
Aceratagallia sanguinolenta, 163
Acerentomidae, 58
Achalarus lyciades, 289, 290
Acheta domesticus, 99
Achilidae, 159, 164, 165
Achyra rantalis, 311
Acilius semisulcatus, 191, 192
Aclerda, 172
Aclerdidae, 172
Acmaeodera pulchella, 204
Acrididae, 92, 94-96
Acridinae, 94
Acrolophus, 305
Acrolophus popeanellus, 305
Acroneuria, 88
Acronicta, 299
  americana, 299
Acrosternum hilare, 155
Actia luna, 292
Acyrthosiphon pisum, 168
Adaina ambrosiae, 312
Adalia bipunctata, 215, 216
Adela, 304
Adelges abietis, 169
  cooleyi, 169
  piceae, 169
Adelgidae, 160, 169-170

Adelphocoris lineolatus, 148
  rapidus, 148
Adephaga, 176, 177, 188-192
ADFRONTAL AREAS: Two narrow regions surrounding the sides of the frons on the head of lepidopterous larvae (Fig. 234); they are lacking in similar beetle larvae.
Admiral, Lorquin's, 276
  red, 275
  Weidemeyer's, 276
  white, 276
Admirals, 276
Aecothea fenestralis, 347, 348
Aedes, 328, 329
  aegypti, 329, 330
  communis, 330
  dorsalis, 329
  nigromaculis, 329
  sollicitans, 329, 330
  squamiger, 329-330
  sticticus, 330
  taeniorhynchus, 330
  vexans, 330
Aeolothripidae, 130, 131
Aeolothrips fasciatus, 131
Aeshna constricta, 75
  umbrosa, 75
Aeshnidae, 73, 74-75
Agabus disintegratus, 191, 192
Agapostemon virescens, 388
Agaristidae, 268, 300-301
Agathymus, 286
Agave, 286
Agraulis vanillae, 277
Agriconota bivittata, 227, 228
Agrilus, 203
  anxius, 204
  bilineatus, 204
  ruficollis, 204
Agriotes mancus, 205
Agrius cingulatus, 290
Agromyzidae, 324, 347
Agrotis ipsilon, 298, 299
  orthogonia, 299
Agulla, 243
Alabama argillacea, 300

Alaus melanops, 205
  myops, 205
  oculatus, 205
Alderflies, 239, 242-243
Alderfly, smoky, 242, 243
Aleyrodidae, 159, 167
Alleculidae, 184, 185, 220
Allocapnia pygmaea, 83, 86
Allograpta obliqua, 341, 342
Allonemobius allardi, 99
  fasciatus, 99
Alloperla, 84, 87
  atlantica, 88
Alobates pennsylvanica, 218, 219
Alpine, common, 278
Alpines, 278
Alsophila pometaria, 303, 304
Alticinae, 229-230
Alydidae, 137, 153-154
Alydus, 154
  eurinus, 153, 154
  pilosulus, 154
Alypia octomaculata, 301
Amathes c-nigrum, 298, 299
Amatidae, 302
Amblycorypha oblongifolia, 96
Amblyscirtes vialis, 287, 288
Ambrysus, 139
  mormon, 139
AMETABOLOUS: Immature development with very little change in form; applied to Protura, Collembola, Diplura, and Thysanura.
Ametastegia glabrata, 371
Amitermes wheeleri, 115, 116
Ampedus collaris, 205
Amphicerus bicaudatus, 209, 210
  cornutus, 209
Anabrus simplex, 98
Anaea, 276
  andria, 276
Anagasta kuehniella, 312
Anagrapha falcifera, 300
ANAL: Refers to the last or most posterior part; region bearing the anus.

ANAL CELL: A cell in the anal region of a wing (Fig. 293).
ANAL LOBE: A lobe or localized expansion along the posterior wing margin near the wing base (Fig. 84A).
ANAL LOOP: A series of cells that forms a rounded to elongated or foot-shaped area in the hind wings of dragonflies (Figs. 70A; 76B).
ANAL VEINS: The lowest or posterior veins in the wing (Figs. 237; 293).
ANAMORPHOSIS: Adding a new body segment upon molting (e.g., Protura).
Anaphothrips obscurus, 131
Anarsia lineatella, 308
Anasa tristis, 152, 153
Anatoecus dentatus, 125
Anax junius, 74-75
Anchicera ephippiata, 214
Ancyloxypha numitor, 287
Andrena, 238, 388, 389
Andrenidae, 238, 369, 388-389
Andricus californicus, 376
Anglewing, satyr, 274
  zephyr, 274
Anglewings, 274
Anisembiidae, 111
Anisolabis maritima, 110
Anisomorpha buprestoides, 102
Anisoptera, 69, 71, 74
Anisota stigma, 294
  virginiensis, 294
Anobiidae, 182, 208-209
Anomala undulata, 200
Anopheles, 328, 329
  punctipennis, 330
  quadrimaculatus, 330
Anoplius americanus, 385
  marginalis, 385
Anoplura, 53, 122, 126-129
Anormenis septentrionalis, 166
Ant, Allegheny mound, 382
  Argentine, 381, 382
  black carpenter, 382
  California harvester, 381

393

Florida carpenter, 382
larger yellow, 381
little black, 381
odorous house, 381
pavement, 381
Pharaoh, 381, 382
red harvester, 381
red imported fire, 381
silky, 382
smaller yellow, 381
southern fire, 381, 382
thief, 381
western harvester, 381, 382
ANTENNA (pl., antennae): A pair of segmented, sensory appendages on the head usually between the compound eyes (Figs. 28; 29).
ANTENODAL CROSSVEINS: One to many crossveins meeting the anterior (costal) margin of the front wing between the base and the nodus (Fig. 70); used in identification of dragonflies and damselflies.
ANTEPYGIDAL BRISTLES: Large bristles situated just in front of the sensilium of fleas.
ANTERIOR: Front, or in front of.
ANTERIOR CROSSVEIN: A crossvein usually above the discal cell in the wings of Diptera (Fig. 293); identical to r-m crossvein.
Anthaxia aeneogaster, 204
Antheraea polyphemus, 293
Anthicidae, 185, 224
Anthicus cervinus, 224
Anthocharis midea, 282
 sara, 282
Anthocoridae, 135, 136, 144, 145
Anthomyiidae, 326, 348
Anthomyzidae, 324, 347
Anthonominae, 236
Anthonomus, 236
 grandis, 236
 signatus, 236
Anthophora abrupta, 390
 occidentalis, 390
Anthophoridae, 370, 389-390
Anthrax, 335
Anthrax analis, 338
 tigrinus, 338
Anthrenus scrophulariae, 208
Anthribidae, 188, 232
Antlions, 52, 239, 240, 245-246
Ants, 51, 361-362, 364, 380-382
 army, 381
 driver, 381
 leafcutting, 381
 velvet, 379-380
Anurida maritima, 62
Aonidiella aurantii, 172
 citrinia, 172
Apache degeerii, 164, 165
Apanteles congregatus, 373
 glomeratus, 373
Apantesis, 301
 arge, 301
 virgo, 301, 302
APEX (adj., apical; pl., apices): The tip, end, or outermost part.
Aphelinidae, 365, 375
Aphelinus mali, 375

Aphid, balsam woolly, 169
 bean, 167, 168
 cabbage, 168
 chrysanthemum, 168
 Cooley spruce gall, 169
 corn leaf, 168
 cotton, 168
 eastern spruce gall, 169
 English grain, 168
 green peach, 168
 mealy plum, 168
 melon, 168
 pea, 168
 pine bark, 170
 rose, 168
 spotted alfalfa, 168
 woolly alder, 169
 woolly apple, 168, 169
Aphididae, 159, 167-168
Aphidlions, 245
Aphids, 43, 167-168, 380
 gallmaking, 160, 168-169
 pine, 160, 169-170
 spruce, 160, 169-170
 woolly, 160, 168-169
Aphidwolves, 244
Aphis fabae, 167, 168
 gossypii, 168
Aphodiinae, 198
Aphodius distinctus, 198, 199
 fimetariuo, 198
Aphorista vittata, 217, 218
Aphrophora permutata, 162
 quadrinotata, 162
 saratogensis, 162
Aphytis proclia, 375
Apidae, 369, 370, 389, 390-391
Apinae, 390
Apiomerus crassipes, 145, 146
Apioninae, 234
Apis mellifera, 390-391
Apocrita, 361, 363, 372-391
Apodemia mormo, 283
 palmeri, 283
Apoidea, 362
Apterobittacus apterus, 248
Apterygota, 54, 57
APTERYGOTA: The subclass of primitive, wingless insects in the orders Protura, Collembola, Diplura, and Thysanura.
Aradidae, 138, 147, 148
Aradus acutus, 147, 148
 robustus, 148
Araecerus fasciculatus, 232
Archilestes grandis, 81
Archips argyrospilus, 309
Archytas apicifer, 352, 353
 californiae, 353
Arctic, chryxus, 278
Arctics, 278
Arctiidae, 267, 268, 269, 301-302
Arctiinae, 301
ARCULUS: A crossvein between the radius and cubitus near the base of the wing (Fig. 70); used in identification of dragonflies and damselflies.
Arcynopteryx, 87
Arenivaga, 106
Arge pectoralis, 370
Argia, 80
Argia fumipennis violacea, 80
 sedula, 80
Argidae, 364, 370

Argyresthia, 307
Argyrotaenia velutinana, 309, 310
Arhyssus lateralis, 153
Arilus cristatus, 145
ARISTA (adj., aristate): A large bristle near the tip of the last antennal segment of certain flies (Diptera) (Fig. 294E).
Armyworm, 299
 beet, 299
 fall, 298, 299
 yellowstriped, 299
Armyworms, 299
AROLIUM (pl., arolia): A pad between the tarsal claws in Orthoptera or at the base of each tarsal claw in Hemiptera (Fig. 127C).
Arotes amoenus, 373
Arphia pseudonietana, 95
Arrhenodes minutus, 232, 233
Arthromacra aenea, 220
Arthropoda, 38
Ascalaphidae, 241, 246
Ascia monuste, 282
Asilidae, 318, 319, 321, 337-338
Aspidiotiphagus citrinus citrinus, 375
Aspidiotus nerii, 173
Asterocampa, 276
 celtis, 276, 277
Asterolecaniidae, 172
Asterolecanium, 172
 puteanum, 172
ASYMMETRICAL: Two sides are unlike.
Atalopedes campestris, 287, 288
Atlanticus testaceus, 98
Atlides halesus, 285, 286
Atropidae, 119
Attagenus megatoma, 208
Attalus scincetus, 211
Atteva punctella, 307, 308
Auchenorrhyncha, 156
Aulacaspis rosae, 173
Aulacidae, 365, 378
Aulacus, 378
Aulocara elliotti, 94
AURICLE: In male dragonflies, a dorsolateral appendage on the second abdominal segment.
Author, 39
Autographa californica, 300
Automeris io, 294
Azure, spring, 284

B

Babia quadriguttata, 228, 229
Backswimmers, 140, 141
Baetidae, 65, 67
Baetis, 67
Bagworm, 305
Balsam, Canada, 17
Barber's fluid, 13
Baridinae, 235
BASAL CELL: One or more cells near the base of the wing (Fig. 293).
BASE (adj., basal): A part of an appendage or structure nearest the body; on the

thorax, the part nearest the abdomen; on the abdomen, the part nearest the thorax.
Battus philenor, 280
BEAK: The narrow, elongated, and often somewhat rigid mouthpart complex of sucking insects (e.g., Hemiptera) or certain insects with chewing mouthparts (e.g., weevils, scorpionflies); proboscis (Figs. 30B; 125; 213A; 223B,C).
Beans, Mexican jumping, 309
Beauty, painted, 275
Bee, alkali, 388
 carpenter, 390
 honey, 390-391
Bees, 361-362, 367
 andrenid, 369, 388, 389
 bumble, 51, 369, 390-391
 carpenter, 389-390
 colletid, 387-388
 cuckoo, 362, 388, 389
 digger, 389-390
 halictid, 369, 388
 honey, 369, 390-391
 leafcutting, 369, 389
 mining, 388
 nomadine, 389-390
 plasterer, 387
 silk, 387
 solitary, 362
 sweat, 369, 388
 yellow-faced, 387
Beetle, argus tortoise, 227, 228
 asparagus, 230, 231
 black blister, 223, 224
 black carpet, 208
 black turpentine, 233
 bumble flower, 201
 carpet, 208
 cigarette, 209
 Colorado potato, 230, 231
 confused flour, 26, 219
 convergent lady, 217
 corn flea, 229
 corn sap, 212
 driedfruit, 212
 drugstore, 209
 eyed click, 205
 giant stag, 197
 golden tortoise, 227, 228
 green June, 201
 hairy rove, 196
 Japanese, 200
 larder, 208
 merchant grain, 213
 Mexican bean, 217
 mottled tortoise, 227, 228
 Nuttall blister, 224
 potato flea, 229
 red flour, 219
 redlegged ham, 211
 red milkweed, 225
 redshouldered ham, 211
 red turpentine, 233
 rhinoceros, 201
 sawtoothed grain, 213
 seedcorn, 191
 smaller European elm bark, 233
 southern lyctus, 210
 southern pine, 233
 spinach flea, 229
 spotted asparagus, 230
 spotted blister, 223
 spotted cucumber, 230, 231

squarenecked grain, 213
squash, 217
striped blister, 223, 224
striped cucumber, 230
tenlined June, 199
thirteenspotted lady, 217
threelined potato, 230
threespotted flea, 229
tobacco flea, 229
transverse lady, 215, 216
twicestabbed lady, 215, 216
twospotted lady, 215, 216
western spotted cucumber, 230
western striped cucumber, 230, 231
whitemarked spider, 208, 209
Beetles, 45, 175-237
  antlike flower, 224
  aphodian dung, 198
  bark, 233
  bess, 197
  blister, 223-224
  bombardier, 189
  carrion, 193-194
  casebearing leaf, 229
  cedar, 203
  checkered, 210-211
  click, 45, 204-206
  combclawed, 220
  crawling water, 191
  cryptophagid, 213-214
  darkling, 218-219
  deathwatch, 208-209
  dermestid, 207-208
  drugstore, 208-209
  dung, 198
  earthboring dung, 198
  elephant, 201
  false darkling, 222
  false powderpost, 209, 210
  firecolored, 221-222
  flat bark, 213
  flea, 229
  flower, 201
  fruitworm, 217, 218
  ground, 45, 177, 188-191
  hairy fungus, 220, 221
  handsome fungus, 217, 218
  Hercules, 201
  hister, 196-197
  June, 198-199, 378, 380
  lady, 45, 215-217
  languriid, 214
  leaf, 227-231
  leafmining leaf, 229
  longhorned, 225-227
  longjointed, 220
  longtoed water, 202-203
  marsh, 201
  May, 198
  melandryid bark, 222
  minute treefungus, 221
  narrowwaisted bark, 220, 221
  netwinged, 207
  oedemerid, 221, 222
  pedilid, 224
  picnic, 212
  pill, 201, 202
  pleasing fungus, 214-215
  powderpost, 209-210
  predaceous diving, 191-192
  rhinoceros, 201
  riffle, 202, 203
  rove, 45, 194-196
  sap, 212-213
  scarab, 45, 197-201, 379
  seed, 232
  shining fungus, 194, 215
  shortwinged mold, 196
  skin, 198
  snout, 234-236
  softbodied plant, 201-202
  softwinged flower, 211-212
  soldier, 206, 207
  spider, 208, 209
  stag, 197
  tiger, 188-189
  toothnosed snout, 234
  tortoise, 227
  trogositid, 210
  tumbling flower, 222
  variegated mudloving, 202, 203
  waterpenny, 201, 202
  water scavenger, 45, 193
  whirligig, 192
  whitefringed, 234, 235
Belostoma, 139
  bakeri, 140
  flumineum, 140
  lutarium, 140
Belostomatidae, 135, 139-140
Bembidion patruele, 189, 190
  quadrimaculatum, 189
Bembix americana comata, 386
  americana spinolae, 386, 387
  pallidipicta, 386
Berosus pantherinus, 193
  striatus, 193
Berytidae, 137, 151
Bibio albipennis, 333
Bibionidae, 318, 319, 332-333
Biddies, 76
Billbug, bluegrass, 235
  maize, 234, 235
Billbugs, 235
BILOBED: Divided into two rounded parts.
BIOLUMENESCENCE: Light production by living organisms (e.g., fireflies).
BISEXUAL: Males and females present.
Bittacidae, 247, 248
Bittacomorpha clavipes, 327
  occidentalis, 327
Bittacus apicalis, 248
  stigmaterus, 248
  strigosus, 248
Blaberidae, 103, 105, 106
Blaberus craniifer, 106
Blatta orientalis, 105
Blattaria, 103, 105-106
Blattella, 104
Blattella germanica, 105, 106
Blattellidae, 104, 106
Blattidae, 104, 105-106
Blattodea, 103
Blissus leucopterus, 150, 151
Bloodworms, 331
Blue, acmon, 284
  eastern, 284, 285
  Karner's, 284
  Melissa, 284, 285
  Reakirt's, 284
  saepiolus, 284
  silvery, 284
  western pygmy, 284
  western tailed, 284
Blues, 265, 284
Boatmen, water, 42, 140-141
Bolitotherus cornutus, 219
Bollworm, 298
  pink, 308
Boloria, 272
  bellona, 272
  selene, 272
Bombinae, 390
Bombus, 390
  fervidus, 391
  griseocollis, 391
  pennsylvanicus, 391
Bombyliidae, 321, 322, 338-339
Bombyliopsis abrupta, 352, 353
Bombylius lancifer, 338
  major, 338, 339
Bombyx mori, 259
Booklouse, 120
Booklouse, larger pale, 119
Boopidae, 122, 123
Boreidae, 247, 248-249
Borer, apple twig, 209, 210
  ash, 307-308
  broadnecked root, 225, 226
  bronze birch, 204
  clover stem, 214
  dogwood, 308
  European corn, 311
  flatheaded appletree, 204
  hemlock, 204
  lilac, 307-308
  locust, 226, 227
  peachtree, 308
  peach twig, 308
  poplar, 225, 226
  potato stalk, 235, 236
  raspberry crown, 307
  redheaded ash, 226, 227
  rednecked cane, 204
  red oak, 227
  roundheaded appletree, 225
  shothole, 233
  squash vine, 307
  sugarcane, 312
  turpentine, 204
  twolined chestnut, 204
  wharf, 221, 222
Borers, flatheaded wood, 203
  metallic wood, 203
  roundheaded wood, 225-227
Boreus brumalis, 249
  californicus, 248, 249
  nivoriundus, 249
Bostrichidae, 182, 209, 210
Bots, 353-354
  horse, 354
  rabbit, 323, 353
  rodent, 323, 353
Bourletiella hortensis, 61
Bovicola bovis, 125
  equi, 125
  ovis, 125
BRACE VEIN: A slanted crossvein in the wing; located near the stigma in dragonflies and damselflies (Fig. 72A).
Brachinus, 189
Brachycera, 316, 334-340
Brachymeria ovata, 376
Brachynemurus abdominalis, 246
  sackeni, 246
Brachypanorpa, 248
Brachys, 203
Brachystola magna, 94
Braconidae, 366, 372-373
Braconids, 366, 372-373
Bradysia, 333
Brentidae, 188, 232-233
Brephidium exilis, 284
Brevicoryne brassicae, 168
Bristletails, 54, 62-63
  jumping, 63
Brochymena quadripustulata, 154, 155
  sulcata, 155
Brown, eyed, 277
Bruchidae, 187, 232
Bruchomorpha oculata, 164, 166
Bruchus pisorum, 232
Bucculatrix, 305
  ainsliella, 306
  albertiella, 305
  canadensisella, 306
  thurberiella, 305
Buckeye, 274
Buckeyes, 274
Budworm, spruce, 309
  tobacco, 298
  western spruce, 309
Buenoa margaritacea, 141
Bug, alfalfa plant, 148
  bed, 144, 145
  boxelder, 153
  brown stink, 155
  chinch, 150, 151
  chrysanthemum lace, 147
  clouded plant, 148, 149
  common damsel, 145
  fourlined plant, 149
  giant water, 140
  green stink, 155
  harlequin, 155
  large bigeyed, 151
  large milkweed, 26, 150, 151
  leaffooted, 153
  meadow plant, 148, 149
  minute pirate, 144
  negro, 155
  onespot stink, 154, 155
  rapid plant, 148
  redshouldered stink, 155
  rough stink, 155
  Say stink, 154, 155
  small milkweed, 150, 151
  spined assassin, 146
  spined soldier, 154, 155
  spined stilt, 151, 152
  squash, 152, 153
  sycamore lace, 147
  tarnished plant, 148, 149
  toad, 138, 139
  western bigeyed, 151
  western damsel, 145
  wheel, 145
Bugs, alydid, 153-154
  ambush, 146-147
  ash-gray leaf, 147, 148
  assassin, 145-146
  bed, 42, 144, 145
  broadheaded, 154
  burrower, 156
  coreid, 152-153
  creeping water, 139
  damsel, 42, 144-145
  flat, 147, 148
  flower, 144
  giant water, 139, 140
  lace, 42, 147
  largid, 151
  leaffooted, 152
  lygus, 148
  minute pirate, 144, 145
  negro, 155-156
  plant, 148-149
  red, 152
  rhopalid, 153

Index and Glossary   395

riffle, 143
scentless plant, 153
seed, 42, 150-151
shield, 154
shore, 143, 144
squash, 152
stilt, 151
stink, 42, 154-155
toad, 138-139
true, 133-156
velvet water, 143
Buprestid, golden, 204
Buprestidae, 183, 203-204
Buprestis apricans, 204
aurulenta, 204
Butterflies, 47, 257-265, 271-290
brushfooted, 47, 264, 271-277
goatweed, 276
hackberry, 276
milkweed, 262, 278-279
rearing, 25-26
snout, 279
thistle, 275
Butterfly, alfalfa, 282, 283
goatweed, 276
hackberry, 276
pine, 281, 282
snout, 279
zebra, 277
Byrrhidae, 187, 201, 202
Byrrhus americanus, 202
Byturellus grisescens, 217
Byturidae, 181, 217, 218
Byturus bakeri, 217
rubi, 217, 218

C

Cabbageworm, cross-striped, 310
imported, 281, 282
southern, 281, 282
Cactoblastis cactorum, 312
Caddisflies, 50, 249-257
fingernet, 252-253
large, 255
longhorned, 256-257
netspinning, 253-254
northern, 255, 256
primitive, 252
trumpetnet, 253
tubemaking, 253
Cadelle, 210
Caeciliidae, 121
Caecilius aurantiacus, 121
Caelifera, 89, 92
Calathus ruficollis, 189
Calephelis virginiensis, 283, 284
Caliroa cerasi, 371
Calitys scabra, 210
Callidium antennatum hesperum, 226, 227
Calligrapha philadelphica, 230, 231
Calliphora vicina, 351
vomitoria, 351
Calliphoridae, 318, 325, 326, 351
Callophrys, 285
Callosamia promethea, 293
Calopteron reticulatum, 207
terminale, 207
Calopterygidae, 73, 79-80
Calopteryx aequabilis, 80
maculata, 80
Calosoma calidum, 189, 190
scrutator, 189
Calycopis cecrops, 286

CALYPTER (pl., calypteres): One of two membranous lobes or flaps near the base of fly (Diptera) wings (Fig. 293); squama.
Camnula pellucida, 95
Campodea, 59
Campodeidae, 58, 59
Camponotus abdominalis floridanus, 382
castaneus, 382
pennsylvanicus, 382
Campsomeris plumipes, 380
Campylenchia latipes, 161, 162
Cankerworm, fall, 303, 304
spring, 304
Cantantopinae, 95
Cantharidae, 180, 206, 207
Cantharis bilineatus, 206
divisa, 206
Canthon pilularius, 198, 199
CAPITATE ANTENNA: With a distinct, rounded, apical enlargement (Fig. 29F).
Capnia, 85
gracilaria, 86
Capniidae, 84, 85-86
Carabidae, 177, 178, 188-191
Carcinophoridae, 109, 110
Carpenterworm, 309
Carpophilus dimidiatus, 212
hemipterus, 212
Caryomyia, 333, 334
Casebearer, birch, 307
cigar, 307
larch, 307
pecan cigar, 307
pistol, 307
Cassidinae, 227
CASTE: A group of insects, with a specific appearance and function, that comprises a colony (e.g., ant workers; termite soldiers).
CATERPILLAR: A larva with a cylindrical body, distinct head, thoracic legs, and abdominal prolegs (Figs. 234; 352C); a larva of a butterfly, moth, sawfly, or scorpionfly.
Caterpillar, eastern tent, 295
forest tent, 295
puss, 313, 314
redhumped, 297
saddleback, 313, 314
saltmarsh, 301, 302
variable oakleaf, 297
walnut, 297
western tent, 295
yellownecked, 297
Cathartus quadricollis, 213
Catocala, 300
Catogenus rufus, 213
Catonia, 165
CAUDAL: At the posterior end of the body.
Cecidomyia ocellaris, 333
Cecidomyiidae, 315, 319, 320, 333-334
Cediopsylla inaequalis, 358
simplex, 358
Celastrina argiolus pseudargiolus, 284
Celithemis elisa, 79
CENTIMETER (cm): 10 millimeters; 0.39 inches; 2.56 centimeters per inch.

Cephidae, 363, 364, 372
Cephus cinctus, 372
pygmaeus, 372
Ceraclea maculata, 256
Cerambycidae, 187, 225-227
Cerambycinae, 227
Ceratina, 389
Ceratogastra ornata, 374
Ceratoma trifurcata, 230
Ceratomia catalpae, 290
undulosa, 291
Ceratophyllidae, 357, 359-360
Ceratophyllus niger, 360
Ceratopogonidae, 320, 331
Cerceris clypeata, 387
Cercopidae, 157, 158, 162, 163
CERCUS (pl., cerci): One of two appendages near the tip of the abdomen (Figs. 28; 65A; 87A; 106); varies from a short stub to a long filament.
Cercyonis pegala, 277, 278
Ceroplastes ceriferus, 172
CERVICAL GILLS: Gills in the neck or cervix region between the head and thorax (e.g., stoneflies).
Cetoniinae, 201
Ceuthophilus, 98
maculatus, 98
Chaetocnema pulicaria, 229
Chaetopsylla, 360
Chafer, rose, 199
Chafers, 198-199
shining leaf, 200-201
Chagas disease, 145
Chalcididae, 365, 376
Chalcidoidea, 362
Chalcids, 376
seed, 375, 376
Chalcophora liberta, 204
virginiensis, 204
Chalybion californicum, 385
Chaoboridae, 320, 328
Chaoborus, 328
flavicans, 328
Chauliodes, 241
rastricornis, 242
Chauliognathus pennsylvanicus, 206, 207
Checkerspot, Baltimore, 273
chalcedon, 273
nycteis, 273
Checkerspots, 273
Cheleutoptera, 101
Chelisoches morio, 110
Chelisochidae, 109, 110
Chelonus texanus, 373
Chelymorpha cassidea, 227, 228
Chermidae, 169
Cheumatopsyche analis, 254
campyla, 254
Chilocorus stigma, 215, 216
Chimarra aterrima, 252
obscura, 253
Chionaspis furfura, 173
pinifoliae, 173
Chironomidae, 320, 331-332
Chironomini, 331
Chironomus attenuatus, 331
plumosus, 331
Chlaenius sericeus, 189
tricolor, 190
Chlamisinae, 229
Chlorochroa ligata, 155
sayi, 154, 155

Chloroperlidae, 84, 85, 87-88
Chloropidae, 324, 346-347
Chlosyne nycteis, 273
Choristoneura fumiferana, 309
occidentalis, 309
rosaceana, 309
Chorthippus curtipennis, 94, 95
Chortophaga viridifasciata, 95
CHRYSALIS (pl., chrysalids): A butterfly pupa (Fig. 235B).
Chrysididae, 366, 379
Chrysis coerulans, 378, 379
Chrysobothris, 203
femorata, 204
Chrysochus auratus, 230
cobaltinus, 230
Chrysomela scripta, 230, 231
Chrysomelidae, 187, 227-231
Chrysomelinae, 230
Chrysopa carnea, 245
coloradensis, 245
oculata, 244, 245
rufilabris, 245
Chrysopidae, 240, 241, 244-245
Chrysops, 335
aestuans, 336
callidus, 336
noctifer, 336
proclivis, 336
vittatus, 335, 336
Cicada, dogday, 161
Linnaeus' 17-year, 161
Linne's, 161
minor, 161
orchard, 161
periodical, 161
Riley's 13-year, 161
Cicadas, 43, 160-161
Cicadellidae, 157, 158, 163-164
Cicadidae, 157, 160-161
Cicindela oregona, 189
punctulata, 189
repanda, 189
sexguttata, 189
tranquebarica, 189
Cicindelidae, 188
Cicindelinae, 178
Ciidae, 186, 221
Cimbex americana, 370, 371
Cimbicidae, 364, 370, 371
Cimex lectularius, 144, 145
Cimicidae, 138, 144, 145
Cingilla catenaria, 303, 304
Circulifer tenellus, 163
Cis fuscipes, 221
Citheronia regalis, 294
Citheroniidae, 292
Citheroniinae, 294
Cixiidae, 159, 164, 165
Cixius, 165
CLASS: The next lower division of a phylum or subphylum and composed of a group of related orders (e.g., Class Insecta).
Clastoptera proteus, 162
CLAVATE ANTENNA: Enlarging gradually toward the tip (Fig. 29E).
CLAVUS: The elongated, anal (posterior) area of the front wing in Hemiptera that lies next to the scutellum when the wing is folded (Figs. 124; 126A,B).
Clearwing, common, 291
Cleoninae, 235
Cleridae, 181, 182, 210-211

Climacia areolaris, 243, 244
CLOSED CELL: A cell enclosed by veins on all sides and does not reach the wing margin (Fig. 32).
CLOSED COXAL CAVITY: Front coxal cavity: enclosed posteriorly by the sides of the prothorax (Fig. 164A); middle coxal cavity: enclosed by the meso- and metasternum and not touched by the lateral (pleural) portions of the mesothorax (Fig. 164C).
CLUBBED ANTENNA: Enlarged toward the tip; enlargement ranges from gradual (clavate, Fig. 29E) to abrupt (capitate, Fig. 29F; lamellate, Fig. 29G).
Clubtails, 75-76
CLYPEUS: A sclerite above and adjacent to the labrum on the front of the face (Fig. 30).
Clytrinae, 229
Cnephia pecuarum, 332
Coccidae, 171-172
Coccids, giant, 170
Coccinella transversoguttata richardsoni, 215
Coccinellidae, 187, 215-217
Coccoidea, 159, 170-175
Coccus hesperidum, 172
Coccygomimus pedalis, 374
Cochineal insect, 172
Cochineal insects, 172
Cochliomyia hominivorax, 351
Cockroach, American, 103, 105, 106
  brownbanded, 106
  Cuban, 106
  death's-head, 106
  German, 102, 103, 105, 106
  Madeira, 106
  oriental, 102, 103, 105
  Pennsylvania wood, 106
  smokybrown, 106
  Surinam, 106
Cockroaches, 46, 102-106
  rearing, 26
  wood, 106
COCOON: A silken covering around a pupa.
Coelopa, 344
  frigida, 344
  vanduzeei, 344
Coelopidae, 324, 344
Coenagrionidae, 72, 73, 80-81
Coenonympha california, 278
  ochracea, 278
Colaspis brunnea, 230, 231
Colaspis, grape, 230, 231
Coleomegilla maculata fuscilabris, 216, 217
Coleophora, 307
  fuscedinella, 307
  laricella, 307
  laticornella, 307
  malivorella, 307
  serratella, 307
Coleophoridae, 270, 306-307
Coleoptera, 45, 175-237
Colias, 282
  cesonia, 282, 283
  eurydice, 282
  eurytheme, 282, 283
  philodice, 282, 283

Collecting equipment and supplies, 3-12
  aspirators, 3, 7
  beating trays and umbrellas, 3, 6-7
  Berlese funnel, 3
  dredges, 6
  killing jars, 3-4
  killing liquids for immatures, 3
  lights, 3, 8
  nets, 3, 4-6
  sifters and sieves, 3, 7-8
  traps, 8-10, 316, 362
Collecting techniques, 10-12
Collembola, 54, 55, 59-62
Colletes, 387
Colletidae, 369, 370, 387-388
COLLOPHORE: A ventral tube on the first abdominal segment of springtails.
Collops, 211
  bipunctatus, 211
  quadrimaculatus, 211, 212
  vittatus, 211
Coloradia pandora, 294
Columbicola columbae, 125
Comma, 274
  green, 274
COMPOUND EYE: One of two protruding eyes of most adult insects and nymphs that consists of a few to thousands of eye units (Fig. 28).
Compsilura concinnata, 352
Conchuela, 155
Conehead, swordbearing, 97
Conenose, bloodsucking, 146
  western bloodsucking, 146
Coniopterygidae, 240, 241, 243-244
Conocephalinae, 97
Conocephalus brevipennis, 97
  fasciatus, 97
Conoderus vespertinus, 205
Conopidae, 322, 342-343
Conotelus obscurus, 212
  stenoides, 212
Conotrachelus, 235
  nenuphar, 235, 236
Contarinia sorghicola, 334
Copaeodes minima, 287
Copris fricator, 198, 199
Cordulegaster, 76
  dorsalis, 76
  maculata, 76
Cordulegastridae, 73, 76
Cordulia shurtleffi, 78
Corduliidae, 73, 77
Coreidae, 137, 152-153
Corimelaena, 155
  pulicaria, 155
Corimelaenidae, 155
Corisciidae, 154
CORIUM: Usually the larger of two thickened, basal portions of the front wings of Heteroptera (Figs. 124; 126); extends along anterior margin.
Corixidae, 135, 140-141
Corizidae, 153

CORNICLE: One of a pair of tubular, dorsal projections near the tip of an aphid's abdomen (Fig. 153).
Corrodentia, 118
Corydalidae, 241
Corydalus, 241
  cornutus, 242
Corythucha, 147
  ciliata, 147
  marmorata, 147
  pruni, 147
Cosmopepla bimaculata, 155
Cossidae, 266, 309
COSTA: The most anterior, longitudinal wing vein, usually forming the anterior margin (Fig. 32).
COSTAL CELL: The wing area between the costa and subcosta (Fig. 340A).
Cotalpa lanigera, 200
Cotinus nitida, 201
  texana, 201
COXA (pl., coxae): The first (basal) leg segment (Fig. 31).
Crabroninae, 387
Crambinae, 312
Crambus, 312
  teterrellus, 312
Crescent, field, 273
  mylitta, 273
  nycteis, 273
  pearl, 274
  phaon, 273
Crescents, 273
Cricket, Allard's ground, 99
  blackhorned tree, 100
  fall field, 99
  fourspotted tree, 120
  house, 99
  Jerusalem, 98
  larger pygmy mole, 93
  minute pygmy mole, 92, 93
  Mormon, 98
  northern mole, 100
  snowy tree, 99
  spotted camel, 98
  spring field, 99
  striped ground, 99
Crickets, 46, 89-93, 98-100
  camel, 90, 92
  cave, 90, 98
  field, 99
  ground, 99
  house, 99
  mole, 90, 100
  pygmy mole, 89, 92-93
  pygmy sand, 92-93
  rearing, 25
  tree, 99-100
Crioceriinae, 230
Crioceris asparagi, 230, 231
  duodecimpunctata, 230
CROCHETS: Hooked spines on the tips of prolegs of butterfly and moth larvae (Fig. 234).
CROSSVEIN: A wing vein that connects adjacent longitudinal veins (Fig. 32).
Cryptocephalinae, 229
Cryptocephalus, 228, 229
Cryptocercidae, 105
Cryptocercus punctulatus, 105
Cryptophagidae, 185, 213-214
Cryptophagus acutangulus, 214

Cryptorhynchinae, 235
Cryptotermes brevis, 115
Cryptus albitarsis, 374
Ctenicera inflata, 205
  lobata tarsalis, 205
  pruinina, 205
CTENIDIUM (pl., ctenidia): A row of stout bristles or spines (gc, prc, Fig. 336); used commonly in flea identification.
Ctenocephalides canis, 358
  felis, 358
Ctenophthalmus pseudagyrtes, 359
Ctenucha, 302
  virginica, 302
Ctenucha, Virginia, 302
Ctenuchas, 302-303
Ctenuchidae, 268, 302-303
CUBITO-ANAL CROSSVEIN: A crossvein connecting the cubitus and an anal vein (Fig. 32).
CUBITUS: A longitudinal vein below (posterior to) the media (Fig. 32).
Cuclotogaster heterographus, 124, 125
Cucujidae, 183, 184, 213
Cucujus clavipes, 213
Culex, 328, 329
  pipiens, 330
  quinquefasciatus, 330
  tarsalis, 330
Culicidae, 320, 328-330
Culicoides variipennis variipennis, 331
CUNEUS: In certain Heteroptera, the apical, separate portion of the corium along the anterior wing margin (Figs. 124; 126B).
Curculio, 236
  caryae, 236
  sulcatulus, 236
Curculio, plum, 235, 236
  rhubarb, 235, 236
  rose, 234
Curculionidae, 188, 234-236
Curculioninae, 236
Cuterebra, 353
Cuterebridae, 323, 353
CUTICLE: The outer, nonliving layer of the skin of arthropods.
Cutworm, black, 298, 299
  pale western, 299
  spotted, 298, 299
  variegated, 299, 300
Cutworms, 299
Cybister fimbriolatus, 191
Cycloneda munda, 216, 217
  sanguinea, 217
Cyclorrhapha, 315, 316, 340-354
Cydia pomonella, 309, 310
Cydnidae, 138, 156
Cyladinae, 234
Cylas formicarius elegantulus, 234
Cylindrarctus longipalpis, 196
Cynipidae, 365, 376-377
Cynipids, 376-377
Cynthia, 275
Cyphoderris monstrosa, 98
Cyphon collaris, 201
Cyrpoptus belfragei, 164, 165
Cyrtacanthacridinae, 94, 95

Index and Glossary   397

D

Dachnonypha, 259
Dactylopiidae, 172
Dactylopius coccus, 172
  confusus, 172
  opuntiae, 172
Daktulosphaira vitifoliae, 170
Damselflies, 48, 69-73, 79-82
  broadwinged, 79-80
  narrowwinged, 80-81
  spreadwinged, 81-82
Damselfly, blue, 81
Danaidae, 258, 262, 263, 278-279
Danaus gilippus, 279
  plexippus, 258, 279
Darner, green, 74, 75
Darners, 74-75
Dascillidae, 183, 201-202
Dascillus davidsoni, 201, 202
Dasineura leguminicola, 334
Dasychira plagiata, 295
Dasymutilla bioculata, 380
  gloriosa, 380
  occidentalis, 379, 380
  sackenii, 380
Datana, 296
  integerrima, 297
  ministra, 297
  perspicua, 297
Datana, sumac, 297
Daubers, mud, 379, 385-386
  organpipe mud, 385, 386
Decticinae, 98
Deloyala guttata, 227, 228
Delphacidae, 157, 158, 164
Delphinia picta, 343
Dendroctonus brevicomis. 233
  frontalis, 233
  terebrans, 233
  valens, 233
Dendroides cyanipennis, 221
Derbidae, 159, 164
Dermaptera, 44, 54, 108-111
Dermestes caninus, 208
  lardarius, 208
Dermestidae, 183, 207-208
Desmia funeralis, 311
Devil, hickory horned, 294
Diabrotica longicornis, 230
  undecimpunctata howardi, 230, 231
  undecimpunctata undecimpunctata, 230
Diacrisia virginica, 301
Dialeurodes citri, 167
Dialictus, 388
Diamanus montanus, 359
Diaperis, 218
Diaphania hyalinata, 311
  nitidalis, 311
Diapheromera femorata, 101, 102
Diaspididae, 172-173
Diatraea saccharalis, 312
Diaulota, 195
Dibrachys cavus, 375
Dicerca, 203
  divaricata, 204
Dichomeris marginella, 308
Dictyopharidae, 159, 165
Dictyoptera, 45, 46, 56, 102-107
Dictyopterus aurora, 207
Didea fuscipes, 342
Didymops, 76
  transversa, 76
Dineutus americanus, 192
  assimilis, 192

ciliatus, 192
  discolor, 192
Dioryctria zimmermani, 312
Diplazon laetatorius, 374
Diplolepis rosae, 376, 377
Diplotaxis tenebrosa, 199
Diplura, 54, 58-59
Diplurans, 54, 58-59
Diprionidae, 364, 370, 371
Diptera, 40, 53, 315-355
DISCAL CELL: An elongated, medium to large cell in the basal or central wing area (Figs. 237; 293).
DISCOIDAL CELL: A cell near the center of the wing of Hymenoptera (Fig. 340A).
Disonycha triangularis, 229
  xanthomelas, 229
Dissosteira carolina, 94, 95
DISTAL: The portion farthest from the body or point of attachment.
Ditrysia, 259
DIURNAL: Active during the daytime.
DIVERGE: Separate or spread apart.
Dixa, 328
Dixidae, 320, 328
Dobsonflies, 239, 241-242
Dobsonfly, 52, 242
Dolichoderinae, 381
Dolichopodidae, 322, 339, 340
Dolichopus, 339, 340
Dolophilodes distinctus, 252
Donacia, 229, 231
Donaciinae, 229
Doodlebugs, 245
Doratopsylla blarinae, 359
Dorcus parallelus, 197
DORSAL: Refers to the top, back, or upper side of an object.
DORSOLATERAL: Between the top and side.
DORSUM (adj., dorsal): The top or upper side; the back.
Doru aculeatum, 110
Dorylinae, 381
Draeculacephala mollipes, 163
Dragonflies, 48, 69-79
Drepana arcuata, 303
Drepanidae, 268, 303
Drilidae, 176
Drosophila, 346
  melanogaster, 346
Drosophilidae, 324, 346
Dryopidae, 179, 181, 202-203
Dustywings, 240, 243-244
Dutch elm disease, 233
Dynastes, 201
  granti, 201
  tityus, 201
Dynastinae, 201
Dysdercus, 152
  suturellus, 152
Dysentery, 316
Dytiscidae, 178, 191-192
Dytiscus fasciventris, 192

E

Eacles imperialis, 294
Earwig, black, 110
  European, 110
  handsome, 109

little, 109, 110
  ringlegged, 110
  seaside, 110
  shore, 110
  spinetailed, 110
Earwigs, 108-111
Earworm, corn, 298
Echidnophaga, 356, 357
  gallinacea, 358
Echinophthiriidae, 126, 128
Echmepteryx hageni, 119
ECTOGNATHOUS: Protruding mouthparts (e.g., Thysanura, Pterygota).
ECTOPARASITE: Externally feeding parasite.
Ectopsocidae, 121
Ectopsocus meridionalis, 120, 121
Edge, hoary, 289, 290
Efferia pogonias, 337, 338
Elateridae, 176, 183, 204-206
ELBOWED ANTENNA: An antenna with a long first segment and the remaining segments attached to the first at a distinct angle (Fig. 29D).
Eleodes, 218
Elfins, 285
Ellipes minuta, 92, 93
Ellychnia corrusca, 206
Elmidae, 182, 202, 203
ELYTRON: The hardened or thickened front wing of beetles, earwigs, and some homopterans (Fig. 157A).
Embolium: A narrow strip of the corium, separated by a suture, along the anterior margin of the front wing of some Heteroptera (Fig. 126A).
Emesaya brevipennis, 145, 146
Empididae, 321, 322, 339
Empoasca fabae, 163
EMPODIUM: A bristle or pad between the claws of the last tarsal segment (Fig. 298).
Enallagma civile, 81
Enaphalodes rufulus, 227
Encephalitis, 328
Enchenopa binotata, 161
Endelomyia aethiops, 371
Endomychidae, 186, 187, 217, 218
Endomychus biguttatus, 217, 218
Endria inimica, 163
Engraver, southern pine, 233
Enicospilus medarius, 374
Ennomos magnarius, 303
  subsignarius, 303
Enochrus ochraceus, 193
Enoclerus nigripes, 211
  sphegeus, 211
Ensifera, 90-91, 96
ENTOGNATHOUS: Mouthparts withdrawn in the head (e.g., Protura, Diplura, Collembola).
Entomobrya multifasciata, 60, 62
Entomobryidae, 61, 62
Entotrophi, 58
Eosentomidae, 58
Epargyreus clarus, 289, 290

Ephemera simulans, 69
Ephemerella, 68
Ephemerellidae, 65, 66, 68
Ephemeridae, 65, 68, 69
Ephemeroptera, 42, 46, 47, 64-69
Ephydra, 345
  cinerea, 345, 346
Ephydridae, 323, 325, 345-346
Epicauta albida, 223
  maculata, 223
  pennsylvanica, 223, 224
  vittata, 223, 224
Epilachna, 215
  borealis, 217
  varivestis, 217
EPISTERNUM (pl., episterna): The side of the thorax (pleuron) anterior to the pleural suture (Fig. 28).
Epitedia wenmanni, 359
Epitheca, 78
  cynosura, 78
  princeps, 78
Epitrix cucumeris, 229
  hirtipennis, 229
Erebia epipsodea, 278
Erebus odora, 298
Eriococcidae, 174
Eriosoma lanigerum, 168, 169
Eriosomatidae, 168
Eristalis, 341
  tenax, 341, 342
Erotylidae, 184, 187, 214-215
Erynnis, 289
  juvenalis, 289
Estigmene acrea, 301, 302
Eubaeocera apicalis, 194
Euborellia annulipes, 110
Eucanthus lazarus, 198
Euchaetias egle, 301
Eudonia, 310
Eulophidae, 375
Eumastacidae, 91, 93
Eumenes, 384
  fraternus, 384
Eumenidae, 238, 368, 384
Eumicrosoma beneficum, 378
Eumolpinae, 230
Eumorpha achemon, 291
Euparius marmoreus, 232
Eupeodes volucris, 342
Euphoria, 201
  inda, 200, 201
Euphydryas, 273
  chalcedona, 273
  phaeton, 273
Euphyes vestris, 287
Euptoieta claudia, 272
Euptychia cymela, 277, 278
  hermes, 278
Eurema lisa, 282, 283
  nicippe, 282
Eurosta, 343
  solidaginis, 343
Eurytomidae, 365, 375-376
Eurytomids, 375-376
Euschistus servus, 155
  variolarius, 154, 155
Euspilotus assimilis, 197
Eustrophinus bicolor, 222
Euthochtha galeator, 153
Evania appendigaster, 377
Evaniidae, 365, 377
Everes amyntula, 284
  comyntas, 284, 285
Evergestinae, 310
Evergestis rimosalis, 310
Exitianus exitiosus, 163

F

FACE: The front of the head.
Face, California dog, 282
  dog, 282, 283
FACET: The outer (external) surface of each eye unit of a compound eye.
FAMILY: A group of related genera, tribes, or subfamilies; a subdivision of an order, suborder, or superfamily; family names end in -idae (e.g., Acrididae).
Fannia, 325, 349
  canicularis, 349
  scalaris, 349
FECES: Excrement.
Felicola subrostratus, 125
FEMUR (pl., femora): The leg segment between the trochanter and tibia (Fig. 31); usually broad, sometimes long.
Fenusa pusilla, 371
  ulmi, 371
Fever, relapsing, 127
  typhoid, 316
  yellow, 328
FILAMENT: Very slender and long; threadlike.
Filariasis, 328
Filbertworm, 309
FILE: A minute, rough ridge near the base of the front wing (ventral side) of crickets and katydids; used for sound production.
FILIFORM ANTENNA: Threadlike antenna composed of slender, elongated segments (Fig. 29B).
Firebrats, 63
Fireflies, 206, 207
Fishflies, 239, 241-242
FLABELLATE ANTENNA: Antenna with broad, fanlike projections (Fig. 160E).
Flatidae, 158, 165, 166
Flea, California ground squirrel, 359
  cat, 358
  chigoe, 357, 358
  dog, 358
  European mouse flea, 359
  Franklin's ground squirrel, 359
  gray squirrel, 360
  human, 358
  oriental rat, 358
  pocket gopher, 360
  snow, 62
  sticktight, 357, 358
  western chicken, 360
Fleahopper, cotton, 149
  garden, 148, 149
Fleas, 53, 355-360
  bat, 360
  carnivore, 360
Flies, 315-355
  anthomyiid, 348
  anthomyzid, 347
  bee, 321, 338-339
  bigheaded, 340-341
  black, 332
  black scavenger, 344
  blow, 318, 325, 351
  bot, 323, 353-354
  chloropid, 346
  crane, 319, 326-327
  dance, 321, 339
  deer, 320, 335-336
  flesh, 351-352
  flower, 322
  fruit, 325, 343-344
  heleomyzid, 347-348
  horse, 318, 320, 335-336
  house, 325, 349
  humpbacked, 320, 340
  lauxaniid, 345
  leafminer, 347
  lonchopterid, 321
  longlegged, 322, 339, 340
  louse, 350-351
  March, 318, 332
  marsh, 344
  milichiid, 345
  moth, 327-328
  muscoid, 316
  mydas, 337
  otitid, 343
  ox warble, 353
  phantom crane, 319, 327
  picture-winged, 343
  robber, 318, 319, 337-338
  sand, 327
  seaweed, 344
  shore, 325, 345-346
  snipe, 336
  soldier, 318, 320, 334-335
  stable, 325, 349
  stiletto, 336-337
  tachina, 325, 352-353
  thickheaded, 342-343
  tsetse, 316
  vinegar, 346
  warble, 323, 353-354
  winter crane, 326
Fly, black blow, 351
  black cherry fruit, 343-344
  black horse, 335, 336
  black soldier, 335
  cherry fruit, 343
  cluster, 351
  drone, 341, 342
  face, 350
  false stable, 350
  frit, 347
  Hessian, 334, 378
  horn, 349
  horse bot, 354
  house, 349, 350
  latrine, 349
  little house, 349
  nose bot, 354
  range crane, 326
  sheep bot, 354
  stable, 349, 350
  striped horse, 335, 336
  throat bot, 354
  western cherry fruit, 344
Folsomia elongata, 62
FONTANELLE: A small, rounded depression on top of the head between the compound eyes of termites (Fig. 108).
Forester, eightspotted, 301
Forficula auricularia, 110
Forficulidae, 109, 110
Formica exsectoides, 382
  fusca, 382
Formicidae, 364, 380-382
Formicinae, 381-382
Foxella ignota, 360
Frankliniella occidentalis, 132
  tritici, 132

FRASS: Dry excrement of wood-boring insects consisting chiefly of wood particles or other plant material.
Frenatae, 260
FRENULUM: One or more bristles at the base of the anterior (costal) margin of the hind wing in many Lepidoptera (Fig. 237); projects beneath the front wing to unite the wings in flight.
Fritillaries, 272-273
  greater, 272
  lesser, 272
Fritillary, aphrodite, 272
  eastern meadow, 272
  great spangled, 272
  gulf, 277
  regal, 272
  silverbordered, 272
  variegated, 272
Froghoppers, 162
FRONS: The central area on the front of the face.
FRONTAL SUTURE: In Diptera, a groove shaped like an upside down "U" with the arms extending downward on each side of the face (Fig. 291A).
FRONTO-ORBITAL BRISTLES: In Diptera, the bristles extending along the front margin of a compound eye (Fig. 291A).
Fruitminer, apple, 306
Fruitworm, eastern raspberry, 217, 218
  tomato, 298
  western raspberry, 217
Fucellia rufitibia, 348
Fulgoridae, 158, 164-165
FURCULA: The forked, springing structure of most springtails located ventrally on the fourth abdominal segment (Figs. 63A,D,F,G).

G

Galerita janus, 190
Galerucinae, 230
Galgupha, 155
  atra, 156
GALL: An excessive, localized growth of plant tissue caused by the presence of insect larvae or other organisms (Figs. 313B-D; 323C; 359B,C).
Gall, mossyrose, 377
  pinecone willow, 333, 334
Galleria mellonella, 312
Galleriinae, 312
Galls, goldenrod, 343
  hickory, 334
  oak, 376, 377
GASTER: The large, rounded part of the abdomen of ants.
Gasterophilidae, 323, 354
Gasterophilus haemorrhoidalis, 354
  intestinalis, 354
  nasalis, 354
Gasteruptiidae, 365, 377-378

Gasteruption, 377, 378
Gelastocoridae, 135, 138-139
Gelastocoris oculatus, 138, 139
Gelechiidae, 270, 308
Gelis tenellus, 374
GENAL COMB: A series of stout teeth on the lower side of the head of some fleas (Fig. 336).
GENITALIA: The external sexual organs and associated structures.
GENUS (pl., genera): A group of closely related species; the first name used in the latinized scientific name of a species (e.g., *Musca* of *Musca domestica*); first letter of a generic name is capitalized and the word is in italics or underlined.
Geocoris, 150
  bullatus, 151
  pallens, 151
  punctipes, 150, 151
Geometer, chainspotted, 303, 304
  chickweed, 303
  notchwing, 303
Geometridae, 267, 271, 303-304
Geomydoecus, 125
Geotrupes splendidus, 198, 199
Geotrupinae, 198
Gerridae, 136, 141-142
Gerris, 142
  marginatus, 142
  remigis, 142
GILL: A thin structure used for breathing (gas exchange) by aquatic animals; in insects, generally feathery, forked, or platelike and located on the thorax, abdomen, or in the hindgut (Figs. 68; 79B; 80B; 311C).
Glaucopsyche lygdamus, 284
Glipa octopunctata, 223
  oculata, 223
Gliricola porcelli, 123
Glischrochilus, 212
  fasciatus, 212
  quadrisignatus, 213
  sanguinolentus, 213
GLOSSA (pl., glossae): One of two lobes at the tip of the labium and between the paraglossae (Figs. 85C; 350).
Glossinidae, 316
Glowworms, 206
Gnat, eye, 346, 347
  southern buffalo, 332
Gnathamitermes perplexus, 117
Gnats, buffalo, 354
  darkwinged fungus, 319, 333
  fungus, 333
Gnorimoschema, 308
Gomphidae, 73, 75-76
Gomphocerinae, 94
Gomphus, 75
  exilis, 75
  externus, 76
Goniocotes gallinae, 124, 125
Gossyparia spuria, 174
Gracillaria negundella, 306
  sassafrasella, 306
  syringella, 306
Gracillariidae, 269, 306

Index and Glossary 399

Graphium marcellus, 280
Graphocephala coccinea, 163
Graphognathus leucoloma, 234, 235
Grasshopper, American, 95
  American bird, 95
  awlshaped pygmy, 93
  bigheaded, 94
  blacksided pygmy, 93
  Carolina, 94, 95
  clearwinged, 95
  differential, 95
  eastern lubber, 94
  greenstriped, 95
  lubber, 94
  meadow, 94, 95
  migratory, 96
  obscure pygmy, 93
  pasture, 94, 95
  prairie bird, 95
  redlegged, 95-96
  redwinged, 95
  Rocky Mountain, 90
  sanded pygmy, 93
  sedge pygmy, 93
  slender pygmy, 93
  spotted bird, 95
  twostriped, 95
Grasshoppers, 46, 89-96
  bandwinged, 95
  bird, 95
  longhorned, 90, 96
  lubber, 94
  monkey, 93
  pygmy, 46, 89, 93, 238
  rearing, 26
  shieldbacked, 98
  slantfaced, 94
Graybacks, 74
Greenbug, 168
GREGARIOUS: Living in groups.
GRUB: A stout, usually sluggish larva with a distinct head and thoracic legs (Figs. 35D; 178D).
Grub, common cattle, 353
  northern cattle, 353-354
Grubs, white, 198, 200
Gryllacrididae, 92, 98
Gryllidae, 92, 99-100
Gryllinae, 99
Grylloblatta campodeiformis, 108
Grylloblattidae, 108
Grylloblattodea, 56, 107-108
Gryllotalpidae, 91, 100
Gryllus, 99
  pennsylvanicus, 99
  veletis, 99
Gyrinidae, 178, 192
Gyrinus, 192
  borealis, 192
  maculiventris, 192
Gyropidae, 122, 123
Gyropus ovalis, 123

H

Haematobia irritans, 349-350
Haematopinidae, 127, 128
Haematopinus asini, 127
  eurysternus, 127
  suis, 127, 128
Haematopis grataria, 303
Hagenius brevistylus, 76
Hairstreak, banded, 286
  coral, 286

gray, 286
great blue, 285, 286
  redbanded, 286
Hairstreaks, 265, 284, 285-286
Halictidae, 368, 369, 388
Halictophagidae, 238
Halictophagus, 238
Halictus farinosus, 388
  ligatus, 388
Haliplidae, 177, 191
Haliplus, 191
  immaculicollis, 191
  triopsis, 191
Halisidota, 301
  caryae, 301
  tessellaris, 301, 302
Halobates, 142
HALTER (pl., halteres): A small, flattened structure enlarged on the end and located on each side of the metathorax in flies (Diptera), representing the hind wings (Fig. 292).
Halticus, 148
  bractatus, 148, 149
HAMULI: Small hooks; in dragonflies and damselflies, the usually forked, ventral appendage on the second abdominal segment of males (Fig. 71B).
Hangingflies, 248
Haploembia solierii, 111
Harkenclenus titus, 286
Harmolita grandis, 376
  tritici, 376
Harpalus caliginosus, 177, 190
  pensylvanicus, 190
Harrisina americana, 313
Hastaperla brevis, 88
HEAD: The anterior of the three body regions.
Head, dog's, 282, 283
Heartworm, dog, 328
Hebridae, 136, 143
Helichus lithophilus, 202, 203
Helicobia rapax, 352
Heliconians, 264, 277
Heliconiidae, 277
Heliconius charitonius, 277
Heliothis virescens, 298
  zea, 298
Heliothrips haemorrhoidalis, 131, 132
Hellgrammites, 242
Hellula rogatalis, 310
Helodidae, 182, 201
Hemaris thysbe, 291
HEMELYTRON (pl., hemelytra): The partially thickened front wing of Heteroptera.
Hemerobiidae, 240, 241, 244
Hemerobius pacificus, 244
  stigmaterus, 244
Hemiargus isola, 184
Hemileuca maia, 294
  nevadensis, 294
Hemileucinae, 294
HEMIMETABOLOUS: One of two major types of development where the immatures (nymphs) usually resemble the adults; incomplete metamorphosis; sometimes restricted only to Ephemeroptera, Odonata, and Plecoptera.

Hemiptera, 42, 43, 44, 46, 53, 55, 133-175
Hepialidae, 261, 314
Hepialus, 314
Heptagenia flavescens, 67
Heptageniidae, 65, 67, 68
Hermetia illucens, 334, 335
Hesperia, 287
  leonardus, 287
Hesperiidae, 259, 262, 287-289
Hesperocorixa atopodonta, 141
  interrupta, 140, 141
  laevigata, 141
Hesperoperla pacifica, 88
Hesperophylax designatus, 255, 256
Hetaerina americana, 80
Heterocampa manteo, 297
Heterocera, 260
Heteroceridae, 186, 202, 203
Heterodoxus spiniger, 123
Heteroneura, 260
Heteroptera, 42, 44, 53, 55, 133-156
Heterothripidae, 130, 132
Heterothrips, 132
Hexagenia bilineata, 69
  limbata, 69
Hippelates, 347
  pusio, 346, 347
Hippoboscidae, 318, 350-351
Hippodamia convergens, 216, 217
  parenthesis, 216, 217
  tredecimpunctata tibialis, 217
Hispinae, 229
Hister abbreviatus, 196
Histeridae, 181, 196-197
Holcostethus abbreviatus, 155
  limbolarius, 155
Hololepta quadridentata, 197
Holomenopon leucoxanthum, 124
HOLOMETABOLOUS: One of two major types of development where the immatures (larvae) do not resemble the adults and a pupal stage occurs; complete metamorphosis.
Homaemus bijugis, 154, 155
  parvulus, 155
Homoneura, 260
Homoptera, 42, 43, 44, 46, 53, 55, 133, 156-175
HONEYDEW: Anal secretions (often sugary) secreted by certain Homoptera (e.g., aphids, treehoppers, others).
Hoplia, 198
Hoplopleura, 128
  acanthopus, 128
  hesperomydis, 128
Hoplopleuridae, 127, 128
Hornet, baldfaced, 383
Hornets, 366, 382-384
Horntail, blue, 372
Horntails, 363, 370, 372
Hornworm, sweetpotato, 290
  tobacco, 292
  tomato, 292
HOST: The plant on which an insect feeds; an organism that is externally or internally parasitized.
Hoyer's solution, 17
HUMERAL: Near the base of the wing on the anterior margin; humeral angle (Fig. 239A); humeral crossvein (Fig. 32); humeral suture, a line of weakness where the termite wing is broken off (Fig. 110B).
Hunter, fiery, 189, 190
  masked, 145, 146
Hyalophora cecropia, 293
  euryalus, 293
Hyalopsocus striatus, 120, 121
Hyalopterus pruni, 168
Hybomitra lasiophthalma, 336
Hydrellia, 345
  griseola, 345
Hydrometra, 144
  martini, 143, 144
Hydrometridae, 136, 143, 144
Hydrophilidae, 180, 193
Hydrophilus triangularis, 193
Hydroporus undulatus, 192
Hydropsyche occidentalis, 254
  orris, 254
  simulans, 254
Hydropsychidae, 251, 253-254
Hydroptila angusta, 255
  hamata, 254, 255
Hydroptilidae, 251, 254-255
Hylaeus, 387
  modestus, 387, 388
Hylemya antiqua, 348
  brassicae, 348
  platura, 348
Hyles lineata, 291
Hymenoptera, 51, 52, 361-392
Hymenorus niger, 220
Hypera, 235
  postica, 235, 236
  punctata, 235, 236
Hyperaspis signata, 217
  undulata, 217
Hyperinae, 235
HYPERMETAMORPHOSIS: A modification of holometabolous development (complete metamorphosis) where certain larval instars assume quite different forms and behavior (e.g., blister beetles).
HYPERPARASITE: A parasite that uses another parasite as a host.
Hyphantria cunea, 301
Hypoderma, 353
  bovis, 353-354
  lineatum, 353, 354
Hypogastrura nivicola, 60, 62
Hypogastruridae, 61, 62
HYPOPLEURAL BRISTLES: A nearly vertical row of bristles usually located just above the hind coxae of flies (Diptera) (Fig. 303).
HYPOPLEURON (pl., hypopleura): In flies (Diptera), a sclerite found just above the hind coxa (Fig. 292).
Hypoprepia miniata, 301
Hystrichopsyllidae, 356, 358-359

I

Icebugs, 56, 107-108
Icerya purchasi, 170
Ichneumonidae, 366, 373-374

Ichneumons, 51, 366, 373-374
Ichthyura, 297
  inclusa, 297
Icosta albipennis, 350
  americana, 350
Idiocerus, 164
  pallidus, 163
Ilnacora malina, 148
IMAGO: Adult insect.
IMMATURE: Not fully developed; a larva or a nymph.
Inchworms, 303
Incisitermes minor, 115
  snyderi, 115
Incurvariidae, 270, 304
Indian lac insect, 171
Insect, adult, 35
  definition, xi
  development, 33-35
  food, 25-26
  habitats, 27-29
  larva, nymph, 34-35
  metamorphosis, 34
  observing, 25
  pupa, 35
  range, 2
  rearing, 25-26
  structure, 30-33
Insecta, 38
Insects, cochineal, 172
  scale, 170-175
INSTAR: The immature insect; the first instar is the insect after hatching but before its first molt, the second instar is the insect after the first molt but before its second molt, etc.
INTERANTENNAL SUTURE: A groove running between the bases of the antennae of fleas (Fig. 338B).
INTERCALARY VEINS: The short, extra, longitudinal veins along the lower wing margin of mayflies (Fig. 65C).
Ioscytus politus, 144
Ips calligraphus, 233
  grandicollis, 233
Iridomyrmex humilis, 381, 382
Ischnopsyllidae, 357, 360
Ischnura verticalis, 81
Ischyrus quadripunctatus, 214
Isia isabella, 302
Isocapnia grandis, 86
Isomira sericea, 220
Isoperla, 87
  bilineata, 87
  patricia, 87
Isoptera, 48, 49, 56, 112-117
Isotoma viridis, 60, 62
Isotomidae, 61, 62
Issidae, 158, 164, 165-166
Ithycerinae, 234
Ithycerus noveboracensis, 234
Itoplectis conquisitor, 374

J

Jalysus spinosus, 151, 152
Japygidae, 58, 59
Japyx, 59
Jointworm, wheat, 376
Jointworms, 375-376
JUGAL LOBE: A lobe near the base of the wing along the posterior margin (Fig. 340A).
Jugatae, 260
JUGUM: A lobelike extension at the base of the front wing that overlaps the hind wing in certain moths (Fig. 238).

K

Kalotermitidae, 114-115
Katydid, blacklegged meadow, 97
  broadwinged, 96
  common meadow, 97
  forktailed bush, 96
  gladiator meadow, 97
  northern true, 97
  oblongwinged, 96
  shortwinged meadow, 97
  slender meadow, 97
Katydids, 46, 89-92, 96-98
  bush, 96
  coneheaded, 96, 97
  false, 96
  meadow, 97
  primitive, 98
  roundheaded, 96
  shieldbacked, 98
  true, 97
Ked, sheep, 350
KEEL: An elevated ridge.
Kermes, 172
Kermesidae, 172
Kerriidae, 171
Keys, 1
Killer, cicada, 386, 387
Killers, cicada, 385
Kinnaridae, 159

L

Labia minor, 109, 110
LABIAL PALP (pl., palpi or palps): One of two short, antennalike structures on the labium (Fig. 30A).
Labidura riparia, 110
Labiduridae, 109-110
Labidus coecus, 381
Labiidae, 109, 110
LABIUM: The lowest (most posterior) mouthpart (Fig. 30).
LABRUM: The uppermost (most anterior) mouthpart (Fig. 30).
Laccifer lacca, 171
Lacewing, goldeneye, 244, 245
Lacewings, 52, 239, 240, 244-245
  brown, 244
  green, 244-245
Lachesilla nubilis, 120, 121
  pedicularia, 121
Lachesillidae, 121
Lady, painted, 275
  west coast, 275
Laemobothriidae, 123, 124
Laemobothrion atrum, 123
  chloropodis, 123
  glutinans, 123
  maximum, 123
  tinnunculi, 123
Lagriidae, 184, 220
LAMELLATE ANTENNA: Flattened or layered plates toward the tip of the antenna (Fig. 29G).
Lamiinae, 225
Lampyridae, 176, 180, 206, 207
Languria mozardi, 214
Languriidae, 184, 214
Laphria, 337
  flavicollis, 337, 338
Largidae, 138, 151
Largus succinctus, 151, 152
Larrinae, 385
LARVA: The immature insect, between the egg and pupal stage, with holometabolous development (Fig. 35); sometimes used for certain immature forms of termites and other hemimetabolous insects.
Lasiocampidae, 268, 295
Lasioderma serricorne, 209
Laspeyresia, 309
  saltitans, 309
LATERAL: Refers to the side of an object.
Lauxaniidae, 324, 345
Leafcutter, maple, 304
Leaffolder, grape, 311
Leafhopper, beet, 163
  clover, 163
  gray lawn, 163
  painted, 163
  potato, 163
Leafhoppers, 43, 158, 163-164, 238
LEAF MINER: An insect living and feeding between the upper and lower layers of a leaf.
Leafminer, birch, 371
  elm, 371
  lilac, 306
  spinach, 348
Leafperforator, cotton, 305
Leafroller, boxelder, 306
  fruittree, 309
  obliquebanded, 309
  redbanded, 309, 310
Leafworm, cotton, 300
Lebia atriventris, 190
  grandis, 190
Lecanium, European fruit, 172
Lecanium corni, 172
  nigrofasciatum, 171, 172
Lema trilineata, 230
Lepidopsocidae, 119
Lepidoptera, 46, 47, 55, 257-315
Lepidosaphes ulmi, 171, 173
Lepisma saccharina, 63
Lepismatidae, 63
Leptinotarsa decemlineata, 230, 231
Leptoceridae, 251, 256-257
Leptocoris trivittatus, 153
Leptogaster flavipes, 337, 338
Leptoglossus clypealis, 153
  occidentalis, 153
  oppositus, 152, 153
  phyllopus, 153
Leptophlebia cupida, 68
  nebulosa, 68
Leptophlebiidae, 65, 66, 67-68
Leptopsylla segnis, 359
Leptopsyllidae, 357, 359
Leptopterna dolabrata, 148, 149
Lepturinae, 227
Lepyronia quadrangularis, 162, 163
Lerodea eufala, 289
Lestes dryas, 81
  unguiculatus, 82
Lestidae, 73, 81-82
Lethe eurydice, 277
Lethocerus, 139
  americanus, 140
  griseus, 139, 140
  uhleri, 140
Leucophaea maderae, 106
Leucorrhinia intacta, 79
Leucotrichia pictipes, 254, 255
Leuctridae, 84, 86
Libellula, 78
  luctuosa, 78, 79
  pulchella, 79
Libellulidae, 72, 73, 78-79
Libytheana bachmanii, 279
Libytheidae, 263, 279
Lice, bird body, 124
  chewing, 53, 122-126
  feather chewing, 124-125
  human, 127-128
  jumping plant, 166
  mammal chewing, 125
  rodent, 123
  smallmammal sucking, 128
  smooth sucking, 128
  sucking, 53, 126-129
  wrinkled sucking, 127
Ligyrocoris diffusus, 151
Limacodidae, 266, 313-314
Limenitis, 276
  archippus, 276
  arthemis, 276
  astyanax, 276
  lorquini, 276
  weidemeyerii, 276
Limnephilidae, 251, 255, 256
Limnephilus rhombicus, 255, 256
  submonilifer, 255, 256
Limnogonus hesione, 142
Limonius californicus, 205
Linognathidae, 126, 128
Linognathus ovillus, 128
  setosus, 128
  vituli, 128
Liparidae, 295
Lipeurus caponis, 125
Liposcelidae, 119, 120
Liposcelis, 120
  divinatorius, 120
Litaneutra minor, 106
Lithocolletis, 306
  hamadryadella, 306
  hamamelielia, 306
  lucetiella, 306
Lithosiinae, 301
Lixus, 235
  concavus, 235
Locust, desert, 90
  migratory, 90
Locusta migratoria, 90
LOCUSTS: Grasshoppers that regularly migrate in large groups.
Locusts (cicadas), 160
Locusts, grouse, 46, 89, 93
Lonchoptera, 340
Lonchopteridae, 321, 340
LONGITUDINAL: Running lengthwise to the body or an appendage.
Looper, alfalfa, 300
  cabbage, 300
  celery, 300
  omnivorous, 304

Index and Glossary 401

Loopers, 300, 303
Louse, body, 127, 128
  cat, 125
  cattle chewing, 125
  chicken body, 123, 124
  chicken fluff, 124, 125
  chicken head, 124, 125
  crab, 127, 128
  dog chewing, 125
  dog large body, 123
  dog sucking, 128
  head, 128
  hog, 127
  horse chewing, 125
  horse sucking, 127
  large duck, 124
  longnosed cattle, 128
  oval guineapig, 123
  shaft, 123, 124
  sheep chewing, 125
  shortnosed cattle, 127, 128
  slender pigeon, 125
  wing, 125
Lovebug, 332
Loxostege cerealis, 311
  sticticalis, 311
Lucanidae, 179, 197
Lucanus elaphus, 197
Lucidota atra, 206, 207
Lucilia illustris, 351
Lycaena helloides, 285
  phlaeas americana, 285
  xanthoides dione, 285
  xanthoides xanthoides, 285
Lycaenidae, 264, 265, 284-286
Lycidae, 180, 207
Lycomorpha pholus, 302
Lyctidae, 183, 209-210
Lyctocoris campestris, 144
Lyctus cavicollis, 209
  linearis, 209
  planicollis, 210
Lycus, 207
Lygaeidae, 135, 137, 138, 150-151
Lygaeus kalmii, 150, 151
Lygus, 148
  lineolaris, 148, 149
Lymantria dispar, 295-296
Lymantriidae, 269, 271, 295-296
Lyonetiidae, 269, 305-306
Lysiphlebus testaceipes, 373
Lytta cyanipennis, 224
  immaculata, 224
  nuttalli, 224

M

Machilidae, 63
Machilids, 63
Machilis variabilis, 63
Macrocentrus ancylivorus, 373
Macrodactylus subspinosus, 199, 200
Macrolepidoptera, 260
Macromia, 76, 77
  illinoiensis, 76
  magnifica, 76, 77
Macromiidae, 73, 76-77
Macronema zebratum, 254
Macrosiphoniella sanborni, 168
Macrosiphum avenae, 168
  rosae, 168
MAGGOT: A larva without legs or a distinct head (Fig. 35E).
Maggot, apple, 344
  cabbage, 348

  cherry, 343
  onion, 348
  seedcorn, 348
  sugarbeet root, 343
  wheat stem, 346, 347
Maggots, rattailed, 341
Magicicada, 160
  septendecim, 161
  tredecim, 161
Malachius aeneus, 212
Malacosoma, 295
  americanum, 295
  californicum, 295
  disstria, 295
Malaria, 328
Malenka californica, 85
Mallophaga, 53, 122-126
MANDIBLE: One of two jaws forming part of the mouth (Fig. 30A); can be elongated to form various types of piercing and sucking mouths (Fig. 30B).
Manduca quinquemaculata, 292
  sexta, 292
Mantid, Carolina, 107
  Chinese, 102, 107
  European, 102, 106, 107
  minor ground, 106
Mantidae, 104, 106-107
Mantids, 46, 102, 103-104, 100-107
  praying, 103-104, 106-107
  rearing, 26
Mantis religiosa, 106
Mantispa brunnea, 243, 245
  interrupta, 245
Mantispidae, 240, 243, 245
Mantispids, 52, 239, 240, 243, 245
Mantodea, 103-104, 106-107
Marava pulchella, 109
Marbles, 282
Margarodes, 170
  laingi, 170
  meridionalis, 170
Margarodidae, 170
MARGINAL CELL: A cell along the outer part of the anterior (coastal margin (Fig. 340A).
Mark, question, 274
Marmara pomonella, 306
Masaridae, 364, 368, 384
MAXILLA (pl., maxillae): One of two mouthparts behind (posterior to) the mandibles (Fig. 30).
MAXILLARY PALP (pl., palpi or palps): A small antennalike structure on each maxilla (Fig. 30A).
Mayetiola destructor, 334
Mayflies, 47, 64-69
  rearing, 26
Mealworm, dark, 219
  yellow, 219
Mealybug, cactus, 172
  citrus, 173
  grape, 174
  longtailed, 173
Mealybugs, 173-174
Measurers, water, 143, 144
Mecoptera, 52, 55, 247-249
Mecynotarsus, 224
MEDIA: A longitudinal wing vein between the radius and cubitus (Fig. 32).

MEDIAN: In the middle; along the body's midline.
Megachile latimanus, 389
  pacifica, 389
Megachilidae, 369, 370, 389
Megacyllene robiniae, 226, 227
Megalodacne fasciata, 214
  heros, 214
Megalomus moestus, 244
Megaloptera, 239, 241-243
Megalopyge crispata, 313, 314
  opercularis, 313, 314
Megalopygidae, 265, 313
Megarhyssa, macrurus icterosticta, 374
  macrurus macrurus, 374
Megaselia rufipes, 340
Megathymidae, 263, 286-287
Megathymus, 286
  yuccae, 287
Melalgus confertus, 209, 210
Melandryidae, 185, 222
Melanolestes abdominalis, 145
  picipes, 145
Melanophila fulvoguttata, 204
Melanopleurus belfragei, 151
Melanoplinae, 94, 95
Melanoplus bivittatus, 95
  differentialis, 89, 95
  femurrubrum, 95-96
  sanguinipes, 95, 96
  spretus, 90
Melanorhopala clavata, 147
Melanotus communis, 205, 206
  similis, 206
Meleoma emuncta, 245
Melissopus latiferreanus, 309
Melittia satyriniformis, 307, 308
Meloe, 223
  angusticollis, 224
Meloidae, 185, 223-224
Melolonthinae, 198-200
Melonworm, 311
Melophagus ovinus, 350
Melyridae, 181, 211-212
Membracidae, 157, 161-162
MEMBRANE (adj., membranous): A very thin piece of tissue, often transparent; the apical part of the front wing (hemelytron) of Heteroptera (Figs. 124; 126).
Menacanthus stramineus, 123, 124
Menopon gallinae, 123, 124
Menoponidae, 123, 124
MENTUM: The lower part of the labium (Figs. 73D; 80C; 82C).
Merchant, hop, 274
Meringis parkeri, 359
Meromyza americana, 346, 347
Merothripidae, 130, 132
Merothrips morgani, 132
Merragata hebroides, 143
MESONOTUM: The dorsal portion of the mesothorax (Fig. 28).
MESOSCUTELLUM: See Scutellum.
MESOSCUTUM: The middle region of the mesonotum.
MESOSTERNUM: The ventral portion of the mesothorax (Figs. 157B; 348).
MESOTHORAX: The middle of the three thoracic segments (Fig. 28).

Mesovelia mulsanti, 143
Mesoveliidae, 137, 143
Metalmarks, 265, 283-284
METAMORPHOSIS: The change in form during development.
METANOTUM: The dorsal portion of the metathorax (Fig. 28).
METASTERNUM: The ventral portion of the metathorax (Figs. 157B; 348).
Metasyrphus americanus, 342
METATHORAX: The most posterior of the three thoracic segments (Fig. 28).
Metcalfa pruinosa, 165, 166
METER (m): 100 centimeters or 1,000 millimeters; 39.37 inches.
Metriona bicolor, 227, 228
Metrobates, 142
  hesperius, 142
Metylophorus novaescotiae, 121
Microbembex monodonta, 386
Microcaddisflies, 254-255
Microcentrum rhombifolium, 96
Microcoryphia, 63
Microlepidoptera, 260
Microphotus angustus, 206
Microrhopala vittata, 229
Microtus, 359, 360
Microvelia, 143
  pulchella, 142, 143
MIDDORSAL STRIPE: A longitudinal stripe running down the middle of the back or dorsum.
Midge, clover seed, 334
  sorghum, 334
Midges, 331-332
  biting, 331
  dixid, 328
  gall, 319, 333-334
  phantom, 328
Milichiidae, 324, 345
MILLIMETER (mm): 0.001 meter or 0.1 centimeter; 0.039 inches or about 1/25 inch.
Miners, leafblotch, 269
Minettia lupulina, 345
Miridae, 135, 138, 148-149
Mochlonyx, 328
MOLTING: Process leading to and including the shedding of the cuticle.
Monarch, 258, 279
MONILIFORM ANTENNA: Beadlike rounded segments form the antenna (Fig. 29C).
Monobia quadridens, 384
Monochamus maculosus, 225, 226
  titillator, 225
Monomorium minimum, 381
  pharaonis, 381, 382
Monophadnoides geniculatus, 372
Monotrysia, 259
Mordella albosuturalis, 223
  atrata, 223
  marginata, 223
Mordellidae, 184, 222-223
Mordellistena pustulata, 223
Mordwilkoja vagabunda, 168, 169
Mormidea lugens, 155

402   Index and Glossary

MORPHOLOGY: The study of form or structure.
Mosquito, California saltmarsh, 329-330
  common malaria, 330
  floodwater, 330
  northern house, 330
  saltmarsh, 329, 330
  southern house, 330
  yellowfever, 329, 330
Mosquitoes, 328-330
  rearing larvae, 26
Moth, American dagger, 299
  Angoumois grain, 308
  artichoke plume, 312
  atlas, 292
  buck, 294
  cactus, 312
  carpet, 305
  casemaking clothes, 305
  ceanothus silk, 293
  cecropia, 293
  codling, 309, 310
  crinkled flannel, 313, 314
  cynthia, 293
  grape plume, 313
  greater wax, 26, 312
  gypsy, 295, 296
  hickory tussock, 301
  honey locust, 294
  hummingbird, 291
  imperial, 294
  Indian meal, 312
  io, 294
  lichen, 302
  luna, 292
  meal, 311, 312
  Mediterranean flour, 312
  milkweed tiger, 301
  Nantucket pine tip, 310
  pale tussock, 301, 302
  pandora, 294
  pine tussock, 295
  polyphemus, 293
  promethea, 293
  ragweed plume, 312
  regal, 294
  striped footman, 301
  virgin tiger, 301, 302
  webbing clothes, 305
  western tussock, 296
  whitemarked tussock, 296
  yellowcollared scape, 302, 303
  Zimmerman pine, 312
Moths, 47, 257-262, 264-271, 290-314
  bagworm, 266, 305
  bogus yucca, 304
  carpenterworm, 266, 309
  casebearer, 270, 306-307
  clearwing, 307-308
  clothes, 305
  dagger, 299
  ermine, 271, 307
  flannel, 313
  footman, 301
  forester, 300-301
  gelechiid, 270, 308
  geometrid, 47, 267, 303-304
  giant silkworm, 261, 267, 292-294
  grass, 310, 312
  handmaid, 296
  hawk, 290
  hepialid, 314
  hooktip, 268, 303
  hummingbird, 290
  leafrolling, 271, 309-310

  leaf skeletonizer, 266, 313
  noctuid, 47, 261, 269
  notodontid, 266, 296-297
  olethreutid, 309
  owlet, 47, 297
  plume, 312-313
  pyralid, 310-312
  royal, 266, 292, 294
  slug caterpillar, 266, 313-314
  snout, 310-312
  sphinx, 261, 266, 290-292
  tent caterpillar, 268, 295
  tiger, 269, 301-302
  tineid, 305
  tussock, 269, 295-296
  wasp, 302-303
  yucca, 304
MOTTLED: Blotchy; irregular spots.
Mounting and preserving insects, 13-21
  arrangement of insects, 22-23
  boxes, 22-24
  care of the collection, 23-24
  clearing, 20
  inflation of larvae, 20-21
  labeling, 19-20
  liquid preservation, 16
  microscope slide preparation, 16-17
  mounting, 14, 15-16
  pinning, 13-15
  points, 15
  relaxing, 13
  wing mounting, 20
  wing spreading, 17-19
Mourningcloak, 275
Murgantia histrionica, 155
Mus, 359
Musca autumnalis, 350
  domestica, 349, 350
Muscidae, 317, 325, 326, 349-350
Muscina stabulans, 350
Mutillidae, 364, 368, 379-380
Mycetophagidae, 186, 220, 221, 333
Mycetophagus punctatus, 221
Mycetophilidae, 319
Mydas clavatus, 337
Mydidae, 320, 337
Myodocha serripes, 151
Myodopsylla insignis, 360
Myopa, 342, 343
Myotis, 360
Myrmeleon crudelis, 246
  immaculatus, 246
Myrmeleontidae, 241, 245-246
Myrmicinae, 381
Mystacides sepulchralis, 256
Myzinum quinquecinctum, 379
Myzus persicae, 168

N

Nabicula subcoleoptrata, 145
Nabidae, 137, 138, 144-145
Nabis alternatus, 145
  americoferus, 145
Nacerdes melanura, 221, 222
NAIAD: An aquatic nymph that uses gills to breathe and does not generally resemble the adult (e.g., mayflies, dragonflies, and stoneflies).
Names, 38-39
  common, 39
  scientific, 38-39

Napthalene, 23
NASUTES: Certain termite soldiers with a long snout through which a sticky, toxic fluid is sprayed.
Nathalis iole, 282, 283
Naucoridae, 135, 139
Neacoryphus bicrucis, 151
Neatus tenebroides, 219
Necrobia, 210
  ruficollis, 211
  rufipes, 211
Nectopsyche exquisita, 256
Neelidae, 60, 61
Neelus minimus, 61
Nematocera, 316, 326-334
Nemeobiidae, 283
Nemobiinae, 99
Nemouridae, 84, 85
Neochlamisus gibbosa, 228, 229
Neoclytus acuminatus, 226, 227
Neoconocephalus ensiger, 97
Neocurtilla hexadactyla, 100
Neodiprion, 370
  lecontei, 370, 371
  sertifer, 370, 371
  taedae linearis, 370
Neohaematopinus, 129
Neoheegeria leucanthemi, 131
  verbasci, 131
Neoheterocerus pallidus, 202, 203
Neomida bicornis, 219
Neoneura aaroni, 81
Neoperla clymene, 87, 88
Neophasia menapia, 281, 282
Neopyrochroa flabellata, 222
Neotridactylus apicialis, 93
Nepa apiculata, 140
Nephrotoma ferruginea, 326
Nepidae, 135, 140
Nerthra martini, 139
Neurocolpus nubilus, 148, 149
Neuroctenus simplex, 148
Neuroptera, 52, 239-247
Nicrophorus, 193-194
  americanus, 194
  marginatus, 194
  orbicollis, 194
  tomentosus, 194
Nigronia, 239, 241
  fasciata, 242
Nitidula bipunctata, 213
Nitidulidae, 183, 212-213
NOCTURNAL: Active at night.
Noctuidae, 268, 269, 297-300
Noctuids, 297-300
NODE: A raised bump or knob; in ants, located at the narrowed, anterior end of the abdomen (Fig. 343).
NODUS: A stout crossvein near the middle of the anterior (costal) margin of dragonfly and damselfly wings (Fig. 70).
Nomenclature, 38-39
Nomia melanderi, 388
Notodontidae, 140, 267, 296-297
Notonecta, 141
  indica, 141
  insulata, 141
  irrorata, 141
  kirbi, 141
  undulata, 140, 141
Notonectidae, 135, 141
NOTOPLEURAL SUTURE: A groove between the notum

and the pleural sclerites (Fig. 157B).
NOTOPLEURON (pl., notopleura): In flies (Diptera), the dorsal area on the thorax at the end of the transverse suture (Fig. 303).
Notoptera, 107
Notoxus, 224
  constricta, 224
  monodon, 224
NOTUM (pl., nota): The dorsal body surface, generally restricted to the thorax.
NYMPH: The immature stage of a hemimetabolous insect (one with incomplete metamorphosis); it usually resembles the adult (Fig. 34).
Nymph, wood, 277, 278
Nymphalidae, 263, 264, 271-277
Nymphalis antiopa, 275
  californica, 276
  milberti, 275
Nymphula ekthlipsis, 310, 311
Nymphulinae, 310
Nysius niger, 150, 151
Nyssoninae, 386-387

O

Oakworm, spiny, 294
OCELLAR TRIANGLE: A somewhat raised triangular area on the head that encloses the ocelli (Fig. 291A).
OCELLUS (pl., ocelli): A simple eye (Fig. 30A).
Ochlodes sylvanoides, 288, 289
Ochthera mantis, 346
Odonata, 48, 69-82
Odontota dorsalis, 228, 229
  scapularis, 229
Oebalus pugnax, 154, 155
Oecanthinae, 99
Oecanthus fultoni, 99
  nigricornis, 100
  quadripunctatus, 100
Oecetis cinerascens, 256, 257
  inconspicua, 257
Oeclidius, 159
Oedemeridae, 185, 221, 222
Oedipodinae, 95
Oeneis chryxus, 278
Oestridae, 323, 353-354
Oestrus ovis, 354
Okanagana bella, 161
Olethreutidae, 309
Olethreutinae, 271, 309
Oliarus humilis, 164, 165
Oligothrips oreios, 132
Oligotoma, 111
  nigra, 112
  saundersii, 112
Oligotomidae, 111, 112
Olla abdominalis, 216, 217
OMNIVOROUS: Feeding on both animal and plant materials.
Omoglymmius americanus, 188
Omophron, 188
  tessellatum, 190
Omophronidae, 188
Omosita colon, 212, 213
Omus, 188
Oncopeltus fasciatus, 150, 151
Onthophagus hecate, 198

Index and Glossary   403

Onychiuridae, 60, 61
Onychiurus folsomi, 60, 61
OOTHECA (pl., oothecae): The hardened covering of an egg mass produced by a cockroach or mantid (Figs. 102; 103B).
OPEN CELL: A wing cell that extends to the wing margin (Fig. 32).
OPEN COXAL CAVITY: Front coxal cavity: surrounded posteriorly by the mesosternum (Fig. 164B); middle coxal cavity: touched by lateral (pleural) portions of the mesothorax (Fig. 164D).
Ophiogomphus, 75
Ophiogomphus rupinsulensis, 76
Ophiomyia simplex, 347
Ophyra leucostoma, 350
Opisocrostis bruneri, 359
Opuntia, 172
ORAL VIBRISSAE: A pair of stout bristles on the lower part of the face of flies (Diptera) (Fig. 291A).
Orange, sleepy, 282
Orangedog, 280
Orangetip, falcate, 282
Orangetips, 265, 281-282
Orchelimum gladiator, 97
nigripes, 97
vulgare, 97, 98
Orchopeas howardii, 360
leucopus, 360
ORDER: A subdivision of a class or subclass; contains related superfamilies or families.
Orgyia, 301
leucostigma, 296
vetusta, 296
Orius insidiosus, 144, 145
tristicolor, 144
Orphulina speciosa, 95
Orthezia, greenhouse, 171
Orthezia insignis, 171
Ortheziidae, 170-171
Ortholomus scolopax, 150, 151
Orthoptera, 40, 45, 46, 55, 56, 89-101, 103, 107
Orthosoma brunneum, 225
Oryzaephilus mercator, 213
surinamensis, 213
Oscinella frit, 347
OSMETERIUM (pl., osmeteria): A fleshy, often forked gland that can be everted from the head of certain caterpillars (e.g., swallowtail larvae).
Osmia lignaria, 389
Osphya varians, 222
Ostomidae, 210
Ostominae, 210
Ostrinia nubilalis, 311
Otiocerus, 164
Otiorhynchinae, 235
Otiorhynchus ovatus, 234, 235
Otitidae, 325, 343
OVIPAROUS: Capable of laying eggs.
OVIPOSIT: To deposit or lay eggs.
OVIPOSITOR: The external egg-laying structure of certain female insects (Figs. 28; 95D; 98A; 121; 356).
Owlflies, 240, 246

P

Pachybrachis, 228, 229
Pachydiplax longipennis, 79
Pachypsylla, 166
Pachysphinx modesta, 292
PAEDOGENESIS: Egg or larval production by insect larvae.
Paleacrita vernata, 304
Panchlora nivea, 106
Pangaeus bilineatus, 156
Panoquina ocola, 289
Panorpa, 248
debilis, 248
helena, 248
nebulosa, 248
nuptialis, 248
Panorpidae, 247, 248
Panorpodidae, 247, 248
Pantomorus cervinus, 235
Paonias excaecatus, 291, 292
Papilio cresphontes, 280
glaucus, 280
multicaudatus, 280
palamedes, 281
polyxenes asterius, 280, 281
rutulus, 281
troilus, 280, 281
Papilionidae, 259, 262, 263, 279-281
Papilioninae, 263
Parabacillus hesperus, 102
Paraclemensia acerifoliella, 304
Paradejeania rutilioides, 352
Paradichlorobenzene, 23-24
PARAGLOSSA (pl., paraglossae): One of two lobes at the tip of the labium and on the outside of the glossae (Figs. 85C; 350).
Paragus bicolor, 342
tibialis, 342
Paraleptophlebia debilis, 68
Paraleuctra occidentalis, 86
sara, 86
Paranaemia vittigera, 217
Paraneotermes simplicicornis, 115
Paraphlepsius irroratus, 164
Parapoynx badiusalis, 310
PARASITE: An organism that lives and feeds on or in the body of another (the host).
Parasite, minute egg, 374, 375
Parasites, minute egg, 374-375
twistedwinged, 237-238
Paratendipes albimanus, 332
Paratrioza cockerelli, 166
Parcoblatta, 104, 106
Parcoblatta pennsylvanica, 106
Paria fragariae, 230, 231
Parnassians, 280, 281
Parnassiidae, 280
Parnassiinae, 259, 263, 280
Parnassius, 280
clodius, 280, 281
phoebus, 281
Parnopes edwardsii, 379
Paromius longulus, 151
PARTHENOGENESIS: Development of an egg without fertilization.
Pasimachus depressus, 190
Passalidae, 179, 197
Passalus, 197
Pearls, ground, 170
Pearslug, 371
PECTINATE: Projections or branches like the teeth of a comb; certain antennae (Fig. 29K), tarsal claws.
Pectinophora gossypiella, 308
PEDICEL: The narrowed anterior end of the abdomen that may form a stalk (Fig. 343); the second antennal segment.
Pedicia albivitta, 326
Pediculidae, 126, 127-128
Pediculus humanus capitis, 128
humanus humanus, 126, 127, 128
Pedilidae, 185, 224-225
Pedilus lugubris, 224, 225
Pegomya hyoscyami, 348
Pelecinidae, 366, 377, 378
Pelecinus polyturator, 377, 378
Pelecomalium testaceum, 195
Pelidnota punctata, 200
Pelocoris, 139
femoratus, 139
Peltodytes, 191
duodecimpunctatus, 191
edentulus, 191
pedunculatus, 191
Peltoperla arcuata, 87
Peltoperlidae, 83, 86-87
Pemphigidae, 160, 168-169
Pemphigus populitransversus, 168, 169
Pemphredoninae, 385
Pennisetia marginata, 307
Pentacora ligata, 144
Pentatomidae, 138, 154-155
Penthe obliquata, 222
pimelia, 222
Pepsis, 385
Perdita, 388, 389
Peridroma saucia, 299, 300
Periplaneta americana, 105, 106
fuliginosa, 106
Peripsocidae, 121
Peripsocus quadrifasciatus, 121
Perlesta placida, 88
Perlidae, 84, 87, 88
Perlodidae, 84, 85, 87
Peromyscopsylla, 359
hesperomys, 359
Peromyscus, 128, 359, 360
Petaluridae, 73, 74
PETIOLE: The narrow stem or stalk attaching the abdomen to the thorax in the Hymenoptera (Fig. 367C,D); a nodelike first segment in ants; a second nodelike segment, if present, is the postpetiole and both together form the pedicel (Fig. 343).
Phaenicia sericata, 351
Phalacridae, 186, 215
Phalacrus simplex, 215
Phaneropterinae, 96
Phasganophora capitata, 88
Phasmatidae, 102
Phasmatodea, 101
Phasmatoptera, 101
Phasmida, 56, 101-102
Pheidole bicarinata vinelandica, 381
Phengodidae, 176
PHEROMONE: A rapidly vaporizing, glandular substance released by an insect that causes a specific response in others of the same species (e.g., sex attractants, alarm substances).
Philaenus spumarius, 162
Philanthinae, 387
Philanthus ventilabris, 387
Phileurus, 201
Philonthus cyanipennis, 195
lomatus, 195
Philopotamidae, 251, 252-253
Philopteridae, 123, 124-125
Philopterus, 124
Philotes, 284
Phlaeothripidae, 130-131
Phlebotomus, 327
Phloeodes pustulosus, 219
Phoebis sennae, 282, 283
Pholeomyia indecora, 345
Pholisora catullus, 289
Phoridae, 320, 321, 340
Phormia regina, 351
Photinus pyralis, 206
scintillans, 206
Photuris pennsylvanicus, 206, 207
Phrygea cinerea, 255
Phryganeidae, 251, 255
Phthiraptera, 122, 126
Phthorimaea operculella, 308
Phyciodes, 273
campestris, 273
mylitta, 273
phaon, 273
tharos, 274
Phycitinae, 312
Phylliidae, 101
Phyllobaenus pallipennis, 211
Phyllocnistis liquidambarisella, 306
populiella, 306
Phyllodecta, 230
Phyllophaga, 198
ephilida, 199
errans, 199
fervida, 199, 200
fusca, 199, 200
rugosa, 199
Phylloscelis, 165
Phylloxera, 170
Phylloxera, grape, 170
Phylloxerans, 160, 170, 333
Phylloxeridae, 160, 170
PHYLUM (pl., phyla): One of the major divisions of the plant and animal kingdoms.
Phymata americana, 147
fasciata, 147
pennsylvanica, 147
Phymatidae, 136, 146-147
Physocephala, 342
Physoconops, 343
Phytomyza, 347
PHYTOPHAGOUS: Plant feeders.
Pickleworm, 311
Pieridae, 263, 264, 265, 281-283
Pieris protodice, 281, 282
rapae, 281, 282
Piesma, 148
cinerea, 147, 148
Piesmatidae, 136, 147, 148
Piesmatids, 147, 148
Pineus strobi, 170

Pinkeye, 347
Pipunculidae, 322, 340-341
Pipunculus, 340
Pisenus humeralis, 222
Plagiognathus politus, 149
Plague, bubonic, 358
　sylvatic, 359
Planipennia, 239, 240, 243-246
Planococcus citri, 173
Planthoppers, 133, 156, 164-166, 238
　acanaloniid, 165, 166
　achilid, 165
　cixiid, 165
　delphacid, 164
　derbid, 164
　dictyopharid, 165
　flatid, 166
　fulgorid, 164-165
　issid, 165
Plantlice 167-168
Plathemis lydia, 79
Platycerus depressus, 197
Platycotis vittata, 161
Platydema ellipticum, 219
　excavatum, 219
　ruficorne, 219
Platygaster, 378
Platygastridae, 367, 378
Platypedia areolata, 161
　minor, 161
Platyptilia carduidactyla, 312
Platysoma carolinum, 196, 197
Plebejus acmon, 284
　melissa, 284, 285
　saepiolus, 284
Plecia nearctica, 332
Plecoptera, 49, 82-88
PLEURON: The lateral region of a body segment (generally a thoracic segment).
Plodia interpunctella, 312
PLUMOSE: Featherlike (e.g., antennae, Fig. 29M).
Poanes hobomok, 289
Podabrus tomentosus, 206, 207
Podalonia luctuosa, 386
　violaceipennis, 386
Podisus maculiventris, 154, 155
Podosesia syringiae, 307-308
Podura aquatica, 60, 61
Poduridae, 60, 61
Poecilocapsus lineatus, 149
Pogonomyrmex barbatus, 381
　californicus, 381
　occidentalis, 381, 382
Polistes, 238, 384
　apachus, 384
　exclamans, 384
　fuscatus aurifer, 384
　fuscatus fuscatus, 383, 384
Polistinae, 384
Polites coras, 288, 289
　themistocles, 288, 289
Pollenia rudis, 351
Polycentropodidae, 252, 253
Polycentropus cinereus, 253
POLYEMBRYONY: One embryo divides into many embryos at the beginning of cell division in an egg; occurs in some parasitic wasps.
Polygenis gwyni, 360
Polygonia, 274
　comma, 274
　faunus, 274
　interrogationis, 274
　satyrus, 274
　zephyrus, 274

POLYMORPHISM: Variation in shape, size or winged condition.
Polyphaga, 177, 178, 193-236
Polyphagidae, 105, 106
Polyphylla crinita, 199
　decimlineata, 199
Polypsocidae, 118, 121
Polypsocus corruptus, 121
Pompeius verna, 288, 289
Pompilidae, 368, 385
Pompilus luctuosus, 385
Pontania, 371
Popilius disjunctus, 197
Popillia japonica, 200-201
POSTERIOR: Rear or back end.
POSTERIOR CELL: One of several cells extending to the hind (posterior) wing margin (Fig. 293).
POSTSCUTELLUM: In flies (Diptera), the area just behind or beneath the mesoscutellum (Figs. 301B; 303A).
POSTVERTICAL BRISTLES: A pair of converging or diverging bristles behind the ocelli on the head of flies (Diptera) (Fig. 291A).
Potamyia flava, 254
Precis coenia, 274
PREDATOR (adj., predaceous): An animal that kills and feeds on another, the prey, and is usually larger and/or stronger than the prey.
Prioninae, 225
Prionoxystus robiniae, 309
Prionus, California, 225
　tilehorned, 225
Prionus californicus, 225
　imbricornis, 225
　laticollis, 225, 226
Prionyx atratus, 386
Pristaulacus, 378
Pristiphora abbreviata, 372
　erichsonii, 371, 372
PROBOSCIS: The elongated, extended mouthparts of certain insects (Figs. 30B; 125; 213A; 223B,C; 233).
Prociphilus tessellatus, 169
Prodoxidae, 304
Prodoxinae, 304
Prodoxus, 304
PRODUCED: Projecting; extended.
PROLEGS: One or more pairs of fleshy legs on the abdomen of caterpillars (Fig. 234) and sawfly larvae (Fig. 352C).
Promachus vertebratus, 338
Prominents, 296
PRONOTUM: The dorsal surface or sclerite of the prothorax (Fig. 28).
Prophalangopsidae, 92, 98
PROSTERNUM: The ventral surface or sclerite of the prothorax (Fig. 28).
PROTHORAX: The anterior of the three thoracic segments (Fig. 28).
Protochauliodes, 241
　minimus, 242
Protoneura cara, 81
Protoneuridae, 73, 81

Protura, 54, 57-58
Proturans, 54, 57-58
Psalididae, 110
Pselaphidae, 176, 187, 196
Psephenidae, 183, 201, 202
Psephenus, 202
　herricki, 201, 202
Pseudaletia unipuncta, 299
Pseudatomoscelis seriatus, 149
Pseudaulacaspis pentagona, 173
Pseudocaeciliidae, 119, 120, 121
PSEUDOCELLI: Skin pores on springtails.
Pseudococcidae, 173-174
Pseudococcus longispinus, 173
　maritimus, 173
Pseudodorus clavatus, 341, 342
Pseudolucanus capreolus, 197
Pseudolynchia canariensis, 351
Pseudomasaris vespoides, 384
Pseudophyllinae, 97
Pseudosermyle straminea, 102
Pseudoxenos, 238
Psilocephala aldrichi, 336, 337
　haemorrhoidalis, 337
Psithyrus, 390
Psocid, cereal, 120
　cosmopolitan grain, 121
Psocidae, 119, 120, 121
Psocids, 118-121
Psocoptera, 50, 55, 118-121
Psocus, 121
　leidyi, 121
Psorophora, 329
Psychidae, 266, 271, 305
Psychoda alternata, 327, 328
Psychodidae, 319, 327-328
Psychomyia flavida, 253
Psychomyiidae, 252, 253
Psylla, pear, 166
Psylla alni americana, 166
　floccosa, 166
　pyricola, 166
Psyllid, cottony alder, 166
　potato, 166
　tomato, 166
Psyllidae, 159, 166
Psyllids, 166
Psyllipsocidae, 119, 120
Psyllipsocus ramburii, 120
Pteromalidae, 365, 366, 375
Pteromalids, 376
Pteromalus puparum, 375, 376
Pteronarcyidae, 83, 86, 87
Pteronarcys californica, 86, 87
　dorsata, 86
Pterophoridae, 264, 312-313
Pterophorus periscelidactylus, 313
Pterophylla camellifolia, 97
PTEROPLEURAL BRISTLES: A cluster of bristles on the pteropleuron (Fig. 303).
PTEROPLEURON (pl., pteropleura): A sclerite on the side of the thorax just below the base of the wing (Figs. 292; 303).
PTERYGOTA: The subclass of winged insects.
Pterygota, 64
Pthirus pubis, 127, 128
PTILINUM: An inflatable structure pushed through the frontal suture on the head of certain flies (Diptera) to assist in emerging from their puparium, and then withdrawn into the head.

Ptilinus ruficornis, 209
Ptilodactyla serricollis, 202
Ptilodactylidae, 181, 202
Ptilostomis ocellifera, 255
Ptinidae, 181, 208, 209
Ptinus fur, 208, 209
Ptychopteridae, 319, 327
PUBESCENCE (adj., pubescent): Fine, short, downy hairs.
Pulex irritans, 358
Pulicidae, 356, 357-358
PULVILLUS (pl., pulvilli): A lobe or pad beneath each tarsal claw (Fig. 298).
Pulvinaria innumerabilis, 172
PUNCTURE: A minute pit or depression.
PUPA (pl., pupae): The non-feeding and usually inactive stage between the larval and adult stage in holometabolous insects (those with complete metamorphosis) (Fig. 36).
PUPARIUM (pl., puparia): A hardened case around a fly (Diptera) pupa formed from the larval skin (Figs. 36D; 330D).
PUPATE: To change from the larval stage to the pupal stage.
Purple, banded, 276
　redspotted, 276
Pycnoscelus surinamensis, 106
Pyralidae, 265, 266, 310-312
Pyralinae, 311
Pyralis farinalis, 311, 312
Pyraustinae, 311
Pyrgus communis, 289, 290
Pyrochroidae, 185, 221-222
Pyromorphidae, 313
Pyrrhocoridae, 138, 152
Pytho niger, 220, 221

Q

Quadraceps, 124
Quadraspidiotus perniciosus, 171, 173
Queen, 279

R

RADIAL SECTOR: The posterior (lower) of the two main branches of the radius (Fig. 32).
RADIO-MEDIAL (r-m) CROSS-VEIN: A crossvein connecting the radius and the media (Figs. 32; 293).
RADIUS: The longitudinal wing vein between the subcosta and the media (Fig. 32).
Ranatra, 140
　brevicollis, 140
　fusca, 140
　nigra, 140
Raphidiidae, 240, 243
Raphidiodea, 240, 243
Raphidioptera, 239
RAPTORIAL LEGS: Usually enlarged and often spined front legs used to grasp prey (Figs. 103A; 128B,C).

Index and Glossary 405

Rasahus biguttatus, 145
 thoracicus, 145
Rattus, 359
RECURRENT VEIN: One of two wing veins used in the identification of certain wasps (Figs. 340A; 345B,D).
RECURVED: Curved back (either upward or downward) upon itself (Fig. 86A).
Reduviidae, 134, 137, 145-146
Reduvius personatus, 145, 146
Repipta taurus, 145, 146
RETICULATED: Netlike.
Reticulitermes flavipes, 116
 hesperus, 116
 tibialis, 116
 virginicus, 116
Rhabdophaga strobiloides, 333, 334
Rhagio mystaceus, 336, 337
Rhagionidae, 321, 336
Rhagoletis cingulata, 343
 fausta, 343
 indifferens, 344
 pomonella, 344
Rhagovelia, 143
 distincta, 143
 obesa, 142, 143
Rhamphomyia rava, 339
Rhaphidophoridae, 98
Rhinotermitidae, 114, 115-116
Rhipiceridae, 179, 203
Rhipiphoridae, 238
Rhodobaenus tredecimpunctatus, 235
Rhopalidae, 137, 153
Rhopalocera, 260
Rhopalopsyllidae, 356, 360
Rhopalosiphum maidis, 168
Rhyacionia frustrana, 310
Rhyacophila fuscula, 252
 lobifera, 250, 252
Rhyacophilidae, 250, 251, 252
Rhynchites bicolor, 234
Rhynchitinae, 234
Rhynchophorinae, 235
Rhysodidae, 178, 188
Ricinidae, 123, 124
Ricinus, 124
Riker mounts, 19, 22
Ringlet, California, 278
 ochre, 278
Ringlets, 278
Riodinidae, 264, 265, 283-284
Rock crawler, Rocky Mountain, 108
Rock crawlers, 56, 107-108
Romalea microptera, 94
Romaleinae, 90, 94
Rootworm, northern corn, 230
 southern corn, 230, 231
 strawberry, 230, 231
Roseslug, 371
ROSTRUM: Beak or snout.
Ruby spot, American, 80
 common, 80
RUDIMENTARY: Poorly developed, greatly reduced in size.
Rutelinae, 200-201

S

Sabulodes caberata, 304
Saissetia coffeae, 172
Saldidae, 135, 137, 143, 144

Saldula pallipes, 143, 144
Salmonfly, California, 86, 87
 Pacific, 85
Salpingidae, 185, 220, 221
Samia cynthia, 293
Sandalus, 203
Saperda calcarata, 225, 226
 candida, 225
Saprinus pennsylvanicus, 196, 197
Sarcophaga, 352
 haemorrhoidalis, 352
Sarcophagidae, 326, 351-352
Saturniidae, 267, 292-294
Saturniinae, 292-293
Satyr, Carolina, 278
 little wood, 277, 278
Satyridae, 259, 263, 264, 277-278
Satyrium behrii, 286
 calanus falacer, 286
Satyrs, 264, 277-278
 wood, 277-278
Sawflies, 51, 361-362, 370-372
 argid, 370
 cimbicid, 364, 370, 371
 conifer, 364, 370, 371
 stem, 363, 372
Sawfly, birch, 370
 California pear, 372
 dock, 371
 elm, 370, 371
 European pine, 370, 371
 European wheat stem, 372
 larch, 371, 372
 loblolly pine, 370
 raspberry, 372
 redheaded pine, 370, 371
 wheat stem, 372
Sawyer, southern pine, 225
 spotted pine, 225, 226
Scaeva pyrastri, 342
SCALE: One of numerous, flattened pieces of cuticle on the body and wings of butterflies and moths; the wing stub of termites; a type of homopteran.
Scale, brown soft, 172
 California red, 172
 cottonycushion, 170
 cottony maple, 172
 euonymus, 173
 European elm, 174
 hemispherical, 172
 oleander, 173
 oystershell, 171, 173
 pine needle, 173
 rose, 173
 San Jose, 171, 173
 scurfy, 171
 tamarisk manna, 173
 terrapin, 172
 white peach, 173
 yellow, 172
Scale insects, 43, 170-175
Scales, armored, 172-173
 ensign, 170-171
 gall-like, 172
 lac, 171
 margarodid, 170
 pit, 172
 soft, 171-172
SCALLOPED: Edge formed of rounded depressions, concavities, or hollows.
Scaphidiidae, 180, 194
Scaphidium quadriguttatum, 194

Scapteriscus, 100
 acletus, 100
 vicinus, 100
Scarabaeidae, 179, 197-201
Scarabaeinae, 198
Scarites subterraneus, 190
Scatophaga furcata, 348
 stercoraria, 348
Scatophagidae, 348
Scatophaginae, 348
SCAVENGER: An animal that feeds on decaying organic materials from dead plants and/or animals.
Scelionidae, 367, 378
Sceliphron caementarium, 385, 386
Scepsis fulvicollis, 302, 303
Schistocerca americana, 95
 emarginata, 95
 gregaria, 90
Schizaphis graminum, 168
Schizura concinna, 297
Sciaridae, 319, 333
SCIENTIFIC NAME: The latinized name of a species or subspecies, the species name consisting of the generic and specific names followed by the author's name.
Sciomyzidae, 324, 344
SCLERITE: A hardened (sclerotized) plate surrounded by membranous areas or sutures.
SCLEROTIZED: Hardened; in insects, the outer skin layer (cuticle) is hardened.
Scobicia declivis, 209
Scolia bicincta, 379, 380
 dubia dubia, 380
Scoliidae, 366, 368, 380
Scolops, 165
 pallidus, 165
 sulcipes, 165
Scolytidae, 186, 187, 233
Scolytus multistriatus, 233
 rugulosus, 233
Scopaeus concavus, 195
 exiguus, 195
Scoparia, 310, 311
Scopariinae, 310, 311
Scorpionflies, 247-249
Scorpionfly, 247
Screwworm, 351
Scudderia furcata, 96
Scutelleridae, 154
Scutellerinae, 154
SCUTELLUM: In Hemiptera and Coleoptera, the triangular sclerite (mesoscutellum) behind the pronotum (Figs. 124; 157A); often well-developed on mesonotum of Diptera (Figs. 301; 303).
Scydmaenidae, 176
Scymnus, 215
Searcher, fiery, 189
SEGMENT: A subdivision of the body or an appendage usually between flexible areas such as joints.
Sehirus cinctus, 156
Semidalis angusta, 243, 244
 vicina, 244
Senotainia rubriventris, 352

SENSILIUM: A small dorsal area near the tip of a flea's abdomen (Fig. 336).
Sepedon fuscipennis, 344
Sepsidae, 324, 344
Sepsis punctum, 344
Serica, 198
 anthracina, 200
 sericea, 200
 vespertina, 200
SERRATE: Toothed or sawlike edges (e.g., antenna, Fig. 29J).
Sesiidae, 264, 307-308
SETA (pl., setae): A hair or bristle.
Sharpshooter, watercress, 163
Shieldbearer, shortlegged, 98
Sialidae, 241, 242-243
Sialis californica, 242, 243
 infumata, 242, 243
 mohri, 242, 243
 velata, 243
Sialodea, 239
Sibine stimulea, 313, 314
Sickness, African sleeping, 316
Sigara alternata, 141
Silkworm, 359
Silpha, 194
 americana, 194
 lapponica, 194
 ramosa, 194
Silphidae, 179, 181, 193-194
Silverfish, 63
Silverspots, 272-273
Simuliidae, 319, 332
Simulium venustum, 332
 vittatum, 332
Sinea diadema, 146
 spinipes, 146
Siphlonuridae, 65, 66-67
Siphlonurus alternatus, 66
 occidentalis, 67
 spectabilis, 67
Siphonaptera, 53, 355-360
Sirex areolatus, 372
 cyaneus, 372
Siricidae, 363, 372, 374
Sisyridae, 240, 241, 243, 244
Sitophilus granarius, 235
 oryzae, 234, 235
Sitotroga cerealella, 308
Skeletonizer, birch, 306
 grapeleaf, 313
 oak, 306
Skimmers, belted, 76-77
 common, 78-79
 greeneyed, 77-78
 river, 76-77
Skipper, checkered, 289, 290
 dun, 287
 eufala, 289
 field, 287, 288
 hobomok, 289
 least, 287
 leonardus, 287
 longtailed, 289
 ocola, 289
 Peck's, 288, 289
 roadside, 287, 288
 silverspotted, 289, 290
 tawny edged, 288, 289
Skipperling, southern, 287
Skippers, 261, 262, 287-289
 branded, 287-289
 giant, 286-287
 pyrginae, 289-290
Slaterocoris stygicus, 149

Sminthuridae, 60, 61
Sminthurinus elegans, 61
Snakeflies, 239, 243
SOCIAL INSECTS: Insects living in a colony where castes and a division of labor occur (e.g., termites, ants, certain wasps and bees).
Solenopsis invicta, 381
  molesta, 381
  xyloni, 381, 382
SOLITARY INSECTS: Insects of the same species that live independently from each other.
Somatochlora tenebrosa, 77, 78
Spanishfly, 223
Spanworm, elm, 303
Spanworms, 303
Sparnopolius lherminierii, 338, 339
SPATULATE: Like a spatula; flat, thin, and narrow.
SPECIES: Groups of interbreeding natural populations that are similar in structure and function and are reproductively isolated from other populations.
Species, 38-39
Speyeria, 272
  aphrodite, 272
  cybele, 272
  idalia, 272
  leto, 272
  zerene, 272-273
Sphaeridium scarabaeoides, 193
Sphecidae, 238, 367, 385-387
Sphecinae, 385-386
Sphecius speciosus, 386, 387
Sphenophorus maidis, 234, 235
  parvulus, 235
  zeae, 235
Sphex ichneumoneus, 386
Sphingicampa bicolor, 294
Sphingidae, 266, 290-292
Sphinx, achemon, 291
  blinded, 291, 292
  catalpa, 290
  great ash, 291, 292
  modest, 292
  waved, 291
  whitelined, 291
Sphinx chersis, 291, 292
Spilodiscus biplagiatus, 197
SPIRACLE: The breathing pore that opens into the respiratory system of tracheal tubes.
Spittlebug, diamondbacked, 162, 163
  dogwood, 162
  meadow, 162
  Saratoga, 162
Spittlebugs, 158, 162
Spodoptera, 299
  exigua, 299
  frugiperda, 298, 299
  ornithogalli, 299
Spoggosia claripennis, 352, 353
Spongilla, 244
Spongillaflies, 240, 243, 244
Springtails, 54, 59-62
SPUR: A movable spine.
SPURIOUS VEIN: An extra, incomplete wing vein between the radius and media

used in the identification of flies (Diptera) (Fig. 300A).
Stagmomantis carolina, 107
Stainer, cotton, 152
Stainers, 152
STALKED: With a stem or stalk; veins that fuse to form a single vein for part of their length (R_4 and R_5, Fig. 241A).
Staphylinidae, 176, 180, 194-196
Staphylinus cinnamopterus, 195
  maculosus, 195-196
  maxillosus, 196
Stegobium paniceum, 209
Stenacron interpunctatum, 67
  tripunctatum, 67
Stenelmis quadrimaculata, 202, 203
Stenodema vicinum, 149
Stenolophus comma, 190, 191
  lecontei, 191
Stenopelmatidae, 98
Stenopelmatus fuscus, 98
Stenoponia americana, 359
Stenotus binotatus, 149
Stenus comma, 195, 196
STERNITE: A ventral plate of the sternum (Figs. 157; 158).
STERNOPLEURAL BRISTLES: In flies (Diptera), bristles found on a sclerite (sternopleuron) just above the base of the middle leg (Fig. 292).
Sternorrhyncha, 156
STERNUM (pl., sterna): The ventral (lower) part of the body (Fig. 28).
Sthenopis, 314
Stictia carolina, 387
Stictocephala bubalus, 162
  diceros, 162
  festina, 162
  inermis, 162
STIGMA (pl., stigmata): A darkened or thickened area of the wing along the anterior (costal) margin near the tip (Figs. 70; 340A).
Stilbus apicalis, 215
  nitidus, 215
Stobaera tricarinata, 164
Stomoxys calcitrans, 349, 350
Stoneflies, 49, 82-88
  common, 88
  giant, 86
  green, 87
  roachlike, 86
Stonefly, giant, 86
Strangalia famelica, 226, 227
Strategus, 201
Stratiomyidae, 318, 320, 334-335
Stratiomys, 334
  maculosa, 335
  meigenii, 334, 335
Strawworm, wheat, 376
Strepsiptera, 40, 53, 237-238
Striders, water, 141-142
  broadshouldered, 142-143
  smaller, 142, 143
STRIDULATE: To produce a sound by rubbing two structures together.
Strophiona laeta, 227
Strophopteryx fasciata, 85
Strymon melinus, 286

STYLE: A very slender or bristle-like process at the tip of an antenna (Fig. 294D) or on the abdomen of certain Thysanura (Fig. 64B).
Stylopidae, 238
Stylops, 238
SUBANTENNAL SUTURE: A groove on the face extending downward from the base of an antenna (Fig. 351).
SUBAPICAL: Just before the apex or tip.
SUBCOSTA: The longitudinal wing vein between the costa and radius (Fig. 32).
SUBFAMILY: A group of related genera or tribes; a major family subdivision; subfamily names end in -inae (e.g., Vespinae).
SUBGENITAL PLATE: A flattened or curved sternite below the genitalia (Figs. 94; 101C).
SUBIMAGO: The winged, preadult stage of a mayfly.
SUBMARGINAL CELL: One or more wing cells below the marginal cell, used in the identification of Hymenoptera (Fig. 340A).
SUBMEDIAN CELL: A cell toward the base of the wing, used in identification of Hymenoptera (Fig. 340A).
SUBORDER: A major subdivision of an order (e.g., Heteroptera).
SUBPHYLUM: A major subdivision of a phylum (e.g., Mandibulata).
SUBSPECIES: Local populations of species usually occupying different geographical areas and showing some biological differences from each other; the scientific name of a subspecies consists of the generic, specific, and subspecific name, and the name of the author.
Subspecies, 38-39
SUBTRIANGLE: One or more cells on the inner (toward the wing base) side of the triangle in dragonflies (Figs. 70A; 74A).
Sulphur, clouded, 282, 283
  cloudless, 282, 283
  dainty, 282, 283
  little, 282, 283
Sulphurs, 265, 281-283
Supella, 104
  longipalpa, 106
SUPERFAMILY: A group of closely related families; superfamily names end in -oidea (e.g., Chalcidoidea).
SUTURE: A groove or narrow membranous area between sclerites.
Suwallia pallidula, 88
Swallowtail, black, 280, 281
  giant, 280
  palamede, 281

  pipevine, 280
  spicebush, 280, 281
  tiger, 280
  twotailed, 280
  western, 281
  zebra, 280
Swallowtails, 262, 279-281
Sympetrum, 78
  rubicundulum, 79
Symphoromyia, 336
  atripes, 336
  hirta, 336
Symphyta, 361, 363, 370-372
Synanthedon exitiosa, 308
  scitula, 308
Synclita obliteralis, 310
Synecdoche impunctata, 164, 165
Syrphidae, 322, 341-342
Syrphus opinator, 342
  ribesii, 341, 342
  torvus, 342
Systena blanda, 229, 230

T

Tabanidae, 318, 320, 335-336
Tabanus, 335
  atratus, 335, 336
  lineola, 335, 336
  nigrovittatus, 336
  punctifer, 336
  quinquevittatus, 336
  sulcifrons, 336
Tachardiella, 171
Tachinidae, 325, 326, 352-353
Tachinus fimbriatus, 196
Tachopteryx thoreyi, 74
Tachyporus jocosus, 195, 196
Taenionema pacificum, 85
Taeniopterygidae, 84, 85
Taeniopteryx maura, 85
  nivalis, 85
Taeniothrips inconsequens, 132
  simplex, 132
Tanaoceridae, 92, 93
Tanaocerids, 93
Tanaocerus koebelei, 93
Tanypteryx hageni, 74
Tapeworm, dog, 358
Tapinoma sessile, 381
TARSAL CLAW: A claw at or near the tip of the last tarsal segment (Fig. 31).
TARSUS (pl., tarsi): The last leg segment, connected to the tibia, that consists of one or more small subdivisions (Fig. 31).
TAXONOMY: The science of describing, naming, and classifying organisms based on their similarities and differences.
Tegeticula, 304
Telephanus velox, 213
Telmatoscopus albipunctatus, 327
Temnochila virescens, 210
TENACULUM: A ventral structure on the third abdominal segment of springtails that holds the furcula in place.
Tenebrio molitor, 219
  obscurus, 219
Tenebrionidae, 184, 218-219

Index and Glossary 407

Tenebroides corticalis, 210
   mauritanicus, 210
Tenebroidinae, 210
Tenodera aridifolia sinensis, 107
Ten spot, 79
Tenthredinidae, 364, 371-372
Tentmaker, poplar, 297
Tephritidae, 323, 325, 343-344
Teratembiidae, 111
Terebrantia, 129, 130, 131
TERGITE: A sclerite of the tergum (Fig. 28).
TERGUM (pl., terga): The dorsal (upper) surface of a body (Fig. 28).
TERMINAL: At the end (often the posterior end).
Termite, eastern subterranean, 116
   Pacific dampwood, 115
   roughheaded powderpost, 115
   western drywood, 115
   western subterranean, 116
   Wheeler's desert, 115, 116
Termites, 48, 112-117
Termitidae, 113, 114, 115, 116-117
TERRESTRIAL: Living on land.
Tetanops myopaeformis, 343
Tetramorium caespitum, 381
Tetraopes femoratus, 225
   melanurus, 225
   tetrophthalmus, 225
Tetrigidae, 91, 93
Tetrix arenosa, 93
   subulata, 93
Tettigidea lateralis, 93
Tettigoniidae, 92, 96-98
Thanasimus dubius, 211
Therevidae, 321, 336-337
Therioaphis maculata, 168
Therion morio, 336
Thermobia domestica, 63
THORAX: The central region of the three main insect body regions (Fig. 28); bears the legs and wings.
Thorybes pylades, 289, 290
Thripidae, 130, 131-132
Thrips, 47, 129-132
   banded, 131
   flower, 132
   gladiolus, 132
   grass, 131
   greenhouse, 131, 132
   mullein, 131
   onion, 132
   pear, 132
   western flower, 132
Thrips tabaci, 132
Thyanta accerra, 155
Thylacitinae, 234-235
Thyreocoridae, 138, 155-156
Thyridopteryx ephemeraeformis, 305
Thysanoptera, 47, 48, 54, 129-132
Thysanura, 54, 62-63
TIBIA (pl., tibiae): The fourth leg segment, between the femur and the tarsus (Fig. 31); often long and slender.
TIBIAL SPUR: A large spine of the tibia usually toward the tip (Figs. 31; 341C).
Tibicen canicularis, 161
   linnei, 161
   pruinosa, 161

Timberworm, oak, 232, 233
Timema, 102
Timemidae, 102
Tinea pellionella, 305
Tineidae, 270, 305
Tineola bisselliella, 305
Tingidae, 136, 147
Tiphia intermedia, 379
Tiphiidae, 368, 379
Tipula abdominalis, 326, 327
   acuta, 326
   simplex, 326
   trivittata trivittata, 326
Tipulidae, 319, 326-327
Tmesiphorus costalis, 196
Tomocerus flavescens, 62
Tomosvaryella, 340
   subvirescens, 341
Tortoiseshell, California, 276
   Milbert's, 275
Tortoiseshells, 275-276
Tortricidae, 270, 271, 309-310
TRANSVERSE: Running across; at right angles to the longitudinal axis.
TRANSVERSE SUTURE: A groove or line across a sclerite or between two sclerites (Figs. 292; 304B; 347D; 348A).
Trap, Malaise, 9, 316, 362
Treaders, marsh, 143, 144
   water, 142, 143
Treehopper, buffalo, 162
   twomarked, 161
Treehoppers, 43, 161-162, 238
Tremex, pigeon, 372, 374
Tremex columba, 372
Trepobates, 142
   pictus, 142
Triaenodes tarda, 256, 257
Trialeurodes vaporariorum, 166, 167
TRIANGLE: A cell or group of cells near the base of the dragonfly wing (Fig. 70A).
Triatoma, 145
   protracta, 146
   sanguisuga, 146
TRIBE: A group of related genera forming a subdivision of a subfamily; tribe names end in -ini (e.g., Bembicini).
Tribolium castaneum, 219
   confusum, 219
Trichiotinus affinis, 200
   piger, 201
Trichobaris trinotata, 235, 236
Trichocera, 326
Trichoceridae, 318, 326
Trichocerids, 326
Trichocorixa calva, 141
Trichodectes canis, 125
Trichodectidae, 123, 125
Trichodes nutalli, 211
   ornatus, 211
Trichogramma minutum, 374, 375
Trichogrammatidae, 365, 374-375
Tricholipeurus, 125
Trichophaga tapetzella, 305
Trichoplusia ni, 300
Trichoptera, 50, 249-257
Tridactylidae, 91, 92-93
Trimenopon hispidum, 124
Trimenoponidae, 122, 124
Trinoton querquedulae, 124

Triplax festiva, 214
Tritoma humeralis, 214, 215
   sanguinipennis, 215
TROCHANTER: The second leg segment, between the coxa and femur (Fig. 31); sometimes offset or divided (Figs. 157B; 344A).
Troctidae, 120
Trogidae, 198
Trogiidae, 119
Troginae, 198
Trogium pulsatorium, 119
Trogoderma ornata, 208
Trogositidae, 183, 210
Trogus pennator, 374
Tropiduchidae, 158
Tropisternus lateralis, 193
Trox, 198
   monachus, 198
True bugs, 42
TRUNCATE: Cut off square (squared-off) at the end.
Trypherus latipennis, 206
Trypoxylon, 385
   clavatum, 385
   politum, 385
TUBERCLE: A small bump or knoblike protuberance.
Tuberworm, potato, 308
Tubulifera, 129, 130
Tularemia, 335
Tullbergia granulata, 61
Tumblebugs, 198
Tunga, 356, 357
   penetrans, 358
Tungidae, 358
TYMPANUM (pl., tympana): An auditory (hearing) organ; a vibrating membrane (e.g., on grasshoppers [Fig. 88], cicadas, some moths).
Typhus, epidemic, 127
   murine (endemic), 358
Typocerus velutinus, 227

U

Uleiota dubius, 213
Ulolodes macleayana hageni, 246
Uloma impressa, 219
   punctulata, 219
Unaspis euonymi, 173
Underwings, 297, 300
Urbanus proteus, 289

V

Vanessa, 275
   annabella, 275
   atalanta, 275
   cardui, 275
   virginiensis, 275
VEIN: A thickened longitudinal line or crossline in the wing (Fig. 32).
Velia, 143
Veliidae, 135, 136, 142
VENTRAL: Referring to the lowerside or underside of an object; underneath.
VENTROMESAL: On the underside or near the middle.
Vermipsyllidae, 356, 360

VESICLE: A sac or bladder, sometimes extensible.
Vespidae, 238, 366, 367, 368, 382-384
Vespinae, 383-384
Vespula maculata, 383
   maculifrons, 383
   pensylvanica, 383
   vulgaris, 384
Viceroy, 276
Viceroys, 276
Viehoperla zipha, 87
VIVIPAROUS: Giving birth to living young rather than laying eggs (e.g., aphids).

W

Walkingstick, 102
   gray, 102
   northern, 102
   twostriped, 102
   western shorthorned, 102
Walkingsticks, 56, 101-102
Wasp, golden paper, 384
   mossyrose gall, 376, 377
Wasps, 51, 361-362
   aphid, 385, 386
   cuckoo, 366, 379
   ensign, 377
   gall, 376-377
   paper, 384
   pelecinid, 378
   potter, 379, 384
   sand, 385, 386
   scelionid, 378
   solitary, 361
   spider, 385
   threadwaisted, 385-386
   tiphiid, 379
Waterscorpions, 140
Webspinners, 49, 111-112
Webworm, ailanthus, 307
   alfalfa, 311
   beet, 311
   bluegrass, 312
   cabbage, 310
   fall, 301
   garden, 311
   juniper, 308
Webworms, sod, 312
Weevil, alfalfa, 235, 236
   boll, 236
   clover leaf, 235, 236
   cocklebur, 235
   coffee bean, 232
   Fuller rose, 235
   granary, 235
   New York, 234
   pea, 232
   pecan, 236
   rice, 234, 235
   strawberry, 236
   strawberry root, 234, 235
   sweetpotato, 234
Weevils, 45, 234-236
   acorn, 236
   brentid, 232-233
   fungus, 232
   grain, 235
   nut, 236
White, great southern, 282
Whiteflies, 167
Whitefly, citrus, 167
   greenhouse, 166, 167
Whites, 265, 281-282

Widow, the, 79
Wing, common sooty, 289
  Juvenal's dusky, 289
  little glassy, 288, 289
  northern cloudy, 289, 290
Winthemia rufopicta, 353
Wireworm, Great Basin, 205
  sugarbeet, 205-206
  tobacco, 205
  wheat, 205
Wireworms, false, 218
Witch, black, 298
Wohlfahrtia vigil, 352
Woollybear, banded, 302
  yellow, 301
Worms, measuring, 303

X

Xenopsylla cheopis, 358
Xenos, 238
Xyleborus saxesensi, 233
Xylocopa, 389
  virginica, 390
Xyloryctes jamaicensis, 201

Y

Yellow, sleepy, 282
Yellowjacket, eastern, 383
  western, 383
Yellowjackets, 51, 366, 382-384
Yersiniops, 106
Yoraperla brevis, 83, 87
Yponomeuta, 307
Yponomeutidae, 270, 271, 307
Yucca, 286

Z

Zapada cinctipes, 85
Zeugloptera, 259
Zootermopsis, 115
  angusticollis, 115
  nevadensis, 115
Zoraptera, 49, 50, 55, 117-118
Zorapterans, 49, 117-118
Zorotypidae, 117
Zorotypus, 117
  hubbardi, 117
  snyderi, 117
Zygaenidae, 265, 266, 313
Zygoptera, 69, 71, 79

NOTES

NOTES

NOTES

NOTES

NOTES